C_0-SEMIGROUPS AND APPLICATIONS

NORTH-HOLLAND MATHEMATICS STUDIES 191

(Continuation of the Notas de Matemática)

2003
ELSEVIER
Amsterdam – Boston – London – New York – Oxford –Paris
San Diego – San Francisco – Singapore – Sydney – Tokyo

C_0-SEMIGROUPS AND APPLICATIONS

Ioan I. VRABIE
Faculty of Mathematics
"Al. I. Cuza" University of Iaşi and
"O. Mayer" Institute of Mathematics
of the Romanian Academy, Iaşi, Romania

2003
ELSEVIER
Amsterdam – Boston – London – New York – Oxford – Paris
San Diego – San Francisco – Singapore – Sydney – Tokyo

ELSEVIER SCIENCE B.V.
Sara Burgerhartstraat 25
P.O. Box 211, 1000 AE Amsterdam, The Netherlands

First edition 2003

Library of Congress Cataloging in Publication Data
A catalog record from the Library of Congress has been applied for.

British Library Cataloguing in Publication Data
A catalogue record from the British Library has been applied for.

ISBN: 0-444-51288-8
ISSN: 0304-0208

♾ The paper used in this publication meets the requirements of ANSI/NISO Z39.48-1992 (Permanence of Paper).
Transferrred to Digital Printing 2009.

To my wife Veronica

Contents

Preface

This book is an entirely rewritten English version of the lecture notes of an advanced course I taught during the last eleven years at the Faculty of Mathematics of "Al. I. Cuza" University of Iaşi. Lecture notes appeared in 2001 in Romanian. The idea was to give a unified and systematic presentation of a fundamental branch of operator theory: the linear semigroups. The existence of several very good books on this topic such as: Ahmed [**2**], Belleni-Morante [**24**], Butzer and Berens [**32**], Davies [**45**], Engel and Nagel et al [**51**], Goldstein [**61**], Haraux [**68**], Hille and Phillips [**70**], McBride [**89**], and Pazy [**101**] made this task very hard to accomplish. Nevertheless, I decided to accept it, simply because there are several particular topics which have not found their place into a monograph until now, mainly because they are very new. This book, although containing the main parts of the classical theory of C_0-semigroups, as the Hille-Yosida theory, illustrated by a wealth of applications of both traditional and non-standard mathematical models, also includes some new, or even unpublished results. We refer here to: the characterization in terms of real regular values of both differentiable and analytic semigroups, the study of elliptic and parabolic systems with dynamic boundary conditions, the study of linear and semilinear differential equations with distributed measures, as well as a finite-dimensional like treatment of semilinear hyperbolic equations, mainly due to the author. As far as I know, some other topics appear for the first time in a book form here: the equations of linear thermoelasticity, the equations of linear viscoelasticity and the characterization of generators of equicontinuous and of compact semigroups, being the most important ones. Besides, the last part of the book contains detailed solutions to all the problems included at the end of each chapter.

There are some interesting topics which, although useful, were not discussed in this book. In this respect I would like to mention the spectral mapping theorems and a thorough study of the asymptotic behavior of solutions. Moreover, in order to avoid some slight complications, most of the results in this book refer only to C_0-semigroups of contractions, although they hold true for general C_0-semigroups, i.e. of type (M, ω). However, I assume that the interested readers will be able to fill in this gap, if necessary.

I believe that someone who has some acquaintance with functional analysis and differential equations can read the book. Therefore, I hope that it will be found useful not only by graduate students and researchers in Mathematics to whom it is primarily addressed, but also by physicists and engineers interested in deterministic mathematical models expressed in terms of differential equations.

I am greatly thankful to my former professors and students, as well as to my colleagues and friends who helped me to clarify many ideas and to organize the presentation. More specifically, I am grateful to professor Viorel Barbu for the courses he taught, which had a decisive influence on my further evolution, and for his unceasing interest in my efforts. The discussions with Professor Dorin Ieşan were of great help to me in order to clarify some aspects concerning the examples in Mechanics presented in Sections 4.8, 4.9 and 10.5. The writing of this book was facilitated by a very careful reading of the manuscript followed by many suggestions and comments by Professors Ovidiu Cârjă, Mihai Necula and Constantin Zălinescu, by Dr. Corneliu Ursescu, Senior Researcher at The "Octav Mayer" Institute of Mathematics of the Romanian Academy, as well as by Dr. Silvia-Otilia Corduneanu. Both Professor Cătălin Lefter and my former student Eugen Vărvărucă read the entire Romanian version of the manuscript and made several useful remarks I took into account in the presentation. Dr. Ioana Sîrbu from SUNY at Buffalo was of great help to make the English read smoothly.

It is a great pleasure to express my appreciation to all of them.

Iaşi, November 12th, 2002 Ioan I. Vrabie

CHAPTER 1

Preliminaries

The aim of this chapter is to give a brief presentation of some auxiliary notions and results which are needed for a good understanding of the whole book. In the first two sections, we define and study the class of vector-valued measurable functions as well as the integral of such functions with respect to a σ-finite and complete measure. In the third section, we recall the definition of the spaces $\mathcal{L}^p(\Omega, \mu; X)$ and $L^p(\Omega, \mu; X)$, with (Ω, Σ, μ) a σ-finite and complete measure space, and X a Banach space, and we recall their most remarkable properties. Also here, we present some properties of $W^{k,p}(a, b\,; X)$ and $A^{k,p}(a, b\,; X)$. The fourth section is devoted to a short presentation of the space $BV([\,a, b]; X)$ of functions of bounded variation from $[\,a, b]$ to X, while in the fifth section, we collect several results referring to Sobolev spaces, exactly in form they will be used later in the book. The sixth section contains some basic facts concerning unbounded linear operators in Banach spaces, with main emphasis on self-adjoint and respectively skew-adjoint operators acting in Hilbert spaces. In the seventh section, we include several spectral analysis results with regards to unbounded, closed linear operators on Banach spaces, while in the last two sections, we introduce and study the Dunford integral in order to offer an elegant way to define the value of an analytic scalar function at such an operator.

1.1. Vector-Valued Measurable Functions

Let X be a real Banach space and (Ω, Σ, μ) a σ-finite and complete measure space. We recall that $\mu : \Sigma \to \mathbb{R}_+$ is *σ-finite* if there exists a family $\{\Omega_n;\, n \in \mathbb{N}\} \subset \Sigma$ such that $\mu(\Omega_n) < +\infty$ for each $n \in \mathbb{N}$ and $\Omega = \cup_{n\in\mathbb{N}}\Omega_n$. The measure μ is called *complete* if each subset of a null μ-measure set is measurable (belongs to Σ).

Definition 1.1.1. A function $x : \Omega \to X$ is called:

(i) *countably-valued* if there exist two families: $\{\Omega_n;\, n \in \mathbb{N}\} \subset \Sigma$ and $\{x_n;\, n \in \mathbb{N}\} \subset X$, with $\Omega_k \cap \Omega_p = \emptyset$ for each $k \neq p$, $\Omega = \cup_{n\geq 0}\Omega_n$, and such that $x(\omega) = x_n$ for all $\omega \in \Omega_n$;

(ii) *almost separably-valued* if there exists a μ-null set Ω_0 such that $x(\Omega \setminus \Omega_0)$ is separable;

(iii) *strongly measurable* if there exists a sequence of countably-valued functions convergent to x μ-a.e. on Ω;

(iv) *weakly measurable* if, for each $x^* \in X^*$, the function $x^*(x) : \Omega \to \mathbb{R}$ is measurable[1].

Definition 1.1.2. A subset Λ in X^* is called *determining set* for X if for each $x \in X$ we have

$$\|x\| = \sup\{|x^*(x)|;\ x^* \in \Lambda\}.$$

Remark 1.1.1. If Λ is a determining set for X then its elements have the norm at most equal to 1. This is a consequence of the definition of the usual sup-norm on X^*.

Theorem 1.1.1. *Each separable Banach space has at least one countable determining set.*

Proof. Let $\{x_n;\ n \in \mathbb{N}\}$ a dense subset in X. Since for each $x \in X$ we have

$$\|x\| = \sup\{|x^*(x)|;\ x^* \in X^*,\ \|x^*\| = 1\},$$

it follows that there exists a family $\{x^*_{n,m};\ n, m \in \mathbb{N}\}$ in the unit closed ball in X^*, such that, for each $n \in \mathbb{N}$,

$$\lim_{m \to \infty} |x^*_{n,m}(x_n)| = \|x_n\|.$$

As $\|x^*_{n,m}\| = 1$ for $n, m \in \mathbb{N}$, $\|x_n\| = \sup\{|x^*_{n,m}(x_n)|;\ m \in \mathbb{N}\}$ for each $n \in \mathbb{N}$ and $\{x_n;\ n \in \mathbb{N}\}$ is dense in X, we deduce that

$$\|x\| = \sup\{|x^*_{n,m}(x)|;\ n, m \in \mathbb{N}\}$$

for each $x \in X$. The proof is complete. □

Theorem 1.1.2. *If X admits a countable determining set Λ and $x : \Omega \to X$ is weakly measurable, then $\|x\| : \Omega \to \mathbb{R}_+$ is measurable.*

Proof. Since the supremum of a countable family of real measurable functions is a measurable function and

$$\|x(\omega)\| = \sup\{|x^*(x(\omega))|;\ x^* \in \Lambda\}$$

for each $\omega \in \Omega$, where each function $x^*(x)$ is measurable, it follows that $\|x\|$ has the same property and this achieves the proof. □

---

[1]Some authors prefer the term *scalarly measurable* instead of weakly measurable, keeping the latter term for those functions $x : \Omega \to X$ with the property that, for each weakly open subset D in X, $x^{-1}(D) \in \Sigma$.

Theorem 1.1.3. (Pettis) *A function $x : \Omega \to X$ is strongly measurable if and only if it is weakly measurable and almost separably-valued.*

Proof. *Necessity.* As x is strongly measurable there exists a sequence $(x_n)_{n\in\mathbb{N}}$ of countably-valued functions and a μ-null set Ω_0, such that

$$\lim_{n\to\infty} x_n(\omega) = x(\omega) \tag{1.1.1}$$

for each $\omega \in \Omega \setminus \Omega_0$. But each function in the sequence is at most countably-valued, and thus $\cup_{n\geq 0}\{x_n(\omega);\ \omega \in \Omega\}$ is at most countable and dense in $x(\Omega \setminus \Omega_0)$. Hence x is almost separably-valued.
From (1.1.1) we conclude that, for each $x^* \in X^*$ and $\omega \in \Omega \setminus \Omega_0$, we have

$$\lim_{n\to\infty} x^*(x_n(\omega)) = x^*(x(\omega)).$$

Taking into account that the functions $x^*(x_n)$ are almost countably-valued, and thus measurable, it follows that $x^*(x)$ is measurable.
Sufficiency. Since x is almost separably-valued, we may assume with no loss of generality that X is separable. Indeed, if X is not separable, let us consider the μ-null set Ω_0 such that $x(\Omega \setminus \Omega_0)$ is separable and let Y be the closed linear subspace spanned by $x(\Omega \setminus \Omega_0)$. Obviously this is separable and, in addition, x coincides μ-a.e. with a function y defined on Ω and taking values in Y. It is easy to see that y is strongly measurable if and only if x enjoys the same property. Similarly, x is weakly measurable if and only if y is weakly measurable, since, by virtue of the Hahn-Banach theorem (see Theorem 2.7.1, p. 29 in Hille and Phillips [**70**]), each linear bounded functional on Y coincides with the restriction of a linear bounded functional on X.
So, let $\{x_n;\ n \in \mathbb{N}^*\}$ be a dense subset in X and let $\varepsilon > 0$. We define

$$\Omega_+ = \{\omega \in \Omega;\ x(\omega) \neq 0\} \quad \text{and} \quad \Omega_n^\varepsilon = \{\omega \in \Omega_+;\ \|x(\omega) - x_n\| \leq \varepsilon\}$$

for $n \in \mathbb{N}^*$. From Theorems 1.1.1 and 1.1.2, it follows that both Ω_+ and Ω_n^ε are measurable. Since $\{x_n;\ n \in \mathbb{N}^*\}$ is dense in X, we deduce that, for each $\varepsilon > 0$,

$$\bigcup_{n\geq 1} \Omega_n^\varepsilon = \Omega_+. \tag{1.1.2}$$

Indeed, if we assume by contradiction that this is not the case, then there exist $\varepsilon > 0$ and $\omega \in \Omega_+$ such that $\|x(\omega) - x_n\| > \varepsilon$ for each $n \in \mathbb{N}^*$. But the inequalities above show that $x(\omega)$ does not belong to the closure of the set $\{x_n;\ n \in \mathbb{N}^*\}$ which coincides with X. This contradiction can be eliminated only if (1.1.2) holds. Let us define now

$$E_1^\varepsilon = \Omega_1^\varepsilon \quad \text{and} \quad E_n^\varepsilon = \Omega_n^\varepsilon \setminus \bigcup_{k=1}^{n-1} \Omega_k^\varepsilon \quad \text{for} \quad n = 2, 3, \ldots$$

and let us observe that all the sets E_n^ε are measurable and

$$\bigcup_{n\geq 1} E_n^\varepsilon = \Omega_+ \quad \text{and} \quad E_k^\varepsilon \cap E_p^\varepsilon = \emptyset \quad \text{for} \quad k \neq p.$$

Let $x_\varepsilon : \Omega \to X$ be defined by

$$x_\varepsilon(\omega) = \begin{cases} x_n & \text{if } \omega \in E_n^\varepsilon \\ 0 & \text{if } \omega \in \Omega \setminus \Omega_+. \end{cases}$$

Obviously x_ε is countably-valued and $\|x(\omega) - x_\varepsilon(\omega)\| \leq \varepsilon$ for each $\omega \in \Omega$. The proof is complete. □

Remark 1.1.2. The definition of x_ε in the proof of Theorem 1.1.3 shows that a function $x : \Omega \to X$ is strongly measurable if and only if there exists a sequence of countably-valued functions from Ω to X which is uniformly μ-a.e. convergent on Ω to x.

1.2. The Bochner Integral

As in the preceding section, let X be a real Banach space, (Ω, Σ, μ) a σ-finite and complete measure space and let $x : \Omega \to X$ be a countably-valued function. Then there exist $\{\Omega_n;\ n \in \mathbb{N}\} \subset \Sigma$ and $\{x_n;\ n \in \mathbb{N}\} \subset X$, satisfying $\Omega_k \cap \Omega_p = \emptyset$ for each $k \neq p$, $\Omega = \cup_{n\geq 0}\Omega_n$, and such that $x(\omega) = x_n$ for each $n \in \mathbb{N}$ and each $\omega \in \Omega_n$. Obviously, the two families $\{\Omega_n;\ n \in \mathbb{N}\}$ and $\{x_n;\ n \in \mathbb{N}\}$ which define a countably-valued function are not unique. For this reason, in all that follows, a pair of sets $(\{\Omega_n;\ n \in \mathbb{N}\}, \{x_n;\ n \in \mathbb{N}\})$ enjoying the above properties is called *a representation of the countably-valued function* x. Inasmuch as Ω has σ-finite measure, each countably-valued function $x : \Omega \to X$ admits at least one representation with the property that, for each $n \in \mathbb{N}$, $\mu(\Omega_n) < +\infty$. Such a representation is called *σ-finite representation.*

Definition 1.2.1. Let $x : \Omega \to X$ be a countably-valued function and let $\mathcal{R} = (\{\Omega_n;\ n \in \mathbb{N}\}, \{x_n;\ n \in \mathbb{N}\})$ be one of its σ-finite representations. We say that $\mathcal{R}$ is *Bochner integrable* (B-integrable) on Ω with respect to μ, if

$$\sum_{n=0}^{\infty} \mu(\Omega_n)\|x_n\| < +\infty.$$

Remark 1.2.1. If $\mathcal{R}$ and $\mathcal{R}'$ are two σ-finite representations of a countably-valued function $x : \Omega \to X$, the series $\sum_{n=0}^{\infty} \mu(\Omega_n)x_n$ and $\sum_{n=0}^{\infty} \mu(\Omega'_n)x'_n$ are either both convergent, or both divergent, in the norm topology of X, and, in the former case, they have the same sum. Accordingly, $\mathcal{R}$ is B-integrable on Ω with respect to μ if and only if $\mathcal{R}'$ enjoys the same property.

This remark enables us to introduce:

Definition 1.2.2. The countably-valued function $x : \Omega \to X$ is *Bochner integrable* on Ω with respect to μ if it has a σ-finite representation

$$\mathcal{R} = (\{\Omega_n;\ n \in \mathbb{N}\}, \{x_n;\ n \in \mathbb{N}\})$$

which is B-integrable on Ω with respect to μ. In this case, the vector

$$\sum_{n=0}^{\infty} \mu(\Omega_n)x_n = \int_\Omega x(\omega)\, d\mu(\omega) = \int_\Omega x\, d\mu,$$

which does not depend on the choice of $\mathcal{R}$ (see Remark 1.2.1), is called *the Bochner integral* on Ω of the function x with respect to μ.

Definition 1.2.3. A function $x : \Omega \to X$ is *Bochner integrable* on Ω with respect to μ if it is strongly measurable and there exists a sequence of countably-valued functions $(x_k)_{k\in\mathbb{N}}$, Bochner integrable on Ω with respect to μ, such that

$$\lim_{k\to\infty} \int_\Omega \|x(\omega) - x_k(\omega)\|\, d\mu(\omega) = 0.$$

Proposition 1.2.1. *If $x : \Omega \to X$ is Bochner integrable on Ω with respect to μ and $(x_k)_{k\in\mathbb{N}}$ is a sequence with the properties in Definition* 1.2.3, *then there exists*

$$\lim_{k\to\infty} \int_\Omega x_k\, d\mu$$

in the norm topology of X. In addition, if $(y_k)_{k\in\mathbb{N}}$ is another sequence of countably-valued functions with the property that

$$\lim_{k\to\infty} \int_\Omega \|x(\omega) - y_k(\omega)\|\, d\mu(\omega) = 0,$$

then

$$\lim_{k\to\infty} \int_\Omega x_k(\omega)\, d\mu(\omega) = \lim_{k\to\infty} \int_\Omega y_k(\omega)\, d\mu(\omega).$$

Proof. Let $\varepsilon > 0$ and let $k(\varepsilon) \in \mathbb{N}$ be such that

$$\int_\Omega \|x - x_k\|\, d\mu \leq \varepsilon$$

for each $k \geq k(\varepsilon)$. Let us observe that

$$\left\|\int_\Omega x_k\,d\mu - \int_\Omega x_p\,d\mu\right\| \leq \int_\Omega \|x_k - x_p\|\,d\mu$$
$$\leq \int_\Omega \|x_k - x\|\,d\mu + \int_\Omega \|x - x_p\|\,d\mu.$$

This inequality and the preceding one show that, for each $k \geq k(\varepsilon)$ and $p \geq k(\varepsilon)$, we have

$$\left\|\int_\Omega x_k\,d\mu - \int_\Omega x_p\,d\mu\right\| \leq \varepsilon.$$

Accordingly, $\left(\int_\Omega x_k\,d\mu\right)_{k\in\mathbb{N}}$ is a Cauchy sequence and consequently it is convergent.

Let now $(y_k)_{k\in\mathbb{N}}$ be another sequence with the specified property. We have

$$\left\|\int_\Omega x_k\,d\mu - \int_\Omega y_k\,d\mu\right\| \leq \int_\Omega \|x_k - y_k\|\,d\mu$$
$$\leq \int_\Omega \|x_k - x\|\,d\mu + \int_\Omega \|x - y_k\|\,d\mu,$$

which completes the proof. □

Definition 1.2.4. Let $x : \Omega \to X$ be a Bochner integrable function on Ω. The vector

$$\lim_{k\to\infty}\int_\Omega x_k\,d\mu = \int_\Omega x(\omega)\,d\mu(\omega) = \int_\Omega x\,d\mu$$

which, according to Proposition 1.2.1, exists and does not depend on the choice of the sequence $(x_k)_{k\in\mathbb{N}}$ in Definition 1.2.3, is called *the Bochner integral* of the function x on Ω with respect to μ.

Theorem 1.2.1. (Bochner) *A function $x : \Omega \to X$ is Bochner integrable on Ω with respect to μ if and only if x is strongly measurable and the real function $\|x\|$ is integrable on Ω with respect to μ.*

Proof. *Necessity.* Let x be Bochner integrable on Ω with respect to μ and let $(x_k)_{k\in\mathbb{N}}$ be a sequence as in Definition 1.2.3. Then, by Lebesgue theorem (see Proposition 14, p. 126 in Dinculeanu [**47**]), it follows that, at least on subsequence (denoted for simplicity again by $(x_k)_{k\in\mathbb{N}}$), we have

$$\lim_{k\to\infty} x_k(\omega) = x(\omega)$$

a.e. for $\omega \in \Omega$. So x is strongly measurable. The fact that $\|x\|$ is integrable on Ω with respect to μ follows from the obvious inequality

$$\int_\Omega \|x\|\,d\mu \leq \int_\Omega \|x - x_k\|\,d\mu + \int_\Omega \|x_k\|\,d\mu < +\infty.$$

Sufficiency. Let x be a strongly μ-measurable function on Ω such that $\|x\|$ is μ-integrable on Ω. Since Ω has σ-finite measure, there exists a family $\{\Omega_n;\ n \in \mathbb{N}\}$ such that $\mu(\Omega_n) < +\infty$ and $\cup_{n\geq 0}\Omega_n = \Omega$. In addition, we may assume with no loss of generality that $\Omega_k \cap \Omega_p = \emptyset$ for each $k \neq p$. Let $\varepsilon > 0$. From Remark 1.1.2, it follows that, for each $n \in \mathbb{N}$ for which $\mu(\Omega_n) \neq 0$, there exists a countably-valued function $x_n^\varepsilon : \Omega_n \to X$ such that

$$\|x(\omega) - x_n^\varepsilon(\omega)\| \leq \frac{\varepsilon}{2^{n+1}\mu(\Omega_n)} \tag{1.2.1}$$

a.e. for $\omega \in \Omega_n$. Let $E = \cup_{n\geq 0} E_n$ where E_n is the set of all elements in Ω_n for which the above inequality does not hold. Obviously E is negligible. Let $x_\varepsilon : \Omega \to X$ be defined by

$$x_\varepsilon(\omega) = \begin{cases} x_n^\varepsilon(\omega) & \text{if } \omega \in \Omega_n \setminus E_n \\ 0 & \text{if } \omega \in E. \end{cases}$$

It is easy to see that x_ε is a countably-valued function. In addition, the function x_ε is integrable because it is measurable, bounded from above by the function $\|x_\varepsilon - x\| + \|x\|$, $\|x\|$ is integrable and, in view of (1.2.1), $\|x_\varepsilon - x\|$ is integrable too, inasmuch as

$$\int_\Omega \|x_\varepsilon - x\|\, d\mu \leq \sum_{n=0}^{\infty} \int_{\Omega_n} \|x_\varepsilon - x\|\, d\mu \leq \sum_{n=0}^{\infty} \mu(\Omega_n) \frac{\varepsilon}{2^{n+1}\mu(\Omega_n)} = \varepsilon.$$

Since ε is arbitrary, from the inequality above, it follows that x is Bochner integrable on Ω with respect to μ and this achieves the proof. □

The next consequence shows that each Bochner integrable function x can be approximated, as in the proof of Theorem 1.1.1, by countably-valued functions whose values belong to the range of x except 0.

Corollary 1.2.1. *Let $x : \Omega \to X$ be Bochner integrable function on Ω. Then, for each $\varepsilon > 0$, there exists a partition $\{\Omega_n;\ n \in \mathbb{N}\} \subset \Sigma$ of the set $\Omega_+ = \{\omega \in \Omega;\ x(\omega) \neq 0\}$, such that, for each choice of the elements $\omega_n \in \Omega_n$, $n = 0, 1, \ldots$, the function $x_\varepsilon : \Omega \to X$, defined by*

$$x_\varepsilon(\omega) = \begin{cases} x(\omega_n) & \text{for } \omega \in \Omega_n \\ 0 & \text{for } \omega \in \Omega \setminus \Omega_+, \end{cases}$$

is countably-valued, Bochner integrable on Ω and

$$\int_\Omega \|x(\omega) - x_\varepsilon(\omega)\|\, d\mu(\omega) \leq \varepsilon.$$

In addition, the inequality above holds for each refinement of the partition considered.[2]

Proof. Let $\{\Omega_n;\ n \in \mathbb{N}\}$ be the class of all subsets of Ω on which the x_n^ε's, defined as in the proof of Theorem 1.2.1, are constant. Redefining the functions x_n^ε as suggested in the statement of Corollary 1.2.1, we observe that, by multiplying the right-hand side of (1.2.1) by 2 if necessary, the inequality thus obtained holds true. It is easy to see that the partition $\{\Omega_n;\ n \in \mathbb{N}\}$ has all the required properties for $\varepsilon' = 2\varepsilon$. The proof is complete. □

We conclude this section with a result which will prove useful in the sequel. We recall that an operator acting between two Banach spaces X and Y is called *closed* if its graph is closed.

Theorem 1.2.2. (Hille) *Let $A : D(A) \subseteq X \to Y$ be a linear closed operator and let $x : \Omega \to D(A)$. Then*

$$\int_\Omega Ax(\omega)\, d\mu(\omega) = A \int_\Omega x(\omega)\, d\mu(\omega)$$

whenever both sides of the above equality are well-defined.

Proof. Let $\varepsilon > 0$. We apply Corollary 1.2.1 to both x and Ax in order to obtain two partitions of $\Omega_+ = \{\omega \in \Omega;\ \|x(\omega)\| > 0\}$, the first one defining an ε-approximation for x and the second one for Ax. Let $\{\Omega_n;\ n \in \mathbb{N}\}$ be a refinement for these partitions. We define $x_\varepsilon : \Omega \to X$ by

$$x_\varepsilon(\omega) = \begin{cases} x(\omega_n) & \text{for } \omega \in \Omega_n \\ 0 & \text{for } \omega \in \Omega \setminus \Omega_+. \end{cases}$$

Then we have

$$\int_\Omega \|x(\omega) - x_\varepsilon(\omega)\|\, d\mu(\omega) \le \varepsilon \ \text{ and } \ \int_\Omega \|Ax(\omega) - Ax_\varepsilon(\omega)\|\, d\mu(\omega) \le \varepsilon.$$

But

$$\int_\Omega x_\varepsilon(\omega)\, d\mu(\omega) = \sum_{n=0}^{\infty} x_\varepsilon(\omega_n)\mu(\Omega_n \cap \Omega) = \lim_{m\to\infty} \sum_{n=0}^{m} x_\varepsilon(\omega_n)\mu(\Omega_n \cap \Omega),$$

and

$$\int_\Omega Ax_\varepsilon(\omega)\, d\mu(\omega) = \sum_{n=0}^{\infty} Ax_\varepsilon(\omega_n)\mu(\Omega_n \cap \Omega)$$

[2] By a *refinement* of the partition $\{\Omega_n;\ n \in \mathbb{N}\} \subset \Sigma$ of the set Ω we mean a partition $\{E_k;\ k \in \mathbb{N}\} \subset \Sigma$ of the same set with the property that for each $k \in \mathbb{N}$ there exists $n \in \mathbb{N}$ such that $E_k \subseteq \Omega_n$.

$$= \lim_{m\to\infty} \sum_{n=0}^{m} Ax_\varepsilon(\omega_n)\mu(\Omega_n \cap \Omega).$$

Since A is closed, it follows that

$$\int_\Omega x_\varepsilon(\omega)\, d\mu(\omega) \in D(A) \quad \text{and} \quad A\left(\int_\Omega x_\varepsilon(\omega)\, d\mu(\omega)\right) = \int_\Omega Ax_\varepsilon(\omega)\, d\mu(\omega).$$

Let $(\varepsilon_k)_{k\in\mathbb{N}}$ be a sequence convergent to 0. By Proposition 1.2.1, we have

$$\lim_{k\to\infty} \int_\Omega x_{\varepsilon_k}(\omega)\, d\mu(\omega) = \int_\Omega x(\omega)\, d\mu(\omega)$$

and

$$\lim_{k\to\infty} A\int_\Omega x_{\varepsilon_k}(\omega)\, d\mu(\omega) = \lim_{k\to\infty} \int_\Omega Ax_{\varepsilon_k}(\omega)\, d\mu(\omega) = \int_\Omega Ax(\omega)\, d\mu(\omega).$$

We conclude the proof by using once again the closedness of A. □

1.3. Basic Function Spaces

Let X be a real Banach space, (Ω, Σ, μ) a σ-finite measure space and $p \geq 1$. We denote by $\mathcal{L}^p(\Omega, \mu\,; X)$ the set of all functions $f : \Omega \to X$ with the property that f is strongly measurable on Ω and $\|f\|^p$ is integrable on Ω with respect to μ. Let us define $\|\cdot\|_{\mathcal{L}^p(\Omega,\mu\,;X)} : \mathcal{L}^p(\Omega, \mu\,; X) \to \mathbb{R}_+$ by

$$\|f\|_{\mathcal{L}^p(\Omega,\mu\,;X)} = \left(\int_\Omega \|f\|^p\, d\mu\right)^{1/p}$$

for each $f \in \mathcal{L}^p(\Omega, \mu\,; X)$ and let us observe that this is a seminorm on $\mathcal{L}^p(\Omega, \mu\,; X)$. We define the relation "$\sim$" by $f \sim g$ if $f(\omega) = g(\omega)$ μ-a.e. $\omega \in \Omega$. Clearly "$\sim$" is an equivalence on $\mathcal{L}^p(\Omega, \mu\,; X)$. Let $L^p(\Omega, \mu\,; X)$ be the quotient space $\mathcal{L}^p(\Omega, \mu\,; X)/\sim$ and let us remark that if $f \sim g$ then $\|f\|_{\mathcal{L}^p(\Omega,\mu\,;X)} = \|g\|_{\mathcal{L}^p(\Omega,\mu\,;X)}$. So, $\|\cdot\|_{L^p(\Omega,\mu\,;X)} : L^p(\Omega, \mu\,; X) \to \mathbb{R}_+$, given by

$$\|\hat{f}\|_{L^p(\Omega,\mu\,;X)} = \left(\int_\Omega \|f\|^p d\mu\right)^{1/p}$$

for each $\hat{f} \in L^p(\Omega, \mu\,; X)$, is well-defined (i.e. it does not depend on the choice of $f \in \hat{f}$) and, in addition, is a norm on $L^p(\Omega, \mu\,; X)$, in respect to which this is a Banach space.
Next, let $\mathcal{L}^\infty(\Omega, \mu\,; X)$ be the space of all functions $f : \Omega \to X$ satisfying

$$\|f\|_{\mathcal{L}^\infty(\Omega,\mu\,;X)} = \inf\{\alpha \in \overline{\mathbb{R}};\ \|f(\omega)\| \leq \alpha, \quad \text{a.e.} \quad \omega \in \Omega\} < +\infty.$$

The mapping $\|\cdot\|_{\mathcal{L}^\infty(\Omega,\mu\,;X)} : \mathcal{L}^\infty(\Omega,\mu\,;X) \to \mathbb{R}_+$, defined as above, is a seminorm. Let $L^\infty(\Omega,\mu\,;X) = \mathcal{L}^\infty(\Omega,\mu\,;X)/\sim$, where "$\sim$" is the μ-a.e. equality on Ω and let $\|\cdot\|_{L^\infty(\Omega,\mu\,;X)} : L^\infty(\Omega,\mu\,;X) \to \mathbb{R}_+$, given by

$$\|\hat{f}\|_{L^\infty(\Omega,\mu\,;X)} = \|f\|_{\mathcal{L}^\infty(\Omega,\mu\,;X)}$$

for each $\hat{f} \in L^\infty(\Omega,\mu\,;X)$. Obviously $\|\cdot\|_{L^\infty(\Omega,\mu\,;X)}$ is well-defined and, in addition, is a norm on $L^\infty(\Omega,\mu\,;X)$, with respect to which this is a Banach space.

For simplicity, in that follows, we denote by f both a fixed element in $\mathcal{L}^p(\Omega,\mu\,;X)$, and its corresponding equivalence class in $L^p(\Omega,\mu\,;X)$.

The next properties are either well-known, or follow directly from their specific counterparts corresponding to the case $X = \mathbb{R}$.

Theorem 1.3.1. *Let (Ω,Σ,μ) be a finite measure space.*[3]

(i) *If $f \in L^p(\Omega,\mu\,;\mathbb{R})$ and $g \in L^q(\Omega,\mu\,;\mathbb{R})$, where $p \in (1,+\infty)$ and $\frac{1}{p}+\frac{1}{q}=1$, then*

$$\int_\Omega |fg|\,d\mu \le \left(\int_\Omega |f|^p d\mu\right)^{1/p} \left(\int_\Omega |g|^q d\mu\right)^{1/q} \quad \text{(Hölder's inequality).}$$

(ii) *If $f \in L^1(\Omega,\mu\,;\mathbb{R})$ and $g \in L^\infty(\Omega,\mu\,;\mathbb{R})$, then*

$$\int_\Omega |fg|\,d\mu \le \|f\|_{L^1(\Omega,\mu\,;\mathbb{R})}\|g\|_{L^\infty(\Omega,\mu\,;\mathbb{R})}.$$

(iii) *If X is an arbitrary Banach space, and $1 \le p \le r \le +\infty$, then $L^r(\Omega,\mu\,;X) \subset L^p(\Omega,\mu\,;X)$, with continuous imbedding;*

(iv) *If, in addition, Ω is a compact subset in $\mathbb{R}^n$, μ is the Lebesgue measure on Ω and $p \in [1,+\infty]$, then $C(\Omega;X) \subset L^p(\Omega,\mu\,;X)$, with continuous imbedding*[4].

We state, again without proof, the following result which gives a simple but precise description of the topological dual of an L^p space, description well-known in the classical case $X = \mathbb{R}$.

Theorem 1.3.2. *If either X is reflexive, or X^* is separable, then, for each $p \in [1,+\infty)$,*

$$(L^p(\Omega,\mu\,;X))^* \approx L^q(\Omega,\mu\,;X^*),$$

where $\frac{1}{p}+\frac{1}{q}=1$ if $p>1$ and $q=\infty$ if $p=1$.

---

[3]We recall that (Ω,Σ,μ) is a finite measure space if $\mu(\Omega) < +\infty$.

[4]Here $C(\Omega;X)$ is endowed with the uniform convergence topology, or equivalently with the sup-norm topology.

See Dinculeanu [**48**], Corollary 1, p. 252. Some extensions and variants of Theorem 1.3.2 can be found in Edwards [**50**], Theorem 8.18.2, p. 588, Remarks, p. 589 and Theorem 8.20.5, p. 607.
A remarkable consequence of Theorem 1.3.2 is stated below.

Corollary 1.3.1. *If X is reflexive then, for each $p \in (1, +\infty)$, $L^p(\Omega, \mu; X)$ is reflexive. If X is separable, then, for each $p \in [1, +\infty)$, $L^p(\Omega, \mu; X)$ is separable.*

The next density result will prove useful in what follows. We notice that in the theorem below, μ is the Lebesgue measure on $\mathbb{R}^n$.

Theorem 1.3.3. *Let X be Banach space and let $\varphi \in C^\infty(\mathbb{R}^n; \mathbb{R})$ with $\varphi(\omega) = 0$, for each $\omega \in \mathbb{R}^n \setminus B(0,1)$ and $\int_{\mathbb{R}^n} \varphi(\omega)\, d\omega = 1$. Let $\varepsilon > 0$ and let $\varphi_\varepsilon : \mathbb{R}^n \to \mathbb{R}$ be given by*

$$\varphi_\varepsilon(\omega) = \frac{1}{\varepsilon^n} \varphi\left(\frac{\omega}{\varepsilon}\right)$$

for each $\omega \in \mathbb{R}^n$. Let $p \in [1, +\infty)$, $f \in L^p(\mathbb{R}^n, \mu; X)$ and let $f_\varepsilon : \mathbb{R}^n \to X$ be given by

$$f_\varepsilon(x) = \int_{\mathbb{R}^n} \varphi_\varepsilon(x - \omega) f(\omega)\, d\omega$$

for each $x \in \mathbb{R}^n$. Then $f_\varepsilon \in C^\infty(\mathbb{R}^n; X) \cap L^\infty(\mathbb{R}^n, \mu; X)$ and

$$\lim_{\varepsilon \to 0} f_\varepsilon = f$$

in the norm of $L^p(\mathbb{R}^n, \mu; X)$.

See Barbu [**21**], Lemma 1.1, p. 14.

Definition 1.3.1. A function φ with the properties in Theorem 1.3.3 is called *mollifier*, and the function f_ε is called *the ε-mollified of f*.

We present next some basic results concerning vector-valued distributions of one variable.
Let (a, b) be an open interval and let $\mathcal{D}(a, b)$ the set of C^∞ real functions defined on (a, b) with compact support in (a, b). We recall that the *support* of a function $\varphi : (a, b) \to \mathbb{R}$ is the set $\operatorname{supp} \varphi = \overline{\{t \in (a, b) ;\ \varphi(t) \neq 0\}}$. By definition, a sequence $(\varphi_n)_{n \in \mathbb{N}}$ in this space is *convergent* to φ if there exists a compact set $K \subset (a, b)$ such that $\varphi_n(t) = 0$ for each $n \in \mathbb{N}$ and each $t \in (a, b) \setminus K$ and, for each $k \in \mathbb{N}$, we have

$$\lim_{n \to \infty} \varphi_n^{(k)}(t) = \varphi^{(k)}(t)$$

uniformly for $t \in K$. Let X be a Banach space. We denote by $\mathcal{D}'(a, b; X)$ the set of linear continuous operators from $\mathcal{D}(a, b)$ to X and we call the elements of this space *X-valued distributions on* (a,b). If $f \in \mathcal{D}'(a,b;X)$ and $k \in \mathbb{N}$, we denote by $f^{(k)}$ the kth-order derivative of f in the sense of X-valued distributions on (a,b), i.e.

$$f^{(k)}(\varphi) = (-1)^k f(\varphi^{(k)})$$

for each $\varphi \in \mathcal{D}(a,b)$, where $\varphi^{(k)}$ is the classical k^{th}-order derivative of φ. Let $k,p \in \mathbb{N}$ and let us denote by $W^{k,p}(a,b;X)$ the set of all X-valued distributions f on (a,b) satisfying

$$f^{(m)} \in L^p(a,b;X)$$

for each $m = 0,1,\ldots,k$, where the set (a,b) is endowed with the Lebesgue measure.
We denote by $A^{k,p}(a,b;X)$ the set of all $u : [\,a,b\,] \to X$, whose m^{th}-order derivatives $u^{(m)}$ are absolutely continuous on $[\,a,b\,]$ for $m = 0,1,\ldots,k-1$, and belong to $L^p(a,b;X)$, for $m = 0,1,\ldots,k$. By convention, $u^{(0)} = u$.

Theorem 1.3.4. *If X is reflexive then each function $u \in A^{1,p}(a,b;X)$ is a.e. differentiable on (a,b) and for each $t \in [\,a,b\,]$*

$$u(t) = u(a) + \int_a^t u'(s)\,ds.$$

See Barbu and Precupanu [**22**], Theorem 3.4, p. 53.

Theorem 1.3.5. *If X is a Banach space, $1 \leq p \leq \infty$ and $f \in L^p(a,b;X)$, then $f \in W^{k,p}(a,b;X)$ if and only if there exists $u \in A^{k,p}(a,b;X)$ with $f(t) = u(t)$ a.e. on (a,b).*

See Brezis [**28**], Proposition A.6, p. 154.

1.4. Functions of Bounded Variation

In this section we recall some basic concepts and results concerning vector-valued functions of bounded variation.
Let $\mathcal{P}[\,a,b\,]$ be the set of all partitions of the interval $[\,a,b\,]$. We recall that, if $g : [\,a,b\,] \to X$, then, for each $\mathcal{P} \in \mathcal{P}[\,a,b\,]$, $\mathcal{P} : a = t_0 < t_1 < \cdots < t_k = b$, the number

$$\mathrm{Var}_{\mathcal{P}}(g,[\,a,b\,]) = \sum_{i=1}^{k} \|g(t_i) - g(t_{i-1})\|$$

is called the *variation of the function g with respect to the partition $\mathcal{P}$*. If

$$\sup_{\mathcal{P}\in\mathcal{P}[\,a,b\,]} \mathrm{Var}_{\mathcal{P}}(g,[\,a,b\,]) < +\infty,$$

then g is said to be of *bounded variation*[5] and the number

$$\operatorname{Var}(g,[a,b]) = \sup_{\mathcal{P}\in\mathcal{P}[a,b]} \operatorname{Var}_{\mathcal{P}}(g,[a,b])$$

is called *the variation* of the function g on the interval $[a,b]$. Similarly, we define $\operatorname{Var}(g,(a,b])$, $\operatorname{Var}(g,[a,b))$ and $\operatorname{Var}(g,(a,b))$. Of course, in each of the above three cases, we have to consider only partitions $\mathcal{P}$ satisfying either $a < t_0$ or $t_k < b$ or both, as required by the type of the interval considered. We denote by $BV([a,b];X)$ the vector space of all functions of bounded variation from $[a,b]$ to X. The mapping $g \mapsto \operatorname{Var}(g,[a,b])$ is a seminorm on $BV([a,b];X)$, while $g \mapsto \|g(a)\| + \operatorname{Var}(g,[a,b])$ is a norm, denoted by $\|\cdot\|_{BV([a,b];X)}$ and, in respect to which, $BV([a,b];X)$ is a Banach space. In the proposition below, we collect without proof some simple but useful properties of the functions of bounded variation.

Proposition 1.4.1. *Let $g \in BV([a,b];X)$. Then:*

(i) *if $[c,d] \subseteq [a,b]$ then $g \in BV([c,d];X)$ and*

$$\operatorname{Var}(g,[c,d]) \le \operatorname{Var}(g,[a,b]);$$

(ii) *the function $t \mapsto \operatorname{Var}(g,[a,t])$ is nondecreasing on $[a,b]$;*

(iii) *for each $c \in (a,b)$ we have*

$$\operatorname{Var}(g,[a,b]) = \operatorname{Var}(g,[a,c]) + \operatorname{Var}(g,[c,b]).$$

The next extension of Lemma 16, p. 140 in Dunford and Schwartz [**49**] will prove useful in the sequel.

Lemma 1.4.1. *For each $g \in BV([a,b];X)$, $t \in [a,b)$, and $s \in (a,b]$, we have*

$$\lim_{h\downarrow 0} \operatorname{Var}(g,(t,t+h]) = 0, \tag{1.4.1}$$

and

$$\lim_{h\downarrow 0} \operatorname{Var}(g,[s-h,s)) = 0. \tag{1.4.2}$$

Proof. Since the proofs of both (1.4.1) and (1.4.2) are quite similar, we confine ourselves only to the proof of (1.4.1). Let us assume by contradiction that this is not the case. Then there exist $\varepsilon > 0$ and $h_1 \in (0,b-t)$ such that

$$\operatorname{Var}(g,(t,t+h_1]) > \varepsilon.$$

So, there exists at least one $\mathcal{P}_1 : t < t_0^1 < t_1^1 < \cdots < t_{k(1)}^1 = t+h_1$ such that

$$\operatorname{Var}_{\mathcal{P}_1}(g,(t,t+h_1]) > \varepsilon.$$

---

[5]Some authors, as for instance Hille and Phillips [**70**], p. 59, refer to this kind of functions as to *functions of strong bounded variation*.

Repeating the same reasoning on $(t, t_0^1]$, we find an $h_2 \in (0, t_0^1 - t]$, and a partition $\mathcal{P}_2 : t < t_0^2 < t_1^2 < \cdots < t_{k(2)}^2 = t_0^1$ of $(t, t_0^1]$, such that, for the very same $\varepsilon > 0$, we have

$$\mathrm{Var}_{\mathcal{P}_2}(g, (t, t + h_2]) > \varepsilon.$$

Clearly, $\mathcal{P}_{1,2} : t_0^2 < t_1^2 \ldots t_{k(2)}^2 < t_0^1 < t_1^1 < \cdots < t_{k(1)} = t + h_1$ is a partition of $(t, t + h_1]$, and

$$\mathrm{Var}_{\mathcal{P}_{1,2}}(g, (t, t + h_1]) > 2\varepsilon.$$

By an inductive argument, we get that $\mathrm{Var}\,(g, (t, t + h_1]) = +\infty$, which contradicts the hypotheses. This contradiction can be eliminated only if (1.4.1) holds, and this achieves the proof. □

Proposition 1.4.2. *If $g \in BV([a, b]; X)$, then g is piecewise continuous on $[a, b]$, i.e. there exists an at most countable subset E of $[a, b]$, such that g is continuous on $[a, b] \setminus E$ and, at each $t \in E \cap [a, b)$ $(s \in E \cap (a, b])$, there exists the one-sided limit $g(t + 0)$ $(g(s - 0))$.*

Proof. Let $t \in [a, b)$. Then, for each $t < s < \tau < b$, we have

$$\|g(\tau) - g(s)\| \leq \mathrm{Var}\,(g, [s, \tau]) \leq \mathrm{Var}\,(g, (t, \tau]).$$

By virtue of Lemma 1.4.1, this inequality shows that $\theta \mapsto g(\theta)$ satisfies the Cauchy Criterion for the existence of the one-sided limit $g(t + 0)$. Since the existence of $g(s - 0)$ follows by a very similar argument, the proof is complete. □

We say that a subset $\mathcal{G}$ in $BV([a, b]; X)$ is of *equibounded variation* if there exists $M > 0$ such that, for each $g \in \mathcal{G}$, we have $\mathrm{Var}\,(g, [a, b]) \leq M$. Similarly, a sequence $(g_k)_{k \in \mathbb{N}}$ in $BV([a, b]; X)$ is of *equibounded variation* if the set of its terms enjoys this property.

Theorem 1.4.6. (Helly-Bray) *Let $(g_k)_{k \in \mathbb{N}}$ be a sequence of functions with equibounded variation in $BV([a, b])$, and let us assume that, for each t in $[a, b]$, we have*

$$\lim_{k \to \infty} g_k(t) = g(t).$$

Then $g \in BV([a, b])$ and, for each $f \in C([a, b])$, we have

$$\lim_{k \to \infty} \int_a^b f(s)\, dg_k(s) = \int_a^b f(s)\, dg(s).$$

See Graves [**63**], Theorem 2.3, p. 283. An extension of this result which we shall need later is presented below. Let $X^{\odot}$ be a closed subspace in X^* with the property that, for each $x \in X$, we have

$$\|x\| = \sup\{|(x, x^{\odot})|\,;\; x^{\odot} \in X^{\odot},\; \|x^{\odot}\| \leq 1\}.$$

In other words, the closed unit ball in $X^{\odot}$ is a determining set for X. On X we consider $\sigma(X, X^{\odot})$ which, from now on, we shall refer as to the weak-$\odot$ topology which, in general, is weaker than the weak topology on X.

Theorem 1.4.7. *Let $(g_k)_{k\in\mathbb{N}}$ be a sequence in $BV([a,b];X)$ of equibounded variation, let $g : [a,b] \to X$, and let us assume that*

$$\lim_{k\to\infty} (g_k(t), x^{\odot}) = (g(t), x^{\odot})$$

for each $x^{\odot} \in X^{\odot}$ and $t \in [a,b]$. Then $g \in BV([a,b];X)$ and, for each $f \in C([a,b];X^{\odot})$, we have

$$\lim_{k\to\infty} \int_a^b (dg_k(s), f(s)) = \int_a^b (dg(s), f(s)).$$

Proof. First, let us observe that, whenever $\lim_{k\to\infty}(x_k, x^{\odot}) = (x, x^{\odot})$ for each $x^{\odot} \in X^{\odot}$, we have $\|x\| \leq \liminf_{k\to\infty} \|x_k\| \leq \limsup_{k\to\infty} \|x_k\|$ and this clearly shows that $g \in BV([a,b];X)$. Next, let us observe that, by the classical Helly-Bray Theorem 1.4.6, for each $\varphi_i \in C([a,b])$ and $x_i^{\odot} \in X^{\odot}$ with $i = 1, 2, \ldots, n$, we have

$$\lim_{k\to\infty} \int_a^b \left(dg_k(s), \sum_{i=1}^n \varphi_i(s) x_i^{\odot} \right) = \lim_{k\to\infty} \sum_{i=1}^n \int_a^b \varphi_i(s)\, d(g_k(s), x_i^{\odot})$$

$$= \sum_{i=1}^n \int_a^b \varphi_i(s)\, d(g(s), x_i^{\odot}) = \int_a^b \left(dg(s), \sum_{i=1}^n \varphi_i(s) x_i^{\odot} \right).$$

So, the conclusion holds for any function $f \in C([a,b];X^{\odot})$ of the form $f(t) = \sum_{i=1}^k \varphi_i(t) x_i^{\odot}$. Since the set of functions of the form as above is dense in $C([a,b];X)$, this achieves the proof. □

1.5. Sobolev Spaces

Let Ω be a nonempty and open subset in $\mathbb{R}^n$ and $\varphi : \Omega \to \mathbb{R}$. As in the one-dimensional case, the set $\operatorname{supp}\varphi$, defined by

$$\operatorname{supp}\varphi = \overline{\{x \in \Omega;\ \varphi(x) \neq 0\}},$$

is called *the support* of the function φ.
Let $\mathcal{D}(\Omega)$ be the set of C^∞ functions from Ω to $\mathbb{R}$ with compact supports included in Ω. Let $\alpha \in \mathbb{N}^n$ be a multi-index, $\alpha = (\alpha_1, \alpha_2, \ldots, \alpha_n)$ and $\varphi \in \mathcal{D}(\Omega)$. We define

$$D^\alpha \varphi = \frac{\partial^{\alpha_1+\alpha_2+\cdots+\alpha_n} \varphi}{\partial x_1^{\alpha_1} \partial x_2^{\alpha_2} \ldots \partial x_n^{\alpha_n}}.$$

One may easily see that $\mathcal{D}(\Omega)$ is a vector space (even an algebra) over $\mathbb{R}$. We endow this space with a convergence structure as follows.

Definition 1.5.1. We say that the sequence $(\varphi_n)_{n\in\mathbb{N}}$ is *convergent in* $\mathcal{D}(\Omega)$ to φ and we write $\varphi_n \overset{\mathcal{D}(\Omega)}{\longrightarrow} \varphi$, if

(i) there exists a compact subset $K \subset \Omega$ such that, for each $n \in \mathbb{N}$, $\operatorname{supp}\varphi_n \subset K$;

(ii) for each multi-index α we have $\lim_{n\to\infty} D^\alpha\varphi_n = D^\alpha\varphi$ uniformly on Ω, or equivalently on K.

Definition 1.5.2. By a *distribution* on $\mathcal{D}(\Omega)$, we mean a real-valued, linear continuous functional defined on $\mathcal{D}(\Omega)$.

We denote by $\mathcal{D}'(\Omega)$ the set of all distributions on $\mathcal{D}(\Omega)$. If $u \in \mathcal{D}'(\Omega)$ and $\varphi \in \mathcal{D}(\Omega)$ we denote $(u, \varphi) = u(\varphi)$. Similarly, we may consider the space $\mathcal{D}(\Omega; \mathbb{C})$ of all C^∞ functions from Ω to $\mathbb{C}$ and we may define complex-valued distributions on $\mathcal{D}(\Omega; \mathbb{C})$.

Definition 1.5.3. Let $\alpha \in \mathbb{N}^n$ be a multi-index and $u : \Omega \to \mathbb{R}$ a locally integrable function. By definition *the derivative of order α of the function u in the sense of distributions over* $\mathcal{D}(\Omega)$ is the distribution $\mathcal{D}^\alpha u$ defined by

$$(\mathcal{D}^\alpha u, \varphi) = (-1)^{|\alpha|} \int_\Omega u D^\alpha \varphi \, d\omega \tag{1.5.1}$$

for each $\varphi \in \mathcal{D}(\Omega)$, where $|\alpha| = \alpha_1 + \alpha_2 + \cdots + \alpha_n$ is *the length of the multi-index* α.

Let us observe that, if the function u is a.e. differentiable of order α on Ω in the classical sense and $D^\alpha u$ is locally integrable, then $\mathcal{D}^\alpha u$ can be identified with $D^\alpha u$ by means of the equality

$$(\mathcal{D}^\alpha u, \varphi) = \int_\Omega D^\alpha u \varphi \, d\omega$$

for each $\varphi \in \mathcal{D}(\Omega)$, equality obtained by integrating $|\alpha|$-times (1.5.1) by parts. Let $m \in \mathbb{N}$, $1 \le p \le +\infty$ and let us define

$$W^{m,p}(\Omega) = \{u \in L^p(\Omega);\ \mathcal{D}^\alpha u \in L^p(\Omega) \text{ for } 0 \le |\alpha| \le m\}.$$

Theorem 1.5.1. *The mapping $\|\cdot\|_{m,p} : W^{m,p}(\Omega) \to \mathbb{R}_+$, defined by*

$$\|u\|_{m,p} = \begin{cases} \left(\displaystyle\sum_{0\le|\alpha|\le m} \|\mathcal{D}^\alpha u\|^p_{L^p(\Omega)} \right)^{1/p} & \text{if } 1 \le p < +\infty \\ \displaystyle\max_{0\le|\alpha|\le m} \|\mathcal{D}^\alpha u\|_{L^\infty(\Omega)} & \text{if } p = \infty \end{cases}$$

for each $u \in W^{m,p}(\Omega)$, *is a norm with respect to which* $W^{m,p}(\Omega)$ *is a real Banach space.*

For the proof see Adams [**1**], Theorem 3.2, p. 45. The space $W^{m,p}(\Omega)$ is called *the Sobolev space* of integer order m and exponent p.
Let $H^{m,p}(\Omega)$ the completion of the space $\{u \in C^m(\Omega);\ \|u\|_{m,p} < +\infty\}$ with respect to $\|\cdot\|_{m,p}$.

Theorem 1.5.2. (Meyers, Serrin) *For any nonempty and open subset* Ω *of* $\mathbb{R}^n$, *any* $m \in \mathbb{N}$ *and* $1 \leq p < +\infty$, *we have*

$$W^{m,p}(\Omega) = H^{m,p}(\Omega).$$

For the proof see Adams [**1**], Theorem 3.16, p. 52. Let us define now the space $W_0^{m,p}(\Omega)$ as being the closure of $\mathcal{D}(\Omega)$ in $W^{m,p}(\Omega)$. We also denote this space by $H_0^{m,p}(\Omega)$. It is easy to see that

$$W_0^{m,p}(\Omega) \subset W^{m,p}(\Omega) \subset L^p(\Omega),$$

both imbeddings being continuous. Moreover, let us observe that, for each $1 \leq p \leq \infty$, we have $W^{0,p}(\Omega) = L^p(\Omega)$. If $1 \leq p < +\infty$, we also have $W_0^{0,p}(\Omega) = L^p(\Omega)$, the last equality being a simple consequence of the fact that $\mathcal{D}(\Omega)$ is dense in $L^p(\Omega)$ if and only if $p < +\infty$. Let $1 \leq q \leq +\infty$ and let us denote by $W^{-m,q}(\Omega)$ the set of all distributions $u \in \mathcal{D}'(\Omega)$ of the form

$$u = \sum_{|\alpha| \leq m} \mathcal{D}^\alpha f_\alpha$$

with $f_\alpha \in L^q(\Omega)$.

Theorem 1.5.3. *For each* $1 \leq p < \infty$, *the topological dual of* $W_0^{m,p}(\Omega)$ *is* $W^{-m,q}(\Omega)$, *where* $1/p + 1/q = 1$ *if* $1 < p < +\infty$ *and* $q = \infty$ *if* $p = 1$.

See Adams [**1**], Theorem 3.8, p. 48. In all that follows we shall denote the spaces $W^{m,2}(\Omega) = H^{m,2}(\Omega)$ and $W_0^{m,2}(\Omega) = H_0^{m,2}(\Omega)$ by $H^m(\Omega)$ and by $H_0^m(\Omega)$ respectively. Clearly these are Hilbert spaces with respect to the inner product

$$\langle u, v\rangle_m = \sum_{|\alpha| \leq m} \int_\Omega \mathcal{D}^\alpha u \mathcal{D}^\alpha v \, d\omega.$$

Theorem 1.5.4. (Sobolev, Rellich, Kondrachov) *Let us assume that* Ω *is a nonempty, open and bounded subset in* $\mathbb{R}^n$ *whose boundary is of class* C^1, $m \in \mathbb{N}$ *and* $p, q \in [1, +\infty)$.

(i) *If* $mp < n$ *and* $q < \frac{np}{n-mp}$, *then* $W^{m,p}(\Omega)$ *is compactly imbedded in* $L^q(\Omega)$.

(ii) *If* $mp = n$ *and* $q \in [1, +\infty)$, *then* $W^{m,p}(\Omega)$ *is compactly imbedded in* $L^q(\Omega)$.

(iii) *If* $mp > n$, *then* $W^{m,p}(\Omega)$ *is compactly imbedded in* $C(\overline{\Omega})$.

Se Adams [**1**], Theorem 5.4, p. 97 and Theorem 6.2, p. 144.
In order to define the Sobolev space of fractional order $s \geq 0$ and exponent 2, we shall use the Fourier transform. More precisely, if $s \geq 0$, we define

$$H^s(\mathbb{R}^n) = \{u \in L^2(\mathbb{R}^n);\ (1 + \|\xi\|^2)^{s/2}\hat{u}(\xi) \in L^2(\mathbb{R}^n)\},$$

where $\hat{u}$ is the Fourier transform of the function u, i.e.

$$\hat{u}(\xi) = \frac{1}{(2\pi)^{n/2}} \int_{\mathbb{R}^n} e^{-i\langle \xi, \omega \rangle} u(\omega)\, d\omega.$$

On the space $H^s(\mathbb{R}^n)$, let us define the norm $\|\cdot\|_s$ by

$$\|u\|_s = \|(1 + \|\xi\|^2)^{s/2}\hat{u}\|_{L^2(\mathbb{R}^n)}.$$

From Plancherel theorem (see Stein and Weiss [**116**], Theorems 2.3, 2.4, p. 17), it follows that if $s = m$ is natural, then $H^s(\mathbb{R}^n) = H^m(\Omega)$ with $\Omega = \mathbb{R}^n$. See, for instance, Lions and Magenes [**85**], Théorème 1.2, p. 7. In order to define the space $H^s(\Omega)$ for every $s \in \mathbb{R}_+$, let us assume that Ω is a nonempty and open subset whose boundary Γ is an $n - 1$-dimensional C^1 manifold. Moreover, let us assume that Ω is locally on one side of Γ. We define both $H^s(\Omega)$, as the set of restrictions to Ω of all elements in $H^s(\mathbb{R}^n)$, and $\|\cdot\|_s : H^s(\Omega) \to \mathbb{R}_+$ as

$$\|u\|_s = \inf\{\|U\|_s;\ U \in H^s(\Omega),\ U_{|\Omega} = u\}.$$

Next, let $\{\chi_i;\ i \in \mathcal{I}\}$ be a family of local charts on Γ, and $\{\theta_i;\ i \in \mathcal{I}\}$ a subordinated finite partition of the unity. For each $u \in L^2(\Gamma)$ we have

$$u = \sum_{i \in \mathcal{I}} \theta_i u.$$

Let $u_i = \chi_i \circ \theta_i u$. We define $H^s(\Gamma)$ as the space of all functions $u \in L^2(\Gamma)$ with the property that $u_i \in H^s(\mathbb{R}^{n-1})$ for each $i \in \mathcal{I}$. One may easily conclude that $H^s(\Gamma)$, endowed with the inner product

$$\langle u, v \rangle_{H^s(\Gamma)} = \sum_{i \in \mathcal{I}} \langle u_i, v_i \rangle_{H^s(\mathbb{R}^{n-1})},$$

is a real Hilbert space. More that this, both $H^s(\Gamma)$ and the inner product are independent of both the choice of the family of local charts and the corresponding partition of unity. By definition, for $s < 0$,

$$H^s(\Gamma) = (H^{-s}(\Gamma))^*.$$

Taking advantage of Riesz theorem, we agree to identify $H^0(\Gamma) = L^2(\Gamma)$ with its own dual. If $u \in C^1(\overline{\Omega})$, we denote by u_ν the *outward normal derivative* of u on Γ, i.e.

$$u_\nu = \langle \nu, \nabla u \rangle,$$

where ν is exterior unit normal at the current point of Γ.

Theorem 1.5.5. *Let Ω be a nonempty and open subset in $\mathbb{R}^n$ whose boundary Γ is of class C^1 and let $\alpha > 1/2$. The function $u \mapsto \left(u_{|\Gamma}\right)$ from $C(\overline{\Omega})$ to $C(\Gamma)$, has a unique linear bounded extension $\mathcal{T}_0 : H^s(\Omega) \to H^{s-1/2}(\Gamma)$. For $s = 1$, the kernel of this application is $H_0^1(\Omega)$. Similarly, if $\alpha > 3/2$, the function $u \mapsto u_\nu$, from $C^1(\overline{\Omega})$ to $C(\Gamma)$, has a unique linear bounded extension $\mathcal{T}_1 : H^s(\Omega) \to H^{s-3/2}(\Gamma)$.*

Theorem 1.5.5 enables us to give a sense to the outward normal derivative of u, whenever $u \in H^s(\Omega)$, with $s > 3/2$. More precisely, for $u \in H^s(\Omega)$, we set $u_\nu = \mathcal{T}_1 u$.
We conclude this section with some general results in the theory of elliptic equations.

Theorem 1.5.6. (Friedrichs) *Let Ω be a nonempty and open subset in $\mathbb{R}^n$ whose boundary Γ is of class C^1. Then there exists $k_1 > 0$ such that, for each $u \in H^1(\Omega)$, we have*

$$\|u\|_{H^1(\Omega)}^2 \leq k_1 \left(\|\nabla u\|_{L^2(\Omega)}^2 + \|u_{|\Gamma}\|_{L^2(\Gamma)}^2 \right).$$

See Nečas [**94**], Théorème 1.9, p. 20.

Theorem 1.5.7. *Let Ω be a nonempty open and bounded subset in $\mathbb{R}^n$ whose boundary Γ is of class C^1. Then $\|\cdot\| : H^1(\Omega) \to \mathbb{R}_+$, defined by*

$$\|u\| = \left(\|\nabla u\|_{L^2(\Omega)}^2 + \|u_{|\Gamma}\|_{L^2(\Gamma)}^2 \right)^{1/2}$$

for each $u \in H^1(\Omega)$, is a norm on $H^1(\Omega)$ equivalent with the usual one. In particular, the restriction of this norm to $H_0^1(\Omega)$, i.e. $\|\cdot\|_0 : H_0^1(\Omega) \to \mathbb{R}_+$, defined by

$$\|u\|_0 = \|\nabla u\|_{L^2(\Omega)},$$

for each $u \in H_0^1(\Omega)$, is a norm on $H_0^1(\Omega)$ (called the gradient norm) equivalent with the usual one. In respect with this norm the application $D : H_0^1(\Omega) \to H^{-1}(\Omega)$, defined by

$$\langle v, Du \rangle_{H_0^1(\Omega), H^{-1}(\Omega)} = \int_\Omega \nabla u \nabla v \, d\omega,$$

is the canonical isomorphism between $H_0^1(\Omega)$ *and its dual* $H^{-1}(\Omega)$. *The restriction of this application to* $H^2(\Omega)$ *coincides with* $-\Delta$, *where* Δ *is the Laplace operator in the sense of distributions over* $\mathcal{D}(\Omega)$.

Proof. The conclusion follows from Theorems 1.5.5 and 1.5.6. □

Remark 1.5.1. Theorem 1.5.7 justifies why, the canonical isomorphism D between $H_0^1(\Omega)$, endowed with the gradient norm $\|\cdot\|_0$, and $H^{-1}(\Omega)$, is by now on denoted by $-\Delta$.

Corollary 1.5.1. *For any* $\lambda > 0$ *and* $f \in H^{-1}(\Omega)$ *the equation* $\lambda u - \Delta u = f$ *has a unique solution* $u \in H_0^1(\Omega)$.

Theorem 1.5.8. *The application* $I - \Delta : H_0^1(\Omega) \to H^{-1}(\Omega)$ *is the canonical isomorphism between* $H_0^1(\Omega)$, *endowed with the usual norm on* $H^1(\Omega)$, *and its dual* $H^{-1}(\Omega)$, *endowed with the usual dual norm. In addition, for each* $u \in H_0^1(\Omega)$ *and each* $v \in L^2(\Omega)$, *we have*

$$\langle u, v\rangle_{L^2(\Omega)} = \langle u, v\rangle_{H_0^1(\Omega), H^{-1}(\Omega)}.$$

1.6. Unbounded Linear Operators

Let X be a Banach space over $\mathbb{K} \in \{\mathbb{R}, \mathbb{C}\}$ whose norm is denoted by $\|\cdot\|$.

Definition 1.6.1. By a *linear operator* in X we mean an ordered pair (D, A), where D is a subspace of X and $A : D \to X$ is a linear application, i.e.:

$$A(\alpha x + \beta y) = \alpha Ax + \beta Ay$$

for each $x, y \in D$ and each $\alpha, \beta \in \mathbb{K}$. The operator (D, A) is called *bounded* if $\sup\{\|Ax\|;\ x \in D,\ \|x\| \leq 1\} < +\infty$. Otherwise, the operator (D, A) is said to be *unbounded*.

Definition 1.6.2. Let (D, A) be a linear operator. The sets $D = D(A)$, $A(D) = R(A)$, $\{(x, y) \in X \times X;\ x \in D,\ y = Ax\} = \mathrm{graph}\,(A)$ and $\{x \in D;\ Ax = 0\} = \ker(A)$ are called: *the domain, the range, the graph* and respectively *the kernel* of the operator (D, A).

In the sequel, from traditional reasons, we shall denote an operator (D, A) by $A : D(A) \subseteq X \to X$ and we shall say that $D(A)$ is its domain, or that A is defined on $D(A)$, $R(A)$ is its image and $\mathrm{graph}\,(A)$ is its graph, instead of the domain of (D, A), the image of (D, A) and respectively the graph of (D, A). Also, we shall write $A = B$ instead of $\mathrm{graph}\,(A) = \mathrm{graph}\,(B)$. We denote by $\mathcal{L}(X)$ the set of all bounded linear operators, from X to X. On this space we define $\|\cdot\|_{\mathcal{L}(X)} : X \to \mathbb{R}_+$ by

$$\|T\|_{\mathcal{L}(X)} = \sup\{\|Tx\|;\ x \in X,\ \|x\| \leq 1\}$$

for each $T \in \mathcal{L}(X)$. It is easy to see that $\|\cdot\|_{\mathcal{L}(X)}$ is a norm on $\mathcal{L}(X)$, called *the operator norm*, with respect to which this is a Banach space and that, for $T \in \mathcal{L}(X)$,

$$\|T\|_{\mathcal{L}(X)} = \sup\{\|Tx\|;\ x \in X,\ \|x\| = 1\}.$$

The norm topology on $\mathcal{L}(X)$ is called *the uniform operator topology*.

Remark 1.6.1. Let $\lambda \in \mathbb{K}$. One may easily verify that $(\lambda I - A)^{-1} \in \mathcal{L}(X)$ if and only if, for each $f \in X$, the equation $\lambda u - Au = f$ has a unique solution $u = u(\lambda, f)$ and the mapping $f \mapsto u(\lambda, f)$ is continuous.

Let X be a Banach space and $A : D(A) \subseteq X \to X$ a linear operator (bounded or not) and let $X(A)$ be the closed linear subspace in X spanned by $D(A)$. Obviously $X(A) = X$ if and only if $D(A)$ is dense in X. We recall that the *adjoint of the linear operator* $A : D(A) \subseteq X \to X$ is the operator $A^* : D(A^*) \subseteq X^* \to X(A)^*$ defined by

$$D(A^*) = \{x^* \in X^*;\ \exists C > 0, \forall y \in D(A),\ |(x^*, Ay)| \leq C\|y\|\}$$

and, for each $x^* \in D(A^*)$, A^*x^* is the unique continuous extension to $X(A)$ of the linear continuous mapping $y \mapsto (x^*, Ay)$ from $D(A)$ to $\mathbb{K}$.

Definition 1.6.3. Let H be a real Hilbert space identified with its own topological dual. The operator $A : D(A) \subseteq H \to H$ is called:

(i) *self-adjoint* if $A = A^*$;
(ii) *skew-adjoint* if $A = -A^*$;
(iii) *symmetric* if $\langle Ax, y\rangle = \langle x, Ay\rangle$ for each $x, y \in D(A)$;
(iv) *skew-symmetric* if $\langle Ax, y\rangle = -\langle x, Ay\rangle$ for each $x, y \in D(A)$.

Remark 1.6.2. If $A : D(A) \subseteq H \to H$ is symmetric (skew-symmetric), then $\operatorname{graph}(A) \subseteq \operatorname{graph}(A^*)$ ($\operatorname{graph}(A) \subseteq \operatorname{graph}(-A^*)$).

Remark 1.6.3. We may easily see that an operator $A : D(A) \subseteq H \to H$ is skew-symmetric if and only if, for each $x \in D(A)$, $\langle Ax, x\rangle = 0$. Indeed, if A is skew-symmetric we have $\langle Ax, x\rangle = -\langle x, Ax\rangle$, which implies $\langle Ax, x\rangle = 0$ for each $x \in D(A)$. Conversely, if $\langle Ax, x\rangle = 0$ for each $x \in D(A)$, we have

$$\begin{aligned} 0 = \langle A(x+y), x+y\rangle &= \langle Ax, x\rangle + \langle Ax, y\rangle + \langle Ay, x\rangle + \langle Ay, y\rangle \\ &= \langle Ax, y\rangle + \langle x, Ay\rangle \end{aligned}$$

for each $x, y \in D(A)$ and thus A is skew-symmetric.

Let A be self-adjoint. Then its graph coincides with the graph of A^* and consequently A is symmetric. Similarly, if A is skew-adjoint, then it is skew-symmetric. The next lemma shows that, for a sufficiently large class of operators, the converse statements are true.

Lemma 1.6.1. *Let H be a Hilbert space and $A : D(A) \subseteq H \to H$ a densely defined operator. We have:*

(i) *if $(I - A)^{-1} \in \mathcal{L}(H)$, then A is self-adjoint if and only if A is symmetric;*

(ii) *if $(I \pm A)^{-1} \in \mathcal{L}(H)$, then A is skew-adjoint if and only if A is skew-symmetric.*

Proof. (i) Let us assume that A is symmetric. In view of Remark 1.6.2, the graph of A is included in the graph of A^*. To complete the proof is suffices to show that $D(A^*) \subseteq D(A)$. Since $D(A)$ is dense in H, it follows that, for each $y \in D(A^*)$, there exists $z \in H$ such that $\langle Ax, y\rangle = \langle x, z\rangle$ for each $x \in D(A)$. Since $I - A$ is surjective, there exists $w \in D(A)$ such that $y - z = w - Aw$. On the other hand, for each $x \in D(A)$, we have

$$\langle Ax - x, y - w\rangle = \langle Ax, y\rangle - \langle Ax, w\rangle - \langle x, y\rangle + \langle x, w\rangle$$

$$= \langle x, z\rangle - \langle Ax, w\rangle - \langle x, y\rangle + \langle x, w\rangle$$

$$= \langle x, y - w + Aw\rangle - \langle Ax, w\rangle - \langle x, y\rangle + \langle x, w\rangle = 0.$$

But $I - A$ is surjective and accordingly, the above equality proves that $y = w$, which implies $y \in D(A)$ and $Ay = A^*y$. So (i) holds. Since (ii) follows by using similar arguments, the proof is complete. □

Let H be a complex Banach space for which there exists a *complex inner product*, i.e. a mapping $[\,\cdot,\cdot\,] : H \times H \to \mathbb{C}$ satisfying:

(i) $[\,x, y\,] = \overline{[\,y, x\,]}$ for each $x, y \in H$;

(ii) $[\,\alpha x + \beta y, z\,] = \alpha[\,x, z\,] + \beta[\,y, z\,]$ for each $x, y, z \in H$ and $\alpha, \beta \in \mathbb{C}$;

(iii) $[\,x, x\,] \geq 0$ and $[\,x, x\,] = 0$ if and only if $x = 0$;

(iv) $\|x\| = \sqrt{[\,x, x\,]}$ for each $x \in H$.

We may easily verify that the mapping $\langle\cdot,\cdot\rangle : H \times H \to \mathbb{R}$ defined by

$$\langle x, y\rangle = Re[\,x, y\,],$$

for each $x, y \in H$, is a real inner product on H. In addition, endowed with this real inner product, H is a real Hilbert space. From now on, a Banach space H endowed with a complex inner product will be called *a complex Hilbert space.*

Lemma 1.6.2. *Let H be a complex Hilbert space, let $A : D(A) \subseteq H \to H$ be a densely defined, $\mathbb{C}$-linear operator and let iA be defined by*

$$\begin{cases} D(iA) = D(A), \\ (iA)x = iAx & \textit{for each } x \in D(iA). \end{cases}$$

Then A^* *is* $\mathbb{C}$*-linear and* $(iA)^* = -iA^*$. *In particular,* A *is self-adjoint if and only if* iA *is skew-adjoint. Analogously,* A *is skew-adjoint if and only if* iA *is self-adjoint.*

Remark 1.6.4. Essentially, Lemma 1.6.2 asserts that the adjoint of A, on the complex Hilbert space H identified with its own topological dual H_c^*, coincides with the adjoint of A, on the real Hilbert space H identified with its own topological dual H_r^*.

Proof. Let $x \in D(A^*)$. For each $y \in D(A)$ and $\lambda \in \mathbb{C}$ we have

$$\begin{aligned}\langle \lambda A^* x, y\rangle &= \langle A^* x, \overline{\lambda} y\rangle = \langle x, A(\overline{\lambda} y)\rangle \\ &= \langle x, \overline{\lambda} A y\rangle = \langle \lambda x, Ay\rangle = \langle A^*(\lambda x), y\rangle.\end{aligned}$$

Since $D(A)$ is dense in H and $y \in D(A)$ is arbitrary, from the above equality it follows that $A^*(\lambda x) = \lambda A^* x$, which proves that A^* is $\mathbb{C}$-linear.
Next, let $x \in D(A^*)$ and $y \in D(A)$. We have

$$\langle -iA^* x, y\rangle = \langle A^* x, iy\rangle = \langle x, A(iy)\rangle = \langle x, iAy\rangle = \langle (iA)^* x, y\rangle.$$

So $x \in D((iA)^*)$, and $\operatorname{graph}(-iA^*) \subseteq \operatorname{graph}((iA)^*)$. Substituting A by iA, we deduce that $\operatorname{graph}(-i(iA)^*) \subseteq \operatorname{graph}(-A^*)$. From the $\mathbb{C}$-linearity, we conclude that $\operatorname{graph}(iA)^* \subseteq \operatorname{graph}(-iA^*)$ and thus it follows that $\operatorname{graph}(iA)^* = \operatorname{graph}(-iA^*)$.
Finally, if A is self-adjoint, from the previous considerations, we conclude that

$$(iA)^* = -iA^* = -iA,$$

which shows that iA is skew-adjoint. Conversely, if iA is skew-adjoint, we have

$$A^* = (-i(iA))^* = i(iA)^* = -i(iA) = A$$

and hence A is self-adjoint. Similarly, if A is skew-adjoint then

$$(iA)^* = -iA^* = iA$$

and so iA is self-adjoint. Conversely, if iA is self-adjoint then

$$A^* = (-i(iA))^* = i((iA)^*) = i(iA) = -A.$$

The proof is complete. □

Lemma 1.6.3. *Let* $A : D(A) \subseteq H \to H$ *be a linear operator. If* A *is densely defined, there exists its adjoint* $A^* : D(A^*) \subseteq H \to H$*, and:*

(i) $\operatorname{graph}(A^*) = \{(x, f) \in H \times H;\ \langle f, y\rangle = \langle x, g\rangle, \forall (y, g) \in \operatorname{graph}(A)\}$;
(ii) $(x, f) \in \operatorname{graph}(A^*)$ *if and only if* $(-f, x) \in (\operatorname{graph}(A))^{\perp}$;
(iii) $\operatorname{graph}(A^*)$ *is closed in* $H \times H$.

Proof. Let $G^* = \{(x,f) \in H \times H;\ \langle f,y\rangle = \langle x,g\rangle,\ \forall (y,g) \in \text{graph}\,(A)\}$. If $(x,f) \in G^*$, we have $\langle f,y\rangle = \langle x,Ay\rangle$ for any $y \in D(A)$ and therefore $|\langle x,Ay\rangle| \leq \|f\|\|y\|$. It follows that $x \in D(A^*)$ and $f = A^*x$, and accordingly $G^* \subseteq \text{graph}\,(A^*)$. Next, let $x \in D(A^*)$ and let $f = A^*x$. Clearly, we have $\langle f,y\rangle = \langle x,Ay\rangle$ for each $x \in D(A)$. But, this last relation shows that $\langle f,y\rangle = \langle x,g\rangle$ for each $(y,g) \in \text{graph}\,(A)$ and hence $\text{graph}\,(A^*) \subseteq G^*$, which proves (i). Obviously (ii) follows from (i), while (iii) from (ii) simply because for each subset B in H its orthogonal complement $B^{\perp}$ is a linear closed subspace. □

We conclude this section with a useful necessary and sufficient condition in order that a densely defined, symmetric operator be self-adjoint.

Theorem 1.6.1. *Let $A : D(A) \subseteq H \to H$ be a linear, densely defined, symmetric operator. Then A is self-adjoint if and only if $(I \pm iA)^{-1} \in \mathcal{L}(H)$.*

See Taylor [**119**], Proposition 8.5, p. 513.

1.7. Elements of Spectral Analysis

Let X be a complex Banach space and $A : D(A) \subseteq X \to X$ a linear closed operator. We recall that *the resolvent set* of A is the set of all $\lambda \in \mathbb{C}$ for which the range of $\lambda I - A$ is dense in X and $(\lambda I - A)^{-1} : R(\lambda I - A) \to X$ is continuous. We denote this set by $\rho(A)$ and we call its elements *regular values of the operator A*. For $\lambda \in \rho(A)$, denote by $R(\lambda; A) = (\lambda I - A)^{-1}$. We also recall that *the spectrum of the operator A*, denoted by $\sigma(A)$, is defined by $\mathbb{C} \setminus \rho(A)$. In the next theorems we assume that X is a complex Banach space.

Theorem 1.7.1. *Let $A : D(A) \subseteq X \to X$ be a linear closed operator. Then, for each $\lambda \in \rho(A)$, $R(\lambda; A) \in \mathcal{L}(X)$.*

Proof. If $\lambda \in \rho(A)$ then $R(\lambda I - A) = D((\lambda I - A)^{-1})$ is dense in X and there exists a constant $c > 0$ such that

$$\|(\lambda I - A)x\| \geq c\|x\| \tag{1.7.1}$$

for each $x \in D(A)$. In order to show that $R(\lambda I - A) = X$, let $y \in X$ and let $(x_n)_{n \in \mathbb{N}}$ such that $\lim_{n \to \infty} (\lambda I - A)x_n = y$. From (1.7.1) it follows that there exists $\lim_{n \to \infty} x_n = x$. Since A is closed, we conclude that $\lambda I - A$ is closed too and consequently $x \in D(\lambda I - A)$ and $(\lambda I - A)x = y$. But $\overline{R(\lambda I - A)} = X$ and accordingly $R(\lambda I - A) = X$. The proof is complete. □

Let $A : D(A) \subseteq X \to X$ be a linear and closed operator. The mapping $R(\cdot\,; A) : \rho(A) \to \mathcal{L}(X)$, defined by $R(\lambda; A) = (\lambda I - A)^{-1}$ for each $\lambda \in \rho(A)$, is called *the resolvent function of A*.

Theorem 1.7.2. *Let $A : D(A) \subseteq X \to X$ be a linear closed operator. Then $\rho(A)$ is an open subset in $\mathbb{C}$ and, on each connected component of $\rho(A)$, the resolvent function $\lambda \mapsto R(\lambda; A)$ is analytic.*

Proof. Let $\mu \in \rho(A)$ and let us consider the series

$$S(\lambda) = R(\mu; A)\left\{I + \sum_{n=1}^{\infty}(\mu - \lambda)^n R(\mu; A)^n\right\}. \tag{1.7.2}$$

Clearly, for each $\lambda \in \mathbb{C}$ for which $|\lambda - \mu|\|R(\mu; A)\|_{\mathcal{L}(X)} < 1$, the series in (1.7.2) is convergent in the operator norm topology to an analytic function. Since

$$\begin{cases} S(\lambda)(\lambda I - A) = S(\lambda)[(\lambda - \mu)I + (\mu I - A)] = I \\ (\lambda I - A)S(\lambda) = [(\lambda - \mu)I + (\mu I - A)]S(\lambda) = I, \end{cases}$$

it follows that $S(\lambda) = R(\lambda; A)$. In conclusion, for each $\mu \in \rho(A)$, there exists a disk centered in μ included in $\rho(A)$, disk on which $R(\,\cdot\,; A)$ is analytic and this achieves the proof. □

Theorem 1.7.3. *Let $A : D(A) \subseteq X \to X$ be a linear operator. If λ, μ are regular values, i.e. $\lambda, \mu \in \rho(A)$, and $R(\lambda; A), R(\mu; A) \in \mathcal{L}(X)$, then*

$$R(\lambda; A) - R(\mu; A) = (\mu - \lambda)R(\lambda; A)R(\mu; A). \tag{1.7.3}$$

The relation (1.7.3) is known as *the Resolvent equation.*

Proof. Let us observe that

$$\begin{aligned} R(\lambda; A) &= R(\lambda; A)(\mu I - A)R(\mu; A) = R(\lambda; A)\{(\mu - \lambda)I \\ &+ (\lambda I - A)\}R(\mu; A) = (\mu - \lambda)R(\lambda; A)R(\mu; A) + R(\mu; A). \end{aligned}$$

The proof is complete. □

Theorem 1.7.4. *Let X be a complex Banach space and $A \in \mathcal{L}(X)$. Then there exists*

$$\lim_{n\to\infty} \|A^n\|_{\mathcal{L}(X)}^{1/n} = r_\sigma(A) \tag{1.7.4}$$

and

$$r_\sigma(A) \le \|A\|_{\mathcal{L}(X)}. \tag{1.7.5}$$

In addition, we have $\{\lambda \in \mathbb{C};\ |\lambda| > r_\sigma(A)\} \subseteq \rho(A)$ and, for each $\lambda \in \mathbb{C}$ with $|\lambda| > r_\sigma(A)$, we have

$$R(\lambda; A) = \sum_{n=1}^{\infty} \lambda^{-n} A^{n-1}. \tag{1.7.6}$$

Definition 1.7.1. The number $r_\sigma(A)$ given by (1.7.4) is called *the spectral radius of the operator A.*

Proof. Let $r = \inf_{n\geq 1} \|A^n\|_{\mathcal{L}(X)}^{1/n}$. In order to prove (1.7.4) it suffices to show that

$$\limsup_{n\to\infty} \|A^n\|_{\mathcal{L}(X)}^{1/n} \leq r.$$

Let $\varepsilon > 0$ and $m \in \mathbb{N}$ such that $\|A^m\|_{\mathcal{L}(X)}^{1/m} \leq r + \varepsilon$. Obviously, each $n \in \mathbb{N}$ can be uniquely written as $n = mp + q$ with $0 \leq q \leq m - 1$. Since, for each $A, B \in \mathcal{L}(X)$, we have $\|AB\|_{\mathcal{L}(X)} \leq \|A\|_{\mathcal{L}(X)}\|B\|_{\mathcal{L}(X)}$, we deduce

$$\|A^n\|_{\mathcal{L}(X)}^{1/n} \leq \|A^m\|_{\mathcal{L}(X)}^{p/n}\|A\|_{\mathcal{L}(X)}^{q/n} \leq (r+\varepsilon)^{mp/n}\|A\|_{\mathcal{L}(X)}^{q/n}.$$

As $\lim_{n\to\infty} mp/n = 1$ and $\lim_{n\to\infty} q/n = 0$, it follows that

$$\limsup_{n\to\infty} \|A^n\|_{\mathcal{L}(X)}^{1/n} \leq r + \varepsilon.$$

Since $\varepsilon > 0$ is arbitrary, the preceding inequality shows that

$$\limsup_{n\to\infty} \|A^n\|_{\mathcal{L}(X)}^{1/n} \leq r,$$

which proves (1.7.4).
In order to prove (1.7.5) it suffices to observe that $\|A^n\|_{\mathcal{L}(X)} \leq \|A\|_{\mathcal{L}(X)}^n$ and then to make use of (1.7.4).
Next, if $|\lambda| > r_\sigma(A)$, there exists $\varepsilon > 0$ such that $|\lambda| \geq r_\sigma(A)+\varepsilon$. But (1.7.4) implies that, for large n, $\|\lambda^{-n}A^n\|_{\mathcal{L}(X)} \leq (r_\sigma(A)+\varepsilon)^{-n}(r_\sigma(A)+\varepsilon/2)^n$. Hence the series in (1.7.6) is convergent. Multiplying this series, either on the left, or on the right, by $(\lambda I - A)$ we obtain the identity in X. So (1.7.6) holds and the proof is complete. □

Corollary 1.7.1. *Let X be a complex Banach space and $A \in \mathcal{L}(X)$. Then $\rho(A)$ is nonempty.*

Theorem 1.7.5. *Let X be a complex Banach space and $A \in \mathcal{L}(X)$. Then*

$$r_\sigma(A) = \sup_{\mu\in\sigma(A)} |\mu|. \tag{1.7.7}$$

Proof. From (1.7.6) it follows that

$$r_\sigma(A) \geq \sup_{\mu\in\sigma(A)} |\mu| = |A|.$$

To prove the converse inequality we recall that, by the inequality above and Theorem 1.7.2, it follows that $R(\lambda; A)$ is analytic for $|\lambda| > |A|$. So, it can be expanded as a Laurent series which converges in the operator norm topology for $|\lambda| > |A|$. Since this expansion is unique, from Theorem 1.7.4, it must coincide with the series on the right-hand side of (1.7.6). Consequently, for each $|\lambda| > |A|$, $\lim_{n\to\infty} \|\lambda^{-n}A^n\|_{\mathcal{L}(X)} = 0$. Then, for each $\varepsilon > 0$, we have $\|A^n\|_{\mathcal{L}(X)} \leq (\varepsilon + |A|)^n$ if n is sufficiently large. By consequence

$$r_\sigma(A) = \lim_{n\to\infty} \|A^n\|_{\mathcal{L}(X)}^{1/n} \leq |A| = \sup_{\mu\in\sigma(A)} |\mu|,$$

which proves (1.7.7). □

Corollary 1.7.2. *For each $\lambda \in \mathbb{C}$ with $|\lambda| < r_\sigma(A)$, the series in* (1.7.6) *diverges.*

Proof. Let r be the smallest positive number with the property that the series $\sum_{n=0}^{\infty} \lambda^{-n}A^{n-1}$ converges for each $\lambda \in \mathbb{C}$ with $|\lambda| > r$. Then, for $|\lambda| > r$, we have $\lim_{n\to\infty} \|\lambda^{-n}A^n\|_{\mathcal{L}(X)} = 0$ and, by repeating the reasoning in the previous proof, we deduce that $\lim_{n\to\infty} \|A^n\|_{\mathcal{L}(X)}^{1/n} \leq r$. Hence $r_\sigma(A) \leq r$ and the proof is complete. □

1.8. Functional Calculus for Bounded Operators

Let X a complex Banach space, $\mathbb{D}$ a domain in $\mathbb{C}$ whose boundary Γ is a finite union of rectifiable Jordan arcs oriented in the positive sense, let $f : \mathbb{D} \to \mathbb{C}$ be an analytic function and $A \in \mathcal{L}(X)$. The problem which we may raise in this context is to give a sense to the expression $f(A)$. A first and very natural idea which comes to mind is to define $f(A)$ by

$$f(A) = f(\lambda)I + \sum_{n=1}^{\infty} \frac{f^{(n)}(\lambda)}{n!}(A - \lambda I)^n,$$

where $\lambda \in \mathbb{D}$ is fixed. One can prove that, for each $A \in \mathcal{L}(X)$ satisfying $\sigma(A) \subset \mathbb{D}$, the series on the right-hand side of the above relation is well-defined and normal convergent, i.e. all the terms are properly defined and the series of the corresponding norms is convergent. Unfortunately, this definition cannot be extended easily to the case in which A is unbounded (when $D(A) \neq X$, which clearly implies that the domain of the operator $(A - \lambda I)^n$ depends on $n \in \mathbb{N}$). So, in this case, the idea to define the operator $f(A)$ by means of the above equality is no longer applicable in its simplest form as suggested by its "A-bounded" counterpart. One way to overcome this difficulty is to find a possibility to do this without making use of the natural powers of $A - \lambda I$. This way has been suggested by the celebrated Cauchy integral formula which asserts that, for each $z \in \mathbb{D}$, we have

$$f(z) = \frac{1}{2\pi i}\int_\Gamma \frac{f(\lambda)}{(\lambda - z)}\,d\lambda.$$

By observing that, at least formally, $\lambda \mapsto R(\lambda; A) = (\lambda I - A)^{-1}$ corresponds to the function $\lambda \mapsto \frac{1}{\lambda - a}$, it is quite natural the define $f(A)$ by

$$f(A) = \frac{1}{2\pi i}\int_\Gamma f(\lambda)R(\lambda; A)\,d\lambda, \tag{1.8.1}$$

whenever, of course, $\sigma(A) \subset \mathbb{D}$. We recall that $\sigma(A)$ is *the spectrum of the operator* A, which is defined by $\mathbb{C} \setminus \rho(A)$, where $\rho(A)$ is *the resolvent set* of A, i.e. the set those $\lambda \in \mathbb{C}$ for which $\lambda I - A$ is invertible with bounded inverse. See Theorem 1.7.1.

Remark 1.8.1. In the case when A is bounded, by Theorem 1.7.4 we know that $\sigma(A)$ is bounded and, so, the set $\mathbb{D}$ can be chosen bounded. Moreover, if f is an analytic function on $\mathbb{C}$, then the domain $\mathbb{D}$ can be chosen simple connected and bounded (for instance, the disk centered at 0 and of radius $r_\sigma(A) + 1$).

Remark 1.8.2. One can easily verify that the integral on the right-hand side of (1.8.1) depends only on f and A, but not on the domain $\mathbb{D}$, with the condition that the spectrum of the operator A is contained in $\mathbb{D}$. This follows by observing that, by virtue of the classical Cauchy theorem, for each $x \in X$ and each $x^* \in X^*$, the integral $\int_\Gamma f(\lambda)(R(\lambda; A)x, x^*)\,d\lambda$ does not depend on $\mathbb{D}$.

Definition 1.8.1. The integral on the right-hand side of (1.8.1) is called *the Dunford integral* associated to f and A.

If $A \in \mathcal{L}(X)$, we denote by $\mathcal{F}(A)$ the set of all complex functions, which are analytic in a certain neighborhood of $\sigma(A)$.

Theorem 1.8.1. (Dunford) *Let $A \in \mathcal{L}(X)$, $f, g \in \mathcal{F}(A)$ and $\alpha, \beta \in \mathbb{C}$. Then:*

(i) *$\alpha f + \beta g \in \mathcal{F}(A)$ and $\alpha f(A) + \beta g(A) = (\alpha f + \beta g)(A)$;*

(ii) *$fg \in \mathcal{F}(A)$ and $f(A)g(A) = (fg)(A)$;*

(iii) *if $f(\lambda) = \sum_{n=0}^{\infty} c_n \lambda^n$ on an open neighborhood $\mathbb{D}$ of $\sigma(A)$, then*

$$f(A) = \sum_{n=0}^{\infty} c_n A^n \text{ in the norm of } \mathcal{L}(X);$$

(iv) *if $(f_n)_{n\in\mathbb{N}}$ are analytic on an open neighborhood $\mathbb{D}$ of $\sigma(A)$ and $\lim_{n\to\infty} f_n = f$ uniformly on $\mathbb{D}$, then $\lim_{n\to\infty} f_n(A) = f(A)$ in the norm of $\mathcal{L}(X)$;*

(v) $f \in \mathcal{F}(A^*)$ *and* $f(A^*) = (f(A))^*$.
(vi) $f(\sigma(A)) = \sigma(f(A))$.

Proof. The fact that $\mathcal{F}(A)$ is a vector space is obvious and so (i) holds. In order to prove (ii), let $\mathbb{D}_1$ and $\mathbb{D}_2$ two open neighborhoods of $\sigma(A)$ whose boundaries Γ_1 and Γ_2 are finite unions of Jordan rectifiable arcs with the property that $\mathbb{D}_1 \cup \Gamma_1 \subset \mathbb{D}_2$ and $\mathbb{D}_2 \cup \Gamma_2$ is included in the common domain of analyticity of both functions f and g. Then, by the resolvent equation (1.7.3) and the Cauchy integral formula, it follows that

$$f(A)g(A) = -\frac{1}{4\pi^2}\int_{\Gamma_1} f(\lambda)R(\lambda;A)\,d\lambda \int_{\Gamma_2} g(\mu)R(\mu;A)\,d\mu$$

$$= -\frac{1}{4\pi^2}\int_{\Gamma_1}\int_{\Gamma_2} f(\lambda)g(\mu)(\mu-\lambda)^{-1}(R(\lambda;A)-R(\mu;A))\,d\lambda\,d\mu$$

$$= \frac{1}{2\pi i}\int_{\Gamma_1} f(\lambda)R(\lambda;A)\left\{\frac{1}{2\pi i}\int_{\Gamma_2}(\mu-\lambda)^{-1}g(\mu)\,d\mu\right\}d\lambda$$

$$-\frac{1}{2\pi i}\int_{\Gamma_2} g(\mu)R(\mu;A)\left\{\frac{1}{2\pi i}\int_{\Gamma_1}(\mu-\lambda)^{-1}f(\lambda)\,d\lambda\right\}d\mu$$

$$= \frac{1}{2\pi i}\int_{\Gamma_1} f(\lambda)g(\lambda)R(\lambda;A)\,d\lambda = (fg)(A),$$

which completes the proof of (ii).
To prove (iii), let us observe that, by virtue of Theorem 1.7.4, there exists $\varepsilon > 0$ such that $B(0, r_\sigma(A)+\varepsilon) \subset \mathbb{D}$, where $r_\sigma(A)$ is the spectral radius of the operator A. See Definition 1.7.1. Since f is analytic on $\mathbb{D}$, the series $f(\lambda) = \sum_{n=0}^{\infty} c_n\lambda^n$ is uniformly convergent on $B(0, r_\sigma(A)+\varepsilon)$. From the Cauchy integral formula and the Laurent expansion of the function $\lambda \mapsto R(\lambda;A)$ (see (1.7.6)), we deduce

$$f(A) = \frac{1}{2\pi i}\int_\Gamma \left(\sum_{k=0}^{\infty} c_k\lambda^k\right) R(\lambda;A)\,d\lambda = \frac{1}{2\pi i}\sum_{k=0}^{\infty} c_k \int_\Gamma \lambda^k R(\lambda;A)\,d\lambda$$

$$= \frac{1}{2\pi i}\sum_{k=0}^{\infty} c_k \sum_{n=1}^{\infty}\int_\Gamma \lambda^{k-n}A^{n-1}\,d\lambda = \sum_{k=0}^{\infty} c_k A^k,$$

which proves (iii).
Obviously, (iv) follows from (1.8.1), while (v) is a consequence of (1.8.1) and Phillips theorem which says that, for each densely defined linear and closed operator $A : D(A) \subset X \to X$, we have

$$\rho(A) = \rho(A^*) \text{ and } R(\lambda;A^*) = R(\lambda;A)^*$$

for each $\lambda \in \rho(A)$. See Yosida [**136**], Theorem 2, p. 225. It remains to show (vi). To this aim, fix $\lambda \in \sigma(A)$ and let us observe that $\lambda \in \mathbb{D}$. Let us define the function $g : \mathbb{D} \to \mathbb{C}$ by

$$g(\mu) = \begin{cases} \dfrac{f(\lambda) - f(\mu)}{\lambda - \mu} & \text{for } \mu \neq \lambda \\ \\ f'(\lambda) & \text{for } \mu = \lambda. \end{cases}$$

Obviously $g \in \mathfrak{F}(A)$ and so, from Theorem 1.8.1 (i), it follows that

$$f(\lambda)I - f(A) = (\lambda I - A)g(A).$$

So, if $f(\lambda)I - f(A)$ has a bounded inverse B, then $g(A)B$, which in accordance with the relation above is the inverse of $(\lambda I - A)$, is bounded too. Hence, if $\lambda \in \sigma(A)$, then $f(\lambda) \in \sigma(f(A))$, implication which is equivalent with $f(\sigma(A)) \subseteq \sigma(f(A))$. Now, let us assume by contradiction that there exists $\lambda \in \sigma(f(A))$ such that $\lambda \notin f(\sigma(A))$. In view of Remark 1.8.1 and Theorem 1.7.2, $f(\sigma(A))$ is a compact set and therefore, in these circumstances, the function $h(\mu) = (f(\mu) - \lambda)^{-1}$ is in $\mathfrak{F}(A)$, because $f(\mu) - \lambda \neq 0$ for each $\mu \in \sigma(A)$ and f is analytic on an open neighborhood of $\sigma(A)$. By virtue of (ii) in Theorem 1.8.1, we have $h(A)(f(A) - \lambda I) = I$, relation which contradicts the supposition $\lambda \in \sigma(f(A))$. This contradiction can be eliminated only if $\sigma(f(A)) \subset f(\sigma(A))$. The proof is complete. $\square$

Theorem 1.8.2. *Let $A \in \mathcal{L}(X)$, $f : \mathbb{D} \to \mathbb{C}$ in $\mathfrak{F}(A)$ and $g \in \mathfrak{F}(f(A))$. Then the function $h : \mathbb{D} \to \mathbb{C}$, $h(\lambda) = g(f(\lambda))$ for each $\lambda \in \mathbb{D}$, is in $\mathfrak{F}(A)$ and $h(A) = g(f(A))$.*

Proof. From (vi) in Theorem 1.8.1 it follows that $h \in \mathfrak{F}(A)$. Let $\mathbb{D}_1$ be an open neighborhood of $\sigma(f(A))$ whose boundary Γ_1 consists from a finite number of Jordan rectifiable arcs and such that $\mathbb{D}_1 \cup \Gamma_1$ is included in the domain of analyticity of the function g. Let $\mathbb{D}_2$ be another open neighborhood of $\sigma(A)$ whose boundary Γ_2 consists from a finite number of Jordan rectifiable arcs and such that $\mathbb{D}_2 \cup \Gamma_2$ is included in the domain of analyticity of the function f with property that $f(\mathbb{D}_2 \cup \Gamma_2) \subseteq \mathbb{D}_1$. Then, for each $\lambda \in \mathbb{D}_1$, we have

$$R(\lambda; f(A)) = \frac{1}{2\pi i} \int_{\Gamma_2} (\lambda - f(\mu))^{-1} R(\mu; A)\, d\mu,$$

this because, by virtue of (i) in Theorem 1.8.2, the operator S defined by the integral on the right-hand side satisfies $(\lambda I - f(A))S = S(\lambda I - f(A)) = I$. Then, from the Cauchy integral formula, we have

$$g(f(A)) = \frac{1}{2\pi i} \int_{\Gamma_1} g(\lambda) R(\lambda; f(A))\, d\lambda$$

$$= -\frac{1}{4\pi^2}\int_{\Gamma_1}\int_{\Gamma_2} g(\lambda)(\lambda - f(\mu))^{-1}R(\mu; A)\,d\mu\,d\lambda$$

$$= \frac{1}{4\pi^2}\int_{\Gamma_2} g(f(\mu))R(\mu; A)\,d\mu = h(A).$$

The proof is complete. □

1.9. Functional Calculus for Unbounded Operators

Let X be a complex Banach space and $A : D(A) \subseteq X \to X$ a closed linear operator. As we already have seen in the preceding section, for $A \in \mathcal{L}(X)$ and f analytic in an open neighborhood $\mathbb{D}$ of $\sigma(A)$ whose boundary Γ consists of a finite union of Jordan rectifiable arcs, oriented in the positive sense, we can define $f(A)$ by

$$f(A) = \frac{1}{2\pi i}\int_{\Gamma} f(\lambda)R(\lambda; A)\,d\lambda, \tag{1.9.1}$$

where Γ is the boundary of $\mathbb{D}$. In the case of an unbounded operator A, one may happen that $\sigma(A) = \mathbb{C}$, and therefore, the integral on the right-hand side of (1.9.1) is meaningless, because $\Gamma = \emptyset$. More than this, even in the case in which $\sigma(A) \neq \mathbb{C}$, $\sigma(A)$ might be unbounded. Consequently, in (1.9.1), one should add the contribution of the point at ∞. This explains why, throughout this section, we shall assume that $\sigma(A) \neq \mathbb{C}$ and we shall show what are the modifications we have to do to (1.9.1) in order to give it a sense in the case in which both A and $\sigma(A)$ are unbounded.
Let $A : D(A) \subseteq X \to X$ be a closed linear operator and let $\mathfrak{F}(A)$ the set of complex functions which are analytic in a neighborhood of the set $\sigma(A)$ and at ∞. Let $\alpha \in \rho(A)$ be fixed and let us define the mapping $P_\alpha : X \to D(A)$ by $P_\alpha = (A - \alpha I)^{-1} = -R(\alpha; A)$. Obviously P_α, which is a bijection from X to $D(A)$, belongs to $\mathcal{L}(X)$ and thus, by using the arguments in the preceding section, we can define $f(P_\alpha)$. We shall show in the sequel that, by means of P_α, we can define an operator which does not depend on α and which is a quite appropriate candidate for $f(A)$. More precisely, let $\mathbb{S} = \mathbb{C} \cup \{\infty\}$ be the complex unit ball and let $\phi_\alpha : \mathbb{S} \to \mathbb{S}$ be the homeomorphism defined by

$$\phi_\alpha(\lambda) = \begin{cases} (\lambda - \alpha)^{-1} & \text{for } \lambda \neq \alpha \text{ and } \lambda \neq \infty \\ 0 & \text{for } \lambda = \infty \\ \infty & \text{for } \lambda = \alpha. \end{cases} \tag{1.9.2}$$

Lemma 1.9.1. *If $\alpha \in \rho(A)$, then $\phi_\alpha(\sigma(A) \cup \{\infty\}) = \sigma(P_\alpha)$ and $f \mapsto h_\alpha$, defined by $h_\alpha(\mu) = f(\phi_\alpha^{-1}(\mu))$, is a bijection from $\mathfrak{F}(A)$ to $\mathfrak{F}(P_\alpha)$.*

Proof. Let $\lambda \in \rho(A)$, $\lambda \neq \alpha$. We prove that $\mu = (\lambda - \alpha)^{-1} \in \rho(P_\alpha)$. To this aim, let us observe that $\mu = (\lambda - \alpha)^{-1} \neq 0$ and

$$(A - \alpha I)(A - \lambda I)^{-1} = \left[(A - \lambda I) + \frac{1}{\mu} I\right](A - \lambda I)^{-1} = \frac{1}{\mu}(A - \lambda I)^{-1} + I.$$

On the other hand, we have

$$\begin{aligned}(A - \alpha I)(A - \lambda I)^{-1} &= P_\alpha^{-1}\left[(A - \alpha I) - \frac{1}{\mu} I\right]^{-1} \\ &= \left\{\left[(A - \alpha I) - \frac{1}{\mu} I\right] P_\alpha\right\}^{-1} = \mu(\mu I - P_\alpha)^{-1},\end{aligned}$$

which shows that

$$(A - \lambda I)^{-1} = \mu^2(\mu I - P_\alpha)^{-1} - \mu I. \tag{1.9.3}$$

So $\mu \in \rho(P_\alpha)$. Conversely, if $\mu \in \rho(P_\alpha)$, $\mu \neq 0$, then

$$(\mu I - P_\alpha)^{-1} P_\alpha = [P_\alpha^{-1}(\mu I - P_\alpha)]^{-1} = (\mu P_\alpha^{-1} - I)^{-1} = \frac{1}{\mu}(A - \lambda I)^{-1},$$

and therefore $\lambda \in \rho(A)$. Obviously $0 \in \sigma(P_\alpha)$ because $P_\alpha^{-1} = A - \alpha I$ is unbounded. Since the last assertion follows from the definition of the function ϕ_α, this achieves the proof. □

Theorem 1.9.1. *Let $A : D(A) \subseteq X \to X$ be a closed linear operator with $\rho(A) \neq \emptyset$ and let $f \in \mathfrak{F}(A)$. Let $\alpha \in \rho(A)$, $P_\alpha = (A - \alpha I)^{-1}$ and $\varphi_\alpha = f(\phi_\alpha^{-1})$, where ϕ_α is defined by means of* (1.9.2). *Let $\mathbb{D}$ be an open neighborhood (possibly unbounded) of $\sigma(A)$ whose boundary Γ is the union of a finite number of Jordan arcs oriented in the positive sense. Then*

$$\varphi_\alpha(P_\alpha) = f(\infty)I + \frac{1}{2\pi i}\int_\Gamma f(\lambda) R(\lambda; A)\, d\lambda. \tag{1.9.4}$$

Proof. We may assume without loss of generality that $\alpha \notin \mathbb{D} \cup \Gamma$. Indeed, if this is not the case, thanks to Cauchy theorem, we can modify the domain $\mathbb{D}$ without altering the value of the integral in (1.9.4) in order to have $\alpha \notin \mathbb{D} \cup \Gamma$. Then $\mathbb{D}_\alpha = \phi_\alpha^{-1}(\mathbb{D})$ is an open set which includes $\sigma(P_\alpha)$ and whose boundary $\Gamma_\alpha = \phi_\alpha^{-1}(\Gamma)$ is a union of a finite number of Jordan rectifiable arcs positively oriented. In addition, $\varphi_\alpha = f(\phi_\alpha^{-1})$ is analytic on $\overline{\mathbb{D}}$. Since $\varphi_\alpha(0) = f(\infty)$ and $0 \in \sigma(P_\alpha)$, from (1.9.3), observing that $d\lambda = -d\mu/\mu^2$, we get

$$\begin{aligned}\frac{1}{2\pi i}\int_\Gamma f(\lambda) R(\lambda; A)\, d\lambda &= \frac{1}{2\pi i}\int_{\Gamma_\alpha} \varphi_\alpha(\mu)[-\mu^{-1} I + R(\mu; P_\alpha)]\, d\mu \\ &= \varphi_\alpha(P_\alpha) - \varphi_\alpha(0) I,\end{aligned}$$

which completes the proof. □

Definition 1.9.1. Let $A : D(A) \subseteq X \to X$ be a closed linear operator with $\rho(A) \neq \emptyset$, let $\mathbb{D}$ be an open neighborhood of $\sigma(A)$ whose boundary is the union of a finite set of Jordan rectifiable arcs positively oriented, let f be an analytic function on $\mathbb{D}$ and at ∞ and let $\alpha \in \rho(A)$. By definition

$$f(A) = \varphi_\alpha(P_\alpha).$$

In view of Theorem 1.9.1, the operator $f(A)$ is well-defined in the sense that it does not depend on $\alpha \in \rho(A)$ and, in addition

$$f(A) = f(\infty)I + \frac{1}{2\pi i}\int_\Gamma f(\lambda)R(\lambda; A)\,d\lambda.$$

From Theorem 1.8.1 we deduce:

Theorem 1.9.2. (Dunford) *Let $A : D(A) \subseteq X \to X$ be a closed linear operator with $\rho(A) \neq 0$, let $f, g \in \mathcal{F}(A)$ and let $\alpha, \beta \in \mathbb{C}$. Then:*

(i) $\alpha f + \beta g \in \mathcal{F}(A)$ *and* $\alpha f(A) + \beta g(A) = (\alpha f + \beta g)(A)$;

(ii) $fg \in \mathcal{F}(A)$ *and* $f(A)g(A) = (fg)(A)$;

(iii) *if* $(f_n)_{n\in\mathbb{N}}$ *are analytic on an open neighborhood* $\mathbb{D}$ *of* $\sigma(A)$ *and* $\lim_{n\to\infty} f_n = f$ *uniformly on* $\mathbb{D}$, *then* $\lim_{n\to\infty} f_n(A) = f(A)$ *in the norm of* $\mathcal{L}(X)$;

(iv) $f \in \mathcal{F}(A^*)$ *and* $f(A^*) = (f(A))^*$;

(v) $f(\sigma(A) \cup \{\infty\}) = \sigma(f(A))$.

Problems

Problem 1.1. Let X be a Banach space and $x : [a, b] \to X$ a function which is weakly continuous from the right on $[a, b]$. Prove that x is strongly measurable. Does this conclusion hold true if $x : \Omega \to X$ is any weakly continuous function, Ω being a locally compact topological space endowed with a positive Radon measure?

Problem 1.2. Prove that, for each $p \in (1, +\infty]$, $W^{1,p}(0, \pi)$ is compactly imbedded in $C([0, \pi])$. Give an example of bounded sequence in $W^{1,1}(0, \pi)$ having no convergent subsequences in $C([0, \pi])$. Show that $W^{1,1}(0, \pi)$ is continuously imbedded in $C([0, \pi])$, but not compactly imbedded.

Problem 1.3. Let $X = L^2(0, \pi)$ and $A : D(A) \subseteq X \to X$ defined by

$$\begin{cases} D(A) = \{u \in H^2(0, \pi)\,;\ u(0) = u'(0) = u(\pi) = u'(\pi) = 0\} \\ Au = u'' \text{ for } u \in D(A). \end{cases}$$

Show that A is symmetric. Find A^* and $D(A^*)$. Show that A is not self-adjoint. This is Exercise 1.8.5, p. 99 in Brezis and Cazenave [**31**].

Problem 1.4. Let $X = L^2(0,\pi)$ and $A : D(A) \subseteq X \to X$ defined by

$$\begin{cases} D(A) = \{u \in H^2(0,\pi)\,;\ u(0) = u'(\pi) = 0\} \\ Au = u'' \text{ for } u \in D(A). \end{cases}$$

Show that A is symmetric. Find A^* and $D(A^*)$. Show that A is self-adjoint.

Problem 1.5. Let $X = L^2(0,\pi)$ and $A : D(A) \subseteq X \to X$ defined by

$$\begin{cases} D(A) = \{u \in W^{1,2}(0,\pi)\,;\ u(0) = 0\} \\ Au = -u' \text{ for } u \in D(A). \end{cases}$$

Find A^*. This is Exercise 1.8.6, p. 99 in Brezis and Cazenave [**31**].

Notes. The results in this chapter are by now classical, and can be found in several well-known monographs and treatises. The first four sections are adaptations upon Adams [**1**], Edwards [**50**], Hille and Phillips [**70**] and Lions and Magenes [**85**], while the other ones, are gathered from Brezis [**29**], Brezis and Cazenave [**31**], Dunford and Schwartz [**49**], and also from Yosida [**136**]. The compactness assertions in Theorem 1.5.4 are due to Rellich and Kondrachov, while the continuous imbeddings where proved by Sobolev. The systematic study of spectral analysis problems for unbounded operators in Hilbert spaces was initiated by Von Neumann in the third decade of the twentieth century. The fundamentals of the functional calculus for unbounded operators in Banach spaces was initiated in the fourth decade of the twentieth century by Dunford, starting from Peano's idea in 1887 (see [**102**]) to define the exponential of a matrix, and using the theory of analytic functions of one complex variable with values in Banach algebras. The problems in this chapter are standard and, in one form or another, can be found in the usual textbooks and monographs.

CHAPTER 2

Semigroups of Linear Operators

The concept of semigroup of linear bounded operators has its roots in the simple remark that the Cauchy functional equation $f(t+s) = f(t)f(s)$ has as continuous nontrivial solutions only functions of the form e^{ta}, with $a \in \mathbb{R}$, combined with the fundamental idea of Peano **[102]**, **[103]** to define the exponential function of a matrix A by $e^A = \sum_{k=0}^{\infty} \frac{1}{k!} A^k$, in order to solve explicitly the first-order linear vector differential equation $u' = Au + f$ by means of the variation of constants formula

$$u(t) = e^{tA}u(0) + \int_0^t e^{(t-s)A} f(s)\, ds.$$

Roughly speaking, the notion of semigroup of linear operators is a quite natural extension of the exponential of a matrix to the exponential of a possible unbounded operator. Taking advantage of the powerful Functional Analysis' machinery, the Theory of Linear Semigroups tremendously emerged between 1930 – 1960 through the major contributions of Stone, Hille, Yosida, Feller, Lumer, Miyadera, Phillips. The aim of this chapter is to introduce the concept of semigroup of linear bounded operators on a Banach space and to present some of its remarkable properties.

2.1. Uniformly Continuous Semigroups

Let X be a Banach space and let $\mathcal{L}(X)$ be the set of all linear bounded operators from X to X. Endowed with the operator norm $\|\cdot\|_{\mathcal{L}(X)}$, defined by

$$\|U\|_{\mathcal{L}(X)} = \sup_{\|x\| \leq 1} \|Ux\|$$

for each $U \in \mathcal{L}(X)$, $\mathcal{L}(X)$ is a Banach space.

Definition 2.1.1. A family $\{S(t);\ t \geq 0\}$ in $\mathcal{L}(X)$ is a *semigroup of linear operators on* X, or simply *semigroup* if:

(i) $S(0) = I$
(ii) $S(t+s) = S(t)S(s)$ for each $t, s \geq 0$.

If, in addition, it satisfies the continuity condition at $t = 0$

$$\lim_{t \downarrow 0} S(t) = I,$$

in the norm topology of $\mathcal{L}(X)$, the semigroup is called *uniformly continuous.*

Example 2.1.1. A first significant example of uniformly continuous semigroup is given by $t \mapsto e^{tA}$, where e^{tA} is the exponential of the matrix tA. Namely, let $A \in \mathcal{M}_{n \times n}(\mathbb{R})$, and let $S(t) = e^{tA}$ for each $t \geq 0$, where

$$e^{tA} = \sum_{n=0}^{\infty} \frac{t^n}{n!} A^n.$$

We can easily see that $\{S(t);\ t \geq 0\}$ is a uniformly continuous semigroup of linear operators. More that this, straightforward computations show that that $t \mapsto S(t)$ is of class C^1 from $[0, +\infty)$ to X, and satisfies the first-order differential equation

$$\frac{d}{dt}(S(t)) = AS(t) = S(t)A, \tag{2.1.1}$$

for each $t \geq 0$. Actually, $\{S(t)\,;\ t \geq 0\}$ is nothing else than the fundamental matrix of the first-order vector differential equation

$$u' = Au$$

which satisfies $S(0) = I$. We shall see later that all uniformly continuous semigroups are of the form e^{tA}, with $A \in \mathcal{L}(X)$ and satisfy (2.1.1)

The next example shows that there exist semigroups which are not uniformly continuous.

Example 2.1.2. Let $X = C_{ub}(\mathbb{R}_+)$ be the space of all bounded and uniformly continuous functions from $\mathbb{R}_+$ to $\mathbb{R}$, endowed with the sup-norm $\|\cdot\|_\infty$, and let $\{S(t);\ t \geq 0\} \subseteq \mathcal{L}(X)$ be defined by

$$[S(t)f](s) = f(t+s)$$

for each $f \in X$ and each $t, s \in \mathbb{R}_+$. One may easily verify that $\{S(t);\ t \geq 0\}$ satisfies (i) and (ii) in Definition 2.1.1, and therefore it is a semigroup of linear operators. As in this specific case, the uniform continuity of the semigroup is equivalent to the equicontinuity of the unit ball in X, property which obviously is not satisfied, the semigroup is not uniformly continuous.

Definition 2.1.2. *The infinitesimal generator*, or *generator* of the semigroup of linear operators $\{S(t)\,;\ t \geq 0\}$ is the operator $A : D(A) \subseteq X \to X$, defined by

$$D(A) = \left\{ x \in X\,;\ \exists \lim_{t \downarrow 0} \frac{1}{t}(S(t)x - x) \right\},$$

and

$$Ax = \lim_{t \downarrow 0} \frac{1}{t} \left(S(t)x - x \right).$$

Equivalently, we say that A *generates* $\{S(t)\,;\ t \geq 0\}$.

Remark 2.1.1. If $A : D(A) \subseteq X \to X$ is the infinitesimal generator of a semigroup of linear operators then $D(A)$ is a vector subspace of X and A is a possibly unbounded linear operator.

Remark 2.1.2. It is easy to see that the infinitesimal generator of the semigroup in Example 2.1.1 is $\mathcal{A} \in \mathcal{L}(\mathbb{R}^n)$, defined by $\mathcal{A}x = Ax$. This remark clarifies the relationship between semigroups of linear operators and first-order linear differential equations.

Example 2.1.3. The generator of the semigroup in Example 2.1.2 is given by

$$D(A) = \left\{ f \in X;\ \exists \lim_{t \downarrow 0} \frac{1}{t} (f(t + \cdot) - f) = f' \text{ strongly in } X \right\},$$

and

$$Af = f'.$$

Let us remark that, if $f \in D(A)$, then $u(t,s) = [S(t)f](s) = f(t+s)$ satisfies the first-order partial differential equation

$$\begin{cases} u_t = u_s \\ u(0,s) = f(s). \end{cases}$$

Accordingly, in this case, we have the following pointwise variant of (2.1.1)

$$\frac{d}{dt}(S(t)x) = AS(t)x = S(t)Ax$$

for each $x \in D(A)$ and each $t \geq 0$.

Proposition 2.1.1. *If $\{S(t)\,;\ t \geq 0\}$ is a uniformly continuous semigroup of linear operators then, for each $t \geq 0$, $S(t)$ is invertible.*

Proof. Inasmuch as

$$\lim_{t \downarrow 0} S(t) - I = 0,$$

in the norm topology of $\mathcal{L}(X)$, there exists $\delta > 0$ such that

$$\|S(t) - I\|_{\mathcal{L}(X)} < 1$$

for each $t \in (0, \delta\,]$. Thus, for each $t \in (0, \delta\,]$, $S(t)$ is invertible. Let $t > \delta$. Then there exist $n \in \mathbb{N}^*$ and $\eta \in [\,0, \delta)$ such that $t = n\delta + \eta$. Therefore $S(t) = S(\delta)^n S(\eta)$, and so $S(t)$ is invertible. The proof is complete. □

Definition 2.1.3. A family of operators $\{G(t);\ t \in \mathbb{R}\}$ in $\mathcal{L}(X)$ is called a *group of linear operators on* X if:

(i) $G(0) = I$
(ii) $G(t+s) = G(t)G(s)$ for each $t, s \in \mathbb{R}$.

If, in addition,

$$\lim_{t \to 0} G(t) = I,$$

in the norm topology of $\mathcal{L}(X)$, the group is called *uniformly continuous.*

Remark 2.1.3. If $\{S(t)\,;\ t \geq 0\}$ is a uniformly continuous semigroup of linear operators then it can be extended to a uniformly continuous group of linear operators. More precisely there exists a group of linear operators $\{G(t)\,;\ t \in \mathbb{R}\}$ such that $G(t) = S(t)$ for each $t \geq 0$. Indeed, by virtue of Proposition 2.1.1, we can define $G(t) : X \to X$ by

$$G(t) = \begin{cases} [S(-t)]^{-1} & \text{if } t < 0 \\ S(t) & \text{if } t \geq 0. \end{cases}$$

One may easily see that $\{G(t)\,;\ t \in \mathbb{R}\}$ is a uniformly continuous group of linear operators which extends $\{S(t)\,;\ t \geq 0\}$. The proof of the uniform continuity of this group is left to the reader.

Corollary 2.1.1. *If* $\{S(t)\,;\ t \geq 0\}$ *is a uniformly continuous semigroup of linear operators then the mapping* $t \mapsto S(t)$ *is continuous from* $[0, +\infty)$ *to* $\mathcal{L}(X)$ *endowed with the operator norm.*

Proof. Let $\{G(t)\,;\ t \in \mathbb{R}\}$ be the uniformly continuous group of linear operators which extends $\{S(t)\,;\ t \geq 0\}$ and let $t > 0$. See Remark 2.1.3. Then

$$\lim_{h \to 0} \|S(t+h) - S(t)\|_{\mathcal{L}(X)} \leq \lim_{h \to 0} \|S(t)\|_{\mathcal{L}(X)} \|G(h) - I\|_{\mathcal{L}(X)} = 0.$$

As at $t = 0$ the continuity follows from Definition 2.1.3, this achieves the proof. □

2.2. Generators of Uniformly Continuous Semigroups

Theorem 2.2.1. *A linear operator* $A : D(A) \subseteq X \to X$ *is the generator of a uniformly continuous semigroup if and only if* $D(A) = X$ *and* $A \in \mathcal{L}(X)$.

Proof. *The "only if" part.* Let $\{S(t)\,;\ t \geq 0\}$ be uniformly continuous. Since

$$\lim_{t \downarrow 0} S(t) = I$$

in the norm topology of $\mathfrak{L}(X)$, there exists $\rho > 0$ such that

$$\left\| \frac{1}{\rho} \int_0^\rho S(t)\, dt - I \right\|_{\mathfrak{L}(X)} < 1.$$

We notice that the integral here is the Riemann integral of the continuous function $S : [0, \rho] \to \mathfrak{L}(X)$, which is defined by a simple analogy with its scalar counterpart. Consequently, the operator $\frac{1}{\rho} \int_0^\rho S(t)\, dt$ is invertible and accordingly $\int_0^\rho S(t) dt$ has the same property. Let $h > 0$. Let us remark that

$$\frac{1}{h}(S(h) - I) \int_0^\rho S(t)\, dt = \frac{1}{h} \int_0^\rho S(t+h)\, dt - \frac{1}{h} \int_0^\rho S(t)\, dt.$$

The change of variable $t + h = s$ in the first integral on the right-hand side yields

$$\begin{aligned} \frac{1}{h}(S(h) - I) \int_0^\rho S(t)\, dt &= \frac{1}{h} \int_h^{\rho+h} S(s)\, ds - \frac{1}{h} \int_0^\rho S(s)\, ds \\ &= \frac{1}{h} \int_\rho^{\rho+h} S(s)\, ds - \frac{1}{h} \int_0^h S(s)\, ds. \end{aligned}$$

Then

$$\frac{1}{h}(S(h) - I) = \left(\frac{1}{h} \int_\rho^{\rho+h} S(s)\, ds - \frac{1}{h} \int_0^h S(s)\, ds \right) \left(\int_0^\rho S(t)\, dt \right)^{-1}.$$

But, the right-hand side of the equality above converges for h tending to 0 by positive values, and thus, the left-hand side enjoys the same property. As the convergence in the uniform operator topology of $\mathfrak{L}(X)$ implies the pointwise convergence, letting h to tend to 0 by positive values, we deduce

$$A = (S(\rho) - I) \left(\int_0^\rho S(t)\, dt \right)^{-1}.$$

Hence $A \in \mathfrak{L}(X)$, which proves the necessity.
The "if" part. Let $A \in \mathfrak{L}(X)$, $t \geq 0$ and let

$$S(t) = \sum_{n=0}^{\infty} \frac{t^n}{n!} A^n,$$

where $A^n = A \cdot A \ldots A$ n times and $A^0 = I$.
We can easily see that $\{S(t)\,;\ t \geq 0\}$ is a semigroup of linear operators. In order to prove that this semigroup is uniformly continuous let us remark that

$$\|S(t) - I\|_{\mathfrak{L}(X)} = \left\| \sum_{n=0}^{\infty} \frac{t^n}{n!} A^n - I \right\|_{\mathfrak{L}(X)}$$

$$= \left\| \sum_{n=1}^{\infty} \frac{t^n}{n!} A^n \right\|_{\mathcal{L}(X)} \leq t \sum_{n=1}^{\infty} \frac{t^{n-1}}{n!} \|A\|^n_{\mathcal{L}(X)}.$$

Since

$$\sum_{n=1}^{\infty} \frac{t^{n-1}}{n!} \|A\|^n_{\mathcal{L}(X)} \leq \|A\| e^{t\|A\|},$$

we conclude that

$$\lim_{t \downarrow 0} S(t) = I,$$

in the norm topology of $\mathcal{L}(X)$, and thus $\{S(t);\ t \geq 0\}$ is a uniformly continuous semigroup. To achieve the proof we have to show that A is the infinitesimal generator of this semigroup. To this aim it suffices to verify that

$$\lim_{t \downarrow 0} \left\| \frac{1}{t}(S(t) - I) - A \right\|_{\mathcal{L}(X)} = 0.$$

But this follows from the obvious inequality

$$\left\| \frac{1}{t}(S(t) - I) - A \right\|_{\mathcal{L}(X)} = t \left\| \sum_{n=2}^{\infty} \frac{t^{n-2}}{n!} A^n \right\|_{\mathcal{L}(X)} \leq t\|A\|^2 e^{t\|A\|},$$

thereby completing the proof. □

Let us consider now the Cauchy problem

$$\begin{cases} u' = Au + f \\ u(a) = \xi, \end{cases} \tag{$\mathcal{CP}$}$$

where $A \in \mathcal{L}(X)$ and $f \in C([0,T];X)$.

Theorem 2.2.2. *For any $(a, \xi, f) \in [0,T) \times X \times C([a,T];X)$, $(\mathcal{CP})$ has a unique solution $u \in C^1([a,T];X)$ given by the so-called variation of constants, or Duhamel, formula*

$$u(t, a, \xi, f) = S(t-a)\xi + \int_a^t S(t-s)f(s)\,ds$$

for each $t \in [a,T]$, where $\{S(t);\ t \geq 0\}$, is the uniformly continuous group generated by A.

Proof. A simple computational argument shows that the function u, defined as above, is of class C^1 and verifies $u'(t,a,\xi,f) = Au(t,a,\xi,f) + f(t)$ for all $t \in [a,T]$, and $u(a,a,\xi,f) = \xi$. The uniqueness follows as in the case $X = \mathbb{R}^n$, by observing that the mapping $x \mapsto Ax$ is Lipschitz continuous. See for instance Corduneanu [**39**], (4.36), p. 65. The proof is complete. □

2.3. C_0-Semigroups. General Properties

In this section we introduce a class of semigroups of linear operators, strictly larger than that of uniformly continuous semigroups, class which proves very useful in the study of many partial differential equations of parabolic or hyperbolic type.

Definition 2.3.1. A semigroup of linear operators $\{S(t)\,;\; t \geq 0\}$ is called *a semigroup of class* C_0, or C_0*-semigroup* if for each $x \in X$ we have

$$\lim_{t \downarrow 0} S(t)x = x.$$

Remark 2.3.1. Each uniformly continuous semigroup is of class C_0 but not conversely as we can state from the example below.

Example 2.3.1. Let $X = C_{ub}(\mathbb{R}_+)$ be the space of all functions which are uniformly continuous and bounded from $\mathbb{R}_+$ to $\mathbb{R}$, endowed with the sup-norm $\|\cdot\|_\infty$, and let $\{S(t)\,;\; t \geq 0\}$ be defined by

$$[S(t)f](s) = f(t+s)$$

for each $f \in X$ and each $t, s \in \mathbb{R}_+$. We know from Example 2.1.2 that $\{S(t);\; t \geq 0\}$ is a semigroup. In addition, this is of class C_0. On the other hand, as we mentioned in Example 2.1.2, it is not uniformly continuous because the unit ball in X is not equicontinuous. See Definition A.2.1.

Theorem 2.3.1. *If* $\{S(t)\,;\; t \geq 0\}$ *is a* C_0*-semigroup, then there exist* $M \geq 1$, *and* $\omega \in \mathbb{R}$ *such that*

$$\|S(t)\|_{\mathcal{L}(X)} \leq Me^{t\omega} \tag{2.3.1}$$

for each $t \geq 0$.

Proof. First, we will show that there exist $\eta > 0$ and $M \geq 1$ such that

$$\|S(t)\|_{\mathcal{L}(X)} \leq M \tag{2.3.2}$$

for each $t \in [\,0, \eta\,]$. To this aim, let us assume by contradiction that this is not the case. Then there exists at least one C_0-semigroup $\{S(t);\; t \geq 0\}$ with the property that, for each $\eta > 0$ and each $M \geq 1$, there exists $t_{\eta,M} \in [\,0, \eta\,]$, such that

$$\|S(t_{\eta,M})\|_{\mathcal{L}(X)} > M.$$

Taking $\eta = 1/n$, $M = n$ and denoting $t_{\eta,M} = t_n$ for $n \in \mathbb{N}^*$, we deduce

$$\|S(t_n)\|_{\mathcal{L}(X)} > n, \tag{2.3.3}$$

where $t_n \in [\,0, 1/n\,]$ for each $n \in \mathbb{N}^*$. Recalling that, for each $x \in X$, $\lim_{n\to\infty} S(t_n)x = x$, it follows that the family $\{S(t_n)\,;\; n \in \mathbb{N}^*\}$ of linear

bounded operators is pointwise bounded, i.e., for each $x \in X$, the set $\{S(t_n)x\,;\ n \in \mathbb{N}^*\}$ is bounded. By the uniform boundedness principle (see Dunford and Schwartz [**49**], Corollary 21, p. 66), it follows that this family is bounded in the uniform operator norm $\|\cdot\|_{\mathcal{L}(X)}$ which contradicts (2.3.3). This contradiction can be eliminated only if (2.3.2) holds.
Next, let $t > 0$. Then there exist $n \in \mathbb{N}^*$ and $\delta \in [\,0, \eta)$, such that $t = n\eta + \delta$. We have

$$\|S(t)\|_{\mathcal{L}(X)} = \|S^n(\eta)S(\delta)\|_{\mathcal{L}(X)} \leq \|S(\eta)\|^n_{\mathcal{L}(X)}\|S(\delta)\|_{\mathcal{L}(X)} \leq MM^n.$$

But $n = \frac{t-\delta}{\eta} \leq \frac{t}{\eta}$ and thus $\|S(t)\|_{\mathcal{L}(X)} \leq MM^{\frac{t}{\eta}} = Me^{t\omega}$, where $\omega = \frac{1}{\eta}\ln M$. The proof is complete. □

Remark 2.3.2. If $\{S(t)\,;\ t \geq 0\}$ is a uniformly continuous semigroup whose generator is A, then (2.3.1) holds with $M = 1$ and $\omega = \|A\|_{\mathcal{L}(X)}$.

Definition 2.3.2. A C_0-semigroup, $\{S(t)\,;\ t \geq 0\}$ is called *of type* (M, ω) with $M \geq 1$ and $\omega \in \mathbb{R}$, if for each $t \geq 0$, we have

$$\|S(t)\|_{\mathcal{L}(X)} \leq Me^{t\omega}.$$

A C_0-semigroup $\{S(t)\,;\ t \geq 0\}$ is called *a C_0-semigroup of contractions*, or of *nonexpansive operators*, if it is of type $(1, 0)$, i.e., if for each $t \geq 0$, we have

$$\|S(t)\|_{\mathcal{L}(X)} \leq 1.$$

We shall use also the term of *contraction semigroup*.

Corollary 2.3.1. *If $\{S(t)\,;\ t \geq 0\}$ is a C_0-semigroup, then the mapping $(t, x) \mapsto S(t)x$ is jointly continuous from $[\,0, +\infty) \times X$ to X.*

Proof. Let $x, y \in X$, $t \geq 0$ and $h \in \mathbb{R}^*$ with $t + h \geq 0$. We distinguish between two cases: $h > 0$, or $h < 0$. If $h > 0$, we have

$$\begin{aligned} &\|S(t+h)y - S(t)x\| \\ &\leq \|S(t+h)y - S(t+h)x\| + \|S(t+h)x - S(t)x\| \\ &\leq \|S(t+h)\|_{\mathcal{L}(X)}\|y - x\| + \|S(t+h)x - S(t)x\| \\ &\leq Me^{(t+h)\omega}\|x - y\| + \|S(t)\|_{\mathcal{L}(X)}\|S(h)x - x\|, \end{aligned}$$

which shows that

$$\lim_{(\tau,y)\to(t+0,x)} S(\tau)y = S(t)x.$$

If $h < 0$, by Theorem 2.3.1, we deduce

$$\begin{aligned} \|S(t+h)y - S(t)x\| &= \|S(t+h)y - S(t+h)S(-h)x\| \\ &\leq \|S(t+h)\|_{\mathcal{L}(X)}\|y - S(-h)x\| \end{aligned}$$

$$\leq Me^{(t+h)\omega}(\|y-x\| + \|S(-h)x - x\|),$$

which implies that

$$\lim_{(\tau,y)\to(t-0,x)} S(\tau)y = S(t)x.$$

The proof is complete. □

Some basic properties of C_0-semigroups are listed below.

Theorem 2.3.2. *Let $A : D(A) \subseteq X \to X$ be the infinitesimal generator of a C_0-semigroup $\{S(t)\,;\ t \geq 0\}$. Then*

(i) *for each $x \in X$ and each $t \geq 0$, we have*

$$\lim_{h\downarrow 0} \frac{1}{h}\int_t^{t+h} S(\tau)x\,d\tau = S(t)x\,;$$

(ii) *for each $x \in X$ and each $t > 0$, we have*

$$\int_0^t S(\tau)x\,d\tau \in D(A) \text{ and } A\left(\int_0^t S(\tau)x\,d\tau\right) = S(t)x - x\,;$$

(iii) *for each $x \in D(A)$ and each $t \geq 0$, we have $S(t)x \in D(A)$. In addition, the mapping $t \mapsto S(t)x$ is of class C^1 on $[0,+\infty)$, and satisfies*

$$\frac{d}{dt}(S(t)x) = AS(t)x = S(t)Ax\,;$$

(iv) *for each $x \in D(A)$ and each $0 \leq s \leq t < +\infty$, we have*

$$\int_s^t AS(\tau)x\,d\tau = \int_s^t S(\tau)Ax\,d\tau = S(t)x - S(s)x.$$

Proof. In order to prove (i), let us observe that

$$\left\|\frac{1}{h}\int_t^{t+h} S(\tau)x\,d\tau - S(t)x\right\| \leq \frac{1}{h}\int_t^{t+h} \|S(\tau)x - S(t)x\|\,d\tau.$$

The conclusion follows from Corollary 2.3.1.
Let $x \in X$, $t > 0$ and $h > 0$. We remark that

$$\frac{1}{h}(S(h) - I)\int_0^t S(\tau)x\,d\tau = \frac{1}{h}\int_0^t S(\tau+h)x\,d\tau - \frac{1}{h}\int_0^t S(\tau)x\,d\tau.$$

The change of variable $\tau + h = s$ in the first integral on the right-hand side yields

$$\frac{1}{h}(S(h) - I)\int_0^t S(\tau)x\,d\tau = \frac{1}{h}\int_h^{t+h} S(s)x\,ds - \frac{1}{h}\int_0^t S(s)x\,ds$$

$$= -\frac{1}{h}\int_0^h S(s)x\,ds + \frac{1}{h}\int_t^{t+h} S(s)x\,ds.$$

From this equality and from (i), we deduce that there exists

$$\lim_{h\downarrow 0}\frac{1}{h}\,(S(h)-I)\int_0^t S(\tau)x\,d\tau = S(t)x - x,$$

which proves (ii).
Next, let $x \in D(A)$, $t \geq 0$ and $h > 0$. We have

$$\left\|\frac{1}{h}\,(S(t+h)x - S(t)x) - S(t)Ax\right\| \leq \|S(t)\|_{\mathcal{L}(X)}\left\|\frac{1}{h}\,(S(h)x - x) - Ax\right\|,$$

inequality which proves that $S(t)x \in D(A)$, $t \mapsto S(t)x$ is differentiable from the right, and that

$$\frac{d^+}{dt}\,(S(t)x) = AS(t)x = S(t)Ax. \tag{2.3.4}$$

On the other hand, for each $t > 0$ and $h < 0$ with $t + h \geq 0$, we have

$$\left\|\frac{1}{h}\,(S(t+h)x - S(t)x) - S(t)Ax\right\|$$

$$\leq \|S(t+h)\|_{\mathcal{L}(X)}\left\|\frac{1}{h}\,(x - S(-h)x) - S(-h)Ax\right\|$$

$$\leq \|S(t+h)\|_{\mathcal{L}(X)}\left(\left\|-\frac{1}{h}\,(S(-h)x - x) - Ax\right\| + \|S(-h)Ax - Ax\|\right).$$

This inequality shows that $t \mapsto S(t)x$ is differentiable from the left as well. From (2.3.4) and the continuity of the function $t \mapsto S(t)Ax$ on $[\,0,+\infty)$, we deduce that $t \mapsto S(t)x$ is of class C^1 on $[\,0,+\infty)$, which completes the proof of (iii).
Since (iv) follows from (iii) by integrating from s to t both sides in (2.3.4), the proof is complete. □

2.4. The Infinitesimal Generator

In this section we shall prove two basic properties of the generator of a C_0-semigroup: the density of the domain and the closedness of the graph. First, let us recall the following:

Definition 2.4.1. An operator $A : D(A) \subseteq X \to X$ is called *closed*, if its graph is closed in $X \times X$.

Theorem 2.4.1. *Let $A : D(A) \subseteq X \to X$ be the infinitesimal generator of a C_0-semigroup $\{S(t);\ t \geq 0\}$. Then $D(A)$ is dense in X, and A is a closed operator.*

Proof. Let $x \in X$ and $\varepsilon > 0$. Then, by virtue of (i) in Theorem 2.3.2, we have

$$\frac{1}{\varepsilon}\int_0^{\varepsilon} S(\tau)x\,d\tau \in D(A)$$

and

$$\lim_{\varepsilon\downarrow 0}\frac{1}{\varepsilon}\int_0^{\varepsilon} S(\tau)x\,d\tau = x.$$

Consequently $D(A)$ is dense in X.
Next, let $(x_n)_{n\in\mathbb{N}^*}$ be a sequence in $D(A)$ such that

$$\lim_{n\to\infty} x_n = x \text{ and } \lim_{n\to\infty} Ax_n = y.$$

From (iv) in Theorem 2.3.2, it follows that

$$S(h)x_n - x_n = \int_0^h S(\tau)Ax_n\,d\tau$$

for each $n \in \mathbb{N}^*$ and each $h > 0$. Passing to the limit in this equality, we obtain

$$S(h)x - x = \int_0^h S(\tau)y\,d\tau.$$

By virtue of (i) in Theorem 2.3.2, it follows that there exists

$$\lim_{h\downarrow 0}\frac{1}{h}\int_0^h S(\tau)y\,d\tau = y.$$

From this relation, and the preceding one, we deduce that $x \in D(A)$ and $Ax = y$. The proof is complete. □

We continue with a uniqueness result.

Theorem 2.4.2. *If $A : D(A) \subseteq X \to X$ is the infinitesimal generator of two C_0-semigroups $\{S(t)\,;\ t \geq 0\}$, and $\{T(t)\,;\ t \geq 0\}$, then $S(t) = T(t)$ for each $t \geq 0$.*

Proof. Let $x \in D(A)$, $t > 0$ and let $f : [\,0, t\,] \to X$ be given by

$$f(s) = S(t-s)T(s)x$$

for each $s \in [\,0, t\,]$. By (ii) in Theorem 2.4.2, it follows f is differentiable on $[\,0, t\,]$, and that

$$\begin{aligned} f'(s) &= -AS(t-s)T(s)x + S(t-s)AT(s)x \\ &= -AS(t-s)T(s)x + AS(t-s)T(s)x = 0 \end{aligned}$$

for each $s \in [\,0, t\,]$. Thus f is constant. Hence we have $f(0) = f(t)$, or equivalently $S(t)x = T(t)x$ for each $x \in D(A)$. Since $D(A)$ is dense in

X, and $S(t)$, $T(t)$ are linear bounded operators, we easily conclude that $S(t)x = T(t)x$ for each $x \in X$, which completes the proof. $\square$

Remark 2.4.1. Theorem 2.3.2 (iii), and Theorem 2.4.2, assert that, for each $\xi \in D(A)$, the function $u : [0, +\infty) \to X$, defined by $u(t) = S(t)\xi$ for each $t \geq 0$, is the unique classical solution of the Cauchy problem

$$\begin{cases} u' = Au \\ u(0) = \xi. \end{cases} \tag{2.4.1}$$

The example below shows that, if $\xi \in X$, but $\xi \notin D(A)$, the function u, defined as above, is not necessarily differentiable on $[0, +\infty)$.

Example 2.4.1. Let $X = C_{ub}(\mathbb{R}_+)$ be the space of all functions which are uniformly continuous and bounded from $\mathbb{R}_+$ to $\mathbb{R}$, endowed with the sup-norm $\|\cdot\|_\infty$, and let us define the linear operator $A : D(A) \subseteq X \to X$ by $D(A) = \{u \in X \,;\ u' \in X\}$, and $Au = u'$ for each $u \in D(A)$. Taking any nondifferentiable function $\xi \in C_{ub}(\mathbb{R}_+)$, we easily observe that (2.4.1) has no classical solution. Indeed, in this case $(\mathcal{CP})$ may be rewritten as

$$\begin{cases} u_t = u_s \\ u(0, s) = \xi(s), \end{cases}$$

whose only classical solution (if there exists any) is given by $u(t, s) = \xi(t+s)$ for each $t \geq 0$ and $s \in \mathbb{R}$.

Let $A : D(A) \subseteq X \to X$ be a linear operator and $n \in \mathbb{N}$. We define the nth-order power of A by:

$$\begin{cases} D(A^0) = X \\ A^0 = I \end{cases}, \quad \begin{cases} D(A^1) = D(A) \\ A^1 = A \end{cases}$$

and

$$\begin{cases} D(A^n) = \{x \in D(A^{n-1});\ A^{n-1}x \in D(A)\} \\ A^n = AA^{n-1} \end{cases}$$

for $n \geq 2$.

Theorem 2.4.3. *Let $A : D(A) \subseteq X \to X$ be the infinitesimal generator of a C_0-semigroup. Then $\cap_{n \geq 0} D(A^n)$ is dense in X.*

Proof. Let us remark that, for each $n \in \mathbb{N}$, $D(A^n)$ is a vector subspace in X. Accordingly, $\cap_{n \geq 0} D(A^n)$ is also a vector subspace in X. Let $x \in X$, and let $\varphi : \mathbb{R} \to \mathbb{R}_+$ be a C^∞ function for which there exists an interval $[a, b] \subset (0, +\infty)$ such that $\varphi(t) = 0$ for each $t \notin [a, b]$. We define

$$x(\varphi) = \int_0^{+\infty} \varphi(t) S(t) x \, dt,$$

and we remark that

$$\lim_{h\downarrow 0}\frac{1}{h}(S(h)-I)x(\varphi)$$

$$=\lim_{h\downarrow 0}\frac{1}{h}\left(\int_0^{+\infty}\varphi(t)S(t+h)x\,dt-\int_0^{+\infty}\varphi(t)S(t)x\,dt\right)$$

$$=\lim_{h\downarrow 0}\frac{1}{h}\left(\int_h^{+\infty}\varphi(t-h)S(t)x\,dt-\int_0^{+\infty}\varphi(t)S(t)x\,dt\right).$$

For $s\in(-\infty,a)$, we have $\varphi(s)=0$, and consequently

$$\lim_{h\downarrow 0}\frac{1}{h}(S(h)-I)x(\varphi)$$

$$=\lim_{h\downarrow 0}\frac{1}{h}\left(\int_0^{+\infty}\varphi(t-h)S(t)x\,dt-\int_0^{+\infty}\varphi(t)S(t)x\,dt\right)$$

$$=\int_0^{+\infty}\lim_{h\downarrow 0}\frac{\varphi(t-h)-\varphi(t)}{h}S(t)x\,dt=-x(\varphi').$$

Accordingly, $x(\varphi)\in D(A)$ and $Ax(\varphi)=-x(\varphi')$. Repeating the above arguments, one may prove by mathematical induction that, for each $n\in\mathbb{N}$, $x(\varphi)\in D(A^n)$ and, in addition, $A^n x(\varphi)=(-1)^n x(\varphi^{(n)})$. Consequently, $x(\varphi)\in\bigcap_{n\geq 0}D(A^n)$.

Next, let φ be a function as above, such that

$$\int_0^{+\infty}\varphi(t)\,dt=1,$$

let $\varepsilon>0$, and let us define $\varphi_\varepsilon:\mathbb{R}\to(0,+\infty)$ by

$$\varphi_\varepsilon(t)=\frac{1}{\varepsilon}\varphi\left(\frac{t}{\varepsilon}\right).$$

Obviously, φ_ε is of class C^∞, $\varphi_\varepsilon(t)=0$ for $t\notin[\,\varepsilon a,\varepsilon b\,]$, and

$$\int_0^{+\infty}\varphi_\varepsilon(t)\,dt=1.$$

Let us remark that

$$\|x(\varphi_\varepsilon)-x\|=\left\|\frac{1}{\varepsilon}\int_0^{+\infty}\varphi\left(\frac{t}{\varepsilon}\right)(S(t)x-x)\,dt\right\|$$

$$\leq\frac{1}{\varepsilon}\int_{\varepsilon a}^{\varepsilon b}\varphi\left(\frac{t}{\varepsilon}\right)\|S(t)x-x\|\,dt\leq\sup_{t\in[\,\varepsilon a,\varepsilon b\,]}\|S(t)x-x\|.$$

This inequality shows that

$$\lim_{\varepsilon\downarrow 0}x(\varphi_\varepsilon)=x$$

and consequently we have $x \in \overline{\bigcap_{n\geq 0} D(A^n)}$. The proof is complete. □

We conclude this section with both a consequence and a completion of (iii) in Theorem 2.3.2.

Corollary 2.4.1. *Let $n \in \mathbb{N}^*$. Then, for each $\xi \in D(A^n)$ and each $t \geq 0$, we have $S(t)\xi \in D(A^n)$, the function $u : [0, +\infty) \to X$, $u(t) = S(t)\xi$ is of class C^n and is a solution of the Cauchy problem*

$$\begin{cases} u^{(n)}(t) = A^n u(t), & t \geq 0 \\ u^{(k)}(0) = A^k \xi, & k = 0, 1, \ldots, n-1. \end{cases}$$

Proof. We proceed by mathematical induction. Let us remark that for $n = 1$ the conclusion follows from (iii) in Theorem 2.3.2. Let us assume that the property in question holds for $n \in \mathbb{N}^*$ and let $\xi \in D(A^{n+1})$. Since $D(A^{n+1}) \subseteq D(A^n)$, the inductive hypothesis yields $u^{(n)}(t) = A^n u(t)$ for each $t \geq 0$. Let $t \geq 0$ and $h \in \mathbb{R}$ with $t + h \geq 0$. We have

$$\frac{1}{h}\left(u^{(n)}(t+h) - u^{(n)}(t)\right) = \frac{1}{h}\left(A^n S(t+h)\xi - A^n S(t)\xi\right)$$

$$= \frac{1}{h}\left(S(t+h)A^n\xi - S(t)A^n\xi\right)$$

because $\xi \in D(A^n)$, while, for each $\tau \geq 0$, A^n and $S(\tau)$ commutes on $D(A^n)$. But $A^n\xi \in D(A)$, and consequently there exists

$$\lim_{h\downarrow 0} \frac{1}{h}\left(S(t+h)A^n\xi - S(t)A^n\xi\right) = AS(t)A^n\xi.$$

Passing to the limit for $h \to 0$ in the preceding equality, and taking into account that $A^n\xi \in D(A)$, we deduce

$$u^{(n+1)}(t) = AS(t)A^n\xi = A^{(n+1)}S(t)\xi.$$

Clearly $u^k(0) = A^k\xi$ for $k = 0, 1, 2, \ldots n$ and this completes the proof. □

Problems

Let $p \in [1, +\infty)$ and let $X = l_p$ be the space of real sequences $(x_n)_{n\in\mathbb{N}^*}$ satisfying $\sum_{n=1}^{\infty} |x_n|^p < +\infty$. This space, endowed with the norm $\|\cdot\|_p$, defined by $\|(x_n)_{n\in\mathbb{N}}\|_p = \left(\sum_{n=1}^{\infty} |x_n|^p\right)^{1/p}$ for each $(x_n)_{n\in\mathbb{N}} \in l_p$, is a real Banach space.

Problem 2.1. Let $p \in [1, +\infty)$, $X = l_p$, $(a_n)_{n\in\mathbb{N}^*}$ a sequence of positive real numbers, and $t \in \mathbb{R}_+$. We define $S(t) : D(A) \subseteq X \to X$ by

$$(S(t)(x_n)_{n\in\mathbb{N}^*})_{k\in\mathbb{N}^*} = \left(e^{-a_k t}x_k\right)_{k\in\mathbb{N}^*}$$

for each $(x_n)_{n\in\mathbb{N}^*} \in l_p$.

(i) Prove that $\{S(t)\,;\ t \geq 0\}$ is a C_0-semigroup of contractions on l_p.
(ii) Find its infinitesimal generator.
(iii) Prove that this semigroup is uniformly continuous if and only if $(a_n)_{n\in\mathbb{N}^*}$ is bounded.

Let c_0 be the space of real sequences vanishing at ∞. Endowed with the norm $\|\cdot\|_\infty$, defined by $\|(x_n)_{n\in\mathbb{N}^*}\|_\infty = \sup_{n\in\mathbb{N}^*} |x_n|$ for each $(x_n)_{n\in\mathbb{N}^*} \in c_0$, this is real Banach space.

Problem 2.2. Let $X = c_0$, $(a_n)_{n\in\mathbb{N}^*}$ a sequence of positive real numbers and $t \in \mathbb{R}_+$. Let us define $S(t) : D(A) \subseteq X \to X$ by

$$(S(t)(x_n)_{n\in\mathbb{N}^*})_{k\in\mathbb{N}^*} = \left(e^{-a_k t} x_k\right)_{k\in\mathbb{N}^*}$$

for each $(x_n)_{n\in\mathbb{N}^*} \in c_0$.

(i) Prove that $\{S(t)\,;\ t \geq 0\}$ is a C_0-semigroup of contractions.
(ii) Find its infinitesimal generator.
(iii) Prove that this semigroup is uniformly continuous if and only if $(a_n)_{n\in\mathbb{N}^*}$ is bounded.

Problem 2.3. Let $X = C_b(\mathbb{R})$ (the space of all continuous and bounded functions from $\mathbb{R}$ to $\mathbb{R}$, which is a real Banach space with respect to the sup-norm), let $t \in \mathbb{R}_+$, and let $S(t) : X \to X$ defined by $[S(t)f](s) = f(t+s)$ for each $f \in X$, and each $s \in \mathbb{R}$. Show that $\{S(t)\,;\ t \geq 0\}$ is a semigroup of linear operators, which is not of class C_0. Find its infinitesimal generator and show that $D(A)$ is not dense in X.

Problem 2.4. Let $X = C_{ub}(\mathbb{R})$ endowed with the norm supremum, let $t \in \mathbb{R}$, $\lambda > 0$, $\delta > 0$ and let us define $G(t) : X \to X$ by

$$[G(t)f](x) = e^{-\lambda t} \sum_{k=0}^{\infty} \frac{(\lambda t)^k}{k!} f(x - k\delta)$$

for each $f \in X$ and each $x \in \mathbb{R}$. Prove that $\{G(t)\,;\ t \in \mathbb{R}\}$ is a uniformly continuous group of isometries, whose infinitesimal generator, $A : X \to X$, is defined by $[Af](x) = \lambda[f(x-\delta) - f(x)]$ for each $f \in X$, and each $x \in \mathbb{R}$. This is Exercise 9, p. 23 in Goldstein [**61**].

Problem 2.5. Let $X = L^p(\mathbb{R}^n)$, let $\mathcal{A}$ be an $n \times n$ matrix with real entries and let us define $G(t) : X \to X$ by $[G(t)f](x) = f\left(e^{-t\mathcal{A}}x\right)$ for each $t \in \mathbb{R}$, $f \in X$ and a.e. for $x \in \mathbb{R}^n$. Prove that $\{G(t)\,;\ t \in \mathbb{R}\}$ is a C_0-group and find its infinitesimal generator. Show that, if $\sum_{i=1}^n a_{ii} = 0$, then the group is of isometries.

Problem 2.6. Show that, with X replaced by $C_{ub}(\mathbb{R}^n)$, i.e. the space of uniformly continuous and bounded functions from $\mathbb{R}^n$ to $\mathbb{R}$, endowed with the sup-norm, the family $\{G(t)\,;\ t\in\mathbb{R}\}$ defined as in Problem 2.5, although a group, is not a C_0-group.

Problem 2.7. Let $\{S(t)\,;\ t\geq 0\}$ be a semigroup of linear operators with the property that, for each $x\in X$, we have $\lim_{t\downarrow 0}S(t)x=x$ in the weak topology of X. Prove that there exists $M\geq 1$, and $\omega\in\mathbb{R}$, such that $\|S(t)\|_{\mathcal{L}(X)}\leq Me^{t\omega}$ for each $t\geq 0$.

Problem 2.8. Let $\{S(t)\,;\ t\geq 0\}$ be a semigroup of linear operators with the property that, for each $x\in X$, we have $\lim_{t\downarrow 0}S(t)x=x$ in the weak topology of X. Prove that $\{S(t)\,;\ t\geq 0\}$ is a C_0-semigroup. This is Dunford theorem. See Pazy [**101**], Theorem 1.4, p. 44, or Engel and Nagel et al, Theorem 5.8, p. 40.

Notes. The main results in Sections 2.1 and 2.2, referring to uniformly continuous groups, were obtained independently by Nathan [**93**], Nagumo [**92**] and by Yosida [**134**], but they have their roots in the pioneering works of Peano [**102**], [**103**] concerning the exponential function of a matrix. Sections 2.3 and 2.4 contain several classical notions and results, which may be found, in one form or another in the monographs and treatises on semigroup theory mentioned in Preface. With some exceptions, the problems included are adapted from Brezis and Cazenave [**31**], Engel and Nagel et al [**51**], Goldstein [**61**] and Pazy [**101**].

CHAPTER 3

Generation Theorems

This chapter begins with the presentation of the most fundamental result within the theory of C_0-semigoups as: the Hille-Yosida generation theorem. This gives a very precise delimitation of the class of linear operators A, acting in a Banach space X, that generate C_0-semigroups containing only operators whose norms do not exceed 1. Next, we prove both the Lumer-Phillips generation theorem, which is a very useful reformulation of the latter, and the Feller-Miyadera-Phillips generation theorem which extends the Hille-Yosida theorem to the general case of arbitrary C_0-semigroups. After presenting some useful consequences and some simple examples, we introduce and study the dual and respectively the sun dual of a C_0-semigroup, and we conclude with the celebrated Stone Generation Theorem.

3.1. The Hille-Yosida Theorem. Necessity

The goal of the next two sections is to prove the most important result in the theory of C_0-semigroups: the Hille-Yosida theorem. More precisely, we shall present a necessary and sufficient condition in order that a linear operator A generate a C_0-semigroup of contractions. See Definition 2.3.2. We recall that, if $A : D(A) \subseteq X \to X$ is a linear operator, *the resolvent set* $\rho(A)$ is the set of all those complex numbers λ, called *regular values*, for which $R(\lambda I - A)$ is dense in X and $R(\lambda; A) = (\lambda I - A)^{-1}$ is continuous from $R(\lambda I - A)$ to X.

Theorem 3.1.1. (Hille-Yosida) *A linear operator $A : D(A) \subseteq X \to X$ is the infinitesimal generator of a C_0-semigroup of contractions if and only if:*

(i) *A is densely defined and closed and*
(ii) *$(0, +\infty) \subseteq \rho(A)$ and for each $\lambda > 0$*

$$\|R(\lambda; A)\|_{\mathcal{L}(X)} \leq \frac{1}{\lambda}.$$

Remark 3.1.1. Since $(\lambda I - A)^{-1} = \lambda^{-1}(I - \lambda^{-1}A)^{-1}$ whenever at least one of the two sides of the equality is well-defined, it follows that (ii) is equivalent with:

(ii') for each $\lambda > 0$ we have $(I - \lambda A)^{-1} \in \mathcal{L}(X)$ and

$$\|(I - \lambda A)^{-1}\|_{\mathcal{L}(X)} \leq 1.$$

Moreover, let us observe that if A is densely defined and satisfies

(jj) for each $\lambda > 0$, $\lambda I - A$ is invertible with continuous inverse and

$$\|(\lambda I - A)^{-1}\|_{\mathcal{L}(X)} \leq \frac{1}{\lambda},$$

then it satisfies (i) and (ii). Indeed, by (jj), we easily deduce that $(I - A)^{-1}$ is closed. So $I - A$ is closed and accordingly A enjoys the same property which proves that (i) holds. As (jj) clearly implies (ii) the proof is complete.

We may now proceed to the proof of Theorem 3.1.1.

Proof. We begin with the necessity. Let $A : D(A) \subseteq X \to X$ be the infinitesimal generator of a C_0-semigroup of contractions $\{S(t)\,;\ t \geq 0\}$. In view of Theorem 2.4.1, A is densely defined and closed. Thus (i) holds. In order to prove (ii), let $\lambda > 0$, $x \in X$, and let us define

$$R(\lambda)x = \int_0^{+\infty} e^{-\lambda t} S(t)x\,dt.$$

We notice that the integral on the right-hand side of the equality above is convergent. Indeed, for each $a, b \geq 0$, $a \leq b$, we have

$$\left\| \int_a^b e^{-\lambda t} S(t)x\,dt \right\| \leq \int_a^b e^{-\lambda t} \|S(t)\|_{\mathcal{L}(X)} \|x\|\,dt$$

$$\leq \int_a^b e^{-\lambda t} \|x\|\,dt = \frac{e^{-\lambda a} - e^{-\lambda b}}{\lambda} \|x\|.$$

Accordingly, we are in the hypotheses of the Cauchy test, and thus the integral is convergent. Clearly $R(\lambda) \in \mathcal{L}(X)$ and

$$\|R(\lambda)x\| \leq \left\| \int_0^{+\infty} e^{-\lambda t} S(t)x\,dt \right\| \leq \int_0^{+\infty} e^{-\lambda t} \|S(t)\|_{\mathcal{L}(X)} \|x\|\,dt \leq \frac{1}{\lambda}\|x\|.$$

Hence

$$\|R(\lambda)\|_{\mathcal{L}(X)} \leq \frac{1}{\lambda}.$$

We prove next that $R(\lambda)$ coincides with $R(\lambda; A)$. To this aim we show that $R(\lambda)$ is both the right and the left inverse of the operator $\lambda I - A$. Let $x \in X$, $\lambda > 0$ and $h > 0$. We have

$$\frac{1}{h}\left(S(h) - I\right) R(\lambda)x$$

$$= \frac{1}{h}\int_0^{+\infty} e^{-\lambda t} S(t+h)x\, dt - \frac{1}{h}\int_0^{+\infty} e^{-\lambda t} S(t)x\, dt$$

$$= \frac{e^{\lambda h} - 1}{h}\int_0^{+\infty} e^{-\lambda t} S(t)x\, dt - \frac{e^{\lambda h}}{h}\int_0^{h} e^{-\lambda t} S(t)x\, dt.$$

As the right-hand side of the above equality converges to $\lambda R(\lambda)x - x$, it follows that $R(\lambda)x \in D(A)$, and

$$AR(\lambda) = \lambda R(\lambda) - I,$$

which proves that

$$(\lambda I - A)R(\lambda) = I.$$

So, $R(\lambda)$ is the right inverse of $\lambda I - A$. Next, let $x \in D(A)$. Let us remark that

$$R(\lambda)Ax = \int_0^{+\infty} e^{-\lambda t} S(t)Ax\, dt = \int_0^{+\infty} e^{-\lambda t} \frac{d}{dt}\left(S(t)x\right)\, dt$$

$$= \lim_{t\to\infty} e^{-\lambda t} S(t)x - x + \lambda \int_0^{+\infty} e^{-\lambda t} S(t)x\, dt = \lambda R(\lambda)x - x.$$

This equality may be equivalently rewritten as

$$R(\lambda)(\lambda I - A) = I,$$

which shows that $R(\lambda)$ is the left inverse of $\lambda I - A$, and this completes the proof of the necessity. □

Remark 3.1.2. Using similar arguments, one may prove that, whenever A generates a C_0-semigroup of contractions, then $\{\lambda \in \mathbb{C};\ \operatorname{Re}\lambda > 0\} \subseteq \rho(A)$ and for each $\lambda \in \mathbb{C}$ with $\operatorname{Re}\lambda > 0$, we have

$$\|R(\lambda; A)\|_{\mathcal{L}(X)} \leq \frac{1}{\operatorname{Re}\lambda}.$$

3.2. The Hille-Yosida Theorem. Sufficiency

In order to prove the sufficiency, some preliminary lemmas are needed. First, let us observe that, by (i) and Theorem 1.7.1, it follows that, for each $\lambda > 0$, $R(\lambda; A) \in \mathcal{L}(X)$.

Definition 3.2.1. Let $A : D(A) \subseteq X \to X$ be a linear operator satisfying (i) and (ii) in Theorem 3.1.1, and let $\lambda > 0$. The operator $A_\lambda : X \to X$, defined by $A_\lambda = \lambda A R(\lambda; A)$, is called *the Yosida approximation* of A.

Lemma 3.2.1. *Let $A : D(A) \subseteq X \to X$ be a linear operator which satisfies* (i) *and* (ii) *in* Theorem 3.1.1. *Then:*

$$\lim_{\lambda\to\infty} \lambda R(\lambda; A)x = x \tag{3.2.1}$$

for each $x \in X$,

$$A_\lambda x = \lambda^2 R(\lambda; A)x - \lambda x \tag{3.2.2}$$

for each $x \in X$, and

$$\lim_{\lambda\to\infty} A_\lambda x = Ax \tag{3.2.3}$$

for each $x \in D(A)$.

Proof. Let $x \in D(A)$ and $\lambda > 0$. We have

$$\|\lambda R(\lambda; A)x - x\| = \|AR(\lambda; A)x\| = \|R(\lambda; A)Ax\| \leq \frac{1}{\lambda}\|Ax\|,$$

and consequently

$$\lim_{\lambda\to\infty} \lambda R(\lambda; A)x = x$$

for each $x \in D(A)$. Since $D(A)$ is dense in X and $\|\lambda R(\lambda; A)\|_{\mathcal{L}(X)} \leq 1$, from the last relation, we deduce (3.2.1). To check (3.2.2), let us remark that we have successively

$$\lambda^2 R(\lambda; A) - \lambda I = \lambda^2 R(\lambda; A) - \lambda(\lambda I - A)R(\lambda; A) = \lambda A R(\lambda; A) = A_\lambda.$$

Finally, if $x \in D(A)$, by (3.2.1), we have

$$\lim_{\lambda\to\infty} A_\lambda x = \lim_{\lambda\to\infty} \lambda A R(\lambda; A)x = \lim_{\lambda\to\infty} \lambda R(\lambda; A)Ax = Ax,$$

which concludes the proof of Lemma 3.2.1. □

Lemma 3.2.2. *Let $A : D(A) \subseteq X \to X$ be a linear operator which satisfies* (i) *and* (ii) *in* Theorem 3.1.1. *Then, for each $\lambda > 0$, A_λ is the infinitesimal generator of a uniformly continuous semigroup $\{e^{tA_\lambda}\,;\ t \geq 0\}$ satisfying*

$$\left\|e^{tA_\lambda}\right\|_{\mathcal{L}(X)} \leq 1 \tag{3.2.4}$$

for each $t \geq 0$. In addition, for each $x \in X$ and each $\lambda, \mu > 0$, we have

$$\left\|e^{tA_\lambda}x - e^{tA_\mu}x\right\| \leq t\|A_\lambda x - A_\mu x\|. \tag{3.2.5}$$

Proof. As $A_\lambda \in \mathcal{L}(X)$, by Theorem 2.2.1, it follows that it generates a uniformly continuous semigroup $\{e^{tA_\lambda}\,;\ t \geq 0\}$. In order to check (3.2.4), let us remark that, by virtue of (3.2.2) and (ii), we have

$$\|e^{tA_\lambda}\|_{\mathcal{L}(X)} = \left\|e^{t\lambda^2 R(\lambda;A)-t\lambda I}\right\|_{\mathcal{L}(X)} \leq \left\|e^{t\lambda^2 R(\lambda;A)}\right\|_{\mathcal{L}(X)} \left\|e^{-t\lambda I}\right\|_{\mathcal{L}(X)}$$

$$\leq e^{t\lambda^2\|R(\lambda;A)\|_{\mathcal{L}(X)}}e^{-t\lambda} \leq e^{t\lambda}e^{-t\lambda} = 1.$$

Since A_λ, A_μ, e^{tA_λ} and e^{tA_μ} commute each to another, we have

$$\|e^{tA_\lambda}x - e^{tA_\mu}x\| = \left\|\int_0^1 \frac{d}{ds}\left(e^{stA_\lambda}e^{(1-s)tA_\mu}x\right)\,ds\right\|$$

$$\leq \int_0^1 t\left\|e^{stA_\lambda}e^{(1-s)tA_\mu}(A_\lambda x - A_\mu x)\right\|\,ds \leq t\|A_\lambda x - A_\mu x\|,$$

which completes the proof. □

We proceed to the proof of the sufficiency of Theorem 3.1.1.
Proof of Theorem 3.1.1 (*continued*). From (3.2.3) and (3.2.5), it follows that, for each $t \geq 0$, there exists a linear operator $S(t) : D(A) \subseteq X \to X$ such that, for each $x \in D(A)$,

$$\lim_{\lambda\to\infty} e^{tA_\lambda}x = S(t)x$$

uniformly on compact subsets in $\mathbb{R}_+$. By (3.2.4) we deduce that

$$\|S(t)x\| \leq \|x\|$$

for each $t \geq 0$ and $x \in D(A)$. Since $D(A)$ is dense in X, it follows that $S(t)$ can be extended by continuity to the whole space X. It is easy to see that the family of linear bounded operators thus obtained is a semigroup, denoted for simplicity again by $\{S(t)\,;\ t \geq 0\}$. Clearly it satisfies

$$\|S(t)\|_{\mathcal{L}(X)} \leq 1.$$

In addition, for each $t > 0$ and $x, y \in X$, we have

$$\|S(t)x - x\|$$

$$\leq \|S(t)x - S(t)y\| + \|S(t)y - e^{tA_\lambda}y\| + \|e^{tA_\lambda}y - y\| + \|y - x\|$$

$$\leq \|S(t)y - e^{tA_\lambda}y\| + \|e^{tA_\lambda}y - y\| + 2\|y - x\|.$$

Let $T > 0$ and $\varepsilon > 0$. Fix $y = x_\varepsilon \in D(A)$, with $\|x - x_\varepsilon\| \leq \varepsilon$, and a sufficiently large λ, such that

$$\|S(t)x_\varepsilon - e^{tA_\lambda}x_\varepsilon\| \leq \varepsilon$$

for each $t \in [0, T]$. By this inequality, we deduce

$$\|S(t)x - x\| \leq 3\varepsilon + \|e^{tA_\lambda}x_\varepsilon - x_\varepsilon\|. \tag{3.2.6}$$

Inasmuch as $\{e^{tA_\lambda}\,;\ t \geq 0\}$ is a uniformly continuous semigroup, for the very same $\varepsilon > 0$, there exists $\delta(\varepsilon) > 0$, such that $\|e^{tA_\lambda} - I\|_{\mathcal{L}(X)} \leq \varepsilon$ for each $t \in (0, \delta(\varepsilon))$. Consequently

$$\|e^{tA_\lambda}x_\varepsilon - x_\varepsilon\| \leq \|e^{tA_\lambda} - I\|_{\mathcal{L}(X)}\|x_\varepsilon\| \leq \varepsilon\|x_\varepsilon\|$$

for each $t \in (0, \delta(\varepsilon))$. Since $\{x_\varepsilon\,;\ \varepsilon > 0\}$ is bounded, this inequality, along with (3.2.6), shows that $\{S(t)\,;\ t \geq 0\}$ is a semigroup of class C_0. To conclude the proof, we have merely to show that the infinitesimal generator, $B : D(B) \subseteq X \to X$, of this semigroup coincides with $A : D(A) \subseteq X \to X$. To this aim, let $x \in D(A)$ and $h > 0$. We have

$$\lim_{\lambda \to \infty} e^{tA_\lambda}A_\lambda x = S(t)Ax$$

uniformly on compact subsets in $\mathbb{R}_+$. Indeed,

$$\begin{aligned}\|e^{tA_\lambda}A_\lambda x - S(t)Ax\| &\leq \|e^{tA_\lambda}A_\lambda x - e^{tA_\lambda}Ax\| + \|e^{tA_\lambda}Ax - S(t)Ax\| \\ &\leq \|e^{tA_\lambda}\|_{\mathcal{L}(X)}\|A_\lambda x - Ax\| + \|e^{tA_\lambda}Ax - S(t)Ax\|.\end{aligned}$$

But this relation, along with (3.2.3) and with the partial conclusions above, proves that

$$S(h)x - x = \lim_{\lambda \to \infty}\left(e^{hA_\lambda}x - x\right) = \lim_{\lambda \to \infty}\int_0^h e^{tA_\lambda}A_\lambda x\,dt = \int_0^h S(t)Ax\,dt.$$

Dividing both sides this equality by h and letting h tend to 0 by positive values, we deduce that $x \in D(B)$ and $Bx = Ax$. Finally, we show that $D(A) = D(B)$. Since B is the infinitesimal generator of a C_0-semigroup of contractions, from the necessity it follows that $1 \in \rho(B)$. Accordingly $I - B$ is invertible and $(I - B)^{-1}X = D(B)$. As $(I - B)D(A) = (I - A)D(A)$ and, by (ii), $(I - A)D(A) = X$, it follows that $(I - B)D(A) = X$, or equivalently $(I - B)^{-1}X = D(A)$. Hence $D(A) = D(B)$, which completes the proof of Theorem 3.1.1. □

3.3. The Feller-Miyadera-Phillips Theorem

As concerns the case of general C_0-semigroups of type (M, ω), we have the following generation theorem due to Feller-Miyadera-Phillips.

Theorem 3.3.1. *The operator $A : D(A) \subseteq X \to X$ is the infinitesimal generator of a C_0-semigroup of type (M, ω) if and only if:*

(i) *A is densely defined and closed and*

(ii) *$(\omega, +\infty) \subseteq \rho(A)$ and, for each $\lambda > \omega$ and each $n \in \mathbb{N}^*$, we have*

$$\|R(\lambda; A)^n\|_{\mathcal{L}(X)} \leq \frac{M}{(\lambda - \omega)^n}.$$

In order to prove Theorem 3.3.1 we need the following renorming lemma.

Lemma 3.3.1. *Let $A : D(A) \subseteq X \to X$ be a linear operator satisfying both $(0, +\infty) \subseteq \sigma(A)$ and*

$$\|\lambda^n R(\lambda; A)^n\|_{\mathcal{L}(X)} \leq M$$

for each $n \in \mathbb{N}$ and $\lambda > 0$. Then there exists a norm $|\cdot|$ on X such that

$$\|x\| \leq |x| \leq M\|x\| \tag{3.3.1}$$

and

$$|\lambda R(\lambda; A)x| \leq |x| \tag{3.3.2}$$

for each $x \in X$ and $\lambda > 0$.

Proof. For $\mu > 0$ let us define $|\cdot|_\mu : X \to \mathbb{R}_+$ by

$$|x|_\mu = \sup_{n\in\mathbb{N}} \|\mu^n R(\mu; A)^n x\|.$$

It is easy to see that

$$\|x\| \leq |x|_\mu \leq M\|x\| \tag{3.3.3}$$

and

$$|\mu R(\mu; A)x|_\mu \leq |x|_\mu.$$

We shall prove next that

$$|\lambda R(\lambda; A)x|_\mu \leq |x|_\mu \tag{3.3.4}$$

for each $\lambda \in (0, \mu\,]$. Indeed, by the resolvent equation (1.7.3), we have

$$R(\lambda; A)x = R(\mu; A)(x + (\mu - \lambda)R(\lambda; A)x)$$

and therefore

$$|R(\lambda; A)x|_\mu \leq \frac{1}{\mu}|x|_\mu + \left(1 - \frac{\lambda}{\mu}\right)|R(\lambda; A)x|_\mu.$$

Consequently $\lambda|R(\lambda; A)x|_\mu \leq |x|_\mu$ which proves (3.3.4). From (3.3.3) and (3.3.4) we deduce that, for each $n \in \mathbb{N}$ and $\lambda \in (0, \mu\,]$, we have

$$\|\lambda^n R(\lambda; A)^n x\| \leq |\lambda^n R(\lambda; A)^n x|_\mu \leq |x|_\mu. \tag{3.3.5}$$

Passing to the sup for $n \in \mathbb{N}$ on the left-hand side of the inequality above, we get $|x|_\lambda \leq |x|_\mu$ for each $\lambda \in (0, \mu\,]$. Now, we can define

$$|x| = \lim_{\mu\to\infty} |x|_\mu.$$

Since (3.3.1) readily follows from (3.3.3), and (3.3.2) from (3.3.5) by taking $n = 1$, the proof is complete □

We may now pass to the proof of Theorem 3.3.1.

Proof. *Necessity.* We consider first the case $\omega = 0$. Let us observe that for any norm on X which is equivalent with the initial one, all the topological properties of both the generator and the semigroup preserve. So, we begin by choosing an equivalent norm with respect to which the C_0-semigroup $\{S(t)\,;\ t \geq 0\}$ of type $(M, 0)$ becomes of type $(1, 0)$, i.e. a semigroup of contractions, and then we apply Hille-Yosida Theorem 3.1.1. We define $|||\cdot||| : X \to \mathbb{R}_+$ by

$$|||x||| = \sup_{t\geq 0} \|S(t)x\|,$$

and we observe that

$$\|x\| \leq |||x||| \leq M\|x\|, \tag{3.3.6}$$

for each $x \in X$. Thus $|||\cdot|||$ and $\|\cdot\|$ are equivalent norms on X. In addition

$$|||S(t)x||| = \sup_{s\geq 0} \|S(s)S(t)x\| \leq \sup_{t\geq 0} \|S(t)x\| = |||x|||$$

and thus $\{S(t)\,;\ t \geq 0\}$ is a C_0-semigroup of contractions on $(X, |||\cdot|||)$. From Hille-Yosida Theorem 3.1.1 it follows that A is closed, densely defined, $(0, +\infty) \subseteq \sigma(A)$ and

$$|||R(\lambda; A)|||_{\mathcal{L}(X)} \leq \frac{1}{\lambda} \tag{3.3.7}$$

for each $\lambda > 0$. By the definition of $|||\cdot|||$, (3.3.6) and (3.3.7) we have

$$\|\lambda^n R(\lambda; A)^n x\| \leq |||\lambda^n R(\lambda; A)^n x||| \leq |||x||| \leq M\|x\|,$$

which show that the conditions (i) and (ii) are necessary.
Sufficiency Let us assume that (i) and (ii) hold. By Lemma 3.3.1 we know that there exists an equivalent norm $|\cdot|$ on X satisfying $\|x\| \leq |x| \leq M\|x\|$ and $|\lambda R(\lambda; A)x| \leq |x|$ for each $x \in X$ and $\lambda > 0$. So, on $(X, |\cdot|)$, A satisfies the hypotheses of Hille-Yosida Theorem 3.1.1 and therefore it generates a C_0-semigroup of contractions $\{S(t)\,;\ t \geq 0\}$. Finally, since

$$\|S(t)x\| \leq |S(t)x| \leq |x| \leq M\|x\|$$

for each $t \geq 0$ and $x \in X$, this achieves the proof in the case $\omega = 0$. For the general case, we have only to observe that, if $\{S(t)\,;\ t \geq 0\}$ is a C_0-semigroup of type (M, ω) and A is its generator, then $\{e^{-\omega t}S(t)\,;\ t \geq 0\}$ is a C_0-semigroup of type $(M, 0)$ and $A - \omega I$ is its generator. □

3.4. The Lumer-Phillips Theorem

Let X be a real, or complex Banach space with norm $\|\cdot\|$, and let us recall that the *duality mapping* $F : X \to 2^{X^*}$ is defined by

$$F(x) = \left\{x^* \in X^*\,;\ (x, x^*) = \|x\|^2 = \|x^*\|^2\right\}$$

for each $x \in X$. In view of the Hahn-Banach theorem, it follows that, for each $x \in X$, $F(x)$ is nonempty.

Definition 3.4.1. A linear operator $A : D(A) \subseteq X \to X$ is *dissipative* if for each $x \in X$ there exists $x^* \in F(x)$ such that $\text{Re}(Ax, x^*) \le 0$.

Theorem 3.4.1. *A linear operator $A : D(A) \subseteq X \to X$ is dissipative if and only if, for each $x \in D(A)$ and $\lambda > 0$, we have*

$$\lambda\|x\| \le \|(\lambda I - A)x\|. \tag{3.4.1}$$

Proof. If A is dissipative, then, for each $x \in D(A)$ and $\lambda > 0$ there exists $x^* \in F(x)$ such that $\text{Re}(\lambda x - Ax, x^*) \le 0$. Therefore

$$\lambda\|x\|^2 \le \text{Re}(\lambda x - Ax, x^*) \le |(\lambda x - Ax, x^*)| \le \|\lambda x - Ax\|\|x\|$$

and this completes the proof of the "only if" part. Next, let $x \in D(A)$ and $\lambda > 0$. Let $y_\lambda^* \in F(\lambda x - Ax)$ and, let us observe that, by virtue of (3.4.1), $\lambda x - Ax = 0$ implies $x = 0$. So, in this case, we clearly have $\text{Re}(x^*, \lambda x - Ax) = 0$. Therefore, let us assume that $\lambda x - Ax \neq 0$. As a consequence, $y_\lambda^* \neq 0$, and thus $z_\lambda^* = y_\lambda^*/\|y_\lambda^*\|$ lies on the unit ball, i.e. $\|z_\lambda^*\| = 1$. So, we have

$$\lambda\|x\| \le \|\lambda x - Ax\| = (\lambda x - Ax, z_\lambda^*)$$

$$= \lambda\text{Re}(x, z_\lambda^*) - \text{Re}(Ax, z_\lambda^*) \le \lambda\|x\| - \text{Re}(Ax, z_\lambda^*).$$

Hence

$$\text{Re}(Ax, z_\lambda^*) \le 0$$

and

$$\text{Re}(z_\lambda^*, x) \ge \|x\| - \frac{1}{\lambda}\|Ax\|.$$

Now, let us recall that the closed unit ball in X^* is weakly-star compact. Thus, the net $(z_\lambda^*)_{\lambda>0}$ has at least one weak-star cluster point $z^* \in X^*$ with $\|z^*\| \le 1$. From the inequalities above, it follows that $\text{Re}(Ax, z^*) \le 0$ and $\text{Re}(x, z^*) \ge \|x\|$. Since $\text{Re}(x, z^*) \le |(x, z^*)| \le \|x\|$, it follows that $(x, z^*) = \|x\|$. Hence $x^* = \|x\|z^* \in F(x)$ and $\text{Re}(Ax, x^*) \le 0$, and this completes the proof. □

Remark 3.4.1. In the case in which $X = H$ is a real Hilbert space with inner product $\langle\cdot,\cdot\rangle$, a linear operator $A : D(A) \subseteq H \to H$ is dissipative if, for each $x \in D(A)$, we have $\langle x, Ax\rangle \le 0$. This is an easy consequence of Definition 3.4.1 combined with the observation that, if we identify H with its own dual, the corresponding duality mapping is given by $F(x) = \{x\}$ for each $x \in H$.

Theorem 3.4.2. (Lumer-Phillips) *Let $A : D(A) \subseteq X \to X$ be a densely defined operator. Then A generates a C_0-semigroup of contractions on X if and only if*

(i) *A is dissipative;*
(ii) *there exists $\lambda > 0$ such that $\lambda I - A$ is surjective.*

Moreover, if A generates a C_0-semigroup of contractions, then $\lambda I - A$ is surjective for any $\lambda > 0$, and we have $\mathrm{Re}(Ax, x^*) \leq 0$ for each $x \in D(A)$ and each $x^* \in F(x)$.

Proof. *Necessity* If A is the infinitesimal generator of a C_0-semigroup of contractions $\{S(t);\ t \geq 0\}$, by Hille-Yosida Theorem 3.1.1, we have $(0, +\infty) \subseteq \rho(A)$ and therefore $\lambda I - A$ is surjective for each $\lambda > 0$. Next, if $x \in D(A)$ and $x^* \in F(x)$, we have

$$|(S(t)x, x^*)| \leq \|x^*\|\|S(t)x\| \leq \|x\|^2.$$

So

$$\mathrm{Re}(S(t)x - x, x^*) \leq \mathrm{Re}(x, x^*) - \|x\|^2 \leq 0.$$

Dividing by $t > 0$ and letting $t \downarrow 0$, we deduce

$$\mathrm{Re}(Ax, x^*) \leq 0$$

and this completes the proof of the necessity.
Sufficiency Since A is dissipative, for each $\lambda > 0$ and $x \in D(A)$, we have

$$\lambda\|x\| \leq \|(\lambda I - A)x\|. \tag{3.4.2}$$

Inasmuch as $\lambda I - A$ is surjective for some λ, say $\lambda = \mu$, from (3.4.2), it follows that $\mu I - A$ has a bounded inverse, and thus it is closed. Hence A is closed. We will prove next that $\lambda I - A$ is surjective for each $\lambda > 0$. To this aim, let us consider the set

$$\Lambda = \{\lambda;\ 0 < \lambda < +\infty,\ \lambda I - A \text{ is surjective}\}.$$

So, it suffices to check that

$$\Lambda = (0, +\infty). \tag{3.4.3}$$

In view of (3.4.2), it follows that $\Lambda \subseteq \rho(A)$. Recalling that, by virtue of Theorem 1.7.2, $\rho(A)$ is open, it follows that, for each $\lambda \in \Lambda$ there exists an open neighborhood of λ, $V \subseteq \mathbb{C}$, which is contained in $\rho(A)$. The intersection of V with the real line is clearly included in Λ, and accordingly Λ is open. To complete the proof of (3.4.3), let $(\lambda_n)_{\in \mathbb{N}}$ be a sequence in Λ with $\lim_{n\to\infty} \lambda_n = \lambda > 0$. Then, for each $n \in \mathbb{N}$ and $y \in X$, there exists $x_n \in D(A)$ so that

$$\lambda_n x_n - Ax_n = y. \tag{3.4.4}$$

By virtue of (3.4.2), it follows that there exists $C > 0$ such that

$$\|x_n\| \leq \frac{1}{\lambda_n}\|y\| \leq C.$$

Also from (3.4.2), we get

$$\begin{aligned}\lambda_m\|x_n - x_m\| &\leq \|\lambda_m(x_n - x_m) - A(x_n - x_m)\| \\ &= |\lambda_n - \lambda_m|\|x_n\| \leq C|\lambda_n - \lambda_m|.\end{aligned}$$

Consequently $(x_n)_{n\in\mathbb{N}}$ is a Cauchy sequence. Let $x = \lim_{n\to\infty} x_n$ and let us observe that, in view of (3.4.4), we have $\lim_{n\to\infty} Ax_n = \lambda x - y$. Since A is closed, if follows that $x \in D(A)$ and $\lambda x - Ax = y$. Thus $\lambda \in \Lambda$ and so Λ is both open and closed. Since it is nonempty (we recall that $\mu \in \Lambda$), we necessarily have $\Lambda = (0, +\infty)$ and this completes the proof. □

3.5. Some Consequences

We begin with the following remarkable consequence.

Corollary 3.5.1. *Let $A : D(A) \subseteq X \to X$ be the infinitesimal generator of a C_0-semigroup and let $\|\cdot\|_{D(A)} : D(A) \to \mathbb{R}_+$ and $|\cdot|_{D(A)} : D(A) \to \mathbb{R}_+$ be defined by $\|x\|_{D(A)} = \|x\| + \|Ax\|$, and respectively by $|x|_{D(A)} = \|x - Ax\|$, for each $x \in D(A)$. Then:*

(i) *$\|\cdot\|_{D(A)}$ is a norm on $D(A)$, called the graph norm, with respect to which $D(A)$ is a Banach space;*
(ii) *$D(A)$ endowed with the norm $\|\cdot\|_{D(A)}$ is continuously imbedded in X;*
(iii) *$A \in \mathcal{L}(D(A), X)$ (where $D(A)$ is endowed with $\|\cdot\|_{D(A)}$);*
(iv) *$|\cdot|_{D(A)}$ is a norm on $D(A)$ equivalent with $\|\cdot\|_{D(A)}$;*
(v) *$I - A$ is an isometry from $(D(A), |\cdot|_{D(A)})$ to $(X, \|\cdot\|)$;*
(vi) *for each $x \in D(A)$, $S(\cdot)x \in C([0, +\infty); D(A)) \cap C^1([0, +\infty); X)$*[1].

Proof. (i) Inasmuch as A is linear, we may easily verify that $\|\cdot\|_{D(A)}$ is a norm on $D(A)$. Let $(x_n)_{n\in\mathbb{N}}$ be a Cauchy sequence in the norm $\|\cdot\|_{D(A)}$. Then both $(x_n)_{n\in\mathbb{N}}$ and $(Ax_n)_{n\in\mathbb{N}}$ are Cauchy sequences in the norm of X. As X is complete and A is closed, it follows that $(x_n)_{n\in\mathbb{N}}$ is convergent to an element $x \in D(A)$. Thus $D(A)$, endowed with the norm $\|\cdot\|_{D(A)}$, is complete and thus a Banach space.
We remark that (ii) and (iii) follow from the inequalities: $\|x\| \leq \|x\|_{D(A)}$ and $\|Ax\| \leq \|x\|_{D(A)}$ for $x \in D(A)$. By Theorem 3.1.1, $\|x\| \leq |x|_{D(A)}$.

---

[1]Here $C([0, +\infty); D(A))$ is the space of all continuous functions from $[0, +\infty)$ in $D(A)$, the latter being endowed with the graph norm $\|\cdot\|_{D(A)}$.

So, $|x|_{D(A)} \leq \|x\|_{D(A)}$ and $\|x\|_{D(A)} \leq 2\|x\| + |x|_{D(A)} \leq 3|x|_{D(A)}$ for each $x \in D(A)$, from where it follows (iv).
Since $(I-A)^{-1} - A(I-A)^{-1} = I$, we have $|(I-A)^{-1}x|_{D(A)} = \|x\|$, for each $x \in X$, which implies that $I - A$ is an isometry from $D(A)$ to X. Thus (v) holds. Finally, (vi) is a consequence of (iii) in Theorem 2.3.2 which shows that, for each $x \in D(A)$, $AS(\,\cdot\,)x \in C([0,+\infty);X)$. From the definition of the graph norm, we finally deduce $S(\,\cdot\,)x \in C([0,+\infty);D(A))$ and so (vi) holds. The proof is complete. □

The following theorem shows that the restriction of a generator of a C_0-semigroup of contractions to its domain, endowed with the graph-norm, is the generator of a C_0-semigroup of contractions too.

Theorem 3.5.1. *Let $A : D(A) \subseteq X \to X$ be the infinitesimal generator of a C_0-semigroup of contractions and let $X_1 = D(A)$ endowed with the graph-norm $|\cdot|_{D(A)} : X_1 \to \mathbb{R}_+$ defined by $|u|_{D(A)} = \|u - Au\|$ for $u \in X_1$. Then the operator $A_{(1)} : D(A_{(1)}) \subseteq X_1 \to X_1$ defined by*

$$\begin{cases} D(A_{(1)}) = \{x \in X_1; Ax \in X_1\} \\ A_{(1)}x = Ax, \ \text{for } x \in D(A_{(1)}), \end{cases}$$

is the infinitesimal generator of a C_0-semigroup of contractions on X_1.

Proof. Let $\lambda > 0$ and $f \in X_1$ and let us consider the equation

$$\lambda u - Au = f.$$

Inasmuch as A generates a C_0-semigroup of contractions, by Theorem 3.1.1, it follows that this equation has a unique solution $u \in D(A)$. As $f \in X_1$, we conclude that $Au \in D(A)$, and thus $u \in D(A_{(1)})$. Thus $\lambda u - A_{(1)}u = f$. On the other hand we have

$$|(\lambda I - A_{(1)})^{-1} f|_{D(A)} = \|(I-A)(\lambda I - A)^{-1} f\|$$

$$= \|(\lambda I - A)^{-1}(I-A)f\| \leq \frac{1}{\lambda}\|f - Af\| = \frac{1}{\lambda}|f|_{D(A)},$$

which shows that $A_{(1)}$ satisfies (ii) in Theorem 3.1.1. Moreover, it follows that $A_{(1)}$ is closed in X_1. Indeed, as $(\lambda I - A_{(1)})^{-1} \in \mathcal{L}(X_1)$, it is closed, and consequently $\lambda I - A_{(1)}$ enjoys the same property which proves that $A_{(1)}$ is closed too.
Next, let $x \in X_1$, $\lambda > 0$ and let $x_\lambda = \lambda x - A_{(1)}x$. Clearly $x_\lambda \in D(A_{(1)})$, and, in addition,

$$\lim_{\lambda \to \infty} |x_\lambda - x|_{D(A)} = 0.$$

Indeed, this simply follows from (3.2.3) in Lemma 3.2.1. Thus $D(A_{(1)})$ is dense in X_1 and, by virtue of Theorem 3.1.1, $A_{(1)}$ generates a C_0-semigroup of contractions on X_1. The proof is complete. □

Remark 3.5.1. An inductive argument shows that, for each infinitesimal generator of a C_0-semigroup of contractions $A : D(A) \subseteq X \to X$, there exists a sequence $(X_n)_{n\in\mathbb{N}}$ of Banach spaces, such that

$$\cdots \subseteq X_{n+1} \subseteq X_n \subseteq \cdots \subseteq X_0 = X,$$

all the inclusions being dense and continuous, and a family of generators of C_0-semigroups of contractions $(A_{(n)})_{n\in\mathbb{N}}$, with $D(A_{(n)}) = X_{n+1}$ and $A_n x = Ax$ for each $x \in X_{n+1}$. Let us remark that, if A is linear continuous, then $X_n = X$ for each $n \in \mathbb{N}$, while whenever A is discontinuous, for each $n \in \mathbb{N}$, X_{n+1} is strictly included in X_n. In addition, for each $n \in \mathbb{N}$, $X_n = D(A^n)$ and the norm $\|\cdot\|_{X_n}$ is equivalent with the norm $\|\cdot\|_{D(A^n)}$. See (iv) and (v) in Corollary 3.5.1.

Theorem 3.5.2. *The linear operator $A : D(A) \subseteq X \to X$ is the generator of a C_0-group of isometries if and only if:*

(i) *A is densely defined and closed and*

(iii) *$\mathbb{R}^* \subseteq \rho(A)$ and for each $\lambda \in \mathbb{R}^*$*

$$\|R(\lambda; A)\|_{\mathcal{L}(X)} \leq \frac{1}{|\lambda|}.$$

Proof. *Necessity.* Let us assume that A is the infinitesimal generator of a C_0-group of isometries $\{G(t)\,;\ t \in \mathbb{R}\}$. Then both operators A and $-A$ generate a C_0-semigroup of contractions $\{G(t)\,;\ t \geq 0\}$, and respectively $\{G(-t)\,;\ t \geq 0\}$. Thus, by Theorem 3.1.1, it follows that A satisfies (i) and (iii).

Sufficiency. If A satisfies (i) and (iii), by Theorem 3.1.1, it follows that both A and $-A$ generate a C_0-semigroup of contractions $\{S(t)\,;\ t \geq 0\}$, and respectively $\{T(t)\,;\ t \geq 0\}$. Then the family of linear bounded operators $\{G(t)\,;\ t \in \mathbb{R}\}$, defined by

$$G(t) = \begin{cases} S(t) & \text{if } t \geq 0 \\ \\ T(-t) & \text{if } t \leq 0, \end{cases}$$

is a C_0-group of isometries, and this achieves the proof. □

3.6. Examples

This section contains two simple examples which illustrate the effectiveness of both Theorems 3.1.1 and 3.5.1. For the sake of simplicity, here we confine ourselves to the 1-dimensional case. We notice that, in the next chapter, we shall reconsider both these examples in the general n-dimensional setting exactly as they are encountered in the study of parabolic, or hyperbolic partial differential equations.

Example 3.6.1. In this Example we shall show that, for $\Omega = (0, \pi)$, the Laplace operator with the Dirichlet boundary condition is the infinitesimal generator of a C_0-semigroup of contractions on $L^2(0, \pi)$. More precisely we have:

Proposition 3.6.1. *The operator*

$$A : D(A) \subseteq L^2(0, \pi) \to L^2(0, \pi),$$

defined by

$$\begin{cases} D(A) = H_0^1(0, \pi) \cap H^2(0, \pi) \\ Au = u'' \text{ for } u \in D(A), \end{cases}$$

is the infinitesimal generator of a C_0-semigroup of contractions.

Proof. We apply Theorem 3.1.1 in the space $X = L^2(0, \pi)$. By the definition of $H^2(0, \pi)$ and by Theorem 1.3.3, it follows that $D(A)$ is dense in $L^2(0, \pi)$, and thus the first part in the condition (i) in Theorem 3.1.1 is satisfied. We shall prove the closedness of A by using (jj) in Remark 3.1.1. In order to check (jj), we begin by showing that, for each $\lambda > 0$, the operator $\lambda I - A$ is bijective. To this aim, let $f \in L^2(0, \pi)$ and let us remark that the equation $(\lambda I - A)u = f$ may be equivalently rewritten as:

$$\begin{cases} \lambda u - u'' = f \\ u(0) = u(\pi) = 0. \end{cases} \tag{3.6.1}$$

At this point, let us recall that the general solution of the nonhomogeneous equation $\lambda u - u'' = f$ is given by

$$u(x) = c_1(x)e^{-\sqrt{\lambda}x} + c_2(x)e^{\sqrt{\lambda}x},$$

where c_1 and c_2 satisfy

$$\begin{cases} c_1'(x)e^{-\sqrt{\lambda}x} + c_2'(x)e^{\sqrt{\lambda}x} = 0 \\ -\sqrt{\lambda}c_1'(x)e^{-\sqrt{\lambda}x} + \sqrt{\lambda}c_2'(x)e^{\sqrt{\lambda}x} = f(x). \end{cases}$$

See (4.50), p. 68 in Corduneanu [**39**]. It then follows

$$u(x) = k_1 e^{-\sqrt{\lambda}x} + k_2 e^{\sqrt{\lambda}x} + \frac{1}{2\sqrt{\lambda}} \int_0^\pi k(x, y) f(y)\, dy,$$

where

$$k(x, y) = \begin{cases} e^{\sqrt{\lambda}(y-x)} & \text{if } 0 \le x \le y \le \pi \\ e^{\sqrt{\lambda}(x-y)} & \text{if } 0 \le y \le x \le \pi. \end{cases}$$

Imposing the boundary conditions $u(0) = u(\pi) = 0$, we obtain a system of linear equations, with the unknowns k_1 and k_2, which has a unique solution. Thus, (3.6.1) has a unique solution $u = (\lambda I - A)^{-1}f$.
Finally, multiplying both sides in (3.6.1) by $u(x)$ and integrating from 0 to π, we obtain

$$\lambda \|u\|^2_{L^2(0,\pi)} + \|u'\|^2_{L^2(0,\pi)} = \langle f, u\rangle_{L^2(0,\pi)},$$

which implies

$$\|u\|_{L^2(0,\pi)} \leq \frac{1}{\lambda}\|f\|_{L^2(0,\pi)}.$$

As $u = (\lambda I - A)^{-1}f$, the last inequality shows that

$$\|(\lambda I - A)^{-1}f\|_{L^2(0,\pi)} \leq \frac{1}{\lambda}\|f\|_{L^2(0,\pi)},$$

for each $f \in L^2(0, \pi)$. Taking the supremum both sides in this inequality for $\|f\|_{L^2(0,\pi)} \leq 1$, we deduce (jj) and implicitly (ii). See Remark 3.1.1. The proof is complete. □

Corollary 3.6.1. *For each $\xi \in H^1_0(0, \pi) \cap H^2(0, \pi)$, the problem*

$$\begin{cases} u_t = u_{xx} & (t,x) \in \mathbb{R}_+ \times (0,\pi) \\ u(t,0) = u(t,\pi) = 0 & t \in \mathbb{R}_+ \\ u(0,x) = \xi(x) & x \in (0,\pi) \end{cases} \tag{3.6.2}$$

has a unique solution $u \in C^1(\mathbb{R}_+; L^2(0, \pi))$.

Proof. The conclusion follows form (iii) in Theorem 2.3.2, by observing that the problem

$$\begin{cases} u' = Au \\ u(0) = \xi, \end{cases}$$

with A defined as in Proposition 3.6.1, may be rewritten as (3.6.2). □

Example 3.6.2. In order to understand the importance of the differential operator we will analyze in this example, let us consider the equation of an isotropic, inextensible and homogeneous string of length π, and having fixed endpoints

$$\begin{cases} u_{tt} - u_{xx} = 0 & (t,x) \in \mathbb{R} \times (0,\pi) \\ u(t,0) = u(t,\pi) = 0 & t \in \mathbb{R} \\ u(0,x) = \xi_1(x) & x \in (0,\pi) \\ u_t(0,x) = \xi_2(x) & x \in (0,\pi). \end{cases} \tag{3.6.3}$$

The equation above may be equivalently rewritten as a first-order (with respect to t) system of partial differential equations in the product space

$H_0^1(0,\pi) \times L^2(0,\pi)$, i.e.

$$\begin{cases} u_t - v = 0 & (t,x) \in \mathbb{R} \times (0,\pi) \\ v_t - u_{xx} = 0 & (t,x) \in \mathbb{R} \times (0,\pi) \\ u(t,0) = u(t,\pi) = 0 & t \in \mathbb{R} \\ u(0,x) = \xi_1(x) & x \in (0,\pi) \\ v(0,x) = \xi_2(x) & x \in (0,\pi). \end{cases}$$

Identifying the real functions of two real variables $u, v : \mathbb{R} \times (0,\pi) \to \mathbb{R}$ with two functions of only one real variable with values in $H_0^1(0,\pi)$, and respectively in $L^2(0,\pi)$, i.e. $u(t,x) = u(t)(x)$ and $v(t,x) = v(t)(x)$ for a.e. $x \in (0,\pi)$, we observe that, in its turn, this system, may be rewritten as a first-order ordinary differential equation of the form

$$\begin{cases} z' = Az \\ z(0) = \xi, \end{cases} \tag{3.6.4}$$

where $z(t) = (u(t), v(t))$ for each $t \in \mathbb{R}$, $\xi = (\xi_1, \xi_2)$, and A is defined as in Proposition 3.6.2 below.

We endow the product space $H_0^1(0,\pi) \times L^2(0,\pi)$ with the inner product $\langle\!\langle \cdot, \cdot \rangle\!\rangle$ defined by

$$\langle\!\langle (u_1, v_1), (u_2, v_2) \rangle\!\rangle = \int_0^\pi u_1'(x) u_2'(x)\, dx + \int_0^\pi v_1(x) v_2(x)\, dx$$

for each $(u_i, v_i) \in H_0^1(0,\pi) \times L^2(0,\pi)$, $i = 1, 2$. We emphasize that, by virtue of Theorem 1.5.7, $H_0^1(0,\pi) \times L^2(0,\pi)$, equipped with this inner product, is a real Hilbert space.

Proposition 3.6.2. *The operator*

$$A : D(A) \subseteq H_0^1(0,\pi) \times L^2(0,\pi) \to H_0^1(0,\pi) \times L^2(0,\pi),$$

defined by

$$\begin{cases} D(A) = (H_0^1(0,\pi) \cap H^2(0,\pi)) \times H_0^1(0,\pi) \\ A(u,v) = (v, u'') \ \text{ for } (u,v) \in D(A), \end{cases}$$

is the infinitesimal generator of a C_0-group of isometries.

Proof. We show that A satisfies the hypotheses of Theorem 3.5.2. Clearly $D(A)$ is dense in $H_0^1(0,\pi) \times L^2(0,\pi)$. In order to prove that A is closed, we shall use again Remark 3.1.1. Let $\lambda \in \mathbb{R}^*$. To check that $\lambda I - A$ is a bijection, let us remark that, for each $(f, g) \in H_0^1(0,\pi) \times L^2(0,\pi)$, the equation $(\lambda I - A)(u, v) = (f, g)$ may be equivalently rewritten as

$$\begin{cases} \lambda u - v = f \\ \lambda v - u'' = g \\ u(0) = u(\pi) = 0. \end{cases} \tag{3.6.5}$$

Multiplying the first equation by λ, and adding side by side both equations, we obtain

$$\begin{cases} \lambda^2 u - u'' = \lambda f + g \\ u(0) = u(\pi) = 0. \end{cases}$$

Repeating the arguments in the proof of Proposition 3.6.1, we conclude that, for each $(f,g) \in H_0^1(0,\pi) \times L^2(0,\pi)$, this problem has a unique solution $u \in H_0^1(0,\pi) \cap H^2(0,\pi)$. Coming back to the first equation in (3.6.5), we easily see that this has a unique solution $v \in H_0^1(0,\pi)$. So, for each $\lambda \in \mathbb{R}^*$, $(\lambda I - A)^{-1}$ is well-defined. By (3.6.5), we have $(\lambda u - v, \lambda v - u'') = (f,g)$. Taking the inner product in $H_0^1(0,\pi) \times L^2(0,\pi)$ both sides of the equality above by (u,v), integrating by parts, and taking into account the boundary conditions $v(0) = v(\pi) = 0$, we obtain

$$|\lambda| \left\|(\lambda I - A)^{-1}(f,g)\right\|^2 = \left|\langle (f,g), (\lambda I - A)^{-1}(f,g)\rangle\right|.$$

By virtue of the Cauchy-Schwarz inequality, this relation proves that, for each $\lambda \in \mathbb{R}^*$, we have

$$\|(\lambda I - A)^{-1}\|_{\mathcal{L}(X)} \leq \frac{1}{|\lambda|}.$$

Hence A satisfies (jj) in Remark 3.1.1, and thus it satisfies (i) and (ii) in Theorem 3.5.2, from where it follows that it generates a C_0-group of isometries. The proof is complete. □

Corollary 3.6.2. *For any* $(\xi_1, \xi_2) \in H^2(0,\pi) \cap H_0^1(0,\pi) \times H_0^1(0,\pi)$, *the problem* (3.6.3) *has a unique solution* u *satisfying*

$$(u, u_t) \in C^1(\mathbb{R}_+; H_0^1(0,\pi) \times L^2(0,\pi)).$$

Proof. The conclusion follows from (iii) in Theorem 2.3.2, recalling that the problem (3.6.3) may be equivalently rewritten as (3.6.4). □

3.7. The Dual of a C_0-Semigroup

Let X be a Banach space and $\{S(t)\,;\ t \geq 0\}$ a semigroup of linear operators on X.

Definition 3.7.1. The family $\{S(t)^*\,;\ t \geq 0\} \subseteq \mathcal{L}(X^*)$, where, for each $t \geq 0$, $S(t)^*$ is the adjoint of the operator $S(t)$, is called *the dual of the semigroup* $\{S(t)\,;\ t \geq 0\}$.

Our goal within this section is to present some basic properties of the dual of a semigroup of linear operators. We begin with the following Functional Analysis result.

Lemma 3.7.1. *If* $T \in \mathcal{L}(X)$ *then* $T^* \in \mathcal{L}(X^*)$ *and* $\|T\|_{\mathcal{L}(X)} = \|T^*\|_{\mathcal{L}(X^*)}$

Proof. First let us observe that, for each $x^* \in X^*$, $x \mapsto (Tx, x^*)$ is a linear continuous functional on X, denoted by y^*. Since $(x, y^*) = (Tx, x^*)$, we have $D(T^*) = X^*$. In addition

$$\|T^*\|_{\mathcal{L}(X^*)} = \sup_{\|x^*\|\leq 1} \|T^*x^*\| = \sup_{\|x^*\|\leq 1} \sup_{\|x\|\leq 1} |(x, T^*x^*)|$$

$$= \sup_{\|x\|\leq 1} \sup_{\|x^*\|\leq 1} |(Tx, x^*)| = \sup_{\|x\|\leq 1} \|Tx\| = \|T\|_{\mathcal{L}(X)},$$

which completes the proof. □

Lemma 3.7.2. *Let $A : D(A) \subseteq X \to X$ be a densely defined linear closed operator. If $\lambda \in \rho(A)$, then $\lambda \in \rho(A^*)$, and*

$$R(\lambda; A^*) = R(\lambda; A)^*.$$

Proof. We have $(\lambda I - A)^* = \lambda I^* - A^*$, where I^* is the identity on X^*. As A is closed, by Theorem 1.7.1, it follows that $R(\lambda; A) \in \mathcal{L}(X)$. From Lemma 3.7.1, we have that $R(\lambda; A)^* \in \mathcal{L}(X^*)$. We shall show next that there exists $R(\lambda; A^*)$ and it coincides with $R(\lambda; A)^*$. We begin by proving that $\lambda I^* - A^*$ is injective. Indeed, if $(\lambda I^* - A^*)x^* = 0$, then

$$0 = (x, (\lambda I^* - A^*)x^*) = ((\lambda I - A)x, x^*)$$

for each $x \in D(A)$. Since $\lambda \in \rho(A)$, we have $R(\lambda I - A) = X$, and thus $x^* = 0$. Hence $\lambda I^* - A^*$ is injective. If $x \in X$ and $x^* \in D(A^*)$, then

$$(x, x^*) = ((\lambda I - A)R(\lambda; A)x, x^*) = (R(\lambda; A)x, (\lambda I^* - A^*)x^*),$$

and consequently

$$R(\lambda; A)^*(\lambda I^* - A^*)x^* = x^*$$

for each $x^* \in D(A^*)$. Finally, if $x^* \in X^*$ and $x \in D(A)$, we have

$$(x, x^*) = (R(\lambda; A)(\lambda I - A)x, x^*) = ((\lambda I - A)x, R(\lambda; A)^*x^*)$$

which shows that

$$(\lambda I^* - A^*)R(\lambda; A)^*x^* = x^*$$

for each $x^* \in X^*$. So, $\lambda \in \rho(A^*)$, and $R(\lambda; A^*) = R(\lambda; A)^*$. □

Let us remark that, by virtue of Lemma 3.7.1, the dual of a C_0-semigroup of contractions is a semigroup of contractions, but, in general, it is not a C_0-semigroup because it may lack the continuity property $\lim_{t\downarrow 0} S(t)^*x^* = x^*$ for each $x^* \in X^*$, as we can easily see from the example below.

Example 3.7.1. Let $S(t)\,;\; t \in \mathbb{R}\} \subset \mathcal{L}(L^1(\mathbb{R}))$ be the translation group, i.e.

$$[S(t)f](s) = f(s-t)$$

for each $t \geq 0$ each $f \in L^1(\mathbb{R})$, and a.e. for $s \in \mathbb{R}$. Recalling that the dual of $L^1(\mathbb{R})$ is $L^\infty(\mathbb{R})$, we easily deduce that, for each $t \in \mathbb{R}$, the dual group $S(t)^* : L^\infty(\mathbb{R}) \to L^\infty(\mathbb{R})$ is defined by

$$[S(t)^*\varphi](s) = \varphi(s+t)$$

for each $\varphi \in L^\infty(\mathbb{R})$, and a.e. for $s \in \mathbb{R}$. Indeed, this follows from

$$(S(t)f, \varphi) = \int_{\mathbb{R}} f(s-t)\varphi(s)\, ds = \int_{\mathbb{R}} f(s)\varphi(s+t)\, ds$$

for each $f \in L^1(\mathbb{R})$, and each $\varphi \in L^\infty(\mathbb{R})$. On the other hand, we can easily see that $\{S(t)^*;\ t \geq 0\}$ does not satisfy the condition

$$\lim_{t \downarrow 0} S(t)^*\varphi = \varphi$$

except if φ is "uniformly continuous" on $\mathbb{R}$ with respect to $\|\cdot\|_{L^\infty(\mathbb{R})}$.

Remark 3.7.1. We notice that the dual semigroup $\{S(t)^*\,;\ t \geq 0\}$, even if not a C_0-semigroup in general, is weakly-$*$ continuous at 0, i.e.

$$\lim_{t \downarrow 0}(x, S(t)^*x^* - x^*) = 0$$

for each $x \in X$. Accordingly, if X is reflexive, the dual semigroup is weakly continuous at $t = 0$, and by virtue of Problem 2.8 it is a C_0-semigroup. We shall give a direct proof of this remarkable result later on.

Lemma 3.7.3. *Let X be reflexive, and let $A : D(A) \subseteq X \to X$ be a densely defined linear closed operator. Then A^* is densely defined and closed.*

Proof. We begin by showing that A^* is densely defined. We proceed by contradiction. To this aim, let us assume that there exist at least one reflexive Banach space, and at least one densely defined linear operator $A : D(A) \subseteq X \to X$, for which $D(A^*)$ is not dense in X^*. This means that there exists at least one element $y^{**} \in X^{**}$, such that $y^{**} \neq 0$, and $(x^*, y^{**}) = 0$ for each $x^* \in D(A^*)$. As X is reflexive, the statement above is equivalent with: there exists $y \in X$ such that $y \neq 0$ and $(y, x^*) = 0$ for each $x^* \in D(A^*)$. Since the graph of A is closed in $X \times X$, it follows that $(0, y) \notin \operatorname{graph}(A)$. By a consequence of the Hahn-Banach separation theorem (see Dunford and Schwartz [**49**], Corollary 13, p. 64), applied to $\operatorname{graph}(A)$ and $(0, y)$ in $X \times X$, it follows that there exist $x_1^*, x_2^* \in X^*$ such that $(x, x_1^*) - (Ax, x_2^*) = 0$ for each $x \in D(A)$, and $(0, x_1^*) - (y, x_2^*) \neq 0$. From the second relation, it follows both $x_2^* \neq 0$ and $(y, x_2^*) \neq 0$. On the other hand, from the first relation we deduce that $x_2^* \in D(A^*)$ and $A^*x_2^* = x_1^*$, which implies $(y, x_2^*) = 0$. This contradiction can be eliminated only if $D(A^*)$ is dense in X^*. In order to prove that A^* is closed, let

$(x_n^*)_{n\in\mathbb{N}}$ be a sequence in $D(A^*)$ with the property that $\lim_{n\to\infty} x_n^* = x^*$, and $\lim_{n\to\infty} A^* x_n^* = y^*$. By the definition of the operator A^*, we have

$$(Ax, x^*) = \lim_{n\to\infty} (Ax, x_n^*) = \lim_{n\to\infty} (x, A^* x_n^*) = (x, y^*)$$

for each $x \in D(A)$. As $D(A)$ is dense, it follows that $x^* \in D(A^*)$ and $A^* x^* = y^*$. So, A^* is closed, and the proof is complete. □

Theorem 3.7.1. *Let X be reflexive, and let $A : D(A) \subseteq X \to X$ be the infinitesimal generator of a C_0-semigroup of contractions $\{S(t)\,;\ t \geq 0\}$. Then $\{S(t)^*\,;\ t \geq 0\}$ is a C_0-semigroup of contractions whose infinitesimal generator is $A^* : D(A^*) \subseteq X^* \to X^*$.*

Proof. We show that A^* satisfies the hypotheses of Theorem 3.1.1. By Lemma 3.7.3, A^* is densely defined and closed. So, A^* satisfies (i) in Theorem 3.1.1. We prove now that $(0, +\infty) \subseteq \rho(A^*)$ and, for each $\lambda > 0$, we have $\|R(\lambda; A^*)\|_{\mathcal{L}(X^*)} \leq 1/\lambda$. As A is the infinitesimal generator of a C_0-semigroup of contractions, we have $(0, +\infty) \subseteq \rho(A)$. But, by Lemma 3.7.2, we have $\rho(A) \subseteq \rho(A^*)$ and, thus, $(0, +\infty) \subseteq \rho(A^*)$. Again by Lemma 3.7.2, we know that $R(\lambda; A^*) = R(\lambda; A)^*$, and by Lemma 3.7.1 it follows that $\|R(\lambda; A)^*\|_{\mathcal{L}(X^*)} = \|R(\lambda; A)\|_{\mathcal{L}(X)}$. Recalling that $\|R(\lambda; A)\|_{\mathcal{L}(X)} \leq 1/\lambda$ for each $\lambda > 0$, we have that $\|R(\lambda; A^*)\|_{\mathcal{L}(X^*)} \leq 1/\lambda$ for each $\lambda > 0$. Hence A^* is the infinitesimal generator of a C_0-semigroup of contractions $\{T^*(t)\,;\ t \geq 0\}$. To conclude the proof, we have merely to check that $T^*(t) = S(t)^*$ for each $t \geq 0$. To this aim, let us we recall that, for each $x^* \in D(A^*)$,

$$T^*(t)x^* = \lim_{\lambda\to+\infty} e^{t(A^*)_\lambda} x^*,$$

where $(A^*)_\lambda$ is the Yosida approximation of the operator A^*. By virtue of Lemma 3.7.1, we have $(A^*)_\lambda = (A_\lambda)^*$. In addition, $e^{t(A_\lambda)^*} = \left(e^{tA_\lambda}\right)^*$, and thus,

$$\lim_{\lambda\to+\infty} e^{t(A^*)_\lambda} x^* = \lim_{\lambda\to+\infty} e^{t(A_\lambda)^*} x^* = S(t)^* x^*$$

for each $x^* \in D(A^*)$. As $D(A^*)$ is dense in X^*, the proof is complete. □

3.8. The Sun Dual of a C_0-Semigroup

Definition 3.8.1. Let $T : D(T) \subseteq X \to X$ a linear operator and let $X^\odot$ a subspace of X. The operator $T^\odot : D(T^\odot) \subseteq X^\odot \to X^\odot$ defined by

$$\begin{cases} D(T^\odot) = \{x \in D(T) \cap X^\odot;\ Tx \in X^\odot\} & \text{and} \\ T^\odot x = Tx & \text{for each } \ x \in D(T^\odot) \end{cases}$$

is called *the part of T in $X^\odot$*.

As concerns the nonreflexive case, we have:

Theorem 3.8.1. *Let $\{S(t)\,;\ t \geq 0\}$ a C_0-semigroup of contractions on X with the infinitesimal generator A and let $\{S(t)^*\,;\ t \geq 0\}$ be the dual semigroup. If A^* is the adjoint of A and $X^\odot$ the closure of $D(A^*)$ in X^*, then the restriction $S(t)^\odot$ of $S(t)^*$ to $X^\odot$ is a C_0-semigroup of contractions whose infinitesimal generator $A^\odot$ is the part of A^* in $X^\odot$.*

Proof. Let $X^\odot = \{x^\odot \in X^*;\ \lim_{t\downarrow 0} \|S(t)^*x^\odot - x^\odot\| = 0\}$. Clearly $X^\odot$ is a subspace in X^* and $D(A^*)$ is dense in $X^\odot$. Indeed, if $x^\odot \in D(A^*)$ and $x \in X$, we have

$$|(x, S(t)^*x^\odot - x^\odot)| = |(S(t)x - x, x^\odot)|$$

$$= \left|\left(A\int_0^t S(s)x\,ds, x^\odot\right)\right| \leq t\|x\|\|A^*x^\odot\|.$$

Accordingly

$$\lim_{t\downarrow 0} |(x, S(t)^*x^\odot - x^\odot)| = 0$$

uniformly for $\|x\| \leq 1$ and therefore

$$\lim_{t\downarrow 0} \|S(t)^*x^\odot - x^\odot\| = 0.$$

Thus, $D(A^*) \subseteq X^\odot$. On the other hand, let us observe that

$$\|S(t)^\odot x^\odot - x^\odot\|$$

$$\leq \|S(t)^\odot x^\odot - S(t)^\odot y^\odot\| + \|S(t)^\odot y^\odot - y^\odot\| + \|y^\odot - x^\odot\|$$

for each $x^\odot, y^\odot \in X^\odot$. Since, by Lemma 3.7.1,

$$\|S(t)^\odot\|_{\mathcal{L}(X^\odot)} \leq \|S(t)^*\|_{\mathcal{L}(X^*)} = \|S(t)\|_{\mathcal{L}(X)} = 1,$$

the inequality above shows that $X^\odot$ is closed and $\{S(t)^\odot\,;\ t \geq 0\}$ is a C_0-semigroup of contractions on $X^\odot$. Let us denote by $A^\odot$ its infinitesimal generator, and let us remark that $D(A^\odot) \subseteq D(A^*)$. Thus $D(A^*)$ is dense in $X^\odot$, $A^\odot$ is the part of A^* in $X^\odot$ and $S(t)^\odot$ is the part of $S(t)^*$ in $X^\odot$, and this achieves the proof. □

Definition 3.8.2. The C_0-semigroup of contractions $\{S(t)^\odot\,;\ t \geq 0\}$ in Theorem 3.8.1, is called *the sun dual* of $\{S(t)\,;\ t \geq 0\}$.

Remark 3.8.1. If X is reflexive, by Lemma 3.7.3, we have $X^\odot = X^*$, and the sun dual of $\{S(t)\,;\ t \geq 0\}$ coincides with its dual $\{S(t)^*\,;\ t \geq 0\}$.

Lemma 3.8.1. *Let $A : D(A) \subseteq X \to X$ be the infinitesimal generator of a C_0-semigroup of contractions. Then, for each $x \in X$, we have*

$$\sup_{\substack{x^* \in D(A^*)\\ \|x^*\|=1}} |(x, x^*)| = \sup_{\substack{x^\odot \in D(A^\odot)\\ \|x^\odot\|=1}} |(x, x^\odot)| = \|x\|.$$

So, both $\{x^* \in D(A^*)\,;\ \|x^*\| = 1\|\}$ *and* $\{x^{\odot} \in D(A^{\odot})\,;\ \|x^{\odot}\| = 1\}$ *are determining sets for* X*, and* $X^{\odot}$ *is weakly-star dense in* X^**. Moreover,*

$$\|S(t)^{\odot}\|_{\mathcal{L}(X^{\odot})} = \|S(t)\|_{\mathcal{L}(X)}.$$

Proof. In view of Lemma 3.2.1 we have

$$(x, x^{\odot}) = \lim_{n\to\infty}(nR(n;A)x, x^{\odot}) = \lim_{n\to\infty}(x, nR(n;A^{\odot})x^{\odot}).$$

As $D(A^{\odot}) \subseteq D(A^*)$, $nR(n;A^{\odot})x^{\odot} \in D(A^{\odot})$ and $\|nR(n;A^{\odot})x^{\odot}\| \leq \|x^{\odot}\|$, this proves the first assertion. The second one is an easy consequence of the first one, while the third one follows from the second one. □

Example 3.8.1. Let $\{S(t)\,;\ t \in \mathbb{R}\} \subset \mathcal{L}(L^1(\mathbb{R}))$ be the translation group, i.e.

$$[S(t)f](s) = f(s-t)$$

for each $t \geq 0$ each $f \in L^1(\mathbb{R})$ and a.e. for $s \in \mathbb{R}$. As we have already seen in Example 3.7.1, the dual group $\{S(t)^*\,;\ t \in \mathbb{R}\} \subset \mathcal{L})L^{\infty}(\mathbb{R}))$ is defined by

$$[S(t)^*\varphi](s) = \varphi(s+t)$$

for each $\varphi \in L^{\infty}(\mathbb{R})$ and a.e. for $s \in \mathbb{R}$. One may easily see that, in this case, $X^{\odot} = C_{ub}(\mathbb{R})$. Indeed, this simply follows from the fact that $\varphi \in X^{\odot}$ if and only if $\varphi \in L^{\infty}(\mathbb{R})$ and

$$\lim_{h\downarrow 0}\|\varphi(\cdot + h) - \varphi(\,\cdot\,)\|_{L^{\infty}(\mathbb{R})} = 0.$$

3.9. Stone Theorem

We conclude this chapter with a result concerning the generation of a C_0-group on a complex Hilbert space H. The next theorem, due to Stone [**117**], concerns the case of C_0-group of unitary operators on Hilbert spaces. We recall that an operator $U \in \mathcal{L}(H)$ is called *unitary* if

$$UU^* = U^*U = I.$$

Theorem 3.9.1. (Stone) *The necessary and sufficient condition in order that* $A : D(A) \subseteq H \to H$ *be the infinitesimal generator of a* C_0*-group of unitary operators on* H *is that* iA *be self-adjoint.*

Proof. *Necessity.* If A is the infinitesimal generator of a C_0-group of unitary operators $\{G(t)\,;\ t \geq 0\}$, then it is densely defined and for each $x \in D(A)$ we have

$$-Ax = \lim_{t\downarrow 0}\frac{1}{t}(G(-t)x - x) = \lim_{t\downarrow 0}\frac{1}{t}(G^*(t)x - x) = A^*x$$

which implies $A = -A^*$. In view of Lemma 1.6.2 it follows that iA is self-adjoint and this proves the necessity.

Sufficiency. If iA is self-adjoint then, by Lemma 1.6.2, we have $A = -A^*$. By (iii) in Lemma 1.6.3, it follows that both A and A^*, which obviously are densely defined, are closed operators. From (ii) in Lemma 1.6.3, we deduce that, for each $\lambda \in \mathbb{R}$, $\lambda \neq 0$, we have

$$(R(\lambda I - A))^{\perp} = \{x \in D(A^*);\ \lambda x - A^* x = 0\}.$$

Inasmuch as A is skew-adjoint, we have $\langle Ax, x\rangle = 0$ for each $x \in D(A)$. Since $\lambda x - A^* x = 0$ if and only if $\lambda x + Ax = 0$, taking the inner product both sides by x, we deduce $\lambda \|x\|^2 = 0$, or $x = 0$. It then follows that $(R(\lambda I - A))^{\perp} = \{0\}$, and accordingly, $R(\lambda I - A)$ is dense in H. Let $\lambda \neq 0$, $f \in H$, let $(f_n)_{n\in\mathbb{N}}$ and let $(x_n)_{n\in\mathbb{N}}$ in $D(A)$ such that

$$\begin{cases} \lambda x_n - Ax_n = f_n & \text{for each } n \in \mathbb{N} \\ \lim_{n\to\infty} f_n = f & \text{in } H. \end{cases}$$

Then we have $\lambda(x_n - x_p) - A(x_n - x_p) = f_n - f_p$, from where, taking into account that $\langle Ax, x\rangle = 0$ for each $x \in D(A)$, we deduce

$$\lambda \|x_n - x_p\|^2 = \langle f_n - f_p, x_n - x_p\rangle \leq \|f_n - f_p\| \|x_n - x_p\|$$

for each $n, p \in \mathbb{N}$. As $\lambda \neq 0$ and $(f_n)_{n\in\mathbb{N}}$ is convergent, it follows that $(x_n)_{n\in\mathbb{N}}$ is a Cauchy sequence, and thus convergent to an element $x \in H$. Since A is a closed operator, $\lambda I - A$ enjoys the same property, and therefore $x \in D(A)$, and $\lambda x - Ax = f$. Hence $\lambda I - A$ is surjective. In addition, if $\lambda x - Ax = 0$, it follows that $\lambda \|x\|^2 = 0$, which proves that $\lambda I - A$ is injective. Thus, there exists $(\lambda I - A)^{-1}$. By the equality $|\lambda| \|x\|^2 = \langle f, x\rangle$ for $x \in D(A)$ with $\lambda x - Ax = f$, we deduce that

$$\|R(\lambda; A)\|_{\mathcal{L}(X)} \leq \frac{1}{|\lambda|}$$

for each $\lambda \in \mathbb{R}^*$. So, A verifies the hypotheses Theorem 3.5.2, and thus it generates a C_0-group of operators $\{G(t)\,;\ t \in \mathbb{R}\}$. Since $\{G(t)\,;\ t \geq 0\}$ is the semigroup generated by A and $\{G(-t)\,;\ t \geq 0\}$ is the semigroup generated by $-A = A^*$, we have $G^{-1}(t) = G(-t) = G^*(t)$ for each $t \in \mathbb{R}$. So, the group contains only unitary operators, and this achieves the proof. □

Remark 3.9.1. As we shall see in the next chapter, Stone Theorem 3.9.1 is very useful in establishing existence and uniqueness results for homogeneous first-order and second-order hyperbolic problems, as well as for the linear Schrödinger equation.

Problems

Problem 3.1. Let $X = l_2$, $(a_n)_{n\in\mathbb{N}^*}$ a sequence of positive real numbers and let $A : D(A) \subseteq X \to X$ be defined by

$$\begin{cases} D(A) = \{(x_n)_{n\in\mathbb{N}^*} \in l_2\,;\ (a_k x_k)_{k\in\mathbb{N}^*} \in l_2\} \\ (A(x_n)_{n\in\mathbb{N}^*})_{k\in\mathbb{N}^*} = -(a_k x_k)_{k\in\mathbb{N}^*} \text{ for } (x_n)_{n\in\mathbb{N}^*} \in D(A). \end{cases}$$

Show that A generates a C_0-semigroup of contractions on X.

Problem 3.2. Let $X = c_0$, $(a_n)_{n\in\mathbb{N}^*}$ a sequence of positive real numbers and let $A : D(A) \subseteq X \to X$ be defined by

$$\begin{cases} D(A) = \{(x_n)_{n\in\mathbb{N}^*} \in c_0\,;\ (a_k x_k)_{k\in\mathbb{N}^*} \in c_0\} \\ (A(x_n)_{n\in\mathbb{N}^*})_{k\in\mathbb{N}^*} = -(a_k x_k)_{k\in\mathbb{N}^*} \text{ for } (x_n)_{n\in\mathbb{N}^*} \in D(A). \end{cases}$$

Show that A generates a C_0-semigroup of contractions on X.

Problem 3.3. Let $X = L^2(0,\pi)$ and $A : D(A) \subseteq X \to X$ defined by

$$\begin{cases} D(A) = \{u \in H^2(0,\pi)\,;\ u(0) = u'(\pi) = 0\} \\ Au = u'' \text{ for } u \in D(A). \end{cases}$$

Show that A generates a C_0-semigroup of contractions on X.

Problem 3.4. Let $X = C_\pi([\,0,\pi\,])$ the space of all continuous real functions defined on $[\,0,\pi\,]$ and satisfying $u(0) = u(\pi)$, endowed with the usual sup norm. Show that $A : D(A) \subseteq X \to X$, defined by

$$\begin{cases} D(A) = \{u \in X\,;\ u' \in X \\ Au = -u' \text{ for } u \in D(A), \end{cases}$$

generates a C_0-semigroup on X.

Problem 3.5. Prove that if $A : D(A) \subseteq X \to X$ is a densely defined linear operator and both A and A^* are dissipative, then A generates a C_0-semigroup of contractions. This is Corollary 4.4, p. 15 in Pazy [**101**].

We recall that a linear operator A is closable if the closure of its graph is the graph of a linear operator $\overline{A}$ called the closure of A.

Problem 3.6. Let $A : D(A) \subseteq X \to X$ be a linear, dissipative operator. Then:

(i) If for some $\lambda > 0$, $\lambda I - A$ is surjective, then for each $\lambda > 0$, $\lambda I - A$ is surjective.
(ii) If A is closable then $\overline{A}$, the closure of A, is also dissipative.
(iii) If $D(A)$ is dense in X, then A is closable.

This is Theorem 4.5, p. 15 in Pazy [**101**]

Problem 3.7. Let $A : D(A) \subseteq X \to X$ be a linear dissipative operator with $I - A$ surjective. Show that, whenever X is reflexive, $D(A)$ is dense in X. This is Theorem 4.6, p. 16 in Pazy [**101**]. See also Brezis and Cazenave [**31**], Exercise 1.8.2, p. 99.

Problem 3.8. Let $X = C([0,1])$ the space of real continuous functions on $[0,1]$ endowed with the sup norm and let $A : D(A) \subseteq X \to X$ be defined by

$$\begin{cases} D(A) = \{u;\ u \in C^1([0,1]),\ u(0) = 0\} \\ Au = -u' \text{ for each } u \in D(A). \end{cases}$$

Show that A is dissipative, $I - A$ is surjective, but $D(A)$ is not dense in X. This is Example 4.7, p. 16 in Pazy [**101**].

Notes. The main result in Sections 3.1 and 3.2 Theorem 3.1.1, was proved independently by Hille [**69**] and Yosida [**135**] in 1948. The sufficiency part, as presented in Section 3.2, is due to Yosida. The Generation Theorem 3.5.1 in Section 3.3, which extends the necessary and sufficient condition in order for an operator to generate a C_0-semigroup of contractions to a necessary and sufficient condition in order for an operator to generate a C_0-semigroup of type (M, ω), has been obtained, also independently, Feller [**55**] in 1953, Miyadera [**91**] and Phillips [**104**] in 1952. Theorem 3.4.2 in Section 3.4 was proved by Phillips [**106**], in the case in which $X = H$ is a Hilbert space, and by Lumer and Phillips [**87**] in the general case. At this point, we notice that the dissipativity condition is very suitable to build up a powerful semigroup theory for nonlinear operators. See for instance Barbu [**17**], Showalter [**112**] and Vrabie [**127**]. The dual of a semigroup was defined and studied by Feller [**53**] in 1952 in the case of some parabolic problems and by Phillips [**105**] in 1955 in the general case. A different point of view concerning this problem, leading to analogous concepts and results, can be found in Butzer and Berens [**32**]. The main result in Section 3.9, Theorem 3.9.1, was obtained by Stone [**117**] in 1932, and this is the first generation result of a C_0-group of linear continuous operators by an unbounded operator. Concerning the general case of C_0-groups of possible non unitary operators, we mention the following result due to Liu [**86**]:

Theorem (Liu) *Let $A : D(A) \subseteq H \to H$ be the infinitesimal generator of a C_0-semigroup. Then A generates a C_0-group on H if and only if there exists $\gamma_0 > 0$ such that:*

$$\{\lambda \in \mathbb{C};\ \operatorname{Re}\lambda \geq \gamma_0\} \subseteq \rho(-A),$$

$$\sup\{\|(\lambda I + A)^{-1}\|_{\mathcal{L}(X)};\ \operatorname{Re}\lambda \geq \gamma_0\} < +\infty$$

and

$$\lim_{\operatorname{Re}\lambda\to+\infty} \|(\lambda I + A)^{-1}x\| = 0 \quad \text{for each} \quad x \in H.$$

There are however situations in which an operator A does not satisfy the hypotheses of the Hille-Yosida Theorem 3.1.1 but we still can introduce a rather powerful notion of generalized solution for the differential equation $u' = Au$. This happens for instance if A satisfies all the hypotheses of Theorem 3.1.1 except the density of $D(A)$. In these cases, the simple observation that whenever $\{S(t)\,;\ t \geq 0\}$ is a C_0-semigroup on X, the family of operators $\{T(t)\,;\ t \geq 0\}$ with $T(t) = \int_0^t S(\tau)\,d\tau$ for $t \geq 0$ satisfies:

$$T(r)T(t) = \int_0^r (T(\tau + t) - T(\tau))\,d\tau \quad \text{and} \quad T(0) = 0$$

suggests the definition of a more general concept, i.e. that of *a C_0-integrated semigroup* as being a family $\{T(t)\,;\ t \geq 0\} \subseteq \mathcal{L}(X)$ satisfying the two conditions above and the continuity condition $\lim_{s\downarrow 0} T(s)x = T(t)x$ for each $x \in X$ and $t \geq 0$. The generator $A : D(A) \subseteq X \to X$ of a C_0-integrated semigroup $\{T(t)\,;\ t \geq 0\}$ is defined by $x \in D(A)$ and $Ax = y$ if and only if $t \mapsto T(t)x$ is C^1 and $T'(t)x - x = T(t)y$ for each $t \geq 0$. From now on, the whole classical machinery can be adapted to handle this case which has interesting applications in the study of the wave equation in $L^2(\mathbb{R}^n)$, as well as in that of the equations of population dynamics. See Ahmed [**2**], Arendt [**8**], Kellermann and Hieber [**73**], Thieme [**120**], [**121**], and the references therein. As already mentioned, with few exceptions, the problems here included are in fact known results or examples, and are gathered from Pazy [**101**].

CHAPTER 4

Differential Operators Generating C_0-Semigroups

This chapter is devoted to the presentation of some partial differential operators which, defined on suitably chosen function spaces, generate either C_0-semigroups of contractions, or groups of isometries. First, we refer to the Laplace operator on a domain Ω in $\mathbb{R}^n$ subjected to the Dirichlet boundary condition which generates C_0-semigroups of contractions on each of the spaces $H^{-1}(\Omega)$, $L^p(\Omega)$, with p in $[1, +\infty)$, and $C_0(\overline{\Omega})$. We also included some remarkable examples of infinitesimal generators of C_0-groups of isometries as for instance: the Maxwell, the directional derivative, the Schrödinger, the wave, as well as the Airy operators. In the last sections, we present two infinitesimal generators of C_0-semigroups of contractions in the linear thermoelasticity and the linear viscoelasticity respectively.

4.1. The Laplace Operator with Dirichlet Boundary Condition

Let us consider the heat equation in a domain Ω in $\mathbb{R}^3$

$$\begin{cases} u_t = \Delta u & (t,x) \in Q_\infty \\ u = 0 & (t,x) \in \Sigma_\infty \\ u(0,x) = u_0(x) & x \in \Omega, \end{cases}$$

where Δ is the Laplace operator, $Q_\infty = \mathbb{R}_+ \times \Omega$ and $\Sigma_\infty = \mathbb{R}_+ \times \Gamma$. This partial differential equation can be rewritten as an ordinary differential equation of the form

$$\begin{cases} u' = Au \\ u(0) = u_0 \end{cases} \qquad (\mathcal{ODE})$$

in a suitably chosen infinite-dimensional Banach space X, in order that the unbounded linear operator $A : D(A) \subseteq X \to X$ generate a C_0-semigroup of contractions.

We present next several such classical possible choices of the space X.

Example 4.1.1. *The $H^{-1}(\Omega)$ Setting.* Let Ω be a nonempty and open subset in $\mathbb{R}^n$, let $X = H^{-1}(\Omega)$, and let us define $A : D(A) \subseteq X \to X$ by:

$$\begin{cases} D(A) = H_0^1(\Omega) \\ Au = \Delta u, \end{cases}$$

for each $u \in D(A)$. In that follows, $H_0^1(\Omega)$ is endowed with the usual norm on $H^1(\Omega)$, defined by

$$\|u\|_{H^1(\Omega)} = \left(\|u\|_{L^2(\Omega)}^2 + \|\nabla u\|_{L^2(\Omega)}^2\right)^{1/2}.$$

Theorem 4.1.1. *The operator A, defined as above, is the generator of a C_0-semigroup of contractions. In addition, A is self-adjoint and $\|\cdot\|_{D(A)}$ is equivalent with the norm of the space $H^1(\Omega)$.*

Proof. By virtue of Theorem 1.5.8, we know that $I - \Delta$ is the canonic isomorphism between $H_0^1(\Omega)$, endowed with the usual norm of $H^1(\Omega)$, and its dual $H^{-1}(\Omega)$. Let us denote by $F = (I - \Delta)^{-1}$ which is an isometry between $H^{-1}(\Omega)$ and $H_0^1(\Omega)$. Consequently

$$\langle u, v\rangle_{H^{-1}(\Omega)} = \langle Fu, Fv\rangle_{H_0^1(\Omega)} \tag{4.1.1}$$

for each $u, v \in H^{-1}(\Omega)$. Let $u, v \in H_0^1(\Omega)$. We have

$$\begin{aligned} \langle u, Fv\rangle_{H_0^1(\Omega)} &= \int_\Omega \nabla u \nabla(Fv)\, d\omega + \int_\Omega uFv\, d\omega \\ &= \int_\Omega u(-\Delta(Fv))\, d\omega + \int_\Omega uFv\, d\omega \\ &= \int_\Omega u(I - \Delta)F(v)\, d\omega = \langle u, v\rangle_{L^2(\Omega)}. \end{aligned} \tag{4.1.2}$$

From (4.1.1), taking into account that $F(I - \Delta) = I$, we deduce

$$\langle -\Delta u, v\rangle_{H^{-1}(\Omega)} = \langle u - \Delta u, v\rangle_{H^{-1}(\Omega)} - \langle u, v\rangle_{H^{-1}(\Omega)}$$

$$= \langle F(u - \Delta u), Fv\rangle_{H_0^1(\Omega)} - \langle u, v\rangle_{H^{-1}(\Omega)} = \langle u, Fv\rangle_{H^{-1}(\Omega)} - \langle u, v\rangle_{H^{-1}(\Omega)}.$$

From (4.1.2), we have

$$\langle \Delta u, v\rangle_{H^{-1}(\Omega)} = \langle u, v\rangle_{H^{-1}(\Omega)} - \langle u, v\rangle_{L^2(\Omega)}.$$

Therefore A is symmetric. But $(I - A)^{-1} \in \mathcal{L}(H^{-1}(\Omega))$, and therefore, from Lemma 1.6.1, it follows that A is self-adjoint. Taking $v = u$ in the above equality, we obtain

$$\langle Au, u\rangle_{H^{-1}(\Omega)} = \|u\|_{H^{-1}(\Omega)}^2 - \|u\|_{L^2(\Omega)}^2 \leq 0. \tag{4.1.3}$$

Corollary 1.5.1 shows that, for $\lambda > 0$, we have $(\lambda I - A)^{-1} \in \mathcal{L}(H^{-1}(\Omega))$, while (4.1.3) implies that, for $\lambda > 0$, $\langle \lambda u - Au, u\rangle_{H^{-1}(\Omega)} \geq \lambda\|u\|_{H^{-1}(\Omega)}^2$.

Hence $\|R(\lambda; A)\|_{\mathcal{L}(H^{-1}(\Omega))} \leq \frac{1}{\lambda}$. Since $H_0^1(\Omega)$ is dense în $H^{-1}(\Omega)$, we are in the hypotheses of Theorem 3.1.1, from where it follows that A generates a C_0-semigroup of contractions on $H^{-1}(\Omega)$. Finally, by (iv) in Corollary 3.5.1 and (4.1.3), it follows that $\|\cdot\|_{D(A)}$ is equivalent with the norm of the space $H^{-1}(\Omega)$. The proof is complete. □

Example 4.1.2. *The $L^2(\Omega)$ Setting.* Let Ω be a nonempty and open subset in $\mathbb{R}^n$, let $X = L^2(\Omega)$, and let us consider the operator A on X, defined by:

$$\begin{cases} D(A) = \{u \in H_0^1(\Omega);\ \Delta u \in L^2(\Omega)\} \\ Au = \Delta u, \end{cases}$$

for each $u \in D(A)$.

Theorem 4.1.2. *The linear operator A, defined above, is the infinitesimal generator of a C_0-semigroup of contractions. Moreover, A is self-adjoint, and $(D(A), \|\cdot\|_{D(A)})$ is continuously included in $H_0^1(\Omega)$. If Ω is bounded with C^1 boundary, then $(D(A), \|\cdot\|_{D(A)})$ is compactly imbedded in $L^2(\Omega)$.*

Proof. Since $C_0^\infty(\Omega)$ is dense in $L^2(\Omega)$, and $C_0^\infty(\Omega) \subseteq D(A)$, it follows that A is densely defined. Let $\lambda > 0$ and $f \in L^2(\Omega)$. Since $L^2(\Omega)$ is continuously imbedded in $H^{-1}(\Omega)$, and $-\Delta : H_0^1(\Omega) \to H^{-1}(\Omega)$ is the duality mapping with respect to the gradient norm on $H_0^1(\Omega)$, we have:

$$\langle Au, v\rangle_{L^2(\Omega)} = -\langle \nabla u, \nabla v\rangle_{L^2(\Omega)} = \langle v, \Delta u\rangle_{H_0^1(\Omega), H^{-1}(\Omega)}. \tag{4.1.4}$$

By Theorem 4.1.1, we know that, for any $\lambda > 0$ and any $f \in L^2(\Omega)$ (notice that $L^2(\Omega) \subset H^{-1}(\Omega)$), the equation

$$\lambda u - \Delta u = f$$

has a unique solution $u_\lambda \in H_0^1(\Omega) \subset L^2(\Omega)$. So, $\Delta u_\lambda = \lambda u_\lambda - f$ is in $L^2(\Omega)$, which shows that $u_\lambda \in D(A)$, and $\lambda u_\lambda - Au_\lambda = f$. Taking the $L^2(\Omega)$ inner product on both sides of the equality above by u_λ, and taking into account that, by (4.1.4), we have $\langle Au, u\rangle_{L^2(\Omega)} \leq 0$ for each $u \in D(A)$, we deduce that

$$\lambda \|u_\lambda\|^2_{L^2(\Omega)} \leq \langle f, u_\lambda\rangle_{L^2(\Omega)} \leq \|f\|_{L^2(\Omega)} \|u_\lambda\|_{L^2(\Omega)},$$

inequality which shows that $\|R(\lambda; A)\|_{\mathcal{L}(X)} \leq \frac{1}{\lambda}$. Finally, from (4.1.4) and Lemma 1.6.1, it follows that A is self-adjoint. Inasmuch as both inclusions $D(A) \subset H_0^1(\Omega) \subset L^2(\Omega)$ are continuous, and the latter is compact whenever Ω is bounded (see Theorem 1.5.4), this achieves the proof. □

Remark 4.1.1. Obviously $H_0^1(\Omega) \cap H^2(\Omega) \subseteq D(A)$. One can prove that, if Ω is bounded and has a C^2 boundary, or $\Omega = \mathbb{R}_+^n$, then the reversed inclusion holds too, and therefore

$$D(A) = H_0^1(\Omega) \cap H^2(\Omega).$$

See Brezis [**29**], Théorème IX.25, p. 181. Moreover, the above equality still holds in the case in which Ω is convex and bounded. See Grisvard [**64**], Theorem 3.1.2.1, p. 139.

Example 4.1.3. *The $L^p(\Omega)$ Setting.* Let Ω be a nonempty and open subset in $\mathbb{R}^n$ and let $1 \leq p < +\infty$.

Lemma 4.1.1. *Let A be the Laplace operator with the Dirichlet boundary condition defined on $H^{-1}(\Omega)$, let $\lambda > 0$ and $R(\lambda; A) : H^{-1}(\Omega) \to D(A)$. Then:*

(i) $R(\lambda; A) \in \mathcal{L}(H^{-1}(\Omega))$ and $\|R(\lambda; A)\|_{\mathcal{L}(H^{-1}(\Omega))} \leq \frac{1}{\lambda}$;
(ii) $R(\lambda; A) \in \mathcal{L}(H^{-1}(\Omega), H_0^1(\Omega))$;
(iii) $R(\lambda; A)_{|H_0^1(\Omega)} \in \mathcal{L}(H_0^1(\Omega))$ *and* $\|R(\lambda; A)_{|H_0^1(\Omega)}\|_{\mathcal{L}(D(A);X)} \leq \frac{1}{\lambda}$;
(iv) $\lim_{\lambda \to +\infty} \|R(\lambda; A)u - u\|_{H^{-1}(\Omega)} = 0$ *for each* $u \in H^{-1}(\Omega)$;
(v) $\lim_{\lambda \to +\infty} \|R(\lambda; A)u - u\|_{H_0^1(\Omega)} = 0$ *for each* $u \in H_0^1(\Omega)$.

Proof. The item (i) follows from Theorem 4.1.1, while (ii) from (v) in Corollary 3.5.1. To check (iii), let us observe that

$$\|R(\lambda; A)x\|_{D(A)} = \|R(\lambda; A)x\|_X + \|AR(\lambda; A)x\|_X$$
$$\leq \frac{1}{\lambda}(\|x\|_X + \|Ax\|_X) = \frac{1}{\lambda}\|x\|_{D(A)}.$$

The item (iv) follows from Lemma 3.2.1, and Theorem 4.1.1. Finally, (v) also follows from Lemma 3.2.1 and Theorem 4.1.1, by observing that the restriction of A to $D(A)$, endowed with the graph-norm, i.e. the operator $A_{|D(A)} : D(A^2) \subseteq D(A) \to D(A)$ satisfies the hypotheses of Theorem 3.1.1. □

Lemma 4.1.2. *Let A be the Laplace operator with the Dirichlet boundary condition in $H^{-1}(\Omega)$ let $\lambda > 0$, and let $R(\lambda; A) : H^{-1}(\Omega) \to D(A)$. Let $1 \leq p < +\infty$. Then, for each $f \in L^p(\Omega) \cap H^{-1}(\Omega)$, we have*

(i) $R(\lambda; A)f \in L^p(\Omega)$
(ii) $\|R(\lambda; A)f\|_{L^p(\Omega)} \leq \frac{1}{\lambda}\|f\|_{L^p(\Omega)}$.

Proof. Let $f \in L^p(\Omega) \cap H^{-1}(\Omega)$. Then $R(\lambda; A)f = u$, where

$$\lambda u - \Delta u = f.$$

Multiplying both sides of this equality by $\|u\|_{L^p(\Omega)}^{2-p}|u|^{p-2}u$, integrating over Ω, using Green's formula and the fact that $u \in H_0^1(\Omega)$, we get

$$\lambda\|u\|_{L^p(\Omega)}^2 \leq \|f\|_{L^p(\Omega)}\|u\|_{L^p(\Omega)},$$

which completes the proof. □

Lemma 4.1.3. *Let A be the Laplace operator with the Dirichlet boundary condition in $H^{-1}(\Omega)$, let $\lambda > 0$, and $1 \leq p < +\infty$. Then there exists a unique $\mathcal{R}_\lambda \in \mathcal{L}(L^p(\Omega))$ so that $\mathcal{R}_\lambda u = R(\lambda; A)u$ for all $u \in H^{-1}(\Omega) \cap L^p(\Omega)$. In addition, $\mathcal{R}_\lambda$ satisfies:*

(i) $\|\mathcal{R}_\lambda u\|_{L^p(\Omega)} \leq \frac{1}{\lambda}\|u\|_{L^p(\Omega)}$;
(ii) *for each $f \in L^p(\Omega)$, $A\mathcal{R}_\lambda f \in L^p(\Omega)$ and $\lambda\mathcal{R}_\lambda f - A\mathcal{R}_\lambda f = f$;*
(iii) *for each $\lambda > 0$, and $\mu > 0$, $\mathcal{R}_\lambda(L^p(\Omega)) = \mathcal{R}_\mu(L^p(\Omega))$.*

Proof. Since $H^{-1}(\Omega) \cap L^p(\Omega)$ is dense in $L^p(\Omega)$, from Lemma 4.1.2, it follows that $R(\lambda; A)$ has a unique extension $\mathcal{R}_\lambda \in \mathcal{L}(L^p(\Omega))$ satisfying (i). Next, let $(f_k)_{k \in \mathbb{N}}$ be a sequence in $\mathcal{D}(\Omega)$ convergent to f in $L^p(\Omega)$. As $\mathcal{R}_\lambda f_k - \lambda A\mathcal{R}_\lambda f_k = f_k$ in $H^{-1}(\Omega)$, we have $\mathcal{R}_\lambda f - \lambda A\mathcal{R}_\lambda f = f$ in $\mathcal{D}'(\Omega)$, from where we get (ii).
Finally, let $f \in H^{-1}(\Omega) \cap L^p(\Omega)$, and $u = \mathcal{R}_\lambda f \in H^{-1}(\Omega) \cap L^p(\Omega)$. For each $\mu > 0$, we have

$$\mu u - \Delta u = f + (\mu - \lambda)\mathcal{R}_\lambda f.$$

Let us denote by g the right-hand side of the equality above, i.e.

$$g = f + (\mu - \lambda)\mathcal{R}_\lambda f,$$

and let us observe that $\mathcal{R}_\lambda f = u = \mathcal{R}_\mu g \in H^{-1}(\Omega) \cap L^p(\Omega)$ and therefore

$$\mathcal{R}_\lambda(H^{-1}(\Omega) \cap L^p(\Omega)) \subseteq \mathcal{R}_\mu(H^{-1}(\Omega) \cap L^p(\Omega)).$$

Analogously,

$$\mathcal{R}_\mu(H^{-1}(\Omega) \cap L^p(\Omega)) \subseteq \mathcal{R}_\lambda(H^{-1}(\Omega) \cap L^p(\Omega)),$$

and so

$$\mathcal{R}_\lambda(H^{-1}(\Omega) \cap L^p(\Omega)) = \mathcal{R}_\mu(H^{-1}(\Omega) \cap L^p(\Omega)).$$

Since $H^{-1}(\Omega) \cap L^p(\Omega)$ is dense in $L^p(\Omega)$, and $\mathcal{R}_\lambda$, $\mathcal{R}_\mu$ are linear continuous operators in $L^p(\Omega)$, we deduce (iii) and this achieves the proof. □

Theorem 4.1.3. *Let $1 \leq p < +\infty$, and $\mathcal{R}_1 \in \mathcal{L}(L^p(\Omega))$ as in Lemma 4.1.3. Then, for each $u \in \mathcal{R}_1(L^p(\Omega))$, we have $\Delta u \in L^p(\Omega)$, and the operator $A : D(A) \subseteq L^p(\Omega) \to L^p(\Omega)$, defined by*

$$\begin{cases} D(A) = \mathcal{R}_1(L^p(\Omega)) \\ Au = \Delta u \quad \text{for} \quad u \in D(A), \end{cases}$$

is the generator of a C_0-semigroup of contractions.

Proof. The fact that, for each $u \in \mathcal{R}_1(L^p(\Omega))$, we have $\Delta u \in L^p(\Omega)$, follows from (ii) in Lemma 4.1.3. So, let $u \in D(A)$, let $\lambda > 0$, and let us denote $f = \lambda u - \Delta u$. From (iii) in Lemma 4.1.3, we know that there exists $g \in L^p(\Omega)$ such that $u = \mathcal{R}_\lambda g$. We then conclude that $g = \lambda u - \Delta u$, and so $f = g$. Then $\lambda \in \rho(A)$ and $R(\lambda; A) = (\lambda I - A)^{-1} = \mathcal{R}_\lambda$. This relation and (i) in Lemma 4.1.3 show that $\|R(\lambda; A)u\|_{L^p(\Omega)} \leq \frac{1}{\lambda}\|u\|_{L^p(\Omega)}$. Thus A satisfies (ii) in Theorem 3.1.1. To complete the proof, we have merely to show $D(A)$ is dense in $L^p(\Omega)$. To this aim, let $u \in \mathcal{D}(\Omega)$ and $f = u - \Delta u \in \mathcal{D}(\Omega)$. Obviously $u = \mathcal{R}_1 f$, and therefore $\mathcal{D}(\Omega) \subseteq D(A)$. Hence $D(A)$ is dense in $L^p(\Omega)$, which achieves the proof. □

Remark 4.1.2. Clearly we have

$$W_0^{1,p}(\Omega) \cap W^{2,p}(\Omega) \subseteq D(A).$$

If $1 < p < +\infty$ and Ω is bounded and has a C^2 boundary, then the reversed inclusion holds too, and we have

$$D(A) = W_0^{1,p}(\Omega) \cap W^{2,p}(\Omega).$$

For the proof see Pazy [**101**], Theorem 3.1, p. 212.

Remark 4.1.3. If $p = 1$ and Ω is bounded and has a C^2 boundary, then

$$D(A) = \{u \in W_0^{1,1}(\Omega);\ \Delta u \in L^1(\Omega)\}.$$

In this case $D(A)$ is not included in $W^{2,1}(\Omega)$. See for instance Brezis and Cazenave [**31**], Remark 1.2.30, p. 28. However, we have $D(A) \subseteq W_0^{1,q}(\Omega)$ for each $1 \leq q < q_n$, where $q_1 = \infty$ and $q_n = n/(n-1)$, for $n \geq 2$. See Maz'ya [**88**].

Example 4.1.4. *The $C_0(\overline{\Omega})$ Setting.* Let Ω be a nonempty and open subset in $\mathbb{R}^n$, let $X = C_0(\overline{\Omega})$ be the space of all continuous functions from $\overline{\Omega}$ to $\mathbb{R}$ vanishing on the boundary, and let us define $A : D(A) \subseteq X \to X$ by:

$$\begin{cases} D(A) = \{u \in C_0(\overline{\Omega}) \cap H_0^1(\Omega),\ \Delta u \in C_0(\overline{\Omega})\} \\ Au = \Delta u, \end{cases}$$

for each $u \in D(A)$.

Theorem 4.1.4. *If Ω has a C^1-boundary, then the operator A, defined as above, is the generator of a C_0-semigroup of contractions on $C_0(\overline{\Omega})$.*

Proof. Since $\mathcal{D}(\Omega)$ is dense in X and $\mathcal{D}(\Omega) \subset D(A)$, it follows that A has dense domain. On the other hand, for each $f \in C_0(\overline{\Omega})$ the problem

$$u - \lambda\Delta u = f$$

has a unique solution $u \in H_0^1(\Omega)$. Moreover, from the classical theory of elliptic problems, this solution $u \in C_0(\overline{\Omega})$. See for instance Gilbarg and Trudinger [**59**], Theorem 8.30, p. 206. Finally, since this solution u satisfies the above equation in any $L^p(\Omega)$, from Theorem 4.1.3 combined with Hille-Yosida Theorem 3.1.1, we deduce that

$$\|u\|_{L^p(\Omega)} \leq \frac{1}{\lambda}\|f\|_{L^p(\Omega)}$$

for each $p \geq 1$. So, the above inequality holds also for $p = \infty$ which shows that

$$\|R(\lambda; A)\|_{C_0(\overline{\Omega})} \leq \frac{1}{\lambda}.$$

The conclusion follows from Theorem 3.1.1, and this completes the proof. □

4.2. The Laplace Operator with Neumann Boundary Condition

Here we consider the heat equation in a domain Ω in $\mathbb{R}^3$

$$\begin{cases} u_t = \Delta u & (t,x) \in Q_\infty \\ u_\nu = 0 & (t,x) \in \Sigma_\infty \\ u(0,x) = u_0(x) & x \in \Omega, \end{cases}$$

where u_ν is the outward normal derivative of u at Γ, $Q_\infty = \mathbb{R}_+ \times \Omega$ and $\Sigma_\infty = \mathbb{R}_+ \times \Gamma$. As in the preceding case, this partial differential equation can be rewritten as an ordinary differential equation of the form

$$\begin{cases} u' = Au \\ u(0) = u_0 \end{cases} \qquad (\mathcal{ODE})$$

in an infinite-dimensional Banach space X, suitably chosen in order that the unbounded linear operator $A : D(A) \subseteq X \to X$ generates a C_0-semigroup of contractions.

Example 4.2.1. *The* $[H^1(\Omega)]^*$ *Setting.* Let Ω be a nonempty and open subset in $\mathbb{R}^n$ with C^1 boundary Γ, let $X = [H^1(\Omega)]^*$, and let us consider the operator $A : D(A) \subseteq X \to X$, defined by:

$$\begin{cases} D(A) = H^1(\Omega) \\ \langle Au, v\rangle_{H^1(\Omega),[H^1(\Omega)]^*} = -\langle \nabla u, \nabla v\rangle_{L^2(\Omega)} \text{ for each } u, v \in H^1(\Omega). \end{cases}$$

Within this section, $\langle\cdot,\cdot\rangle_{[H^1(\Omega)]^*,H^1(\Omega)}$ is the duality between $[H^1(\Omega)]^*$ and $H^1(\Omega)$.

Theorem 4.2.1. *The operator* $A : D(A) \subseteq X \to X$*, defined as above, is the generator of a* C_0*-semigroup of contractions on* X*.*

Proof. Since $H^1(\Omega)$ is densely imbedded in $[H^1(\Omega)]^*$, in view of Hille-Yosida Theorem 3.1.1, we have merely to show that for each $\lambda > 0$, the operator $\lambda I - A : D(A) \subseteq X \to X$, where A is defined as above, is bijective and

$$\|(\lambda I - A)^{-1}\|_{\mathcal{L}(X)} \leq \frac{1}{\lambda}.$$

But this simply follows from the obvious identity

$$\langle \lambda u - Au, u\rangle_{[H^1(\Omega)]^*,H^1(\Omega)} = \lambda\|u\|^2_{L^2(\Omega)} + \|\nabla u\|^2_{L^2(\Omega)}$$

and this achieves the proof. □

Example 4.2.2. *The $L^2(\Omega)$ Setting.* Let Ω be a nonempty and open subset in $\mathbb{R}^n$ with C^1 boundary Γ, let $X = L^2(\Omega)$, and let us consider the operator $B : D(B) \subseteq X \to X$, defined by:

$$\begin{cases} D(B) = \{u \in H^2(\Omega)\,;\ u_\nu = 0 \text{ on } \Gamma\} \\ Bu = \Delta u \text{ for } u \in D(B). \end{cases}$$

Theorem 4.2.2. *The operator $B : D(B) \subseteq X \to X$, defined as above, is the generator of a C_0-semigroup of contractions on X.*

Proof. Let $u \in D(B)$. Then, for each $v \in H^1(\Omega)$, we have

$$\langle Au, v\rangle_{H^1(\Omega),[H^1(\Omega)]^*} = -\langle \nabla u, \nabla v\rangle_{L^2(\Omega)} = \langle \Delta u, v\rangle_{L^2(\Omega)}$$

and thus, $Au = Bu$ for each $u \in D(B)$. In addition,

$$\langle Bu, v\rangle_{L^2(\Omega)} = -\langle \nabla u, \nabla v\rangle_{L^2(\Omega)}$$

for each $u, v \in D(B)$. Thus B is symmetric and, for each $\lambda > 0$, $\lambda I - B$ is bijective from $D(B)$ to $L^2(\Omega)$ and

$$\|(\lambda I - B)^{-1}\|_{\mathcal{L}(X)} \leq \frac{1}{\lambda}.$$

Since $D(B)$ is dense in $X = L^2(\Omega)$, we are in the hypotheses of Hille-Yosida Theorem 3.1.1, and this completes the proof. □

Remark 4.2.1. For the Laplace operator Δ subjected to the Neumann boundary condition, either in $[H^1(\Omega)]^*$, or in $L^2(\Omega)$, 0 is an eigenvalue, and all constant functions are the corresponding eigenfunctions. This is the first major distinction between this operator and the homonymous one with Dirichlet boundary condition.

4.3. The Maxwell Operator

The aim of this section is to present a partial differential operator which generates a C_0-group of unitary operators and which is, without any doubt, one of the most important operators in the Field Theory. Some other examples of infinitesimal generators of C_0-groups of unitary operators, or isometries, will be discussed in the next sections. For the beginning we recall that, if $\varphi \in L^2(\mathbb{R}^3)$ and $\mathbf{F} \in (L^2(\mathbb{R}^3))^3$, then

$$\nabla\varphi = \operatorname{grad}\varphi = \sum_{i=1}^{3} \frac{\partial\varphi}{\partial x_i}\vec{e_i}, \quad \nabla\cdot\mathbf{F} = \operatorname{div}\mathbf{F} = \sum_{\mathbf{i=1}}^{\mathbf{3}} \frac{\partial\mathbf{F_i}}{\partial\mathbf{x_i}} \quad \text{and}$$

$$\nabla\times\mathbf{F} = \operatorname{curl}\mathbf{F} =$$

$$= \left(\frac{\partial\mathbf{F}_3}{\partial x_2} - \frac{\partial\mathbf{F}_2}{\partial x_3}\right)\vec{e_1} + \left(\frac{\partial\mathbf{F}_1}{\partial x_3} - \frac{\partial\mathbf{F}_3}{\partial x_1}\right)\vec{e_2} + \left(\frac{\partial\mathbf{F}_2}{\partial x_1} - \frac{\partial\mathbf{F}_1}{\partial x_2}\right)\vec{e_3}\,.$$

The evolution of the intensity of both the electric field $\mathbf{E}$ and the magnetic field $\mathbf{H}$ in the space $\mathbb{R}^3$ containing no matter, is described by the Maxwell system

$$\begin{cases} \mathbf{E}_t = -c\nabla\times\mathbf{H} & (t,x)\in\mathbb{R}\times\mathbb{R}^3 \\ \mathbf{H}_t = c\nabla\times\mathbf{E} & (t,x)\in\mathbb{R}\times\mathbb{R}^3 \\ \nabla\cdot\mathbf{E} = 0 \ \text{ and } \ \nabla\cdot\mathbf{H} = 0 & (t,x)\in\mathbb{R}\times\mathbb{R}^3 \\ \mathbf{E}(0,x) = \mathbf{E}_0(x) \ \text{ and } \ \mathbf{H}(0,x) = \mathbf{H}_0(x) & x\in\mathbb{R}^3, \end{cases}$$

where $c > 0$ is constant. This system can be rewritten, in a suitably chosen Hilbert space, as

$$\begin{cases} u' = Au \\ u(0) = u_0, \end{cases}$$

where A is the generator of a C_0-semigroup of contractions.

Example 4.3.1. (*The Maxwell Operator*). Let us consider the Hilbert space $H = (L^2(\mathbb{R}^3))^3 \times (L^2(\mathbb{R}^3))^3$, and let us denote by

$$u = (\mathbf{E},\mathbf{H}) = (\mathbf{E_1},\mathbf{E_2},\mathbf{E_3},\mathbf{H_1},\mathbf{H_2},\mathbf{H_3})$$

a generic element in H. Let $H_0 = \{u \in H;\ \nabla\cdot\mathbf{E} = \nabla\cdot\mathbf{H} = \mathbf{0}\}$, where the differential operator ∇ is considered in the sense of distributions, i.e. $\nabla\cdot\mathbf{F} = \mathbf{0}$ if and only if

$$\int_{\mathbb{R}^3} \nabla g\cdot\mathbf{F}\,\mathbf{d}\omega = \mathbf{0}$$

for each $g \in C_0^\infty(\mathbb{R})$. This means that u is in H_0 if and only if it is orthogonal on each element v in H of the form $v = (\nabla\varphi, \nabla\psi)$ with $\varphi, \psi \in H^1(\mathbb{R}^3)$. Let us define the Maxwell operator, $A : D(A) \subseteq H \to H$, by

$$\begin{cases} D(A) = \{(\mathbf{E},\mathbf{H}) \in H;\ (-c\nabla\times\mathbf{H}, \mathbf{c}\nabla\times\mathbf{E}) \in H\} \\ A(\mathbf{E},\mathbf{H}) = (-\mathbf{c}\nabla\times\mathbf{H}, \mathbf{c}\nabla\times\mathbf{E}), \end{cases}$$

for $(\mathbf{E},\mathbf{H}) \in D(A)$. Let us observe that A maps $D(A)$ in H_0, and therefore H_0 is invariant under A, because the divergence of a curl is always 0. This explains why, in all that follows, we shall consider the restriction of A to

H_0, restriction which, for the sake of simplicity, we denote again by A. We emphasize that the operator A is not densely defined in H, but its restriction to H_0 does, as we shall see from the proof of Theorem 4.3.1.

Theorem 4.3.1. *The operator A, defined as above, is the generator of a C_0-group of unitary operators.*

Proof. We show that A satisfies the hypotheses of Stone Theorem 3.9.1. To this aim, let $C_\sigma^\infty(\mathbb{R}^n) = \{\mathbf{F} \in C_0^\infty(\mathbb{R}^n);\ \nabla \cdot \mathbf{F} = 0\}$. Inasmuch as $C_\sigma^\infty(\mathbb{R}^n) \times C_\sigma^\infty(\mathbb{R}^n)$ is included in $D(A)$, and dense in H_0, it follows that A is densely defined. We prove next that A is skew-adjoint. First, let us observe that, for each $\mathbf{E}, \mathbf{H} \in C_\sigma^\infty(\mathbb{R}^n)$, we have

$$\langle A(\mathbf{E}, \mathbf{H}), (\mathbf{E}, \mathbf{H})\rangle = 0. \tag{4.3.1}$$

Indeed, since the integral on $\mathbb{R}^3$ of the divergence of a C^1 function with compact support is 0, we have

$$\langle A(\mathbf{E}, \mathbf{H}), (\mathbf{E}, \mathbf{H})\rangle = \int_{\mathbb{R}^3} \operatorname{div}(\mathbf{H} \times \mathbf{E})\, d\omega = 0$$

for each $(\mathbf{E}, \mathbf{H}) \in C_\sigma^\infty(\mathbb{R}^n)$. Inasmuch as $C_\sigma^\infty(\mathbb{R}^n) \times C_\sigma^\infty(\mathbb{R}^n)$ is dense in H_0, it follows that (4.3.1) holds for each $(\mathbf{E}, \mathbf{H}) \in D(A)$ and, by virtue of Remark 1.6.3, A is skew-symmetric, or equivalently, iA is symmetric. To check that A is skew-adjoint we prove that iA is self-adjoint and, to this aim, we shall use Lemma 1.6.1. More precisely, we shall prove that $1 \in \rho(iA)$. Let us denote by $\hat{\mathbf{H}}$ and $\hat{\mathbf{E}}$ the Fourier transform of $\mathbf{H}$ and respectively of $\mathbf{E}$, i.e.

$$\begin{cases} \hat{E}(\xi) = \dfrac{1}{(2\pi)^{3/2}} \displaystyle\int_{\mathbb{R}^3} e^{-i\langle \xi, x\rangle} \mathbf{E}(x)\, dx, \\[2ex] \hat{H}(\xi) = \dfrac{1}{(2\pi)^{3/2}} \displaystyle\int_{\mathbb{R}^3} e^{-i\langle \xi, x\rangle} \mathbf{H}(x)\, dx. \end{cases}$$

Then the mapping $(\mathbf{E}, \mathbf{H}) \mapsto (\hat{\mathbf{E}}, \hat{\mathbf{H}})$ is an isomorphism from H to a Hilbert space $\hat{H}$ analogously defined. More that this, this isomorphism maps H_0 into a subspace $\hat{H}_0$ in $\hat{H}$, subspace defined by

$$\hat{H}_0 = \{(\hat{\mathbf{E}}, \hat{\mathbf{H}}) \in \hat{H};\ \xi \cdot \hat{\mathbf{E}} = \xi \cdot \hat{\mathbf{H}} = 0\}$$

and it maps the operator A to the operator $\hat{A} : D(\hat{A}) \subseteq \hat{H}_0 \to \hat{H}_0$, defined by

$$\begin{cases} D(\hat{A}) = \{(\hat{\mathbf{E}}, \hat{\mathbf{H}}) \in \hat{H}_0;\ (-c\xi \times \hat{\mathbf{H}}, c\xi \times \hat{\mathbf{E}}) \in \hat{H}_0\} \\ \hat{A}(\hat{E}, \hat{H}) = (-c\xi \times \hat{\mathbf{H}}, c\xi \times \hat{\mathbf{E}}). \end{cases}$$

Let $\hat{\mathbf{v}} \in \hat{H}_0$, $\hat{\mathbf{v}} = (\hat{\mathbf{v}}_1, \hat{\mathbf{v}}_2)$, and let us consider the equation

$$(i\hat{I} - \hat{A})\hat{\mathbf{u}} = \hat{\mathbf{v}}, \tag{4.3.2}$$

where $\hat{I}$ denotes the identity operator on $\hat{H}$. Obviously, $1 \in \rho(iA)$ if and only if $i \in \rho(\hat{A})$. But this last condition holds if and only if, for each $\hat{\mathbf{v}} \in \hat{H}_0$, the equation (4.3.2) has a unique solution $R_1\hat{\mathbf{v}}$ and there exists $K > 0$ such that

$$\|R_1\hat{\mathbf{v}}\| \leq K\|\hat{\mathbf{v}}\|$$

for each $\hat{\mathbf{v}} \in \hat{H}_0$. Clearly, in this case, the solution operator R_1 coincides with $R(1; \hat{A})$. Let us observe that, written on components, (4.3.2) has the form

$$\begin{cases} c\xi \times \hat{\mathbf{u}}_2 + i\hat{\mathbf{u}}_1 = \hat{\mathbf{v}}_1 \\ -c\xi \times \hat{\mathbf{u}}_1 + i\hat{\mathbf{u}}_2 = \hat{\mathbf{v}}_2, \end{cases}$$

system whose unique solution is given by

$$\begin{cases} \hat{\mathbf{u}}_1 = \dfrac{-i\hat{\mathbf{v}}_1 + c\xi \times \hat{\mathbf{v}}_2}{c^2\|\xi\|^2 + 1} \\[2ex] \hat{\mathbf{u}}_2 = \dfrac{i\hat{\mathbf{v}}_2 - c\xi \times \hat{\mathbf{v}}_1}{c^2\|\xi\|^2 + 1} \end{cases}$$

From the equalities above one may observe that $\hat{\mathbf{u}}$ is a linear continuous function of $\hat{\mathbf{v}}$. Therefore $i \in \rho(\hat{A})$, or equivalently $1 \in \rho(iA)$. Analogously we deduce that $-1 \in \rho(iA)$, which, according to Lemma 1.6.1, shows that A is skew-adjoint. The conclusion of Theorem 4.3.1 follows from Stone Theorem 3.9.1. The proof is complete. □

4.4. The Directional Derivative

The aim of this section is to prove that the well-known directional derivative operator is the infinitesimal generator of a C_0-group of isometries in a suitably chosen function space. More precisely, let us consider the transport equation along the direction $a \in \mathbb{R}^3$

$$\begin{cases} u_t + a \cdot \nabla u = 0 & (t, x) \in \mathbb{R} \times \mathbb{R}^3 \\ u(0, x) = u_0(x) & x \in \mathbb{R}^3. \end{cases} \tag{$\mathcal{TE}$}$$

This first-order linear partial differential equation can be rewritten as an ordinary differential equation of the form

$$\begin{cases} u' = Au \\ u(0) = u_0 \end{cases} \tag{$\mathcal{ODE}$}$$

in a Banach space X, chosen so that $A : D(A) \subseteq X \to X$ does generate a C_0-semigroup of contractions. As we shall see later, in this case, A generates even a C_0-group of isometries.

Example 4.4.1. (*The Directional Derivative Operator.*) Let us consider $X = L^p(\mathbb{R}^n)$, with $1 \leq p < +\infty$ and let $a \in \mathbb{R}^n$. We define the operator A in X by

$$\begin{cases} D(A) = \{u \in X;\ a \cdot \nabla u \in X\} \\ Au = -a \cdot \nabla u = -\displaystyle\sum_{i=1}^{n} a_i \frac{\partial u}{\partial x_i}, \end{cases}$$

for $u \in D(A)$[1].

Theorem 4.4.1. *The operator A, defined as above, is the generator of a C_0-group of isometries. Moreover, this group, $G(t) : X \to X$, $t \in \mathbb{R}$, is given by*

$$G(t)f(x) = f(x - ta)$$

for each $f \in X$, $t \in \mathbb{R}$, and $x \in \mathbb{R}^n$.

Lemma 4.4.1. *Let $\lambda > 0$, and $1 \leq p < \infty$. If $u \in L^p(\mathbb{R}^n)$ satisfies*

$$u + \lambda a \cdot \nabla u = 0$$

in $\mathcal{D}'(\mathbb{R}^n)$, then $u \equiv 0$ a.e. $x \in \mathbb{R}^n$.

Proof. Let φ a mollifier, let $(\varepsilon_k)_{k \in \mathbb{N}}$ be a sequence decreasing to 0 and let $u_k = u_{\varepsilon_k}$ the ε_k-mollified of u. See Definition 1.3.1. Thanks to Theorem 1.3.3, we have $u_n \in C^\infty(\mathbb{R}^n) \cap L^\infty(\mathbb{R}^n)$. In addition, since

$$u_k(x) = \int_{\mathbb{R}^n} \varphi(z)u(x - \varepsilon_k z)\, dz = \int_{\mathbb{R}^n} u(y)\varphi_{\varepsilon_k}(x - y)\, dy$$

for each $x \in \mathbb{R}^n$, we have

$$u_k + \lambda a \cdot \nabla u_k = 0$$

for each $k \in \mathbb{N}$, and $x \in \mathbb{R}^n$.
Let $x \in \mathbb{R}^n$, and let us define $h : \mathbb{R} \to \mathbb{R}$ by $h(t) = e^t u_k(x + \lambda t a)$. Clearly h is differentiable on $\mathbb{R}$, and

$$h'(t) = e^t \left(u_k(x + \lambda t a) + \lambda a \cdot \nabla u_k(x + \lambda t a)\right) = 0.$$

Therefore h is constant on $\mathbb{R}$. Letting $t \to -\infty$ and taking into account that u is bounded, we deduce that $h \equiv 0$ on $\mathbb{R}$, which implies $u_k(x) = 0$.

[1] We notice that here $a \cdot \nabla u$ is the directional derivative of u along the direction a in the sense of distributions over $\mathcal{D}(\mathbb{R}^n)$, i.e. $a \cdot \nabla u = \langle a, \nabla u \rangle$, where ∇u is the gradient of u in the sense of distributions over $\mathcal{D}(\mathbb{R}^n)$.

Since x is arbitrary and $\lim_{k\to\infty} u_k = u$ in $L^p(\mathbb{R}^n)$ $(1 \le p < +\infty)$, the proof is complete. □

Lemma 4.4.2. *If $\lambda > 0$, $1 \le p < +\infty$ and $f \in L^p(\mathbb{R}^n)$, then the function*

$$Lf(x) = \frac{1}{\lambda}\int_0^\infty e^{-\frac{s}{\lambda}} f(x - sa)\, ds$$

satisfies

$$Lf + \lambda a \cdot \nabla(Lf) = f \tag{4.4.1}$$

in $\mathcal{D}'(\mathbb{R}^n)$, and

$$\|Lf\|_{L^p(\mathbb{R}^n)} \le \|f\|_{L^p(\mathbb{R}^n)}. \tag{4.4.2}$$

Proof. Let $f \in C_{ub}(\mathbb{R}^n)$, and let us define

$$Mf(x) = \frac{1}{\lambda}\int_0^\infty e^{-\frac{s}{\lambda}} f(x + sa)\, ds.$$

Let $\varphi \in \mathcal{D}(\mathbb{R}^n)$. From Fubini theorem (see Dunford and Schwartz [**49**], Theorem 9, p. 190), it follows that

$$\langle Lf, \varphi\rangle = \int_{\mathbb{R}^n} f M\varphi\, dx.$$

On the other hand

$$M(\lambda a \cdot \nabla\varphi)(x) = \int_0^\infty e^{-\frac{s}{\lambda}} a \cdot \nabla\varphi(x + sa)\, ds$$

$$= \int_0^\infty e^{-\frac{s}{\lambda}} \frac{d}{ds}\left(\varphi(x + sa)\right) ds = -\varphi(x) + M\varphi(x).$$

Accordingly

$$\langle Lf, \varphi\rangle = \int_{\mathbb{R}^n} f M\varphi\, dx = \int_{\mathbb{R}^n} f M(\lambda a \cdot \nabla\varphi)\, dx + \langle f, \varphi\rangle$$

$$= \langle Lf, \lambda a \cdot \nabla\varphi\rangle + \langle f, \varphi\rangle = \langle -\lambda a \cdot \nabla(Lf) + f, \varphi\rangle,$$

equality which proves (4.4.1).
Let $1 < p < +\infty$, and $\frac{1}{p} + \frac{1}{q} = 1$. From Hölder's inequality, we have

$$|Lf(x)| \le \frac{1}{\lambda}\int_0^\infty e^{-\frac{s}{\lambda q}} e^{-\frac{s}{\lambda p}} |f(x - sa)| ds$$

$$\le \lambda^{-1/p}\left(\int_0^\infty e^{-\frac{s}{\lambda}} |f(x - sa)|^p ds\right)^{1/p}.$$

Taking the pth power of both sides in this inequality, integrating on $\mathbb{R}^n$ the inequality thus obtained, and using once again Fubini theorem, we deduce

$$\|Lf\|^p_{L^p(\mathbb{R}^n)} \le \frac{1}{\lambda}\|f\|^p_{L^p(\mathbb{R}^n)} \int_0^\infty e^{-\frac{s}{\lambda}} ds = \|f\|^p_{L^p(\mathbb{R}^n)},$$

from where it follows (4.4.2), in the case $1 < p < +\infty$. Since for $p = 1$, (4.4.2) follows similarly, this achieves the proof. □

Remark 4.4.1. Analogous considerations show that both Lemmas 4.4.1 and 4.4.2 can be extended to the case $\lambda < 0$.

We may now proceed to the proof of Theorem 4.4.1.

Proof. Since $D(A)$ includes $C_0^\infty(\mathbb{R}^n)$ and the latter is dense in X, it follows that $D(A)$ is dense in X. So A satisfies (i) from Theorem 3.5.2. From (4.4.2) and Remarks 3.1.1 and 4.4.1, one may easily deduce that A satisfies (iii) in Theorem 3.5.2 as well. The proof is complete. □

4.5. The Schrödinger Operator

The aim of this section is to present an example of differential operator arising in the mathematical modelling of some phenomena in Quantum Physics, which generates a C_0-group of unitary operators. More precisely, let us consider the well-known Schrödinger equation in the domain Ω in $\mathbb{R}^3$

$$\begin{cases} u_t = \frac{i}{2m}\left[\Delta u - V(x)u\right] & (t,x) \in Q_{\mathbb{R}} \\ u = 0 & (t,x) \in \Sigma_{\mathbb{R}} \\ u(0,x) = u_0(x) & x \in \Omega, \end{cases} \tag{$\mathcal{SE}$}$$

where $Q_{\mathbb{R}} = \mathbb{R} \times \Omega$ and $\Sigma_{\mathbb{R}} = \mathbb{R} \times \Gamma$. In this case too, by an appropriate choice of the Hilbert space H, the partial differential equation ($\mathcal{SE}$) can be rewritten as a first-order ordinary differential equation of the form

$$\begin{cases} u' = Au \\ u(0) = \xi, \end{cases} \tag{$\mathcal{ODE}$}$$

where $A : D(A) \subseteq H \to H$ generates a C_0-group of unitary operators. In this section we consider only the case of a vanishing potential, i.e. ($V \equiv 0$), the general case making the object of Section 5.5. For the sake of simplicity, we shall also assume that $m = 1/2$.

Example 4.5.1. (*The Schrödinger Operator*).
1. *The $H^{-1}(\Omega;\mathbb{C})$ Setting.* Let Ω be a nonempty and open subset in $\mathbb{R}^n$. We consider the space $H = H^{-1}(\Omega;\mathbb{C}) = H^{-1}(\Omega) + iH^{-1}(\Omega)$ endowed with its real Hilbert space structure, and we define the operator A in H by:

$$\begin{cases} D(A) = H_0^1(\Omega;\mathbb{C}) \\ Au = i\Delta u, \end{cases}$$

for each $u \in D(A)$. We notice that, $H_0^1(\Omega;\mathbb{C})$ is equipped with the usual norm of the space $H^1(\Omega;\mathbb{C})$, i.e. $\|u+iv\|_{H_0^1(\Omega;\mathbb{C})} = (\|u\|^2_{H^1(\Omega)} + \|v\|^2_{H^1(\Omega)})^{1/2}$.

Theorem 4.5.1. *The operator A, defined as above, is the generator of a C_0-group of unitary operators.*

Proof. We apply Theorem 3.9.1 combined with Lemma 1.6.2. □

2. *The $L^2(\Omega;\mathbb{C})$ Setting.* Let Ω a nonempty and open subset in $\mathbb{R}^n$, let $H = L^2(\Omega;\mathbb{C}) = L^2(\Omega) + iL^2(\Omega)$, endowed with its real Hilbert structure, and let us consider the operator A in H, defined by:

$$\begin{cases} D(A) = \{u \in H_0^1(\Omega;\mathbb{C});\ \Delta u \in L^2(\Omega;\mathbb{C})\} \\ Au = i\Delta u, \end{cases}$$

for each $u \in D(A)$.

Theorem 4.5.2. *The operator A, defined as above, is the generator of a C_0-group of unitary operators. In addition, $D(A)$ is continuously imbedded in $H_0^1(\Omega;\mathbb{C})$. Moreover, if Ω is bounded, then $D(A)$ is compactly imbedded in $L^2(\Omega;\mathbb{C})$.*

Proof. We apply Theorem 4.1.2, Lemma 1.6.2 and Theorem 3.9.1. □

Remark 4.5.1. If Ω is bounded and has a C^2 boundary, or either $\Omega = \mathbb{R}^n_+$ or Ω is bounded and convex, we have $D(A) = H_0^1(\Omega;\mathbb{C}) \cap H^2(\Omega;\mathbb{C})$. See Remark 4.1.1.

4.6. The Wave Operator

Let us consider the equation of wave propagation in a homogeneous and isotropic body Ω in $\mathbb{R}^3$ with C^1 boundary Γ

$$\begin{cases} u_{tt} = \Delta u & (t,x) \in Q_{\mathbb{R}} \\ u = 0 & (t,x) \in \Sigma_{\mathbb{R}} \\ u(0,x) = u_0, \quad u_t(0,x) = v_0 & x \in \Omega, \end{cases} \tag{$\mathcal{WE}$}$$

where $Q_{\mathbb{R}} = \mathbb{R} \times \Omega$ and $\Sigma_{\mathbb{R}} = \mathbb{R} \times \Gamma$, and let us observe that this can be formally rewritten as a first-order (in t) system

$$\begin{cases} u_t = v & (t,x) \in Q_{\mathbb{R}} \\ v_t = \Delta u & (t,x) \in Q_{\mathbb{R}} \\ u = 0 & (t,x) \in \Sigma_{\mathbb{R}} \\ u(0,x) = u_0(x), \ v(0,x) = v_0(x) & x \in \Omega. \end{cases}$$

Setting $w = (u,v)$, $A(u,v) = (v,\Delta u)$, and $\xi = (u_0, v_0)$, this system can be rewritten in the abstract form

$$\begin{cases} w' = Aw \\ w(0) = \xi, \end{cases} \tag{$\mathcal{ODE}$}$$

in a Hilbert space H, suitably chosen so that $A : D(A) \subseteq H \to H$ generate a C_0-group of unitary operators on H. In what follows, we present two such possible choices of the function space H satisfying all the requirements mentioned above.

Example 4.6.1. (*The Wave Operator*).
1. *The $L^2(\Omega) \times H^{-1}(\Omega)$ Setting.* Let Ω be a nonempty and open subset in $\mathbb{R}^n$, let $H = L^2(\Omega) \times H^{-1}(\Omega)$ and let $A : D(A) \subset H \to H$ be defined by:

$$\begin{cases} D(A) = H_0^1(\Omega) \times L^2(\Omega) \\ A(u,v) = (v, \Delta u), \end{cases}$$

for each $(u,v) \in D(A)$.
In this example, the space H is endowed with the inner product $\langle \cdot, \cdot \rangle$, defined by

$$\langle (u,v),(f,g) \rangle = \langle u, f \rangle_{L^2(\Omega)} + \langle v, g \rangle_{H^{-1}(\Omega)},$$

where $\langle \cdot, \cdot \rangle_{H^{-1}(\Omega)}$ is the inner product in $H^{-1}(\Omega)$ corresponding to the inner product $\langle \cdot, \cdot \rangle_{H_0^1(\Omega)}$ defined by

$$\langle u, v \rangle_{H_0^1(\Omega)} = \langle \nabla u, \nabla v \rangle_{L^2(\Omega)}$$

for each $u, v \in H_0^1(\Omega)$, i.e. the inner product defining the gradient norm $\| \cdot \|_0$ in Theorem 1.5.7.

Theorem 4.6.1. *The operator A, defined as above, is the generator of a C_0-group of unitary operators on H. In addition, the graph-norm of A is equivalent with $\| \cdot \|_{H_0^1(\Omega) \times L^2(\Omega)}$.*

Proof. Obviously $D(A)$ is dense in H. Let $(u,v) \in D(A)$. We have

$$\langle A(u,v),(u,v) \rangle = \langle v, u \rangle_{L^2(\Omega)} + \langle \Delta u, v \rangle_{H^{-1}(\Omega)}.$$

Let $w \in H_0^1(\Omega)$ be the unique solution of the problem $-\Delta w = v$. By virtue of Theorem 1.5.8 we have

$$\langle -\Delta u, v \rangle_{H^{-1}(\Omega)} = \langle u, w \rangle_{H_0^1(\Omega)},$$

and accordingly

$$\begin{aligned} \langle A(u,v),(u,v) \rangle &= \langle v, u \rangle_{L^2(\Omega)} + \langle u, -w \rangle_{H_0^1(\Omega)} \\ &= \langle v, u \rangle_{L^2(\Omega)} + \int_\Omega \nabla u \nabla w \, d\omega \\ &= \langle v, u \rangle_{L^2(\Omega)} + \langle u, -\Delta w \rangle_{(H_0^1(\Omega), H^{-1}(\Omega))} \\ &= \langle v, u \rangle_{L^2(\Omega)} + \langle -v, u \rangle_{H_0^1(\Omega), H^{-1}(\Omega)} = 0. \end{aligned}$$

See Theorem 1.5.8. So, A is skew-symmetric. To complete the proof, it suffices to show that $(I \pm A)^{-1} \in \mathcal{L}(H)$. See Lemma 1.6.1 and Stone

Theorem 3.9.1. To this aim, let us observe first that, for each $(f,g) \in H$, the equation $(I - A)(u,v) = (f,g)$ is equivalent with the system

$$\begin{cases} u = v + f \\ -\Delta u = g - v. \end{cases}$$

The latter can be rewritten under the form

$$\begin{cases} u - \Delta u = f + g \\ v = u - f. \end{cases}$$

Since $f+g \in H^{-1}(\Omega)$, from Theorem 4.1.1, we deduce that the first equation in the above system has a unique solution $u \in H_0^1(\Omega)$. Hence v, given by the second equation, belongs to $L^2(\Omega)$, and therefore $I - A$ is bijective. Since the norm of $H_0^1(\Omega) \times L^2(\Omega)$ is given by $\|(u,v)\| = \|v\|_{L^2(\Omega)} + \|u\|_{H_0^1(\Omega)}$, and the latter, by Theorem 1.5.8, is equivalent with $\|v\|_{L^2(\Omega)} + \|\Delta u - u\|_{H^{-1}(\Omega)}$, which in its turn is nothing else than the graph-norm of A, it follows that $(I - A)^{-1} \in \mathcal{L}(H)$. Similarly, we show that $(I + A)^{-1} \in \mathcal{L}(H)$ and this achieves the proof. □

2. *The $H_0^1(\Omega) \times L^2(\Omega)$ Setting.* Let Ω be a nonempty and open subset in $\mathbb{R}^n$, let us consider the space $H = H_0^1(\Omega) \times L^2(\Omega)$ and let us define $A : D(A) \subset H \to H$ by:

$$\begin{cases} D(A) = \{(u,v) \in H; \Delta u \in L^2(\Omega),\ v \in H_0^1(\Omega)\} \\ A(u,v) = (v, \Delta u), \end{cases}$$

for each $(u,v) \in D(A)$.
In this example, we endow H with $\langle \cdot, \cdot \rangle : H \times H \to \mathbb{R}$, defined by

$$\langle (u,v), (\tilde{u}, \tilde{v}) \rangle = \int_\Omega \nabla u \nabla \tilde{u}\, dx + \int_\Omega v\tilde{v}\, dx,$$

for each $(u,v), (\tilde{u},\tilde{v}) \in H$, in respect to which this is a real Hilbert space. See Theorem 1.5.8.

Theorem 4.6.2. *The operator A, defined as above, is the generator of a C_0-group of unitary operators on H. In addition, $D(A)$ endowed with the graph-norm, is continuously imbedded in $H_0^1(\Omega) \times L^2(\Omega)$.*

Proof. Clearly $D(A)$ is dense in H. In addition, using Green's formula, we deduce that

$$\langle A(u,v), (u,v) \rangle = \langle (v, \Delta u), (u,v) \rangle = \int_\Omega \nabla v \nabla u\, dx + \int_\Omega \Delta u v\, dx$$

$$= \int_\Omega \nabla v \nabla u\, dx - \int_\Omega \nabla u \nabla v\, dx + \int_{\partial\Omega} u_\nu v\, d\sigma = 0$$

for each $(u, v) \in D(A)$. From Remark 1.6.3, it follows that A is skew-symmetric. By Lemma 1.6.1 we conclude that A is skew-adjoint, while from Stone Theorem 3.9.1, it follows that A generates a C_0-group of unitary operators in H. The proof is complete. □

Example 4.6.2. (*The Klein-Gordon Operator*)
1. *The $L^2(\Omega) \times H^{-1}(\Omega)$ Setting.* Let

$$\lambda = \inf\{\|\nabla u\|_{L^2(\Omega)};\ u \in H_0^1(\Omega),\ \|u\|_{L^2(\Omega)} = 1\}^2,$$

let $m > -\lambda$ and $H = L^2(\Omega) \times H^{-1}(\Omega)$. From Corollary 1.5.1, it follows that, for each $u \in H^{-1}(\Omega)$, the equation $m\varphi_u - \Delta\varphi_u = u$ has a unique solution $\varphi_u \in H_0^1(\Omega)$. Then the mapping $\langle\cdot,\cdot\rangle : H \times H \to \mathbb{R}$, defined by

$$\langle(u, v), (\tilde{u}, \tilde{v})\rangle = \langle\nabla\varphi_u, \nabla\varphi_{\tilde{u}}\rangle_{L^2(\Omega)} + m\langle\varphi_u, \varphi_{\tilde{u}}\rangle_{L^2(\Omega)} + \langle v, \tilde{v}\rangle_{L^2(\Omega)},$$

for each $(u, v), (\tilde{u}, \tilde{v}) \in H$, is an inner product on H equivalent with the initial one, in the sense that the corresponding two norms are equivalent. We endow H with this inner product and we define $A : D(A) \subseteq H \to H$ by

$$\begin{cases} D(A) = H_0^1(\Omega) \times L^2(\Omega) \\ A(u, v) = (v, \Delta u - mu) \end{cases}$$

for each $(u, v) \in D(A)$. This is the *Klein-Gordon operator* in the space $L^2(\Omega) \times H^{-1}(\Omega)$. Reasoning as in the proof of Theorem 4.6.2, we obtain:

Theorem 4.6.3. *The operator A, defined as above, is the generator of a group of isometries on H. In addition, $D(A)$ endowed with the graph-norm, is continuously imbedded in $L^2(\Omega) \times H^{-1}(\Omega)$.*

2. *The $H_0^1(\Omega) \times L^2(\Omega)$ Setting.* Let λ be as above, let us consider the space $H = H_0^1(\Omega) \times L^2(\Omega)$ and let us define $A : D(A) \subset H \to H$ by:

$$\begin{cases} D(A) = \{(u, v) \in H; \Delta u \in L^2(\Omega),\ v \in H_0^1(\Omega)\} \\ A(u, v) = (v, \Delta u - mu), \end{cases}$$

for each $(u, v) \in D(A)$. This is the *Klein-Gordon operator* in the space $H_0^1(\Omega) \times L^2(\Omega)$. From the definition of λ and the choice of m, it follows that $\langle\cdot,\cdot\rangle : H \times H \to \mathbb{R}$, defined by

$$\langle(u, v), (\tilde{u}, \tilde{v})\rangle = \langle\nabla u, \nabla\tilde{u}\rangle_{L^2(\Omega)} + m\langle u, \tilde{u}\rangle_{L^2(\Omega)} + \langle v, \tilde{v}\rangle_{L^2(\Omega)}$$

for each $(u, v), (\tilde{u}, \tilde{v}) \in H$, is an inner product on H which is equivalent with the original one, in the sense that the corresponding two norms are equivalent. Using similar arguments as in the proof of Theorem 4.6.2, we get:

---

[2] If Ω is bounded then λ is the first eigenvalue of $-\Delta$ on $H^{-1}(\Omega)$ and $\lambda > 0$.

Theorem 4.6.4. *The operator A, defined as above, is the generator of a group of isometries on H. In addition, $D(A)$ endowed with the graph-norm, is continuously imbedded in $H_0^1(\Omega) \times L^2(\Omega)$.*

Remark 4.6.1. We notice that the *Klein-Gordon equation*, i.e.

$$\begin{cases} a^2 u_{tt} = \Delta u - k^2 u & \text{in } \mathbb{R}_+ \times \Omega \\ u = 0 & \text{on } \mathbb{R}_+ \times \Gamma \\ u(0) = \xi & \text{on } \Omega, \end{cases}$$

where $a > 0$ and $k > 0$, is of interest in Quantum Physics. More precisely, the wave function of the free particles satisfies this equation.

4.7. The Airy Operator

In this section, we shall consider the Airy equation

$$\begin{cases} u_t + u_{xxx} = 0 & (t, x) \in \mathbb{R} \times \mathbb{R} \\ u(0, x) = u_0(x) & x \in \mathbb{R}, \end{cases} \tag{$\mathcal{AE}$}$$

and we will show that it can be rewritten under the abstract form

$$\begin{cases} u' = Au \\ u(0) = \xi, \end{cases} \tag{$\mathcal{ODE}$}$$

where $A : D(A) \subseteq H \to H$ is the infinitesimal generator of a C_0-group of isometries on H, while H is a suitably chosen function space.

Example 4.7.1. (*The Airy Operator*). Let $H = L^2(\mathbb{R})$ and let us consider the operator A on H, defined by:

$$\begin{cases} D(A) = H^3(\mathbb{R}) \\ Au = -u_{xxx} = -\dfrac{d^3 u}{dx^3}, \end{cases}$$

for each $u \in D(A)$.

Theorem 4.7.1. *The operator A, defined as above, is the generator of a C_0-group of unitary operators on H.*

Proof. Obviously A is densely defined. Let $u, v \in C_0^\infty(\mathbb{R})$. Inasmuch as

$$u_{xxx} u = \frac{1}{2} \left[2u_{xx} u - (u_x)^2 \right]_x,$$

it follows that

$$\int_{\mathbb{R}} u_{xxx} u \, dx = 0,$$

which shows that $\langle Au, u \rangle = 0$ for each $u \in C_0^\infty(\mathbb{R})$. Since $C_0^\infty(\mathbb{R})$ is dense in $D(A)$, from Remark 1.6.3, it follows that A is skew-symmetric. One may easily verify that $\pm i \in \rho(A)$, and therefore, by Theorem 1.6.1, it follows that A is skew-adjoint. From Stone Theorem 3.9.1, we conclude that A generates C_0-group of unitary operators on H, and this achieves the proof. □

4.8. The Equations of Linear Thermoelasticity

In this section, we show how the previously developed theory can be used to obtain accurate information on a system of partial differential equations modelling the evolution of three fields: the displacement, the momentum and the temperature of a thermoelastic body.

Example 4.8.1. (*The Equations of Linear Thermoelasticity*). Let $\mathcal{C}$ be a homogeneous body having the referential configuration a nonempty, open and bounded subset Ω in $\mathbb{R}^n$ ($n = 1, 2$ or 3) whose boundary Γ is of class C^1. The state at the time $t \in \mathbb{R}_+$ is characterized by two vector fields: the displacement $u(t,x)$, and the momentum $v(t,x)$, and a scalar field: the temperature $\theta(t,x)$. The system of equations describing the evolution of these three fields is

$$\begin{cases} u_t = \rho^{-1} v & (t,x) \in Q_\infty \\ v_t = \mu\nabla\cdot\nabla u + (\lambda+\mu)\nabla\nabla\cdot u + m\nabla\theta & (t,x) \in Q_\infty \\ \theta_t = c^{-1}k\Delta\theta + c^{-1}\rho^{-1}\Theta m\nabla\cdot v & (t,x) \in Q_\infty \\ u(t,x) = 0,\ \theta(t,x) = 0 & (t,x) \in \Sigma_\infty \\ u(0,x) = u_0(x) \quad v(0,x) = v_0(x) \quad \theta(0,x) = \theta_0(x) & x \in \Omega, \end{cases} \tag{4.8.1}$$

where $Q_\infty = \mathbb{R}_+ \times \Omega$ and $\Sigma_\infty = \mathbb{R}_+ \times \Gamma$. Here $\rho > 0$ is the density of the body, $\Theta > 0$ the referential temperature, and $c > 0$, $k > 0$, m, λ, and μ are constants which characterize the thermoelastic properties of the body. We assume that λ and μ satisfy the strong ellipticity conditions $\mu > 0$ and $\lambda + 2\mu > 0$. We begin by rewriting the system (4.8.1) under the form of an abstract Cauchy problem in a suitably chosen Hilbert space. More precisely, let $H = \left[H_0^1(\Omega)\right]^n \times \left[L^2(\Omega)\right]^n \times L^2(\Omega)$ which, endowed with the inner product $\langle\cdot,\cdot\rangle$, defined by

$$\langle(u,v,\theta),(\widetilde{u},\widetilde{v},\widetilde{\theta})\rangle$$

$$= \int_\Omega \left\{\rho^{-1}v\widetilde{v} + \mu\nabla u : \nabla\widetilde{u} + (\lambda+\mu)(\nabla\cdot u)(\nabla\widetilde{u}) + c^{-1}\theta\widetilde{\theta}\right\} dx,$$

is a real Hilbert space. Here

$$\nabla u : \nabla\widetilde{u} = \sum_{i,j=1}^n \frac{\partial u_i}{\partial x_j}\frac{\partial u_j}{\partial x_i}.$$

We define the operator $A : D(A) \subseteq H \to H$ by

$$D(A) = \left[H^2(\Omega) \cap H_0^1(\Omega)\right]^n \times \left[H_0^1(\Omega)\right]^n \times \left[H^2(\Omega) \cap H_0^1(\Omega)\right],$$

and

$$A(u, v, \theta)$$

$$= \left(\rho^{-1}v, \mu\nabla \cdot \nabla u + (\lambda + \mu)\nabla\nabla \cdot u + m\nabla, c^{-1}k\Delta\theta + c^{-1}\rho^{-1}\Theta m\nabla \cdot v\right)$$

for each $(u, v, \theta) \in D(A)$. Let us observe that the system (4.8.1) can be rewritten in the space H as

$$\begin{cases} z' = Az \\ z(0) = \xi, \end{cases}$$

where A is as above, $z = (u, v, \theta)$ and $\xi = (u_0, v_0, \theta_0)$.

Theorem 4.8.1. *The operator A, defined as above, is the generator of a C_0-semigroup of contractions in H.*

Proof. Since the set $[C_0^\infty(\Omega)]^n \times [C_0^\infty(\Omega)]^n \times C_0^\infty(\Omega)$ is included in $D(A)$ and dense in H, A is densely defined. We shall prove next that, for each $\lambda > 0$, there exists $(\lambda I - A)^{-1} \in \mathcal{L}(H)$, and

$$\|(\lambda I - A)^{-1}\|_{\mathcal{L}(H)} \leq \frac{1}{\lambda}, \tag{4.8.2}$$

from where it will follow, on one hand that A is closed, and on the other hand, that it satisfies the conditions (i) and (ii) in Theorem 3.1.1. Let us observe that $(\lambda I - A)^{-1}(\widetilde{u}, \widetilde{v}, \widetilde{\theta}) = (u, v, \theta)$ if and only if

$$\begin{cases} \lambda u - \rho^{-1}v = \widetilde{u} & x \in \Omega \\ \lambda v - \mu\nabla \cdot \nabla u - (\lambda + \mu)\nabla\nabla \cdot u - m\nabla\theta = \widetilde{v} & x \in \Omega \\ \lambda\theta - c^{-1}k\Delta\theta - c^{-1}\rho^{-1}\Theta m\nabla \cdot v = \widetilde{\theta} & x \in \Omega \\ u = 0, \quad \theta = 0 & x \in \Gamma. \end{cases} \tag{4.8.3}$$

From the strong ellipticity conditions, i.e. $\mu > 0$ and $\lambda + 2\mu > 0$, and from the general existence results concerning systems of elliptic equations, it follows that, for each $(\widetilde{u}, \widetilde{v}, \widetilde{\theta}) \in H$, the system (4.8.3) has a unique solution (u, v, θ). On the other hand, a simple computational argument shows that, for each $(u, v, \theta) \in D(A)$, we have

$$\langle A(u, v, \theta), (u, v, \theta)\rangle = -k\Theta^{-1}\int_\Omega \|\nabla\theta\|^2\,dx \leq 0,$$

which proves that $\lambda I - A$ is a bijection from $D(A)$ to H, and satisfies (4.8.2). The proof is complete. □

4.9. The Equations of Linear Viscoelasticity

In this section, we shall use the abstract results in Chapter 2 in order to get information concerning a system of integro-partial differential equations arising in the modelling of the evolution of the displacement, the history of displacement, and the momentum of a viscoelastic body.

Example 4.9.1. (*The Equations of Linear Viscoelasticity*). Let us consider an one-dimensional homogeneous body of density $\rho > 0$, and having the constitutive equation

$$\sigma(t,x) = cu_x(t,x) - \int_{-\infty}^{t} g(t-\tau)u_x(\tau,x)\,d\tau,$$

where $g : \mathbb{R}_+ \to \mathbb{R}$ is a positive, nonincreasing C^1 function satisfying

$$a = c - \int_0^{\infty} g(s)\,ds > 0. \tag{4.9.1}$$

The referential configuration is the interval $[\,0,\pi\,]$. The state of the body at the time $t \in \mathbb{R}$ is characterized by the displacement $u(t,x)$, the momentum $v(t,x)$ and the history of the displacement $w(t,s,x)$ which is defined by $w(t,s,x) = u(t-s,x)$ for $s \in \mathbb{R}_+$. Assuming that the displacement vanishes at the two endpoints of the interval $[\,0,\pi]$, and taking into account that the equation of motion is $\rho u'' = \sigma_x$, we obtain the system

$$\begin{cases} u_t = \rho^{-1}v & (t,x) \in Q_\infty \\ v_t = cu_{xx} - \displaystyle\int_0^{\infty} g(s)w_{xx}\,ds & (t,x) \in Q_\infty \\ w_t = -w_s & (t,s,x) \in R_\infty \\ u(t,0) = u(t,\pi) = 0 & t \in \mathbb{R}_+ \\ u(0,x) = u_0(x) \quad v(0,x) = v_0(x) & x \in (0,\pi) \\ w(0,s,x) = w_0(s,x) & (s,x) \in Q_\infty, \end{cases} \tag{4.9.2}$$

where $Q_\infty = \mathbb{R}_+ \times (0,\pi)$ and $R_\infty = \mathbb{R}_+ \times Q_\infty$. We shall show next how the system above can be rewritten as a Cauchy problem for an ordinary differential equation in a suitably chosen Hilbert space. To this aim, let

$$H = H_0^1(0,\pi) \times L^2(0,\pi) \times L_g^2(\mathbb{R}_+; H_0^1(0,\pi)),$$

where $L_g^2(\mathbb{R}_+; H_0^1(0,\pi))$ is the g-weighted space $L^2(\mathbb{R}_+; gdt; X)$. On H we consider the inner product $\langle\cdot,\cdot\rangle$, defined by

$$\langle (u,v,w), (\widetilde{u},\widetilde{v},\widetilde{w}) \rangle$$

$$= \int_0^{\pi} \left\{ au_x\widetilde{u}_x + \rho^{-1}v\widetilde{v} + \int_0^{\infty} g(s)[u_x - w_x][\widetilde{u}_x - \widetilde{w}_x]\,ds \right\} dx,$$

with respect to which H is a real Hilbert space. Let us define the operator $A : D(A) \subseteq H \to H$ by

$$D(A) = \{(u, v, w) \in H\,;\ (u, v, w) \text{ satisfies (4.9.3)}\}$$

$$\begin{cases} v \in H_0^1(0,\pi), \quad w_s \in L_g^2(\mathbb{R}_+; H_0^1(0,\pi)), \quad w(0,\cdot) = u \\ cu_{xx} - \displaystyle\int_0^\infty g(s) w_{xx}\, ds \in L^2(0,\pi) \end{cases} \tag{4.9.3}$$

and

$$A(u, v, w) = \left(\rho^{-1} v, cu_{xx} - \int_0^\infty g(s) w_{xx}\, ds, -w_s\right)$$

for each $(u, v, w) \in D(A)$. At this point, let us observe that (4.9.2) can be equivalently rewritten under the abstract form

$$\begin{cases} z' = Az \\ z(0) = \xi, \end{cases}$$

where A is as above, $z = (u, v, w)$ and $\xi = (u_0, v_0, w_0)$.

Theorem 4.9.1. *The operator A, defined as above, is the generator of a C_0-semigroup of contractions on H.*

Proof. Since the set

$$\{(u, v, w) \in C_0^\infty(0,\pi) \times C_0^\infty(0,\pi) \times L_g^2(\mathbb{R}_+; C_0^\infty(0,\pi))\,; w(0,\cdot) = u\}$$

is included in $D(A)$ and dense in H, A is densely defined. We shall show next that, for each $\lambda > 0$, there exists $(\lambda I - A)^{-1} \in \mathcal{L}(H)$ and

$$\|(\lambda I - A)^{-1}\|_{\mathcal{L}(H)} \leq \frac{1}{\lambda}, \tag{4.9.4}$$

from where it will follow, on one hand that A is closed, and on the other hand, that it satisfies the conditions (i) and (ii) in Theorem 3.1.1. Let us observe that $(\lambda I - A)^{-1}(\widetilde{u}, \widetilde{v}, \widetilde{w}) = (u, v, w)$ if and only if

$$\begin{cases} \lambda u - \rho^{-1} v = \widetilde{u} \\ \lambda v - cu_{xx} + \displaystyle\int_0^\infty g(s) w_{xx}\, ds = \widetilde{v} \\ \lambda w + w_s = \widetilde{w} \\ w(0,\cdot) = u. \end{cases} \tag{4.9.5}$$

From the last two conditions, we deduce

$$w(s, x) = u(x) e^{-\lambda s} + \int_0^s e^{-\lambda(s-\tau)} \widetilde{w}(\tau, x)\, d\tau. \tag{4.9.6}$$

If $u \in H_0^1(0,\pi)$, then, again from the last two relations in the system (4.9.5), we get $\lambda w_x = -w_{sx} + \widetilde{w}_x$ and therefore we successively have

$$\lambda g(s) w_x^2 = -\frac{1}{2} g(s)(w_x^2)_s + g(s) w_x \widetilde{w}_x$$

and

$$\lambda \int_0^\infty \int_0^\pi g(s) w_x^2 \, dx \, ds = \frac{1}{2} g(0) \int_0^\pi u_x^2 \, dx + \frac{1}{2} \int_0^\infty \int_0^\pi g'(s) w_x^2 \, dx \, ds$$
$$+ \int_0^\infty \int_0^\pi g(s) w_x \widetilde{w}_x \, dx \, ds.$$

From this equality, using Cauchy-Schwarz inequality, in the form

$$ab = \sqrt{\lambda} a \frac{1}{\sqrt{\lambda}} b \leq \frac{\lambda a^2}{2} + \frac{b^2}{2\lambda},$$

to estimate the last term on the right-hand side, we obtain

$$\lambda \int_0^\infty \int_0^\pi g(s) w_x^2 \, dx \, ds \leq g(0) \int_0^\pi u_x^2 \, dx + \frac{1}{\lambda} \int_0^\infty \int_0^\pi g(s) \widetilde{w}_x^2 \, dx \, ds,$$

which proves that $w \in L_g^2(\mathbb{R}_+; H_0^1(0,\pi))$. Coming back to the last two relations in (4.9.5), we deduce that $w_s \in L_g^2(\mathbb{R}_+; H_0^1(0,\pi))$. Substituting v from the first equation of (4.9.5) and w given by (4.9.6) in the second equation of (4.9.5), we get

$$\lambda^2 \rho u - \left(c - \int_0^\infty e^{-\lambda s} g(s) \, ds \right) u_{xx} = \widetilde{v} + \lambda \rho \widetilde{u} - \int_0^\infty e^{\lambda \tau} \widetilde{w}_{xx} \, d\tau \, ds. \quad (4.9.7)$$

Let us observe that the right-hand side of the above equality belongs to $H^{-1}(0,\pi)$. On the other hand, from (4.9.1) and the fact that $\lambda > 0$, we deduce

$$c - \int_0^\infty e^{-\lambda s} g(s) \, ds > 0.$$

Accordingly, (4.9.7) has a unique solution $u \in H_0^1(0,\pi)$. Consequently, the system (4.9.5) has a unique solution, which shows that $\lambda I - A$ is bijective from $D(A)$ to H. Finally, recalling the definition of the inner product in H and using the fact that g is C^1, and nonincreasing (which implies $g' \leq 0$), we obtain

$$\langle A(u,v,w), (u,v,w) \rangle = \frac{1}{2} \int_0^\infty \int_0^\pi g'(s) |u_x - w_x|^2 \, dx \, ds \leq 0.$$

From this inequality, it follows that $(\lambda I - A)^{-1} \in \mathcal{L}(H)$, and satisfies (4.9.4). This achieves the proof. □

Problems

Problem 4.1. Let $X = C([0,1])$ endowed with the usual sup norm and let us consider the operator $A : D(A) \subseteq X \to X$ defined by

$$\begin{cases} D(A) = \{f \in C^2((0,1))\,;\ \lim_{s\to t} s(1-s)f''(s) = 0 \text{ for } t = 0 \text{ or } t = 1\} \\ Af(s) = s(1-s)f''(s). \end{cases}$$

Show that A generates a C_0-semigroup of contractions on X. See Engel and Nagel et al [**51**], Paragraph II 3.30.(iii), p. 94.

Problem 4.2. Let $X = L^2(0,\pi)$, $t \in \mathbb{R}_+$, and let $S(t) : X \to X$ be defined by

$$[S(t)\xi](x) = \sum_{k=0}^{\infty} a_k(\xi)e^{-k^2 t}\sin kx$$

for each $\xi \in L^2(0,\pi)$, where $a_k(\xi)$ are the Fourier coefficients of ξ with respect to the orthogonal system $\{\sin x, \sin 2x, \ldots, \sin kx, \ldots\}$, i.e.

$$a_k(\xi) = \sqrt{\frac{2}{\pi}}\int_0^{\pi} \xi(y)\sin ky\,dy.$$

(i) Prove that $\{S(t)\,;\ t \geq 0\}$ is a C_0-semigroup of contractions on $L^2(0,\pi)$.
(ii) Find its infinitesimal generator.
(iii) Prove that, for each $\xi \in L^2(0,\pi)$, $u : [0,+\infty) \to L^2(0,\pi)$, defined by $u(t,x) = [S(t)\xi](x)$, satisfies

$$\begin{cases} u_t = u_{xx} & \text{for } (t,x) \in \mathbb{R}_+ \times (0,\pi) \\ u(t,0) = u(\pi,0) = 0 & \text{for } t \in \mathbb{R}_+ \\ u(0,x) = \xi(x) & \text{for } x \in (0,\pi), \end{cases}$$

in the sense of distributions over $(0,\pi)$.

Problem 4.3. Let $p \in [1,+\infty)$, $X = L^p(\mathbb{R})$ and $t \in \mathbb{R}_+$. Let us define $S(t) : X \to X$ by $S(0) = I$ and

$$[S(t)f](x) = \frac{1}{\sqrt{4\pi t}}\int_{-\infty}^{+\infty} e^{-\frac{(x-y)^2}{4t}} f(y)\,dy$$

for $t > 0$. Show that $\{S(t)\,;\ t \geq 0\}$ is a C_0-semigroup of contractions on X. This is the *Gauss-Weierstrass semigroup*. See McBride [**89**], Example 1.8, p. 15.

Problem 4.4. Find the generator of the semigroup in Problem 4.3. Show that, for each $f \in L^p(\mathbb{R})$, the function $u : [0, +\infty) \to L^p(\mathbb{R})$, defined by $u(t,x) = [S(t)f](x)$, satisfies

$$\begin{cases} u_t = u_{xx} & \text{on } \mathbb{R}_+ \times \mathbb{R} \\ u(0,x) = f(x) & \text{on } \mathbb{R}, \end{cases}$$

in the sense of distributions. See McBride [**89**], Example 1.8, p. 15.

Problem 4.5. Let $p \in [1, +\infty)$, $X = L^p(\mathbb{R})$, and let $t \in \mathbb{R}_+$. Let us define $S(t) : X \to X$ by $S(0) = I$, and

$$[S(t)f](x) = \frac{t}{\pi} \int_{-\infty}^{+\infty} \frac{f(y)}{t^2 + (x-y)^2}\, dy$$

for $t > 0$. Show that $\{S(t)\,;\ t \geq 0\}$ is a C_0-semigroup of contractions on X. This is the *Poisson semigroup*. See McBride [**89**], Example 1.10, p. 20.

Problem 4.6. Let $\{S(t)\,;\ t \geq 0\}$ the semigroup in Problem 4.5. Show that, for each $f \in L^p(\mathbb{R})$, the function $u : [0, +\infty) \to L^p(\mathbb{R})$, defined by $u(t,x) = [S(t)f](x)$, satisfies

$$\begin{cases} u_{tt} + u_{xx} = 0 & \text{on } \mathbb{R}_+ \times \mathbb{R} \\ u(0,x) = f(x) & \text{on } \mathbb{R}, \end{cases}$$

in the sense of distributions. See McBride [**89**], Example 1.10, p. 20.

Problem 4.7. Let $p \in [1, +\infty)$, $X = L^p(0,1)$ and $t \in \mathbb{R}_+$. Let us define $S(t) : X \to X$ by $S(0) = I$ and

$$[S(t)f](x) = \frac{1}{\Gamma(t)} \int_0^x (x-y)^{t-1} f(y)\, dy = (J^t f)(x)$$

for $t > 0$. Show that $\{S(t)\,;\ t \geq 0\}$ is a C_0-semigroup of contractions on X. The integral $J^t f$ is *the fractional Riemann-Liouville integral of exponent t of f*. See McBride [**89**], Example 1.12, p. 23.

Problem 4.8. Find the duals of the semigroups in Problems 4.2 and 4.3, in the particular case $p = 2$.

Problem 4.9. Let $H = L^2(\mathbb{R}) \times L^2(\mathbb{R})$ which, endowed with the inner product

$$\langle (u_1, v_1), (u_2, v_2) \rangle = \langle u_1, u_2 \rangle + \langle v_1, v_2 \rangle$$

for each $(u_i, v_i) \in H$, $i = 1, 2$, is a real Hilbert space. Let $t \in \mathbb{R}$, and let us define $G(t) : H \to H$ by

$$[G(t)(u,v)](x) = \begin{pmatrix} \dfrac{u(x+t)+u(x-t)}{2} + \dfrac{v(x+t)-v(x-t)}{2} \\ \\ \dfrac{v(x+t)+v(x-t)}{2} + \dfrac{u(x+t)-u(x-t)}{2} \end{pmatrix}^{\tau}$$

for each $(u, v) \in H$, where B^{τ} denotes the transposed of the matrix B. Prove that $\{G(t)\,;\ t \in \mathbb{R}\}$ is a group of unitary operators on H and find its infinitesimal generator.

Problem 4.10. Let X be a Banach space, $A : D(A) \subseteq X \to X$ the infinitesimal generator of a C_0-semigroup in X, let $a \in L^1(\mathbb{R})$, and let $\varphi \in L^1(-\infty, 0; X)$. Prove that the integro-differential equation with infinite delay

$$\begin{cases} x'(t) = \displaystyle\int_{-\infty}^{t} a(t-s)Ax(s)\,ds \\ \\ x(s) = \varphi(s) \quad \text{for} \quad s \in (-\infty, 0] \end{cases}$$

can be equivalently rewritten in the space $\mathcal{X} = L^1(-\infty, 0; X)$ under the form

$$\begin{cases} u' = \mathcal{A}u \\ u(0) = \varphi, \end{cases}$$

with $\mathcal{A} : D(\mathcal{A}) \subseteq \mathcal{X} \to \mathcal{X}$ defined by

$$D(\mathcal{A}) = \left\{\varphi \in W^{1,1}_{D(A)}(-\infty, 0; X)\,;\ \varphi'(0) = \int_0^{\infty} a(\tau)A\varphi(-\tau)\,d\tau\right\}$$

and

$$[\mathcal{A}\varphi](\theta) = \begin{cases} \varphi'(\theta) & \text{if } \theta < 0 \\ \\ \displaystyle\int_0^{\infty} a(\tau)A\varphi(-\tau)\,d\tau & \text{if } \theta = 0. \end{cases}$$

Here $W^{1,1}_{D(A)}(-\infty, 0; X)$ is the space of all $\varphi \in W^{1,1}(-\infty, 0; X)$ satisfying $\varphi(s) \in D(A)$ a.e. for $s \in (-\infty, 0]$, and $A\varphi \in L^1(-\infty, 0; X)$.

Notes. Except for Section 4.2, which follows an idea in Richtmyer [**110**], and for the last two ones, which are essentially based on Dafermos [**44**], the chapter is adapted from Brezis and Cazenave [**31**]. We note that the results on the Laplace operator with Dirichlet boundary condition are classical and can be found in Barbu [**21**], Brezis [**29**], Grisvard [**64**], Pazy [**101**] and

Taylor [**119**]. As concerns the Laplace operator with Neumann boundary condition, we followed Taylor [**119**]. More details on the Maxwell operator can be found in Taylor [**119**]. Moreover, we add that the wave operator in $\mathbb{R}^n$, i.e. $A : D(A) \subseteq H \to H$ defined by $H = H^1(\mathbb{R}^n) \times L^2(\mathbb{R}^n)$ and

$$\begin{cases} D(A) = H^2(\mathbb{R}^n) \times H^1(\mathbb{R}^n) \\ A(u,v) = (v, \Delta u) \quad \text{for} \quad (u,v) \in D(A), \end{cases}$$

generates a C_0-group $\{G(t)\,;\ t \in \mathbb{R}\}$, satisfying

$$\|G(t)\|_{\mathcal{L}(H)} \leq e^{2|t|}$$

for each $t \in \mathbb{R}^n$. For a direct proof of this generation theorem see Pazy [**101**], Theorem 4.5, p. 222. An alternate proof is to show that A satisfies all the hypotheses of Theorem of Liu [**86**] mentioned in Notes to Chapter 3, with $\gamma_0 = 2$. The idea of treating linear integro-partial differential equations (as those of linear viscoelasticity in Section 8), by using specific semigroup techniques, belongs to Dafermos [**44**]. An extension of this method to the fully nonlinear case is due to Barbu [**19**]. Subsequent developments are due to Aizicovici [**4**], Crandall, Londen and Nohel [**42**], and Hrusa [**71**]. For other examples in population dynamics, which can be analyzed with the tools of the C_0-semigroup theory, see Aniţa [**7**], Engel and Nagel [**51**], Thieme [**121**], Webb [**132**] and the references therein. The problems in this chapter are adapted from Engel and Nagel et al [**51**] and McBride [**89**].

CHAPTER 5

Approximation Problems and Applications

In this chapter we study the continuity properties of the semigroup as a function of its infinitesimal generator, properties which allow, on one hand, to approximate the solutions of a certain differential equation by the solutions of some other more simple equations, and, on the other hand, to give a rigorous meaning to some formulae in Theoretical Physics established by a more or less intuitive manner. We also study the behavior as t tends to $+\infty$ of the solutions of a homogeneous first-order differential equation governed by the infinitesimal generator of a C_0-semigroup, and we show how the celebrated central limit theorem in Probability Theory can be obtained via C_0-semigroup techniques.

5.1. The Continuity of $A \mapsto e^{tA}$

Let X be a real or complex Banach space and let us denote by $\mathcal{L}_p(X)$ the space $\mathcal{L}(X)$ endowed with the pointwise convergence topology. Let us denote also by $\mathcal{G}(X)$ the set of all infinitesimal generators of C_0-semigroups of contractions defined on X. In this section we present some continuity properties of the application $A \mapsto S$, defined on $\mathcal{G}(X)$, and taking values in $C([0,+\infty);\mathcal{L}_p(X))$, where S is the C_0-semigroup of contractions generated by A. We note that all the results here included can be suitably restated in order to handle C_0-semigroups of type (M,ω) as well. We begin with the following fundamental results.

Theorem 5.1.1. (Trotter, Neveu, Kato) *Let $\{A_n;\ n \in \mathbb{N}\} \subset \mathcal{G}(X)$, and let us denote by $\{S_n(t);\ t \geq 0\}$ the C_0-semigroup of contractions generated by A_n, for $n = 0, 1, \ldots$. Then we have:*

(i) *if there exists $\lim_{n\to\infty} S_n(t)x = S_0(t)x$ for each $t \in \mathbb{R}_+$ and each $x \in X$, then there exists $\lim_{n\to\infty} (\lambda I - A_n)^{-1}x = (\lambda I - A_0)^{-1}x$ for each $x \in X$, and uniformly for λ in compact subsets in $(0,+\infty)$;*

(ii) *if there exists* $\lim_{n\to\infty}(\lambda I - A_n)^{-1}x = (\lambda I - A_0)^{-1}x$ *for each* $x \in X$ *and* $\lambda \in (0, +\infty)$, *then there exists* $\lim_{n\to\infty} S_n(t)x = S_0(t)x$ *for each* $x \in X$, *and uniformly for* t *in compact subsets in* $(0, +\infty)$.

Proof. Let us recall that, for each $n \in \mathbb{N}^*$ and each $\lambda > 0$, we have $\lambda \in \rho(A_n)$ and

$$(\lambda I - A_n)^{-1}x = \int_0^\infty e^{-\lambda t} S_n(t)x\,dt.$$

The item (i) follows from the preceding equalities, by observing that, by virtue of the fact that all the semigroups in the family are of contractions, the hypotheses of the Lebesgue dominated convergence theorem are satisfied.

To prove (ii), we shall use Theorem 3.1.1 applied to a suitably defined linear operator on the space of all convergent sequences with elements in X. More precisely, let

$$\mathcal{X} = \{(x_n)_{n\in\mathbb{N}^*};\ \exists \lim_{n\to\infty} x_n\},$$

which, endowed with the norm $\|\cdot\|_\infty$, defined by

$$\|(x_n)_{n\in\mathbb{N}^*}\|_\infty = \sup_{n\geq 1}\|x_n\|,$$

is a Banach space. Let

$$D(\mathcal{A}) = \{(x_n)_{n\in\mathbb{N}^*} \in \mathcal{X};\ x_n \in D(A_n)\ \forall\, n \in \mathbb{N}^* \text{ and } (A_n x_n)_{n\in\mathbb{N}^*} \in \mathcal{X}\},$$

and let $\mathcal{A} : D(\mathcal{A}) \subseteq \mathcal{X} \to \mathcal{X}$ be defined by

$$\mathcal{A}(x_n)_{n\in\mathbb{N}^*} = (A_n x_n)_{n\in\mathbb{N}^*}.$$

We emphasize that, if $(x_n)_{n\in\mathbb{N}^*}$ is in $D(\mathcal{A})$, then

$$\lim_{n\to\infty} x_n = x_0,\ \lim_{n\to\infty} A_n x_n = y_0,\ x_0 \in D(A_0) \text{ and } A_0 x_0 = y_0.$$

Let $\lambda > 0$ and $(f_n)_{n\in\mathbb{N}^*} \in \mathcal{X}$. Then, for each $n \in \mathbb{N}^*$ the equation

$$\lambda x_n - A_n x_n = f_n$$

has a unique solution $x_n \in D(A_n)$ and, by virtue of (ii), the sequence $(x_n)_{n\in\mathbb{N}^*}$ belongs to $\mathcal{X}$. Hence $(0, +\infty) \subset \rho(\mathcal{A})$. By Theorem 3.1.1, we have

$$\|(\lambda I - \mathcal{A})^{-1}(f_n)_{n\in\mathbb{N}^*}\|_\infty \leq \sup_{n\geq 1}\|(\lambda I - A_n)^{-1} f_n\| \leq \frac{1}{\lambda}\|(f_n)_{n\in\mathbb{N}^*}\|.$$

It is easy to see that $D(\mathcal{A})$ is dense in $\mathcal{X}$. This follows from the fact that, for each $n \in \mathbb{N}^*$, A_n is densely defined, combined with the observation that a sequence $((x_n^k)_{n\in\mathbb{N}^*})_{k\in\mathbb{N}^*}$ in $\mathcal{X}$ is convergent in the norm $\|\cdot\|_\infty$ to $(x_n)_{n\in\mathbb{N}^*}$ if and only if all the sequences $(x_n^k)_{k\in\mathbb{N}^*}$ are convergent to x_n uniformly with respect to $n \in \mathbb{N}^*$. From Theorem 3.1.1, it follows that $\mathcal{A}$ generates a C_0-semigroup of contractions $\{S(t);\ t \geq 0\}$. Inasmuch as the operator $\mathcal{A}$ is of "diagonal" type, we conclude that $(\lambda \mathcal{I} - \mathcal{A})^{-1} = ((\lambda I - A_n)^{-1})_{n \in \mathbb{N}^*}$. Hence $\mathcal{A}_\lambda = (A_{n\lambda})_{n \in \mathbb{N}^*}$. Then it follows that $S(t)(x_n)_{n \in \mathbb{N}^*} = (S_n(t)x_n)_{n \in \mathbb{N}^*}$. From the definition of the space $\mathcal{X}$, we deduce $\lim_{n \to \infty} S_n(t)x_n = S_0(t)x_0$. The proof is complete. □

Definition 5.1.1. A *center of the operator A* is any subspace Y of $D(A)$ such that graph A is the closure in $X \times X$ of the graph of the restriction of A to Y, i.e. $\operatorname{graph} A = \overline{\operatorname{graph} A_Y}$.

One may easily see that, in general, the center of a closed operator is not unique.

Definition 5.1.2. The *inferior limit* of a sequence of sets $(D_n)_{n \in \mathbb{N}^*}$ is the set

$$\liminf_{n \to \infty} D_n = \bigcup_{n=1}^{\infty} \bigcap_{m=n}^{\infty} D_m.$$

Theorem 5.1.2. *Let $\{A_n;\ n \in \mathbb{N}\} \subset \mathcal{G}(X)$, and let $\{S_n(t);\ t \geq 0\}$ be the C_0-semigroup of contractions generated by A_n, for $n = 0, 1, \ldots$. Let us assume that there exists a center Y_0 of the operator A_0, such that*

$$Y_0 \subset \liminf_{n \to \infty} D(A_n) \text{ and } \lim_{n \to \infty} A_n y = A_0 y \tag{5.1.1}$$

for each $y \in Y_0$. Then, for each $x \in X$,

$$\lim_{n \to \infty} S_n(t)x = S_0(t)x,$$

uniformly for t in compact subsets in $(0, +\infty)$.

Proof. By virtue of (ii) in Theorem 5.1.1, it suffices to show that, for each $\lambda > 0$ and each $x \in X$, we have

$$\lim_{n \to \infty} (\lambda I - A_n)^{-1} x = (\lambda I - A_0)^{-1} x.$$

Since Y_0 is a center of the operator A_0, $(\lambda I - A_0)Y_0$ is dense in X, and therefore it suffices to prove the equality above merely for $x \in (\lambda I - A_0)Y_0$. Indeed, this follows from the inequality

$$\begin{aligned}
&\|(\lambda I - A_n)^{-1}x - (\lambda I - A_0)^{-1}x\| \\
&\leq \|(\lambda I - A_n)^{-1}x - (\lambda I - A_n)^{-1}x_\varepsilon\| + \|(\lambda I - A_n)^{-1}x_\varepsilon - (\lambda I - A_0)^{-1}x_\varepsilon\| \\
&\quad + \|(\lambda I - A_0)^{-1}x_\varepsilon - (\lambda I - A_0)^{-1}x\| \\
&\leq \frac{2}{\lambda}\|x - x_\varepsilon\| + \|(\lambda I - A_n)^{-1}x_\varepsilon - (\lambda I - A_0)^{-1}x_\varepsilon\|
\end{aligned}$$

and from the fact that, for each $x \in X$ and each $\varepsilon > 0$, there exists $x_\varepsilon \in (\lambda I - A_0)Y_0$ with $\|x - x_\varepsilon\| \le \varepsilon$. Let $x = (\lambda I - A_0)y$ with $y \in Y_0$. We have

$$\|(\lambda I - A_n)^{-1}x - (\lambda I - A_0)^{-1}x\| = \|(\lambda I - A_n)^{-1}(\lambda I - A_0 - \lambda I + A_n)y\|$$
$$= \|(\lambda I - A_n)^{-1}(A_n - A_0)y\| \le \frac{1}{\lambda}\|(A_n - A_0)y\|.$$

By (5.1.1), it follows that

$$\lim_{n\to\infty} \|(\lambda I - A_n)^{-1}x - (\lambda I - A_0)^{-1}x\| = 0,$$

and this achieves the proof. □

Lemma 5.1.1. *Let $A : D(A) \subseteq X \to X$ be a densely defined, linear, closed operator, and let $\lambda \in \rho(A)$. Then $\mu \in \rho(A)$ if and only if the operator $I - (\mu - \lambda)(\mu I - A)^{-1}$ is invertible with the inverse in $\mathcal{L}(X)$, case in which*

$$(\lambda I - A)^{-1} = (\mu I - A)^{-1}\left[I - (\mu - \lambda)(\mu I - A)^{-1}\right]^{-1}. \tag{5.1.2}$$

Proof. Let us assume that $B = I - (\mu - \lambda)(\mu I - A)^{-1}$ is invertible and, in addition, $B^{-1} \in \mathcal{L}(X)$. Then

$$(\lambda I - A)(\mu I - A)^{-1}B^{-1} = [(\lambda - \mu)I + (\mu I - A)]\,(\mu I - A)^{-1}B^{-1}$$
$$= \left[(\lambda - \mu)(\mu I - A)^{-1} + I\right]B^{-1} = I.$$

Since $(\lambda I - A)$ commutes with both $(\mu I - A)^{-1}$ and B^{-1}, by the preceding equality, we deduce that

$$(\mu I - A)^{-1}B^{-1}(\lambda I - A) = I_{|D(A)}.$$

Therefore $\mu \in \rho(A)$, and (5.1.2) holds. Conversely, if $\mu \in \rho(A)$, then

$$(\lambda I - A)(\mu I - A)^{-1} = \lambda(\mu I - A)^{-1} + [(\mu I - A) - \mu I](\mu I - A)^{-1} = B,$$

from where it follows that B is invertible. As

$$B^{-1} = (\mu - \lambda)(\lambda I - A)^{-1} + I,$$

it follows that $B^{-1} \in \mathcal{L}(X)$, and (5.1.2) holds. The proof is complete. □

Theorem 5.1.3. (Trotter, Neveu, Kato) *Let $\{A_n;\ n \in \mathbb{N}\} \subset \mathcal{G}(X)$, and let $\{S_n(t);\ t \ge 0\}$ be the C_0-semigroup of contractions generated by A_n, for $n = 0, 1, \ldots$. Let us assume that, for some fixed $\lambda > 0$, there exists an injective operator $R_\lambda \in \mathcal{L}(X)$, with dense range in X, and such that*

$$\lim_{n\to\infty} (\lambda I - A_n)^{-1}x = R_\lambda x \tag{5.1.3}$$

for $x \in X$. Then there exists $A_0 \in \mathcal{G}(X)$ so that $R_\lambda = (\lambda I - A_0)^{-1}$, and

$$\lim_{n\to\infty} S_n(t)x = S_0(t)x$$

for each $x \in X$, and uniformly for t in compact subsets in $(0, +\infty)$.

Proof. Let $\lambda > 0$, $\mu > 0$ and $n \in \mathbb{N}^*$. Inasmuch as $\lambda, \mu \in \rho(A_n)$, from Lemma 5.1.1 we deduce

$$(\mu I - A_n)^{-1}x = (\lambda I - A_n)^{-1}\sum_{k=0}^{\infty}(\lambda - \mu)^k(\lambda I - A_n)^{-k}x$$

for each $x \in X$. For fixed $\lambda > 0$ and $\varepsilon \in (0, \lambda)$, the series above converges uniformly for $|\lambda - \mu| \leq \lambda - \varepsilon$. From this remark, and from (5.1.3), it follows that, for each $\mu > 0$ with $|\lambda - \mu| < \lambda$, there exists $R_\mu \in \mathcal{L}(X)$ such that

$$\lim_{n\to\infty}(\mu I - A_n)^{-1}x = R_\mu x \tag{5.1.4}$$

for each $x \in X$. We shall show next that the set S, of all numbers μ in $\mathbb{R}_+^*$ for which (5.1.4) holds, coincides with $\mathbb{R}_+^*$. To this aim, we shall prove that S is both open and closed in $\mathbb{R}_+^*$. Clearly S is open. Let $\mu \in \mathbb{R}_+^* \setminus S$ and let $\lambda > 0$ be such that $|\mu - \lambda| < \frac{1}{2}\mu$, or equivalently $\lambda \in (\frac{1}{2}\mu, \frac{3}{2}\mu)$. Since $|\lambda - \mu| < \frac{1}{2}\mu < \lambda$, it follows that $\lambda \in \mathbb{R}_+^* \setminus S$, because otherwise $\mu \in S$ thereby contradicting the assumption $\mu \in \mathbb{R}_+^* \setminus S$. So

$$\left\{\lambda \in \mathbb{R}_+^*;\ |\mu - \lambda| < \frac{1}{2}\mu\right\} \subset \{\lambda \in \mathbb{R}_+^*;\ |\mu - \lambda| < \lambda\} \subset \mathbb{R}_+^* \setminus S.$$

Consequently $\mathbb{R}_+^* \setminus S$ is open and therefore $S = \mathbb{R}_+^*$. From this remark it follows that all the operators R_μ in (5.1.4) are well-defined for each $\mu > 0$. In addition, from the resolvent equation (1.7.3), and from (5.1.4), it follows by passing to the limit that

$$R_\lambda - R_\mu = (\mu - \lambda)R_\lambda R_\mu,$$

and $R_\lambda R_\mu = R_\mu R_\lambda$ for each $\lambda, \mu \in \mathbb{R}_+^*$. Let us denote by $\mathcal{N}$ and $\mathcal{R}$ the kernel, and respectively the range of the operator R_μ. From the fact that $R_\mu = R_\lambda[I - (\mu - \lambda)R_\mu]$, it follows that the range of R_μ includes the range of R_λ. Interchanging λ with μ, we deduce that $\mathcal{R}$ is independent of $\mu \in \mathbb{R}_+^*$. Analogously, one proves that $\mathcal{N}$ is independent of $\mu \in \mathbb{R}_+^*$. As R_λ is injective and with dense range, it follows that $\mathcal{N} = \{0\}$ and $\mathcal{R}$ is dense in X. Let us define the operator $A_0 : D(A_0) \subseteq X \to X$ by $D(A_0) = \mathcal{R}$ and $A_0 = \lambda I - R_\lambda^{-1}$, where $\lambda > 0$ has the specified properties. We have

$$(\lambda I - A_0)R_\lambda = R_\lambda(\lambda I - A_0) = I$$

on $D(A_0)$. In addition,

$$\begin{aligned}(\mu I - A_0)R_\mu &= [(\mu - \lambda)I + (\lambda I - A_0)]R_\mu \\ &= [(\mu - \lambda)I + (\lambda I - A_0)]R_\lambda[I - (\mu - \lambda)R_\mu] \\ &= I + (\mu - \lambda)[R_\lambda - R_\mu - (\mu - \lambda)R_\mu R_\lambda)] = I.\end{aligned}$$

Similarly, we have $R_\mu(\mu I - A_0) = I$ on $\mathcal{R}$. Consequently

$$R_\mu = (\mu I - A_0)^{-1}$$

for each $\mu > 0$. Finally,

$$\|(\mu I - A_0)^{-1}x\| = \lim_{n\to\infty} \|(\mu I - A_n)^{-1}x\| \leq \frac{1}{\mu}\|x\|$$

for each $\mu > 0$ and each $x \in X$, showing that A_0 satisfies the hypotheses of Theorem 3.1.1. So, A_0 generates a C_0-semigroup of contractions. Since the conclusion follows from Theorem 5.1.1, the proof is complete. □

5.2. The Chernoff and Lie-Trotter Formulae

The aim of this section is to establish two approximation formulae for C_0-semigroups of contractions which will prove very useful in applications. We begin with the following fundamental result known as the *Chernoff product formula*

Theorem 5.2.1. (Chernoff) *Let X be a Banach space, real or complex, and let $\{V(t);\ t \geq 0\}$ be a family of contractions on X with $V(0) = I$. Let us assume that there exists $A : D(A) \subseteq X \to X$ which generates a C_0-semigroup of contractions $\{S(t);\ t \geq 0\}$ and, in addition, for each $x \in D(A)$, there exists*

$$\lim_{h\downarrow 0} \frac{1}{h}(V(h) - I)x = Ax.$$

Then, for each $x \in X$, we have

$$\lim_{n\to\infty} V^n(t/n)x = S(t)x, \tag{5.2.1}$$

uniformly for t in compact subsets in $\mathbb{R}_+$.

The proof of Theorem 5.2.1 is essentially based on the next lemma.

Lemma 5.2.1. *If $L \in \mathcal{L}(X)$ satisfies $\|L\|_{\mathcal{L}(X)} \leq 1$, then*

$$\|e^{n(L-I)}x - L^n x\| \leq \sqrt{n}\|Lx - x\|$$

for each $n \in \mathbb{N}^*$, and each $x \in X$.

Proof. From the definition of the operator $e^{n(L-I)}$, it follows that

$$\left\|e^{n(L-I)}x - L^n x\right\| = e^{-n}\left\|\sum_{k=0}^{\infty} \frac{n^k}{k!}(L^k - L^n)x\right\|$$

$$\leq e^{-n}\sum_{k=0}^{\infty} \frac{n^k}{k!}\|L^k x - L^n x\|.$$

Since

$$\|L^k x - L^n x\| \leq \|L^{|k-n|}x - x\| \text{ and } L^m - I = \sum_{i=1}^{m}(L^i - L^{i-1}),$$

we have

$$\|L^m x - x\| \leq \sum_{i=1}^{m} \|L^i x - L^{i-1}x\| \leq m\|Lx - x\|.$$

From these relations and the Cauchy-Schwarz inequality, we deduce

$$\left\|e^{n(L-I)}x - L^n x\right\| \leq e^{-n}\sum_{k=0}^{\infty}\frac{n^k}{k!}|k-n|\|Lx - x\|$$

$$\leq e^{-n}\left\{\sum_{k=0}^{\infty}\frac{n^k}{k!}\right\}^{1/2}\left\{\sum_{k=0}^{\infty}\frac{n^k}{k!}(k-n)^2\right\}^{1/2}\|Lx - x\|.$$

By the relation between the mean and dispersion of a Poisson distributed random variable (see for instance Laha and Rohatgi [**81**], Example 1.2.2, p. 31), we know that

$$\sum_{k=0}^{\infty}\frac{n^k}{k!}(k-n)^2 = ne^n,$$

from where we get the conclusion. □

We may now proceed to the proof of Theorem 5.2.1.

Proof. Let $(a_n)_{n\in\mathbb{N}^*}$ be a sequence in $\mathbb{R}_+$ convergent to 0, and let

$$A_n = \begin{cases} \dfrac{1}{a_n}\left[V(a_n) - I\right] & \text{if } a_n > 0 \\ \\ A & \text{if } a_n = 0. \end{cases}$$

Then, by Theorem 5.1.2, we have that, for each $x \in X$,

$$\lim_{n\to\infty}\left\|e^{tA_n}x - S(t)x\right\| = 0,$$

uniformly for t in compact subsets in $\mathbb{R}_+$. Let $t > 0$ be fixed, and let $a_n = t/n$ for each $n \in \mathbb{N}^*$. From Lemma 5.2.1 with $L = V(t/n)$, we obtain

$$\left\|e^{tA_n}x - V^n(t/n)x\right\| \leq \sqrt{n}\|V(t/n)x - x\| = \frac{t}{\sqrt{n}}\left\|\frac{n}{t}\left[V(t/n)x - x\right]\right\|,$$

which shows the pointwise convergence in (5.2.1). To conclude the proof, we have merely to check that, in (5.2.1), the convergence is uniform on compact subsets in $\mathbb{R}_+$. But this follows from the simple observation that, for each $[a, b] \subset \mathbb{R}_+$, we have

$$\frac{t}{\sqrt{n}}\left\|\frac{n}{t}\left[V(t/n)x - x\right]\right\| \leq \sup_{t\in[a,b]}\left\{\frac{t}{\sqrt{n}}\left\|\frac{n}{t}\left[V(t/n)x - x\right]\right\|\right\},$$

inequality whose right-hand side tends to 0 for n tending to ∞. □

The next two fundamental results are simple consequences of Theorem 5.2.1.

Theorem 5.2.2. *Let $\{S(t);\ t \geq 0\}$ be a C_0-semigroup of contractions and let $A : D(A) \subseteq X \to X$ be its infinitesimal generator. Then, for each $x \in X$, we have*

$$\lim_{n\to\infty}\left(I - \frac{t}{n}A\right)^{-n} x = S(t)x,$$ [1]

uniformly for t in compact subsets in $\mathbb{R}_+$.

Proof. One applies Theorem 5.2.1 to the family $V(t) = (I - tA)^{-1}$ for $t \geq 0$. □

Definition 5.2.1. An operator $L : D(L) \subseteq X \to X$ is called *closable*, if the closure of its graph in $X \times X$ is the graph of a linear operator. If L is a closable operator, *its closure*, denoted by $\bar{L}$, is the operator whose graph is the closure of the graph of L.

Theorem 5.2.3. *Let $A : D(A) \subseteq X \to X$ and $B : D(B) \subseteq X \to X$ be the generators of two C_0-semigroups of contractions $\{S(t);\ t \geq 0\}$, and respectively $\{T(t);\ t \geq 0\}$. Let us assume that $A + B$ is closable and its closure, $\overline{A+B} : D(\overline{A+B}) \subseteq X \to X$, is the infinitesimal generator of a C_0-semigroup of contractions $\{U(t);\ t \geq 0\}$. Then, for each $x \in X$, we have*

$$\lim_{n\to\infty}\{S(t/n)T(t/n)\}^n x = U(t)x,$$ [2]

uniformly for t in compact subsets in $\mathbb{R}_+$.

Proof. One applies Theorem 5.2.1 to the family $V(t) = S(t)T(t)$ for $t \geq 0$ and to the operator $\overline{A+B} : D(\overline{A+B}) \subseteq X \to X$. □

---

[1] This is the famous *Hille's exponential formula.*

[2] This is the celebrated *Lie-Trotter formula.*

5.3. A Perturbation Result

In this section we prove a sufficient condition in order for the sum of an infinitesimal generator of a C_0-semigroup of contractions with a densely defined linear operator to also be the infinitesimal generator of a C_0-semigroup of contractions.

Theorem 5.3.1. *Let $A : D(A) \subseteq X \to X$, $B : D(B) \subseteq X \to X$ be two linear operators with $D(A) \subseteq D(B)$, and such that*

$$\|\lambda x - (A + tB)x\| \geq \lambda \|x\| \tag{5.3.1}$$

for all $\lambda > 0$, $t \in [0, 1]$ and $x \in D(A)$, and

$$\|Bx\| \leq \alpha \|Ax\| + \beta \|x\| \tag{5.3.2}$$

for each $x \in D(A)$, where $\alpha \in [0, 1)$ and $\beta > 0$. If there exists s in $[0, 1]$ such that $A + sB$ is closed, and $(0, +\infty) \subseteq \rho(A + sB)$, then we have $(0, +\infty) \subseteq \rho(A + tB)$ for each $t \in [0, 1]$. In this case, for each $t \in [0, 1]$, $A + tB$ is the infinitesimal generator of a C_0-semigroup of contractions.

Proof. Let $s \in [0, 1]$ be such that, for each $\lambda > 0$, $\lambda I - (A + sB)$ is invertible, and $(\lambda I - (A + sB))^{-1} \in \mathcal{L}(X)$. To complete the proof, it suffices to show that there exists $\delta > 0$ such that, for each $t \in [0, 1]$ with $|t - s| \leq \delta$, we have $(0, +\infty) \subseteq \rho(A + tB)$. From (5.3.1), it follows that $\|R(\lambda; A + sB)\|_{\mathcal{L}(X)} \leq \frac{1}{\lambda}$. We shall show next that $BR(\lambda; A + sB) \in \mathcal{L}(X)$. From (5.3.2), it follows that

$$\begin{aligned}\|Bx\| &\leq \alpha \|Ax\| + \beta \|x\| \leq \alpha \|(A + sB)x\| + \alpha s \|Bx\| + \beta \|x\| \\ &\leq \alpha \|(A + sB)x\| + \alpha \|Bx\| + \beta \|x\|\end{aligned}$$

for each $x \in D(A)$. Accordingly

$$\|Bx\| \leq \frac{\alpha}{1 - \alpha} \|(A + sB)x\| + \frac{\beta}{1 - \alpha} \|x\|.$$

Since $R(\lambda; A + sB) : X \to D(A)$ and

$$(A + sB)R(\lambda; A + sB) = \lambda R(\lambda; A + sB) - I,$$

from the last inequality, it follows that

$$\begin{aligned}&\|BR(\lambda; A + sB)x\| \\ &\leq \frac{\alpha}{1 - \alpha} \|(\lambda R(\lambda; A + sB) - I)x\| + \frac{\beta}{1 - \alpha} \|R(\lambda; A + sB)x\| \\ &\leq \frac{2\lambda\alpha + \beta}{\lambda(1 - \alpha)} \|x\|\end{aligned}$$

for each $x \in X$. Consequently $BR(\lambda; A+sB) \in \mathcal{L}(X)$, and

$$\|BR(\lambda; A+sB)\|_{\mathcal{L}(X)} \leq \frac{2\lambda\alpha+\beta}{\lambda(1-\alpha)}.$$

At this point let us observe that

$$\lambda I - (A+tB) = \lambda I - (A+sB) + (s-t)B$$

$$= (I + (s-t)BR(\lambda; A+sB))(\lambda I - (A+sB)).$$

Then $\lambda I - (A+tB)$ is invertible if and only if $I + (s-t)BR(\lambda; A+sB)$ enjoys the same property. But the latter is invertible for each $t \in [0,1]$ with

$$|t-s| < \lambda(1-\alpha)(2\alpha\lambda+\beta)^{-1} \leq \|BR(\lambda; A+sB)\|^{-1}_{\mathcal{L}(X)}.$$

Taking $\delta < \lambda(1-\alpha)(2\alpha\lambda+\beta)^{-1}$, let us observe that $\lambda I-(A+tB)$ is invertible for each $t \in [0,1]$ with $|t-s| \leq \delta$. Since $[0,1]$ can be covered by a finite union of intervals whose length is less that δ, the proof is complete. □

5.4. The central limit theorem

In order to illustrate the power of Chernoff Theorem 5.2.1, in this section, we will give a proof of one of the most famous result in Probability Theory and Statistics: the central limit theorem. We begin by recalling that a random variable ξ defined on a probability space (Ω, Σ, p) has a *normal distribution of mean* 0 *and dispersion* $r > 0$, if its repartition, or cumulative probability, $F_\xi(t) = p(\xi \leq t)$ is given by

$$F_\xi(t) = F_{N(0,r)}(t) = \frac{1}{\sqrt{2\pi r}} \int_{-\infty}^{t} \exp\left\{-\frac{s^2}{2r}\right\} ds$$

for each $t \in \mathbb{R}$. One may easily see that the two parameters 0 and r are the mean, and respectively the dispersion of the variable ξ.

Theorem 5.4.1. (The central limit theorem) *Let $(\xi_n)_{n\in\mathbb{N}^*}$ be a sequence of random variables having normal distribution of mean* 0 *and dispersion* 1. *Then, for each $x \in \mathbb{R}$, we have*

$$\lim_{n\to\infty} F_{\eta_n}(x) = F_{N(0,1)}(x),$$

where, for each $n \in \mathbb{N}^*$ and $x \in \Omega$,

$$\eta_n(\omega) = \frac{1}{\sqrt{n}} \sum_{k=1}^{n} \xi_k(\omega).$$

In order to prove Theorem 5.4.1, some preliminaries are needed. Let $\mathfrak{F}$ be the set of all repartition functions and let $X = C_{ub}(\mathbb{R})$. For each $F \in \mathfrak{F}$, we define $\widetilde{F} : X \to X$ by the convolution $\widetilde{F}f = f * dF$, i.e.

$$\widetilde{F}f(t) = \int_{-\infty}^{+\infty} f(t-s)\, dF(s)$$

for each $f \in X$ and $t \in \mathbb{R}$. Obviously $\widetilde{F} \in \mathcal{L}(X)$ and $\|\widetilde{F}\|_{\mathcal{L}(X)} \leq 1$.

Lemma 5.4.1. *Let* $(F_n)_{n\in\mathbb{N}^*}$, *and* F *in* $\mathfrak{F}$. *Then* $\lim_{n\to\infty} F_n(x) = F(x)$ *at each point* x *of continuity for* F *if and only if* $\lim_{n\to\infty} \widetilde{F}_n f = \widetilde{F}f$ *for each* $f \in X$.

Lemma 5.4.2. *For each* $F, G \in \mathfrak{F}$, *we have*

$$\widetilde{F * G} = \widetilde{F}\widetilde{G}.$$

We may now proceed to the proof of Theorem 5.4.1.

Proof. By virtue of Lemma 5.4.1, in order to prove the theorem, it suffices to show that

$$\lim_{n\to\infty} \widetilde{F}_{\eta_n} f = \widetilde{F}_{N(0,1)} f$$

for each f in a dense subset in X. Let G be the common repartition function of all random variables ξ_n, let $G_r(x) = G(\sqrt{r}x)$ for $r > 0$ and $x \in \mathbb{R}$, and let $V(t) = \widetilde{G}_{1/t}$ for $t > 0$, and $V(0) = I$. Finally, let $A : D(A) \subseteq X \to X$ be defined by

$$\begin{cases} D(A) = \{f \in X;\ f', f'' \in X\} \\ Au = \frac{1}{2}u'' \end{cases}$$

for each $u \in D(A)$. One may easily prove that A generates a C_0-semigroup of contractions $\{S(t);\ t \geq 0\}$ on X, and, moreover, that

$$[S(t)f](x) = \frac{1}{\sqrt{2\pi t}} \int_{-\infty}^{\infty} \exp\left\{-\frac{y^2}{2t}\right\} f(x-y)\, dy,$$

i.e. $S(t) = \widetilde{F}_{N(0,t)}$, for each $t > 0$.
We shall show next that

$$\lim_{h\downarrow 0} \frac{1}{h}[V(h)f - f] = Af \tag{5.4.1}$$

for each $f \in D(A)$. Assuming for the moment that (5.4.1) holds, from Chernoff product formula, it follows that

$$\lim_{n\to\infty} V^n(t/n)f = \widetilde{F}_{N(0,t)} f \tag{5.4.2}$$

for each $f \in D(A)$ and $t \geq 0$. Taking $t = 1$, we observe that, by Lemma 5.4.2, we have

$$V^n(1/n) = \widetilde{G}_n^n = \widetilde{H}_n,$$

where $\widetilde{H}_n$ is the repartition function of the sum of n random independent variables, each one having the repartition function G_n, while

$$G_n^n = G_n * G_n * \cdots * G_n.$$

Since G_n is the repartition function of the random variable $\frac{1}{\sqrt{n}}\xi_i$, we have $H_n = F_{\eta_n}$, which implies the conclusion.
Thus, it remains to prove (5.4.1). Let us observe that

$$\int_{-\infty}^{\infty} dG_s(r) = 1.$$

In addition, since $G_s = F_{\xi_i/\sqrt{s}}$, we deduce

$$\int_{-\infty}^{\infty} r\, dG_s(r) = E\left(\frac{\xi_i}{\sqrt{s}}\right) = 0,$$

$$\int_{-\infty}^{\infty} r^2 dG_s(r) = E\left[\left(\frac{\xi_i}{\sqrt{s}}\right)^2\right] = \frac{1}{s}.$$

Then, for each $f \in D(A)$ and each $\delta > 0$, we have

$$\left\{\frac{1}{t}\left[V(t)f - f\right] - Af\right\}(x) = \frac{1}{t}\left[V(t)f - f\right](x) - \frac{1}{2}f''(x)$$

$$= \int_{-\infty}^{\infty}\left[\frac{f(x-r) - f(x)}{t}\right] dG_{1/t}(r) - \frac{1}{2}f''(x)$$

$$= \int_{-\infty}^{\infty}\left[\frac{f(x-r) - f(x) + rf'(x) - \frac{r^2}{2}f''(x)}{t}\right] dG_{1/t}(r)$$

$$= \int_{|r|<\delta} \frac{r^2}{2t}[f''(\theta) - f''(x)]\, dG_{1/t}(r) + \int_{|r|\geq\delta} \frac{r^2}{2t}[f''(\theta) - f''(x)]\, dG_{1/t}(r)$$

$$= J_1 + J_2,$$

where θ is between x and $x - r$. Let $\varepsilon > 0$. Inasmuch as f'' is uniformly continuous on $\mathbb{R}$ there exists $\delta > 0$ such that $|f''(\theta) - f''(x)| < \varepsilon/2$ for each $\theta, x \in \mathbb{R}$ with $|\theta - x| < \delta$. Fix $\delta > 0$ with property above, and let us observe that

$$|J_1| \leq \frac{\varepsilon}{2t}\int_{|r|<\delta} r^2 dG_{1/t}(r) \leq \frac{\varepsilon}{2}$$

for each $t > 0$. Since $|f''(\theta) - f''(x)| \leq 2\|f''\|_\infty$, we have

$$|J_2| \leq \int_{|r|\geq\delta} \frac{2\|f''\|_\infty}{2t} r^2 dG_{1/t}(r)$$

$$= \|f''\|_\infty \int_{|r/\sqrt{t}|\geq\delta/\sqrt{t}} \left(\frac{r}{\sqrt{t}}\right)^2 dG\left(\frac{r}{\sqrt{t}}\right)$$

$$= \|f''\|_\infty \int_{|s|\geq\delta/\sqrt{t}} s^2 dG(s).$$

From this inequality and from the fact that $\int_{-\infty}^{\infty} s^2 dG(s) = 1$, we deduce that

$$\lim_{t\downarrow 0} \int_{|s|\geq\delta/\sqrt{t}} s^2 dG(s) = 0.$$

Therefore, for the very same $\varepsilon > 0$, there exists $t(\varepsilon, f) > 0$ such that, for each $t \in (0, t(\varepsilon, f))$, $|J_2| \leq \varepsilon/2$. Consequently we have

$$\left|\left\{\frac{1}{t}\left[V(t)f - f\right] - Af\right\}(x)\right| \leq |J_1| + |J_2| < \varepsilon$$

for each $x \in \mathbb{R}$ and each $t \in (0, t(\varepsilon, f))$. The proof is complete. □

5.5. Feynman Formula

Our aim here is to give a rigorous interpretation of the celebrated Feynman formula by using the Lie-Trotter formula. Let us consider a nonrelativistic particle without spin, of mass m, moving in $\mathbb{R}^3$ under the influence of a potential V. We notice that the *wave function* of the particle is the solution $u = u(t, x)$ of the Cauchy problem for the Schrödinger equation

$$\begin{cases} u_t = i\left[\frac{1}{2m}\Delta u - V(x)u\right] & (t, x) \in \mathbb{R} \times \mathbb{R}^3 \\ u(0, x) = u_0(x) & x \in \mathbb{R}^3, \end{cases}$$

where

$$\int_{R^3} \|u_0(x)\|^2 \, dx = 1.$$

From a physical point of view, $\|u(t, x)\|^2$ represents the probability density of the position of the particle at the time t, and at the point x. More precisely, $\int_\Omega \|u(t, x)\|^2 \, dx$ is the probability that, at the time t, the particle lies in the Borel set Ω in $\mathbb{R}^3$. Analogously, $\|\tilde{u}(t, \xi)\|^2$ is the momentum of the probability density, where $\tilde{u}$ is the Fourier transform of u, i.e.

$$\tilde{u}(t, \xi) = \frac{1}{(2\pi)^{3/2}} \int_{\mathbb{R}^3} e^{-i\langle\xi,\omega\rangle} u(t, \omega) \, d\omega.$$

Let us consider the Hilbert space $H = L^2(\mathbb{R}^3)$, and let us identify u with a function from $\mathbb{R}$ to H satisfying

$$\begin{cases} u' = (A+B)u \\ u(0) = u_0, \end{cases}$$

where $A : D(A) \subseteq H \to H$ and $B : D(B) \subseteq H \to H$ are defined by

$$\begin{cases} D(A) = H^2(\mathbb{R}^3) \\ Au = (i/2m)\Delta u \end{cases} \tag{5.5.1}$$

and, respectively by

$$\begin{cases} D(B) = \{u \in L^2(\mathbb{R}^3);\ V(\,\cdot\,)u \in L^2(\mathbb{R}^3)\} \\ Bu = -iV(\,\cdot\,)u. \end{cases} \tag{5.5.2}$$

Lemma 5.5.1. *The operator A, defined by* (5.5.1), *is the generator of a C_0-group of unitary operators $\{S(t) : L^2(\mathbb{R}^3) \to L^2(\mathbb{R}^3);\ t \in \mathbb{R}\}$, defined by*

$$[S(t)u](x) = \left(\frac{2\pi it}{m}\right)^{-3/2} \int_{\mathbb{R}^3} \exp\left\{\frac{im\|x-\xi\|^2}{2t}\right\} u(\xi)\,d\xi, \tag{5.5.3}$$

for each $t > 0$ and $u \in L^2(\mathbb{R}^3)$.

Proof. We begin by proving that iA is self-adjoint. Let us observe that, for each $u, v \in H^2(\mathbb{R}^3)$, we have

$$\begin{aligned} \langle iAu, v\rangle = \langle -\frac{1}{2m}\Delta u, v\rangle &= -\int_{\mathbb{R}^3} \frac{1}{2m}\Delta u\overline{v}\,d\omega \\ &= -\int_{\mathbb{R}^3} \frac{1}{2m} u\overline{\Delta v}\,d\omega = \langle u, iAv\rangle, \end{aligned}$$

which shows that iA is symmetric. Since $C_0^\infty(\mathbb{R}^3) \subseteq D(iA)$, iA is densely defined and so, by virtue of (i) Lemma 1.6.1, in order to complete the proof, it suffices to check that $(I - iA)^{-1} \in \mathcal{L}(X)$. We begin by showing that the range of $I - iA$ is dense in H. More precisely, we will prove that $C_0^\infty(\mathbb{R}^3)$ is included in the range of $I - iA$. Let $f \in C_0^\infty(\mathbb{R}^3)$, and let us consider the problem $u - iAu = f$, which rewrites equivalently under the form

$$u + \frac{1}{2m}\Delta u = f.$$

Then u, given by

$$u(x) = \frac{1}{(2\pi)^{3/2}} \int_{\mathbb{R}^3} \frac{2m\tilde{f}(\xi)e^{i\langle x,\xi\rangle}}{2m + \|\xi\|^2}\,d\xi,$$

is a solution of the problem above. From Stone Theorem 3.9.1 it follows that A generates a C_0-group of unitary operators. Since formula (5.5.3) follows by direct computation, the proof is complete. □

Lemma 5.5.2. *If $V \in L^2(\mathbb{R}^3)$, then B, defined by (5.5.2), is the generator of a C_0-group of unitary operators, $\{T(t) : L^2(\mathbb{R}^3) \to L^2(\mathbb{R}^3),\ t \in \mathbb{R}\}$, defined by*

$$[T(t)u](x) = e^{-itV(x)}u(x), \tag{5.5.4}$$

for each $u \in L^2(\mathbb{R}^3)$.

Proof. First, let us observe that iB is symmetric, $I - iB$ is invertible and $(I - iB)^{-1} \in \mathcal{L}(H)$. By (i) in Lemma 1.6.1, it follows that iB is self-adjoint, while, by Stone Theorem 3.9.1, B is the infinitesimal generator of a C_0-group of unitary operators, $\{T(t);\ t \in \mathbb{R}\}$, defined by (5.5.4). □

Let $t > 0$, $n \in \mathbb{N}^*$, and let us consider $U_n(t) : L^2(\mathbb{R}^3) \to L^2(\mathbb{R}^3)$ defined by

$$U_n(t) = [S(t/n)T(t/n)]^n .$$

A simple mathematical induction argument shows that

$$U_n(t)u(x)$$

$$= \left(\frac{2\pi it}{nm}\right)^{-3/2} \int_{\mathbb{R}^3} \cdots \int_{\mathbb{R}^3} \exp\{iS(x_0, x_1, \ldots, x_n; t)\}u(x_n)\, dx_1 \ldots dx_n, \tag{5.5.5}$$

where $x_0 = x \in \mathbb{R}^3$ and

$$S(x_0, x_1, \ldots, x_n; t) = \sum_{j=1}^{m} \left[\frac{m|x_j - x_{j-1}|^2}{2(t/n)^2} - V(x_j)\right](t/n). \tag{5.5.6}$$

Let C_x the set of all continuous functions $\omega : \mathbb{R}_+ \to \mathbb{R}^3$ with $\omega(0) = x$. Setting $x_j = \omega(tj/n)$, we observe that $S(x_0, x_1, \ldots, x_n; t)$ is nothing else than a Riemann sum for *the integral of action*

$$S(\omega; t) = \int_0^t \left\{\frac{m}{2}\|\dot{\omega}(s)\|^2 - V(\omega(s))\right\} ds,$$

where $\dot{\omega}$ is derivative of ω with respect to the t variable.
Passing to the limit formally both sides in (5.5.5) for $n \to \infty$, we "obtain" the famous *Feynman integral representation of the wave function*, i.e.

$$\lim_{n\to\infty} U_n(t)u(x) = k \int_{C_x} \exp\{iS(\omega; t)\}u(\omega(t))\, D\omega,$$

where k is constant. We emphasize that, this completely formal manner of approaching the problem has many vulnerable points. First, the constant k is in fact infinite. In addition, the expression $D\omega = \prod_{0\le s\le t} dx_s$, which ought to coincide with $\lim_{n\to\infty} \prod_{i=1}^{n} dx_i$, is meaningless in the context of the classical integration theory. More than this, $S(\omega; t)$ is defined only for a.e. differentiable functions ω. Summing up, none of the participants to

the definition of the expression above, i.e. the constant k, the integrand and the measure has sense in this frame. At this point, let us observe that, in the case in which $A+B$ is closable, and $\overline{A+B}$ generates a C_0-group of unitary operators on $L^2(\mathbb{R}^3)$, the Lie-Trotter formula (see Theorem 5.2.3) allows us to give a rigorous sense of the above expression by observing that, in fact, we can pass to the limit in (5.5.5) for $n \to \infty$ not pointwisely, but in $L^2(\mathbb{R}^3)$, and this limit is the wave function $u(t,x)$.
We conclude this section with a sufficient condition in order that $A+B$ be closable and $\overline{A+B}$ generate a C_0-group of unitary operators in $L^2(\mathbb{R}^3)$.

Lemma 5.5.3. *If $V \in L^p(\mathbb{R}^3)$ with $p \geq 2$, then, for each $\varepsilon > 0$, there exists $C(\varepsilon) > 0$ such that*

$$\|Bu\|_{L^2(\mathbb{R}^3)} \leq \varepsilon \|Au\|_{L^2(\mathbb{R}^3)} + C(\varepsilon)\|u\|_{L^2(\mathbb{R}^3)},$$

for each $u \in H^2(\mathbb{R}^3)$, where A and B are defined by (5.5.1), *and respectively by* (5.5.2).

Proof. If $u \in H^2(\mathbb{R}^3)$ then $(1+\|\xi\|^2)\hat{u}(\xi) \in L^2(\mathbb{R}^3)$. On the other hand, it is obvious that $(1+\|\xi\|^2)^{-1} \in L^p(\mathbb{R}^3)$. Take $q = 2p/(p+2)$ and let us observe that, from Parseval identity and Hölder's inequality, we have

$$\|\hat{u}\|_{L^q(\mathbb{R}^3)} = \left(\int_{\mathbb{R}^3} \|\hat{u}(\xi)\|^q \, d\xi\right)^{1/q}$$

$$= \left(\int_{\mathbb{R}^3} (1+\|\xi\|^2)^{-q}(1+\|\xi\|^2)^q \|\hat{u}(\xi)\|^q \, d\xi\right)^{1/q}$$

$$\leq \left(\int_{\mathbb{R}^3} (1+\|\xi\|^2)^{-p} d\xi\right)^{1/p} \left(\int_{\mathbb{R}^3} (1+\|\xi\|^2)^2 \|\hat{u}(\xi)\|^2 \, d\xi\right)^{1/2}$$

$$\leq C_p(\|\Delta u\|_{L^2(\mathbb{R}^3)} + \|u\|_{L^2(\mathbb{R}^3)}).$$

Since $p \geq 2$ and $1 \leq q \leq 2$, by Hausdorff-Young inequality[3], we get $\|u\|_{L^r(\mathbb{R}^3)} \leq \|\hat{u}\|_{L^q(\Omega)}$ provided r and q satisfy $1/r + 1/q = 1$. Therefore

$$\|u\|_{L^r(\Omega)} \leq C_p(\|\Delta u\|_{L^2(\Omega)} + \|u\|_{L^2(\Omega)}).$$

Substituting in the above inequality $u(x)$ by $u(\rho x)$, we obtain

$$\|u\|_{L^r(\mathbb{R}^3)} \leq C_p(\rho^2 \|\Delta u\|_{L^2(\mathbb{R}^3)} + \|u\|_{L^2(\mathbb{R}^3)}).$$

Let $\varepsilon > 0$. Let us choose $\rho > 0$ such that

$$\|u\|_{L^r(\mathbb{R}^3)}\|V\|_{L^p(\mathbb{R}^3)} \leq \varepsilon \|\Delta\|_{L^2(\mathbb{R}^3)} + C(\varepsilon)\|u\|_{L^2(\mathbb{R}^3)}.$$

---

[3]Hausdorff-Young inequality says that $\|f * g\|_{L^p(\mathbb{R}^n)} \leq \|f\|_{L^q(\mathbb{R}^n)}\|g\|_{L^r(\mathbb{R}^n)}$, provided $1 \leq p, q, r \leq \infty$, $1/p = 1/q + 1/r - 1$, $f \in L^q(\mathbb{R}^n)$ and $g \in L^r(\mathbb{R}^n)$. See for instance Stein and Weiss [**116**], p. 178.

Finally, by Hölder's inequality, we have $\|Vu\|_{L^2(\Omega)} \leq \|V\|_{L^p(\Omega)}\|u\|_{L^r(\Omega)}$ and so, from the preceding inequality, we deduce

$$\|Vu\|_{L^2(\mathbb{R}^3)} \leq \varepsilon\|\Delta u\|_{L^2(\mathbb{R}^3)} + C(\varepsilon)\|u\|_{L^2(\mathbb{R}^3)},$$

which achieves the proof. □

Theorem 5.5.1. *If $V \in L^2(\mathbb{R}^3)$, then $A+B$ if the generator of a C_0-group of unitary operators in $L^2(\mathbb{R}^3)$.*

Proof. Lemmas 5.5.1 and 5.5.2 imply that $i(A+B)$ is symmetric. In view of Theorem 1.6.1, in order to prove that it is self-adjoint, it suffices to show that $I \pm (A+B) \in \mathcal{L}(L^2(\mathbb{R}^3))$. This is a consequence of Lemma 5.5.3 combined with Theorem 5.3.1. We complete the proof with the help of Stone Theorem 3.9.1. □

5.6. The Mean Ergodic Theorem

We conclude with a result concerning the behavior as t tends to $+\infty$ of the means $\frac{1}{t}\int_0^t S(s)\,ds$ of a C_0-semigroup of contractions $\{S(t)\,;\ t \geq 0\}$.

Theorem 5.6.1. *Let $A : D(A) \subseteq X \to X$ be the infinitesimal generator of a C_0-semigroup of contractions, $\{S(t)\,;\ t \geq 0\}$. If $x_1 \in \ker(A)$ and $x_2 \in R(A)$, then there exists a bounded function $\varphi : \mathbb{R}_+ \to X$ with the property*

$$\lim_{t\to\infty} \frac{\varphi(t)}{t} = 0,$$

and such that, for each $t > 0$,

$$\frac{1}{t}\int_0^t S(\tau)(x_1 + x_2)\,d\tau = x_1 + \frac{\varphi(t)}{t}.$$

Proof. Let us observe first that, for each $x_1 \in \ker(A)$ and each $\tau \geq 0$, we have $S(\tau)x_1 = x_1$. As $x_2 \in R(A)$, there exists $x_3 \in D(A)$ such that $x_2 = Ax_3$. Accordingly

$$\frac{1}{t}\int_0^t S(\tau)(x_1 + x_2)\,d\tau = x_1 + \frac{1}{t}\int_0^t S(\tau)x_2\,d\tau,$$

and

$$\left\|\frac{1}{t}\int_0^t S(\tau)x_2\,d\tau\right\| = \left\|\frac{1}{t}\int_0^t S(\tau)Ax_3\,d\tau\right\|$$

$$= \left\|\frac{1}{t}\int_0^t \frac{d}{d\tau}[S(\tau)x_3]\,d\tau\right\| = \left\|\frac{1}{t}(S(t)x_3 - x_3)\right\| \leq \frac{2\|x_3\|}{t}.$$

So, the function $\varphi(t) = S(t)x_3 - x_3$ has all the required properties. The proof is complete. □

Let X be a Banach space and $B \subseteq X$. We denote by

$$B^{\perp} = \{x^* \in X^*\,;\ (x, x^*) = 0 \text{ for all } x \in B\}.$$

Similarly, if $B^* \subseteq X^*$, we denote by

$$(B^*)^{\perp} = \{x^{**} \in X^{**}\,;\ (x^*, x^{**}) = 0 \text{ for all } x^* \in B^*\}.$$

If X is reflexive and identified with its own bidual, then

$$(B^*)^{\perp} = \{x \in X\,;\ (x, x^*) = 0 \text{ for all } x^* \in B^*\}.$$

Lemma 5.6.1. *Let X be reflexive, and let $A : D(A) \subseteq X \to X$ be a densely defined, linear, closed operator. Then*

$$\ker(A)^{\perp} = \overline{R(A^*)} \quad R(A)^{\perp} = \ker A^*,$$

$$\ker(A^*)^{\perp} = \overline{R(A)} \quad R(A^*)^{\perp} = \ker(A).$$

Proof. We shall prove only the first equality, the proofs of the other ones being similar. Let $y^* \in \overline{R(A^*)}$. Then there exists $(x_n^*)_{n\in\mathbb{N}}$ in $D(A^*)$ such that $\lim_{n\to\infty} A^*x_n^* = y^*$. Thus, for each $x \in \ker(A)$, we have

$$(x, y^*) = \lim_{n\to\infty} (x, A^*x_n^*) = \lim_{n\to\infty} (Ax, x_n^*) = 0,$$

which shows that $y^* \in \ker(A)^{\perp}$. So $\overline{R(A^*)} \subseteq \ker(A)^{\perp}$. To prove the converse inclusion, let us assume by contradiction that there exists at least one $y^* \in \ker(A)^{\perp}$ such that $y^* \notin \overline{R(A^*)}$. By the Hahn-Banach theorem, it follows that there exists $x \in X$ (we recall that X is reflexive and identified with X^{**}) such that $(x, y^*) \neq 0$ and $(x, z^*) = 0$ for all $z^* \in \overline{R(A^*)}$. In particular, we have $(Ax, x^*) = (x, A^*x^*) = 0$ for each $x^* \in D(A^*)$. On the other hand, by Lemma 3.7.3, we know that $D(A^*)$ is dense in X^* and therefore $Ax = 0$. So $x \in \ker(A)$ and consequently $(x, y^*) = 0$ which is impossible. This contradiction can be eliminated only if $\ker(A)^{\perp} \subseteq \overline{R(A^*)}$ and this completes the proof. □

Theorem 5.6.2. (The Mean Ergodic Theorem) *Let X be reflexive and let $A : D(A) \subseteq X \to X$ be the generator of a C_0-semigroup of contractions, $\{S(t);\ t \geq 0\}$. Then* $\ker(A)$ *and* $\overline{R(A)}$ *are closed subspaces in X. Moreover,* $\ker(A) \cap \overline{R(A)} = \{0\}$, $X = \ker(A) \oplus \overline{R(A)}$, *and the projection operator $P : X \to \ker(A)$ is well-defined, and satisfies $\|P\|_{\mathcal{L}(X)} \leq 1$. In addition, each $x \in X$ has a unique decomposition $x = Px + (I - P)x \in \ker(A) \oplus \overline{R(A)}$, and*

$$\lim_{t\to\infty} \frac{1}{t} \int_0^t S(\tau)x\, d\tau = Px.$$

Proof. Theorem 5.6.1 implies that $\ker(A) \cap \overline{R(A)} = \{0\}$. Indeed, let $y \in \ker A \cap \overline{R(A)}$. Then there exists $(y_n)_{n\in\mathbb{N}}$ in $R(A)$ such that

$$\lim_{n\to\infty} y_n = y.$$

Since $y \in \ker(A)$ and $y_n \in R(A)$ for each $n \in \mathbb{N}$, from Theorem 5.6.1, we have

$$\lim_{t\to\infty} \frac{1}{t}\int_0^t S(\tau)y\,d\tau = y,$$

and

$$\lim_{t\to\infty} \frac{1}{t}\int_0^t S(\tau)y_n\,d\tau = 0$$

for each $n \in \mathbb{N}$. On the other hand

$$\left\|\frac{1}{t}\int_0^t S(\tau)y\,d\tau - \frac{1}{t}\int_0^t S(\tau)y_n d\tau\right\| \leq \frac{1}{t}\int_0^t \|y - y_n\|\,d\tau = \|y - y_n\|,$$

which implies

$$\lim_{t\to\infty} \frac{1}{t}\int_0^t S(\tau)y\,d\tau = 0.$$

So, $y = 0$, proving that $\ker(A) \cap \overline{R(A)} = \{0\}$.
Let $x \in \ker(A) \oplus \overline{R(A)}$. We have

$$\|Px\| = \lim_{t\to\infty}\left\|\frac{1}{t}\int_0^t S(\tau)x\,d\tau\right\| \leq \|x\|,$$

which shows that $\|P\|_{\mathcal{L}(X)} \leq 1$.
To complete the proof, we have to check that $\ker(A) \oplus R(A)$ is dense in X. So, let us assume by contradiction that there exist $x \in X \setminus \ker(A) \oplus \overline{R(A)}$ and $x^* \in X^*$, satisfying both $x^* \in (\ker(A) + R(A))^\perp$ and $(x, x^*) = 1$. Clearly $x^* \in \ker(A)^\perp$ and $x^* \in R(A)^\perp$. By Lemma 5.6.1, we deduce that $x^* \in \ker(A^*) \cap \overline{R(A^*)}$. On the other hand, since X is reflexive, from Theorem 3.5.1, we know that A^* is the generator of the C_0-semigroup of contractions $\{S^*(t)\,;\,t \geq 0\}$ in X^*. From the considerations above, we have $\ker(A^*) \cap \overline{R(A^*)} = \{0\}$, and therefore $x^* = 0$. This contradiction can be eliminated only if $\ker(A) \oplus \overline{R(A)} = X$, and this achieves the proof. □

In order to get a variant of Theorem 5.6.2 valid in general Banach spaces, we need the following lemma.

Lemma 5.6.2. *Let $u : \mathbb{R}_+ \to X$ be a bounded and continuous function. Then*

$$\lim_{t\to\infty} \frac{1}{t}\int_0^t u(\tau)\,d\tau = \lim_{\lambda\downarrow 0} \lambda\int_0^\infty e^{-\lambda s}u(s)\,ds,$$

in the sense that, either both limits exist and they are equal, or both do not exist.

Proof. Let us assume for the beginning that there exists

$$\lim_{t\to\infty}\frac{1}{t}\int_0^t u(\tau)\,d\tau=\ell,$$

and let us observe that

$$\lambda\int_0^\infty e^{-\lambda s}u(s)\,ds=\lambda^2\int_0^\infty e^{-\lambda s}s\left(\frac{1}{s}\int_0^s u(\tau)\,d\tau\right)\,ds.$$

Then

$$\left|\lambda\int_0^\infty e^{-\lambda s}u(s)\,ds-\ell\right|\leq\lambda^2\int_0^\infty e^{-\lambda s}s\left|\frac{1}{s}\int_0^s u(\tau)\,d\tau-\ell\right|\,ds.$$

Let $\varepsilon>0$, and let us fix a $\delta=\delta(\varepsilon)>0$ such that

$$\left|\frac{1}{s}\int_0^s u(\tau)\,d\tau-\ell\right|\leq\varepsilon$$

for each $s\in[\delta,+\infty)$. From the preceding inequality, it follows

$$\begin{aligned}\left|\lambda\int_0^\infty e^{-\lambda s}u(s)\,ds-\ell\right|&\leq\lambda^2\int_0^\delta e^{-\lambda s}s\left|\frac{1}{s}\int_0^s u(\tau)\,d\tau-\ell\right|\,ds\\&\quad+\lambda^2\int_\delta^\infty e^{-\lambda s}s\left|\frac{1}{s}\int_0^s u(\tau)\,d\tau-\ell\right|\,ds\\&\leq\lambda^2\int_0^\delta e^{-\lambda s}s(M+|\ell|)\,ds+\lambda^2\varepsilon\int_\delta^\infty se^{-\lambda s}\,ds,\end{aligned}$$

where $M>0$ satisfies $\|u(t)\|\leq M$ for each $t\in\mathbb{R}_+$. Therefore, a simple computational argument yields

$$\left|\lambda\int_0^\infty e^{-\lambda s}u(s)\,ds-\ell\right|\leq(M+|\ell|)(1-e^{-\lambda\delta}-\lambda\delta e^{-\lambda\delta})+\varepsilon(\lambda\delta e^{-\lambda\delta}+e^{-\lambda\delta}).$$

Passing to the sup-limit in this inequality for λ tending to 0 by positive values we obtain

$$\limsup_{\lambda\downarrow 0}\left|\lambda\int_0^\infty e^{-\lambda s}u(s)\,ds-\ell\right|\leq\varepsilon.$$

Since $\varepsilon>0$ is arbitrary, this shows that, whenever there exists the first limit in lemma, then there exists the second one too, and both must be equal.

The proof of the fact that the existence of the second limit implies the existence of the first one (and by the preceding proof, the fact that both are equal) is much more subtle and follows from a very deep Theorem of

Wiener which, for the sake of brevity, did not find its place here. However, we refer the interested reader to Widder [**133**], Theorem 14, p. 221. The proof is complete. □

Theorem 5.6.3. *Let X be a Banach space and $A : D(A) \subseteq X \to X$ the generator of a C_0-semigroup of contractions $\{S(t);\ t \geq 0\}$. Then the following conditions are equivalent*:

(i) $X = \ker A \oplus \overline{R(A)}$;

(ii) *for each $x \in X$ there exists* $\lim_{t\to\infty} \frac{1}{t}\int_0^t S(\tau)x\,d\tau$;

(iii) *for each $x \in X$ there exists* $\lim_{\lambda\downarrow 0} \lambda(\lambda I - A)^{-1}x$;

(iv) *for each $x \in D(A)$ there exists* $\lim_{\lambda\downarrow 0} A(\lambda I - A)^{-1}x$.

Proof. As in the proof of Theorem 5.6.2, one shows first that, for each $x \in \ker(A) \oplus \overline{R(A)}$, there exists

$$\lim_{t\to\infty} \frac{1}{t}\int_0^t S(\tau)x\,d\tau.$$

Hence (i) implies (ii). From Lemma 5.6.2 and from the observation that, for each $x \in X$, we have

$$\lambda(\lambda I - A)^{-1}x = \lambda\int_0^\infty e^{-\lambda s}S(s)x\,ds,$$

we deduce that (ii) implies (iii). The fact that (iii) implies (iv) follows from the identity

$$(\lambda I - A)^{-1}Ax = \lambda(\lambda I - A)^{-1}x - x = A(\lambda I - A)^{-1}x, \tag{5.6.1}$$

for each $x \in D(A)$. As $D(A)$ is dense in X, from Lemma 5.6.2 and (5.6.1), it follows that (iv) implies (ii). In order to complete the proof we have merely to show that (ii) implies (i). Let us define $P : X \to X$ by

$$Px = \lim_{t\to\infty} \frac{1}{t}\int_0^t S(\tau)x\,d\tau$$

for each $x \in X$. We have $X = R(P)+R(I-P)$ and $PS(t)x = S(t)Px = Px$, for each $x \in X$ and each $t \geq 0$. In addition, let us remark that $P^2 = P$, $R(P) = \ker(A)$, $R(I - P) = \ker(P)$, and $\overline{R(A)} \subseteq \ker(P)$. To check that $\ker(P) \subseteq \overline{R(A)}$, let $x \in \ker(P)$. Then, by (5.6.1) and (iii) (which is implied by (ii)), it follows that

$$\lim_{\lambda\downarrow 0} A(\lambda I - A)^{-1}x = Px - x = -x,$$

and therefore $x \in \overline{R(A)}$. The proof is complete. □

Problems

We recall that un operator $A : D(A) \subseteq X \to X$ is called *closable* if the closure of its graph is the graph of a linear operator $\overline{A}$, called *the closure of the operator A*.

Problem 5.1. Let $X = L^2(0,\pi)$, and $A : D(A) \subseteq X \to X$, defined by

$$\begin{cases} D(A) = \{u \in X\,;u' \text{ there exists a.e. and } u' = v \text{ a.e. with } v \in X\} \\ Au = u' \text{ for each } u \in D(A). \end{cases}$$

Prove that A is not closable. See Richtmyer [**110**], Example 3, p. 137.

Problem 5.2. Which one of the operators below is closable? In the case of a closable operator, determine the domain of its closure.

(i) $X = l_2$, and $A : D(A) \subseteq X \to X$ defined by:

$$\begin{cases} D(A) = \{(x_n)_{n\in\mathbb{N}^*} \in l_2\,;\ \exists\, m \in \mathbb{N}^*,\ x_n = 0\ \forall\, n > m\} \\ (A(x_n)_{n\in\mathbb{N}^*})_{k\in\mathbb{N}^*} = (2^k x_k)_{k\in\mathbb{N}^*}\,; \end{cases}$$

(ii) $X = l_2$, and $A : D(A) \subseteq X \to X$ defined by:

$$\begin{cases} D(A) = \{(x_n)_{n\in\mathbb{N}^*} \in l_2\,;\ \exists\, m \in \mathbb{N}^*,\ \forall\, n > m\ x_n = 0\} \\ (A(x_n)_{n\in\mathbb{N}^*})_{k\in\mathbb{N}^*} = (\sum_{n=1}^{\infty} x_n, 0, 0, \ldots)\,; \end{cases}$$

(iii) $X = l_2$, and $A : D(A) \subseteq X \to X$ defined by:

$$\begin{cases} D(A) = \{(x_n)_{n\in\mathbb{N}^*} \in l_2\,;\ \exists\, m \in \mathbb{N}^*,\ \forall\, n > m\ x_n = 0\} \\ (A(x_n)_{n\in\mathbb{N}^*})_{k\in\mathbb{N}^*} = (\sum_{n=1}^{\infty} (1/n)x_n, 0, 0, \ldots)\,; \end{cases}$$

(iv) $X = L^2(0,1)$, and $A : D(A) \subseteq X \to X$, defined by:

$$\begin{cases} D(A) = C([\,0,1\,]) \\ (Au)(x) = u(1/2)\sin \pi x\,; \end{cases}$$

(v) $X = L^2(0,1)$, and $A : D(A) \subseteq X \to X$, defined by:

$$\begin{cases} D(A) = C([\,0,1\,]) \\ (Au)(x) = \left(\int_0^1 u(y)\,dy\right)\sin \pi x. \end{cases}$$

See Richtmyer [**110**], Exercise 2, p. 138.

Problem 5.3. Show that if A is closable and densely defined, then A^{**} coincides with the closure of A.

Problem 5.4. Let $A : D(A) \subseteq X \to X$ be the generator a C_0-semigroup of contractions on the Banach space X. Let $n \in \mathbb{N}^*$, and let us recall that the Yosida approximation A_n, of A is defined by

$$A_n = nR(n; A).$$

Using Theorem 5.1.1 prove that

$$S(t)x = \lim_{n\to\infty} e^{tA_n}x.$$

See Engel and Nagel [**51**], 4.10, p 214.

Problem 5.5. Let $X = C_0(\mathbb{R})$, or $X = C_{ub}(\mathbb{R})$ and let us consider the C_0 translation group $[S(t)f] = f(t+s)$, for each $f \in X$ and $t, s \in \mathbb{R}$. For each $n \in \mathbb{N}$ we define

$$A_n = \frac{S(1/n) - I}{1/n}.$$

Prove that:

(i) A_n commute with $S(t)$;
(ii) A_n generates a uniformly continuous semigroup of contractions;
(iii) $\lim_{n\to\infty} A_n f = f$ for each $f \in D(A)$.

Show that

$$f(t+s) = \lim_{n\to\infty} \sum_{k=0}^{\infty} \frac{t^k}{k!}(A_n^k)(f)(s)$$

for each $f \in D(A)$ and uniformly for $s \in \mathbb{R}$, $t \in [0,1]$. Show that, for each $f \in X$ there exists a sequence $(m_n)_{n\in\mathbb{N}}$, tending to ∞ and such that, for each $k \in \mathbb{N}$, $\sum_{k=0}^{m_k} \frac{t^k}{k!}(A_n^k)(f)(0)$ is a polynomial, and

$$f(t) = \lim_{n\to\infty} \sum_{k=0}^{m_k} \frac{t^k}{k!}(A_n^k)(f)(0).$$

This is the well-known Weierstrass approximation theorem. See Engel and Nagel [**51**], 4.11, p. 215.

Notes. Theorems 5.1.1 and 5.1.3 were proved within a particular frame by Neveu [**95**], and generalized by Trotter [**122**]. The lack of an important step in the initial proof, given by Trotter, has been completed by Kato [**72**]. The proof of Theorem 5.1.2 in the form here presented is due to Kisýnski [**74**]. Theorem 5.2.1 was established by Chernoff [**37**], Theorem 5.2.2 is due to Hille [**69**], and Theorem 5.2.3 was proved by Trotter [**123**]. However, the main idea in all these results goes back at the end of the ninetieth century and is due to Lie. For applications of results of this kind in optimal control theory, see Barbu [**20**], pp. 389-404. Theorem 5.3.1 was obtained by Pazy [**100**]. Although equivalent with Pazy's original result, Theorem 5.3.1 is included here in a slightly different formulation as needed in the proof of

Theorem 5.5.1. We note that, under some appropriate extra-assumption, we can take $\alpha = 1$ in (5.3.2). Namely, we have

Theorem (Chernoff) *Let $A : D(A) \subseteq X \to X$ be the infinitesimal generator of a C_0-semigroup of contractions and let $B : D(B) \subseteq X \to X$ be dissipative and such that $D(A) \subseteq D(B)$ and*

$$\|Bx\| \leq \|Ax\| + \beta\|x\|$$

for all $x \in D(A)$, where $\beta \geq 0$ is constant. If the adjoint of B, B^* is densely defined then the closure of $A + B$, $\overline{A+B}$ is the infinitesimal generator of a C_0-semigroup of contractions.

See Pazy [**101**], Theorem 3.4, p. 83. The proof of Theorem 5.4.1 is due to Goldstein [**60**], and the rigorous approach to the Feynman formula [**56**] by means of the Lie-Trotter formula follows Goldstein [**61**] and Pazy [**101**]. For the physical background of this formula see Feynman [**56**] and also Simon [**114**]. The proof of Theorem 5.6.2 is due to Goldstein, Radin and Showalter [**62**]. As we already mentioned, within this chapter, we confined ourselves to the case of C_0-semigroups of contractions although almost all the results can be appropriately restated in order to handle general C_0-semigroups, i.e. of type (M, ω). As concerns the results in Section 5.1, we notice that the only fact we have to add to get the general case is that all the semigroups $\{S_n(t);\ t \geq 0\}$ there considered have to satisfy the stability condition $\|S(t)\|_{\mathcal{L}(X)} \leq Me^{\omega t}$ for each $n \in \mathbb{N}^*$, condition which is automatically satisfied whenever the semigroups are of contractions. As concerns the problems included, they are from Engel and Nagel et al [**51**], Richtmyer [**110**] and Pazy [**101**].

CHAPTER 6

Some Special Classes of C_0-Semigroups

In this chapter we present several remarkable classes of C_0-semigroups which proved useful in the study of parabolic partial differential equations. We mean here equicontinuous, compact and differentiable C_0-semigroups. Each one of these semigroups endows the solution operator $f \mapsto u$, defined by the nonhomogeneous equation $u' = Au + f$, $u(a) = \xi$, by means of the variation of constants formula

$$u(t) = S(t-a)\xi + \int_a^t S(t-s)f(s)\,ds,$$

with additional properties (as regularity, compactness, etc.), comparable with their finite-dimensional counterparts. All these properties will be studied in Chapters 8 and 9, and some of them will be exploited in Chapters 10, 11 and 12 in order to obtain existence results for semilinear, or even fully nonlinear differential equations in abstract Banach spaces.

6.1. Equicontinuous Semigroups

The aim of this section is to study a class of C_0-semigroups which, for $t > 0$, behave exactly as uniformly continuous semigroups. As in the preceding chapter, for the sake of simplicity, here, we confine ourselves only to the case of C_0-semigroups of contractions although all the results remain valid in the general framework of C_0-semigoups of type (M, ω). First, we introduce:

Definition 6.1.1. A C_0-semigroup $\{S(t)\,;\ t \geq 0\}$ is *equicontinuous* if the function $t \mapsto S(t)$ is continuous from $(0, +\infty)$ to $\mathcal{L}(X)$ endowed with the uniform operator norm $\|\cdot\|_{\mathcal{L}(X)}$.

Remark 6.1.1. Each uniformly continuous semigroup is equicontinuous, but the converse statement is not true, as we can state, from the examples analyzed in Problems 6.1 and 6.2.

The next theorem provides a characterization of the infinitesimal generators of equicontinuous C_0-semigroups.

Theorem 6.1.1. *Let $A : D(A) \subseteq X \to X$ be the infinitesimal generator of a C_0-semigroup of contractions, $\{S(t)\,;\ t \geq 0\}$. Then $\{S(t)\,;\ t \geq 0\}$ is equicontinuous if and only if, for each $\alpha \in (0,1)$, we have*

$$\lim_{n\to\infty}\left(I - \frac{t}{n}A\right)^{-n} = S(t)$$

in the usual sup-norm topology of $C([\,\alpha, 1/\alpha\,]; \mathcal{L}(X))$.

The proof of Theorem 6.1.1 rests on the following convergence result.

Lemma 6.1.1. *For each $a \in (0,1)$ and each $b \in (1,+\infty)$ we have*

$$\lim_{n\to\infty}\frac{n^{n+1}}{n!}\int_0^a (ve^{-v})^n dv = 0$$

and

$$\lim_{n\to\infty}\frac{n^{n+1}}{n!}\int_b^{+\infty} (ve^{-v})^n dv = 0.$$

Proof. Since $t \mapsto te^{-t}$ is nondecreasing on $[\,0,1\,]$, it follows that

$$\int_0^a (ve^{-v})^n dv \leq a(ae^{-a})^n.$$

On the other hand, $ve^{-v} < e^{-1}$ for each $v > 0$, $v \neq 1$, and accordingly

$$\lim_{n\to\infty} vn(ve^{-v}e)^n = 0 \tag{6.1.1}$$

for each $v > 0$, $v \neq 1$. Observing that, from Stirling's formula

$$\lim_{n\to\infty}\frac{n!}{\sqrt{2\pi}n^{n+1/2}e^{-n}} = 1$$

(see Nikolsky [**96**], p. 393), it follows

$$\lim_{n\to\infty}\frac{n^n e^{-n}}{n!} = 0, \tag{6.1.2}$$

we obtain

$$0 \leq \lim_{n\to\infty}\frac{n^{n+1}}{n!}\int_0^a (ve^{-v})^n dv \leq \lim_{n\to\infty} an(ae^{-a}e)^n \frac{n^n e^{-n}}{n!} = 0,$$

which proves the first assertion.
Next, let us observe that

$$\frac{n^{n+1}}{n!}\int_b^{+\infty} (ve^{-v})^n dv = e^{-nb}\sum_{k=0}^{n}\frac{(nb)^k}{k!} \tag{6.1.3}$$

for each $n \in \mathbb{N}^*$ and $b \geq 0$. Since $b > 1$, we have

$$\frac{(nb)^k}{k!} \leq \frac{(nb)^n}{n!}$$

for $k = 1, 2, \ldots, n-1$, and accordingly, the last relation implies

$$\frac{n^{n+1}}{n!} \int_b^{+\infty} (ve^{-v})^n dv \leq (n+1)(be^{-b})^n \frac{n^n}{n!}.$$

Consequently, from (6.1.1) and (6.1.2), it follows

$$\lim_{n\to\infty} \frac{n^{n+1}}{n!} \int_b^{+\infty} (ve^{-v})^n dv \leq \lim_{n\to\infty} (n+1)(be^{-b})^n \frac{n^n}{n!} = 0,$$

which achieves the proof. □

Lemma 6.1.2. *Let $A : D(A) \subseteq X \to X$ be the generator of a C_0-semigroup of contractions $\{S(t)\,;\ t \geq 0\}$, let $\lambda > 0$, and let $R(\lambda; A) = (\lambda I - A)^{-1}$. Then the mapping $\lambda \mapsto R(\lambda; A)$ is of class C^∞ on $(0, +\infty)$. In addition, for each $x \in X$, each $t > 0$ and each $n \in \mathbb{N}^*$, we have*

$$\left(I - \frac{t}{n}A\right)^{-n-1} x - S(t)x = \frac{n^{n+1}}{n!} \int_0^{+\infty} (ve^{-v})^n [S(tv)x - S(t)x]\, dv. \quad (6.1.4)$$

Proof. From the proof of the necessity of Theorem 3.1.1, we know that

$$R(\lambda; A)x = \int_0^{+\infty} e^{-\lambda s} S(s)x\, ds. \quad (6.1.5)$$

In addition, by Theorems 1.7.2 and 3.1.1, it follows that $\lambda \mapsto R(\lambda; A)$ is even analytic from $(0, +\infty)$ to $\mathcal{L}(X)$. Differentiating n-times both sides in (6.1.5) with respect to λ and putting $s = vt$, we obtain

$$(R(\lambda; A))^{(n)} x = (-1)^n t^{n+1} \int_0^{+\infty} v^n e^{-\lambda t v} S(tv)x\, dv$$

for $\lambda > 0$ and $x \in X$. On the other hand,

$$(R(\lambda; A))^{(n)} = (-1)^n n! R(\lambda; A)^{n+1},$$

and so, substituting $\lambda = \frac{n}{t}$ in the relations above, we get

$$\left[\frac{n}{t} R\left(\frac{n}{t}; A\right)\right]^{n+1} x = \frac{n^{n+1}}{n!} \int_0^{+\infty} (ve^{-v})^n S(tv)x\, dv.$$

Taking $b = 0$ in (6.1.3), we conclude that

$$\frac{n^{n+1}}{n!} \int_0^{+\infty} (ve^{-v})^n dv = 1. \quad (6.1.6)$$

for each $n \in \mathbb{N}^*$. Therefore (6.1.4) holds and this completes the proof. □

We can now proceed to the proof of Theorem 6.1.1.

Proof. *Necessity.* Let $\alpha \in (0,1)$ and let us fix $\beta \in (0,\alpha)$. Since $\{S(t)\,;\ t \geq 0\}$ is equicontinuous, for each $\varepsilon > 0$ there exists $\delta(\varepsilon) > 0$ such that

$$\|S(t) - S(s)\|_{\mathcal{L}(X)} \leq \varepsilon$$

for each $t, s \in [\,\beta, 1/\beta\,]$ with $|t-s| \leq \delta(\varepsilon)$. On the other hand, for the very same $\varepsilon > 0$, there exists $a = a(\varepsilon)$ and $b = b(\varepsilon)$ with $0 < a < 1 < b < +\infty$ and such that, for each $t \in [\,\alpha, 1/\alpha\,]$ and $v \in [\,a, b\,]$, we have

$$tv \in [\,\beta, 1/\beta\,] \quad \text{and} \quad |t - tv| \leq \delta(\varepsilon).$$

So,

$$\|S(tv) - S(t)\|_{\mathcal{L}(X)} \leq \varepsilon$$

for each $t \in [\,\alpha, 1/\alpha\,]$, and $v \in [\,a, b\,]$. From both (6.1.4) and the last relation, we deduce

$$\left\|\left(I - \frac{t}{n}A\right)^{-n-1} - S(t)\right\|_{\mathcal{L}(X)} \leq 2\frac{n^{n+1}}{n!}\int_0^a (ve^{-v})^n\,dv$$

$$+\varepsilon\frac{n^{n+1}}{n!}\int_a^b (ve^{-v})^n\,dv + 2\frac{n^{n+1}}{n!}\int_b^{+\infty} (ve^{-v})^n\,dv$$

for each $n \in \mathbb{N}^*$. By Lemma 6.1.1, it follows that

$$\limsup_{n\to\infty}\left\|\left(I - \frac{t}{n}A\right)^{-n-1} - S(t)\right\|_{\mathcal{L}(X)} \leq \varepsilon$$

for each $\varepsilon > 0$. Consequently, for each $\alpha \in (0,1)$, we have

$$\lim_{n\to\infty}\left(I - \frac{t}{n}A\right)^{-n-1} = S(t) \tag{6.1.7}$$

in the norm $\|\cdot\|_{\mathcal{L}(X)}$, uniformly for $t \in [\,\alpha, 1/\alpha\,]$. In order to conclude the proof, we have merely to show that

$$\lim_{n\to\infty}\left(I - \frac{t}{n+1}A\right)^{-n-1} = S(t) \tag{6.1.8}$$

in the norm $\|\cdot\|_{\mathcal{L}(X)}$, uniformly for $t \in [\,\alpha, 1/\alpha\,]$. From (6.1.7), we deduce that, for each sequence $(a_n)_{n\in\mathbb{N}}$ of functions from $\mathbb{R}_+^*$ to $\mathbb{R}_+^*$ satisfying

$$\lim_{n\to\infty} a_n(t) = t$$

uniformly on each compact subset in $\mathbb{R}_+^*$, we have

$$\lim_{n\to\infty}\left(I - \frac{a_n(t)}{n}A\right)^{-n-1} = S(t)$$

in the norm $\|\cdot\|_{\mathcal{L}(X)}$, uniformly on each compact in $\mathbb{R}_+^*$. This simply follows from the fact that, for each $\alpha \in (0,1)$, the set of functions

$$\left\{ t \mapsto \left(I - \frac{t}{n}A\right)^{-n-1} ;\ n \in \mathbb{N}^* \right\}$$

is equicontinuous on $[\alpha, 1/\alpha]$ because it is relatively compact in the space $C([\alpha, 1/\alpha]; \mathcal{L}(X))$. See (6.1.7) and Theorem A.2.1. Taking

$$a_n(t) = \frac{nt}{n+1},$$

we obtain (6.1.8), which proves the necessity.

Sufficiency. By Lemma 6.1.2 it follows that $t \mapsto \left(I - \frac{t}{n}A\right)^{-n}$ is continuous from $(0, +\infty)$ to $\mathcal{L}(X)$ with respect to the norm $\|\cdot\|_{\mathcal{L}(X)}$. Since, for each $\alpha \in (0,1)$, we have

$$\lim_{n\to\infty} \left(I - \frac{t}{n}A\right)^{-n} = S(t)$$

in the usual sup-norm topology of $C([\alpha, 1/\alpha]; \mathcal{L}(X))$, it follows that the semigroup is continuous from $(0, +\infty)$ to $\mathcal{L}(X)$, i.e. equicontinuous. The proof is complete. □

Remark 6.1.1. It should be emphasized that the really interesting part of Theorem 6.1.1 is the necessity, which asserts that, if a C_0-semigroup is equicontinuous, then, for each $\alpha \in (0,1)$ and each bounded subset B in X, the family of sequences $\left(\left(I - \frac{t}{n}A\right)^{-n} x\right)_{n\in\mathbb{N}}$ converges to $S(t)x$ "with the very same speed" for all $t \in [\alpha, 1/\alpha]$ and $x \in B$. This property proves useful for the error evaluation in some approximation procedures.

We conclude this section by noticing that, using a very similar proof with that of Theorem 6.1.1, we can give a direct proof of Hille's exponential formula. See Theorem 5.2.2.

6.2. Compact Semigroups

The aim of this section is to study a class of C_0-semigroups of contractions which are specific to parabolic problems in bounded domains.

Definition 6.2.1. A C_0-semigroup $\{S(t);\ t \geq 0\}$ is *compact* if for each $t > 0$, $S(t)$ is a compact operator.

Remark 6.2.1. If a C_0-semigroup $\{S(t);\ t \geq 0\}$ satisfies the condition that $S(t)$ is compact for each $t \geq 0$, then I is necessarily compact, and thus X is finite-dimensional. So, if X is infinite-dimensional and $\{S(t);\ t \geq 0\}$ can be extended to a C_0-group on X, then this semigroup cannot be compact.

Indeed, if we assume by contradiction that the semigroup is compact, then, for $t > 0$, we have $I = S(t)S(-t)$ and therefore I is compact, thereby contradicting the hypothesis that X is infinite-dimensional.

We have the following characterization of the infinitesimal generators of compact C_0-semigroups of contractions.

Theorem 6.2.1. (Pazy) *Let $A : D(A) \subseteq X \to X$ be the generator of a C_0-semigroup of contractions, $\{S(t)\,;\ t \geq 0\}$. Then $\{S(t)\,;\ t \geq 0\}$ is compact if and only if*

(i) *$\{S(t)\,;\ t \geq 0\}$ is equicontinuous, and*
(ii) *for each $\lambda > 0$ the operator $(\lambda I - A)^{-1}$ is compact.*

Proof. *Necessity.* Let $\{S(t)\,;\ t \geq 0\}$ be a compact C_0-semigroup of contractions, let $t > 0$, and let $\lambda > 0$ be such that $t - \lambda > 0$. Then, for each $\varepsilon > 0$, there exists a finite family $\{x_1, x_2, \dots x_{k(\varepsilon)}\}$ in $B(0,1)$ such that, for each $x \in B(0,1)$, there exists $i \in \{1, 2, \dots k(\varepsilon)\}$ with

$$\|S(t-\lambda)x - S(t-\lambda)x_i\| \leq \varepsilon.$$

Since the family $\{S(\,\cdot\,)x_i;\ i = 1, 2, \dots k(\varepsilon)\}$ of continuous functions from $[0, +\infty)$ to X is finite, it is equicontinuous at t. Therefore, for the very same $\varepsilon > 0$, there exists $\delta(\varepsilon) \in (0, \lambda)$, such that

$$\|S(t+h)x_i - S(t)x_i\| \leq \varepsilon$$

for each $i = 1, 2, \dots k(\varepsilon)$, and each $h \in \mathbb{R}$ with $|h| \leq \delta(\varepsilon)$. We have

$$\begin{aligned}
\|S(t+h)x - S(t)x\| &\leq \|S(t+h)x - S(t+h)x_i\| + \|S(t+h)x_i - S(t)x_i\| \\
&\quad + \|S(t)x_i - S(t)x\| = \|S(\lambda+h)(S(t-\lambda)x - S(t-\lambda)x_i)\| \\
&\quad + \|S(t+h)x_i - S(t)x_i\| + \|S(\lambda)(S(t-\lambda)x_i - S(t-\lambda)x)\| \\
&\leq \|S(\lambda+h)\|_{\mathcal{L}(X)}\|S(t-\lambda)x - S(t-\lambda)x_i\| + \|S(t+h)x_i - S(t)x_i\| \\
&\quad + \|S(\lambda)\|_{\mathcal{L}(X)}\|S(t-\lambda)x_i - S(t-\lambda)x\| \leq 3\varepsilon
\end{aligned}$$

for each $x \in B(0,1)$, and each $h \in \mathbb{R}$ with $|h| \leq \delta(\varepsilon)$. Consequently,

$$\|S(t+h) - S(t)\|_{\mathcal{L}(X)} \leq 3\varepsilon$$

for each $h \in \mathbb{R}$ with $|h| \leq \delta(\varepsilon)$, which proves (i).
To check (ii), let us recall that, for each $\lambda > 0$ and $x \in X$, we have

$$(\lambda I - A)^{-1}x = R(\lambda; A)x = \int_0^\infty e^{-s\lambda} S(s)x\, ds.$$

Let $\varepsilon > 0$ and let us define $R_\varepsilon(\lambda; A) : X \to X$ by

$$R_\varepsilon(\lambda; A)x = \int_\varepsilon^\infty e^{-s\lambda} S(s)x\, ds$$

for each $x \in X$. We shall show next that $R_\varepsilon(\lambda; A)$ is a compact operator and that, in addition,

$$\lim_{\lambda \to \infty} \|R(\lambda; A) - R_\varepsilon(\lambda; A)\|_{\mathcal{L}(X)} = 0. \tag{6.2.1}$$

Indeed, let us observe that

$$R_\varepsilon(\lambda; A)x = S(\varepsilon) \int_\varepsilon^\infty e^{-s\lambda} S(s - \varepsilon)x\, ds$$

$$= e^{-\lambda\varepsilon} S(\varepsilon) \int_0^\infty e^{-\tau\lambda} S(\tau)x\, d\tau = e^{-\lambda\varepsilon} S(\varepsilon) R(\lambda; A)x$$

for each $x \in X$. Since $R(\lambda; A)$ is linear continuous and $e^{-\lambda\varepsilon} S(\varepsilon)$ is compact, it follows that $R_\varepsilon(\lambda; A)$ is compact. On the other hand,

$$\|R(\lambda; A) - R_\varepsilon(\lambda; A)\|_{\mathcal{L}(X)} \leq \int_0^\varepsilon e^{\lambda t} \|S(t)\|_{\mathcal{L}(X)} dt \leq \frac{1 - e^{-\lambda\varepsilon}}{\lambda}$$

for each $\varepsilon > 0$, which proves (6.2.1). From Corollary A.1.2 it follows that $R(\lambda; A)$ is compact and this completes the proof of the necessity.

Sufficiency. Let $\{S(t)\,;\ t \geq 0\}$ be a C_0-semigroup of contractions satisfying (i) and (ii). For $t > 0$ and $\lambda > 0$, we define $S_\lambda(t) : X \to X$ by

$$S_\lambda(t)x = \lambda R(\lambda; A) S(t)x$$

for each $x \in X$. We shall prove that $S_\lambda(t)$, which obviously is compact, satisfies

$$\lim_{\lambda \to \infty} \|S_\lambda(t) - S(t)\|_{\mathcal{L}(X)} = 0. \tag{6.2.2}$$

Indeed

$$\|S_\lambda(t) - S(t)\|_{\mathcal{L}(X)} = \left\| \lambda \int_0^\infty e^{-\lambda\tau} (S(t + \tau) - S(t))\, d\tau \right\|_{\mathcal{L}(X)}$$

$$\leq \lambda \int_0^\infty e^{-\lambda\tau} \|S(t + \tau) - S(t)\|_{\mathcal{L}(X)}\, d\tau. \tag{6.2.3}$$

Since the semigroup $\{S(t)\,;\ t \geq 0\}$ is equicontinuous, for each $\varepsilon > 0$ there exists $\delta = \delta(\varepsilon) > 0$ such that, for each $\tau \in (0, \delta)$, we have

$$\|S(t + \tau) - S(t)\|_{\mathcal{L}(X)} \leq \varepsilon.$$

From (6.2.3) and this inequality, it follows

$$\|S_\lambda(t) - S(t)\|_{\mathcal{L}(X)} \leq \lambda \int_0^\delta e^{-\lambda\tau} \|S(t + \tau) - S(t)\|_{\mathcal{L}(X)}\, d\tau$$

$$+\lambda \int_\delta^\infty e^{-\lambda\tau} (\|S(t + \tau)\|_{\mathcal{L}(X)} + \|S(t)\|_{\mathcal{L}(X)})\, d\tau \leq (1 - e^{-\lambda\delta})\varepsilon + 2e^{-\lambda\delta}.$$

Passing to the sup-limit for λ tending to $+\infty$ in this inequality, we obtain

$$\limsup_{\lambda \to \infty} \|S_\lambda(t) - S(t)\|_{\mathcal{L}(X)} \leq \varepsilon$$

for each $\varepsilon > 0$. But this relation implies (6.2.2), and the latter, along with Corollary A.1.2, ensures the compactness of the semigroup $\{S(t)\,;\ t \geq 0\}$. The proof is complete. □

The next simple theorem gives a very precise limitation of the class of Banach spaces on which may exist compact C_0-semigroups.

Theorem 6.2.2. *If $A : D(A) \subseteq X \to X$ is the infinitesimal generator of a C_0-semigroup of contractions for which there exists $\lambda > 0$ such that $R(\lambda; A)$ is compact, then X is separable. In particular, if the semigroup generated by A is compact, then X is separable.*

Proof. First, let us observe that, by virtue of the resolvent equation (1.7.3), we deduce that, for each $\lambda > 0$, $R(\lambda; A)$ is compact. Let $(\lambda_n)_{n \in \mathbb{N}^*}$ be a sequence of numbers, strictly decreasing to 0. Let $n \in \mathbb{N}^*$ be arbitrary. Inasmuch as $R(\lambda_n; A)$ is compact, there exists a finite family D_n in $B(0, n)$ such that, for each $x \in B(0, n)$, there exists $x_n \in D_n$ with

$$\|\lambda_n R(\lambda_n; A)x - \lambda_n R(\lambda_n; A)x_n\| \leq \lambda_n. \tag{6.2.4}$$

Let $x \in X$ and $\varepsilon > 0$. In view of (3.2.1), there exists $n \in \mathbb{N}^*$ such that

$$\begin{cases} \lambda_n \leq \varepsilon \\ \|x\| \leq n \\ \|\lambda_n R(\lambda_n; A)x - x\| \leq \varepsilon. \end{cases}$$

Taking $x_n \in D_n$ satisfying (6.2.4), we deduce

$$\|x - \lambda_n R(\lambda_n; A)x_n\| \leq \|x - \lambda_n R(\lambda_n; A)x\|$$

$$+\|\lambda_n R(\lambda_n; A)x - \lambda_n R(\lambda_n; A)x_n\| \leq 2\varepsilon.$$

This inequality shows that the set

$$D = \cup_n \lambda_n R(\lambda_n; A) D_n$$

is dense in X. Since this set is a countable union of finite sets, it is countable too, and therefore X is separable. Finally, if the semigroup generated by A is compact, thanks to Pazy's Theorem 6.2.1, there exists $\lambda > 0$ such that $R(\lambda; A)$ is compact, and this completes the proof. □

We conclude this section with another characterization of the infinitesimal generators of compact C_0-semigroups.

Theorem 6.2.3. *Let $A : D(A) \subseteq X \to X$ be the infinitesimal generator of a C_0-semigroup of contractions $\{S(t)\,;\ t \geq 0\}$. Then $\{S(t)\,;\ t \geq 0\}$ is compact if and only if*

(i) *for each $\alpha \in (0,1)$ we have*

$$\lim_{n\to\infty}\left(I - \frac{t}{n}A\right)^{-n} = S(t)$$

in the usual sup-norm topology of $C([\,\alpha, 1/\alpha\,]; \mathcal{L}(X))$ and

(ii) *for each $\lambda > 0$ the operator $(\lambda I - A)^{-1}$ is compact.*

Proof. The conclusion follows from Theorems 5.1.1 and 6.2.1. □

6.3. Differentiable Semigroups

Here, we shall study a remarkable class of C_0-semigroups of contractions, $\{S(t)\,;\ t \geq 0\}$, having the property that $S(t)\xi \in D(A)$ for each $\xi \in X$ and each $t > 0$.

Definition 6.3.1. A C_0-semigroup $\{S(t)\,;\ t \geq 0\}$ is called:

(i) *differentiable* at $\tau \geq 0$, if, for each $x \in X$, the function $t \mapsto S(t)x$ is differentiable at τ;

(ii) *differentiable*, if it is differentiable at each $\tau \in (0, +\infty)$.

(iii) *eventually differentiable*, if there exists $\theta > 0$ such that $t \mapsto S(t)x$ is differentiable at each $\tau \in (\theta, +\infty)$.

Remark 6.3.1. From the definition of the infinitesimal generator, it follows that, whenever $\{S(t)\,;\ t \geq 0\}$ is differentiable at $\tau \geq 0$, we necessarily have $S(\tau)X \subseteq D(A)$. So, if $\{S(t)\,;\ t \geq 0\}$ is differentiable at $\tau = 0$, we have $D(A) = X$ and, inasmuch as A is closed, thanks to the closed graph theorem (see Dunford and Schwartz [**49**], Theorem 4, p. 57), we easily deduce that $A \in \mathcal{L}(X)$. Hence, in this case, $\{S(t)\,;\ t \geq 0\}$ is uniformly continuous. Also, if $\{S(t)\,;\ t \geq 0\}$ is differentiable at τ, then it is differentiable at each $t \geq \tau$.

Remark 6.3.2. If $\{S(t)\,;\ t \geq 0\}$ is eventually differentiable, an inductive argument based on (iii) in Theorem 2.3.2 shows that, for each $x \in X$, the function $t \mapsto S(t)x$ is of class C^∞ on $(\theta, +\infty)$, where θ is given by (iii) in Definition 6.3.1.

Definition 6.3.2. A C_0-semigroup $\{S(t)\,;\ t \geq 0\}$ is called:

(i) *uniformly differentiable* at $\tau \geq 0$, if the function $t \mapsto S(t)$, from $(0, +\infty)$ to $\mathcal{L}(X)$, is differentiable at τ;

(ii) *uniformly differentiable* if it is uniformly differentiable at each point $\tau \in (0, +\infty)$.

(iii) *eventually uniformly differentiable*, if there exists $\theta > 0$ such that it is uniformly differentiable at each $\tau \in (\theta, +\infty)$.

Remark 6.3.3. Each C_0-semigroup which is uniformly differentiable at τ is differentiable at τ. Moreover, if a C_0-semigroup is uniformly differentiable at τ, then it is uniformly differentiable at each $t \geq \tau$. We will prove latter that each differentiable C_0-semigroup is uniformly differentiable.

Remark 6.3.4. If $\{S(t)\,;\ t \geq 0\}$ is uniformly differentiable, the function $t \mapsto S(t)$ is of class C^∞ from $(0, +\infty)$ to $\mathcal{L}(X)$. In particular, each uniformly differentiable C_0-semigroup is equicontinuous.

Lemma 6.3.1. *Let $\{S(t)\,;\ t \geq 0\}$ be a C_0-semigroup differentiable at each $t > \theta$, $A : D(A) \subseteq X \to X$ be its infinitesimal generator, and let $n \in \mathbb{N}^*$. Then:*

(i) *for $t > n\theta$, $S(t)X \subseteq D(A^n)$ and $S^{(n)}(t) = A^n S(t)$ is a linear bounded operator;*
(ii) *the mapping $t \mapsto S^{(n-1)}(t)$ is continuous from $(n\theta, +\infty)$ to $\mathcal{L}(X)$ in the uniform operator topology.*

Proof. Let $n = 1$. Since, for each $x \in X$, $t \mapsto S(t)x$ is differentiable at each $t > \theta$, we have $S(t)x \in D(A)$ and $S'(t)x = AS(t)x$ for each $x \in X$ and $t > \theta$. Inasmuch as A is closed and $S(t)$ is bounded it follows that $AS(t)$ is closed and thus, by the closed graph theorem (see Dunford and Schwartz [**49**], Theorem 4, p. 57), we conclude that it is bounded, and this completes the proof of (i) for $n = 1$. To prove (ii), first let us observe that there exists $M \geq 1$ such that $\|S(t)\|_{\mathcal{L}(X)} \leq M$ for each $t \in [\,0, 1\,]$. Then, for each $\theta < t \leq s < \theta + 1$, we have

$$S(s)x - S(t)x = \int_t^s AS(\tau)x\,d\tau = \int_t^s S(\tau - t)AS(t)x\,d\tau,$$

and thus

$$\|S(s) - S(t)\|_{\mathcal{L}(X)} \leq |s - t| M \|AS(t)\|_{\mathcal{L}(X)} \|x\|.$$

So, $t \mapsto S(t)$ is continuous from $(\theta, +\infty)$ to $\mathcal{L}(X)$ in the uniform operator topology, and this proves (ii) for $n = 1$.
Next, we proceed by induction on n. Assume that both (i) and (ii) hold for n and let $t > (n + 1)\theta$. Choose $s > n\theta$ such that $t - s > \theta$. Then, for each $x \in X$, we have

$$S^{(n)}(t)x = S(t - s)A^n S(s)x.$$

Clearly the right-hand side is differentiable at t, and thus $t \mapsto S(t)x$ is $(n + 1)$-times differentiable, and $S^{(n+1)}(t)x = A^{n+1}S(t)x$. From now on, the proof follows exactly the same lines as in the case $n = 1$. □

Corollary 6.3.1. *Let $\{S(t)\,;\ t \geq 0\}$ be a C_0-semigroup. If there exists $\theta \geq 0$ such that $t \mapsto S(t)$ is differentiable at every $t > \theta$, then, for each $n \in \mathbb{N}^*$, $t \mapsto S(t)$ is n-times uniformly differentiable from $((n+1)\theta, +\infty)$ to $\mathcal{L}(X)$ in the uniform operator topology. In particular, if $\{S(t)\,; t \geq 0\}$ is differentiable, then $t \mapsto S(t)$ is infinitely many times differentiable from $(0, +\infty)$ to $\mathcal{L}(X)$ in the uniform operator topology.*

Two characterizations of the class of all linear operators which generate differentiable and thus, uniformly differentiable, C_0-semigroups are listed below.

Theorem 6.3.1. *Let $A : D(A) \subseteq X \to X$ be the infinitesimal generator of a C_0-semigroup of contractions $\{S(t)\,;\ t \geq 0\}$. Then $\{S(t)\,;\ t \geq 0\}$ is differentiable (and thus uniformly differentiable) if and only if, for each $x \in X$ and each $\alpha \in (0,1)$, there exists*

$$\lim_{n\to\infty} A\left(I - \frac{t}{n}A\right)^{-n} x,$$

uniformly for $t \in [\,\alpha, 1/\alpha\,]$.

Theorem 6.3.2. *Let $A : D(A) \subseteq X \to X$ be the infinitesimal generator of a C_0-semigroup of contractions $\{S(t)\,;\ t \geq 0\}$. Then $\{S(t)\,;\ t \geq 0\}$ is differentiable (and thus uniformly differentiable) if and only if, for each $\alpha \in (0,1)$, there exists*

$$\lim_{n\to\infty} A\left(I - \frac{t}{n}A\right)^{-n}$$

uniformly for $t \in [\,\alpha, 1/\alpha\,]$in the norm topology of $\mathcal{L}(X)$).

Proof. *Sufficiency.* Let $x \in X$ and $\alpha \in (0,1)$. From Hille's exponential formula (see Theorem 5.2.2), we have

$$\lim_{n\to\infty} \left(I - \frac{t}{n}A\right)^{-n} x = S(t)x$$

uniformly for $t \in [\,\alpha, 1/\alpha\,]$. As A is a closed operator, from this remark, and from hypothesis, it follows that

$$\lim_{n\to\infty} A\left(I - \frac{t}{n}A\right)^{-n} x = AS(t)x$$

uniformly for $t \in [\,\alpha, 1/\alpha\,]$. But this means that $S(t)x \in D(A)$ for each $x \in X$ and each $t > 0$, which completes the proof of the sufficiency.

Necessity. Using once again the closedness of A, Hille's Theorem 1.2.2 and Lemma 6.1.2, we get

$$A\left(I-\frac{t}{n}A\right)^{-n-1}x - AS(t)x$$

$$= \frac{n^{n+1}}{n!}\int_0^{+\infty}(ve^{-v})^n[AS(tv)x - AS(t)x]\,dv \tag{6.3.1}$$

for each $x \in X$ and each $t > 0$. Let $\alpha \in (0,1)$, and fix $\beta \in (0,\alpha)$. As $\{S(t)\,;\ t \geq 0\}$ is differentiable, it follows that, for each $x \in X$, the mapping $t \mapsto AS(t)x$ is continuous on $(0,+\infty)$. Then, for each $\varepsilon > 0$ there exists $\delta(\varepsilon) > 0$ such that

$$\|AS(t)x - AS(s)x\| \leq \varepsilon$$

for each $t, s \in [\beta, 1/\beta]$ with $|t-s| \leq \delta(\varepsilon)$. Moreover, for the very same $\varepsilon > 0$, there exist $a = a(\varepsilon)$ and $b = b(\varepsilon)$ with $0 < a < 1 < b < +\infty$ such that, for each $t \in [\alpha, 1/\alpha]$ and $v \in [a,b]$, we have

$$tv \in [\beta, 1/\beta] \quad \text{and} \quad |t - tv| \leq \delta(\varepsilon).$$

As a consequence

$$\|AS(tv)x - AS(t)x\| \leq \varepsilon \tag{6.3.2}$$

for each $t \in [\alpha, 1/\alpha]$ and $v \in [a,b]$. From (6.3.1), we deduce

$$\left\|A\left(I-\frac{t}{n}A\right)^{-n-1}x - AS(t)x\right\| \leq \sum_{k=1}^{5}\|\mathcal{J}_k^n(t)\|, \tag{6.3.3}$$

for each $n \in \mathbb{N}^*$ and each $t \in [\alpha, 1/\alpha]$, where

$$\mathcal{J}_1^n(t) = \frac{n^{n+1}}{n!}\int_0^a (ve^{-v})^n \frac{1}{t}\frac{d}{dv}(S(tv))x\,dv,$$

$$\mathcal{J}_2^n(t) = \frac{n^{n+1}}{n!}\int_0^a (ve^{-v})^n AS(t)x\,dv,$$

$$\mathcal{J}_3^n(t) = \frac{n^{n+1}}{n!}\int_a^b (ve^{-v})^n (AS(tv)x - AS(t)x)\,dv,$$

$$\mathcal{J}_4^n(t) = \frac{n^{n+1}}{n!}\int_b^{+\infty} (ve^{-v})^n \frac{1}{t}\frac{d}{dv}(S(tv))x\,dv,$$

$$\mathcal{J}_5^n(t) = \frac{n^{n+1}}{n!}\int_b^{+\infty} (ve^{-v})^n AS(t)x\,dv,$$

for each $n \in \mathbb{N}^*$ and each $t \in [\alpha, 1/\alpha]$. Next, we shall evaluate each of the five terms on the right hand side of (6.3.3). To this aim, let us observe that, for each $\rho > 0$ and each $\tau \in [\rho, +\infty)$, we have

$$\|AS(\tau)x\| \leq \|AS(\rho)x\|. \tag{6.3.4}$$

Indeed, since the semigroup is differentiable, the mapping $\tau \mapsto AS(\tau)x$ is a C^1-solution of the equation $u' = Au$ on the interval $(0, +\infty)$ and then, since the semigroup generated by A is of contractions, we deduce (6.3.4).
We begin to evaluate $\mathcal{I}_1^n(t)$. Integrating by parts, observing that, for each $v \in (0,1)$, $|e^{-v} - ve^{-v}| \leq 1$ and taking into account that the mapping $v \mapsto ve^{-v}$ is nondecreasing on $(0,1)$, we deduce

$$\begin{aligned}\|\mathcal{I}_1^n(t)\| &= \left\| \frac{n^{n+1}}{n!} \int_0^a (ve^{-v})^n \frac{1}{t} \frac{d}{dv}(S(tv))x\, dv \right\| \\ &= \frac{1}{t} \left\| \frac{n^{n+1}}{n!}(ae^{-a})^n S(ta)x - \right. \\ &\left. - \frac{n^{n+1}}{n!} \int_0^a n(ve^{-v})^{n-1}(e^{-v} - ve^{-v})S(tv)x\, dv \right\| \\ &\leq \frac{\|x\|}{\alpha} \left[\frac{n^{n+1}}{n!}(ae^{-a})^n + \frac{n^{n+1}}{(n-1)!}(ae^{-a})^{n-1} \right]\end{aligned}$$

for each $n \in \mathbb{N}^*$ and $t \in [\alpha, 1/\alpha]$. We easily see that

$$\frac{n^{n+1}}{n!}(ae^{-a})^n = \frac{n^n}{n!}e^{-n} \cdot n\left(ae^{1-a}\right)^n \quad \text{and}$$

$$\frac{n^{n+1}}{(n-1)!}(ae^{-a})^{n-1} = \frac{(n-1)^{n-1}}{(n-1)!}e^{-(n-1)} \cdot n^2 \left(\frac{n}{n-1}\right)^{n-1} \left(ae^{1-a}\right)^{n-1}.$$

In view of Stirling's formula (see Nikolsky [**96**], p. 393), we have

$$\lim_{n\to\infty} \frac{n^n}{n!}e^{-n} = 0.$$

In addition, since $a \in (0,1)$, we have $ae^{1-a} < 1$. Therefore, we deduce both

$$\lim_{n\to\infty} \frac{n^{n+1}}{n!}(ae^{-a})^n = 0 \quad \text{and} \quad \lim_{n\to\infty} \frac{n^{n+1}}{(n-1)!}(ae^{-a})^{n-1} = 0. \tag{6.3.5}$$

So, we conclude that

$$\lim_{n\to\infty} \|\mathcal{I}_1^n(t)\| = 0$$

uniformly with respect to $t \in [\alpha, 1/\alpha]$.

Regarding $\mathcal{I}_2^n(t)$, from (6.3.4), we have

$$\|\mathcal{I}_2^n(t)\| \leq \|AS(\alpha)x\|\frac{n^{n+1}}{n!}\int_0^a (ve^{-v})^n\,dv$$

for each $n \in \mathbb{N}^*$, and $t \in [\alpha, 1/\alpha]$. From this inequality and Lemma 6.1.1, we get

$$\lim_{n\to\infty} \|\mathcal{I}_2^n(t)\| = 0,$$

uniformly for $t \in [\alpha, 1/\alpha]$.

In order to evaluate $\mathcal{I}_3^n(t)$, let us observe that, from (6.3.2) and Lemma 6.1.2, we have

$$\|\mathcal{I}_3^n(t)\| \leq \frac{n^{n+1}}{n!}\int_a^b \|(ve^{-v})^n\|AS(tv)x - AS(t)x\|\,dv \leq$$

$$\leq \varepsilon\frac{n^{n+1}}{n!}\int_a^b (ve^{-v})^n\,dv$$

$$\leq \varepsilon\frac{n^{n+1}}{n!}\int_0^\infty (ve^{-v})^n\,dv = \varepsilon$$

for each $n \in \mathbb{N}^*$ and each $t \in [\alpha, 1/\alpha]$. Consequently

$$\limsup_{n\to\infty} \|\mathcal{I}_3^n(t)\| \leq \varepsilon.$$

In order to evaluate $\mathcal{I}_4^n(t)$, let us observe that, in view of (6.3.4), we have

$$\|AS(tv)x\| \leq \|AS(\alpha b)x\|$$

for each $t \in [\alpha, \frac{1}{\alpha}]$ and each $v \in (b, +\infty)$. As a consequence

$$\|\mathcal{I}_4^n(t)\| = \left\|\frac{n^{n+1}}{n!}\int_b^\infty (ve^{-v})^n AS(tv)x\,dv\right\|$$

$$\leq \|AS(\alpha b)x\|\frac{n^{n+1}}{n!}\int_b^\infty (ve^{-v})^n\,dv,$$

for each $n \in \mathbb{N}^*$ and each $t \in [\alpha, 1/\alpha]$. From this relation, (6.3.5) and from Lemma 6.1.1, we get

$$\lim_{n\to\infty} \|\mathcal{I}_4^n(t)\| = 0.$$

Finally, from (6.3.4), we have

$$\|\mathcal{I}_5^n(t)\| \leq \|AS(\alpha)x\|\frac{n^{n+1}}{n!}\int_b^\infty (ve^{-v})^n\,dv$$

for each $n \in \mathbb{N}^*$, and each $t \in [\alpha, 1/\alpha]$ and accordingly

$$\lim_{n\to\infty} \|\mathcal{I}_5^n(t)\| = 0.$$

Therefore

$$\limsup_{n\to\infty}\left\|A\left(I-\frac{t}{n}A\right)^{-n-1}x-AS(t)x\right\|\leq\varepsilon$$

for each $\varepsilon>0$. Consequently, for each $\alpha\in(0,1)$, we have

$$\lim_{n\to\infty}A\left(I-\frac{t}{n}A\right)^{-n-1}x=AS(t)x \tag{6.3.6}$$

uniformly for $t\in[\alpha,1/\alpha]$. To complete the proof, we have merely to show that

$$\lim_{n\to\infty}A\left(I-\frac{t}{n+1}A\right)^{-n-1}x=AS(t)x \tag{6.3.7}$$

uniformly for $t\in[\alpha,1/\alpha]$. To this aim, let us observe that, from (6.3.6) and the Arzelà-Ascoli's Theorem A.2.1, it follows that, for each sequence $(a_n)_{n\in\mathbb{N}}$ of functions from $\mathbb{R}_+^*$ in $\mathbb{R}_+^*$ satisfying

$$\lim_{n\to\infty}a_n(t)=t,$$

uniformly on every compact subset in $\mathbb{R}_+^*$, we have

$$\lim_{n\to\infty}A\left(I-\frac{a_n(t)}{n}A\right)^{-n}x=AS(t)x$$

uniformly on every compact subset in $\mathbb{R}_+^*$. Indeed, this follows from the remark that, for each $x\in X$ and $\alpha\in(0,1)$, the family of functions

$$\left\{t\mapsto A\left(I-\frac{t}{n}A\right)^{-n-1}x;\ n\in\mathbb{N}^*\right\}$$

is relatively compact in $C([\alpha,1/\alpha];X)$, and thus a fortiori equicontinuous on $[\alpha,1/\alpha]$. See (6.3.6) and Arzelà-Ascoli's Theorem A.2.1. To complete the proof, we observe that the choice $a_n(t)=\frac{nt}{n+1}$, for $n\in\mathbb{N}^*$, leads to (6.3.7). □

Since the proof of Theorem 6.3.2 follows exactly the same lines as that of Theorem 6.3.1, we do not give details.

A useful sufficient condition in order that an operator A generate a compact C_0-semigroup of contractions is:

Corollary 6.3.2. *Let $A:D(A)\subseteq X\to X$ be the infinitesimal generator of a differentiable, i.e. uniformly differentiable C_0-semigroup of contractions $\{S(t)\,;\ t\geq0\}$. If for each $\lambda>0$, $(\lambda I-A)^{-1}$ is compact, then the semigroup generated by A is compact. Equivalently, if $D(A)$ endowed with the graph norm is compactly imbedded in X, then the semigroup generated by A is compact.*

Proof. The first part of the conclusion follows from Theorem 6.2.2 and Remark 6.3.4. The second one is a simple consequence of the fact that, for each $\lambda > 0$, $u \mapsto \|\lambda u - Au\|$, defined on $D(A)$, is a norm on $D(A)$, which is equivalent with the graph-norm. □

6.4. Semigroups with Symmetric Generators

In this section we shall prove that all C_0-semigroups of contractions which are generated by symmetric operators in Hilbert spaces are differentiable. So, let H be a Hilbert space. The next two lemmas will prove useful later.

Lemma 6.4.1. *Let $A : D(A) \subseteq H \to H$ be the infinitesimal generator of a C_0-semigroup of contractions. Let $(x_n)_{n\in\mathbb{N}}$ be a sequence in $D(A)$ with the property that $\lim_{n\to\infty} x_n = x$, and $(Ax_n)_{n\in\mathbb{N}}$ is bounded. Then $x \in D(A)$ and $\lim_{n\to\infty} Ax_n = Ax$ in the weak topology of H.*

Proof. Let us observe that, by virtue of Theorem 2.4.1, A is closed. Since graph A is a closed linear subspace in $H \times H$, it is sequentially strongly×weakly closed in $H \times H$. Therefore, if $(Ax_{n_k})_{k\in\mathbb{N}}$ is a weakly convergent subsequence of $(Ax_n)_{n\in\mathbb{N}}$, then $x \in D(A)$, and the weak limit of $(Ax_{n_k})_{k\in\mathbb{N}}$ is $y = Ax$. Since $(Ax_n)_{n\in\mathbb{N}}$ is bounded, and thus sequentially weakly compact, it follows that the set of all its weak sequential limit points is nonempty and, by the argument above, contains only one element y. The proof is complete. □

Lemma 6.4.2. *Let $A : D(A) \subseteq H \to H$ be the infinitesimal generator of a C_0-semigroup of contractions. Then, for each $x \in D(A)$, we have*

$$\langle Ax, x\rangle \leq 0.$$

Proof. This is an easy consequence of Lumer-Phillips Theorem 3.4.2 combined with Remark 3.4.1. □

A remarkable class of uniformly differentiable semigroups is given below.

Theorem 6.4.1. *Let $A : D(A) \subseteq H \to H$ be the infinitesimal generator of a C_0-semigroup of contractions, $\{S(t)\,;\ t \geq 0\}$, let $\xi \in H$, and $u(t) = S(t)\xi$ for $t \geq 0$. If A is self-adjoint, then*

(i) $u \in C([0,+\infty); H) \cap C((0,+\infty); D(A)) \cap C^1((0,+\infty); H)$, *and u is the unique solution of the Cauchy problem*

$$\begin{cases} u' = Au \\ u(0) = \xi \end{cases}$$

in this space;

(ii) $\|Au(t)\| \leq \dfrac{1}{t\sqrt{2}}\|\xi\|$ *for each $t > 0$;*

(iii) *the function* $t \mapsto \sqrt{t}\|Au(t)\|$ *belongs to* $L^2(0,+\infty)$ *and*

$$\int_0^\infty s\|Au(s)\|^2\,ds \le \frac{1}{4}\|\xi\|^2\,;$$

(iv) *if* $\xi \in D(A)$, *then* $\|Au(t)\|^2 \le \frac{1}{2t}\langle -A\xi,\xi\rangle$ *for each* $t>0$, *and*

(v) $t \mapsto Au(t) \in L^2(0,+\infty)$, *and* $\int_0^t \|Au(s)\|^2\,ds \le \frac{1}{2}\langle -A\xi,\xi\rangle$.

Proof. Let $\xi \in D(A^2)$. By virtue of Corollary 2.3.1, it follows that the problem

$$\begin{cases} u' = Au \\ u(0) = \xi \end{cases}$$

has a unique solution u of class C^2 on $\mathbb{R}_+$ which satisfies $u''(t) = Au'(t)$ for each $t \in \mathbb{R}_+$. Taking the inner product of both sides in the above equality by $u'(t)$, integrating from s to t, with $0 \le s \le t$, and using Lemma 6.4.2, we deduce

$$\frac{1}{2}\|u'(t)\|^2 - \frac{1}{2}\|u'(s)\|^2 = \int_s^t \langle Au'(\tau), u'(\tau)\rangle\,d\tau \le 0,$$

which shows that function $t \mapsto \|u'(t)\|$ is nonincreasing. We also have

$$\frac{d}{dt}\|u(t)\|^2 = 2\langle Au(t), u(t)\rangle, \tag{6.4.1}$$

and

$$\frac{d}{dt}\langle Au(t), u(t)\rangle = 2\langle Au(t), u'(t)\rangle = 2\|u'(t)\|^2 \ge 0. \tag{6.4.2}$$

Accordingly, $t \mapsto \langle Au(t), u(t)\rangle$ is nondecreasing. Integrating (6.4.1) from 0 to t, we successively obtain

$$\|u(t)\|^2 - \|\xi\|^2 = 2\int_0^t \langle Au(s), u(s)\rangle\,ds \le 2t\langle Au(t), u(t)\rangle$$

and

$$\begin{aligned} -t\langle Au(t), u(t)\rangle &\le -\int_0^t \langle Au(s), u(s)\rangle\,ds \\ &= -\frac{1}{2}\|u(t)\|^2 + \frac{1}{2}\|\xi\|^2 \le \frac{1}{2}\|\xi\|^2. \end{aligned}$$

Integrating (6.4.2) from 0 to t and recalling that the function $t \mapsto \|u'(t)\|$ is nonincreasing, we obtain

$$\langle Au(t), u(t)\rangle - \langle A\xi, \xi\rangle = 2\int_0^t \|u'(s)\|^2\,ds \ge 2t\|u'(t)\|^2.$$

Since $\langle Au(t), u(t)\rangle \leq 0$, it follows that

$$2t\|u'(t)\|^2 \leq \langle -A\xi, \xi\rangle. \tag{6.4.3}$$

Multiplying both sides in (6.4.2) by t, and integrating, we get

$$2t^2\|u'(t)\|^2 \leq \int_0^t s\langle Au(s), u'(s)\rangle\,ds = \int_0^t s\frac{d}{ds}\langle Au(s), u(s)\rangle\,ds$$

$$= t\langle Au(t), u(t)\rangle - \int_0^t \langle Au(s), u(s)\rangle\,ds.$$

Since $t\langle Au(t), u(t)\rangle \leq 0$, from (6.4.1) and the last inequality, we deduce

$$2t^2\|u'(t)\|^2 \leq \|\xi\|^2.$$

Accordingly,

$$\|Au(t)\| \leq \frac{1}{t\sqrt{2}}\|\xi\|$$

for each $\xi \in D(A^2)$, and $t > 0$. Since $D(A^2)$ is dense in H and the mapping $\xi \mapsto u(t) = S(t)\xi$ is nonexpansive, by the inequality above and Lemma 6.4.1, we obtain (ii). Since (i) and (iii) follow from (ii), while (iv) and (v) are consequences of (6.4.3), this achieves the proof. □

Remark 6.4.1. All the items in conclusion of Theorem 6.4.1 hold also if A is symmetric and generates a C_0-semigroup of contractions on H. Indeed, this follows from the simple remark that, in this case, A is self-adjoint. See Theorem 3.1.1 and Lemma 1.6.1.

Corollary 6.4.1. *Let $A : D(A) \subseteq H \to H$ be a self-adjoint and negatively defined operator, let $\{S(t)\,;\ t \geq 0\}$ be the C_0-semigroup of contractions generated by A, let $\xi \in H$, and $u(t) = S(t)\xi$ for $t \geq 0$. Let $(X_n)_{n\in\mathbb{N}}$ be the sequence of spaces defined as in Remark* 3.5.1. *Then, for each $n \in \mathbb{N}^*$, $u \in C^\infty((0, +\infty); X_n)$, and*

$$\|A^n u(t)\| \leq \left(\frac{n}{t\sqrt{2}}\right)^n \|x\|$$

for each $t > 0$.

Proof. Let X_1 be the Banach space in Theorem 3.5.1, endowed with the graph-norm $|\cdot|_{D(A)}$. Then X_1 is a Hilbert space with respect to the inner product $\langle\cdot,\cdot\rangle_1$ defined by

$$\langle x, y\rangle_1 = \langle x - Ax, y - Ay\rangle$$

for each $x, y \in X_1$. In addition, $A_{(1)}$ is self-adjoint on X_1. Indeed, we have

$$\langle A_{(1)}x, y\rangle_1 = \langle Ax - A^2x, y - Ay\rangle$$

$$= \langle Ax, y\rangle - \langle Ax, Ay\rangle - \langle A^2x, y\rangle + \rangle A^2x, Ay\rangle$$
$$= \langle x, Ay\rangle - \langle x, A^2y\rangle - \langle Ax, Ay\rangle + \langle Ax, A^2y\rangle = \langle x, A_{(1)}y\rangle_1$$

for each $x, y \in D(A_{(1)})$. Thus $A_{(1)}$ is symmetric. Since it generates a C_0-semigroup of contractions on X_1, it follows that $(I - A_{(1)})^{-1} \in \mathcal{L}(X_1)$. See Theorem 3.1.1. By Lemma 1.6.1, it follows that $A_{(1)}$ is self-adjoint. A simple inductive argument shows that, for $n \in \mathbb{N}^*$, X_n is a Hilbert space, and A_n is self-adjoint.

Let $t > 0$ and $n \in \mathbb{N}^*$. From (ii) in Theorem 6.4.1, we deduce that

$$\|Au(t/n)\| \leq \frac{n}{t\sqrt{2}}\|x\|.$$

Inasmuch as $u(2t/n) = S(t/n)u(t/n)x \in X_2$, applying once again (ii) in Theorem 6.4.1, this time to the operator $A_{(1)}$, we obtain

$$\|A^2u(2t/n)\| \leq \left(\frac{n}{t\sqrt{2}}\right)^2 \|x\|.$$

By induction, we conclude that

$$\|A^n u(t)\| \leq \left(\frac{n}{t\sqrt{2}}\right)^n \|x\|.$$

To complete the proof, we have to show that, for each $n \in \mathbb{N}^*$, u belongs to $C^\infty((0, +\infty); X_n)$. But this follows from the simple remark that, for each $x \in H$ and $t > 0$, we have $u(t) \in \cap_{n \geq 0} X_n$, combined with Corollary 2.4.1. □

6.5. The Linear Delay Equation

A remarkable example of a differentiable C_0-semigroup is given below.

Example 6.5.1. Let $r > 0$ and let $X = C([-r, 0]; \mathbb{R}^n)$ which, endowed with the sup-norm, is a Banach space. If $x : [-r, +\infty) \to \mathbb{R}^n$ is continuous, then, for each $t \geq 0$, the function $x_t : [-r, 0] \to \mathbb{R}^n$, defined by

$$x_t(\theta) = x(t + \theta),$$

for each $\theta \in [-r, 0]$, belongs to X. Let $L : X \to \mathbb{R}^n$ be a linear continuous operator, let $\varphi \in X$ and let us consider the Cauchy problem for the linear delay equation

$$\begin{cases} x'(t) = Lx_t & t \geq 0 \\ x(s) = \varphi(s) & s \in [-r, 0]. \end{cases} \tag{6.5.1}$$

By a solution of this problem we mean a function $x \in C([-r, +\infty); \mathbb{R}^n)$ with the property $x_{|[0,+\infty)} \in C^1([0, +\infty); \mathbb{R}^n)$ and satisfying $x(s) = \varphi(s)$, for each $s \in [-r, 0]$, and $x'(t) = Lx_t$, for each $t \geq 0$.

Proposition 6.5.1. *For each* $\varphi \in X$, *the problem* (6.5.1) *has a unique solution* $x : [-r, +\infty) \to \mathbb{R}^n$.

Proof. Let us observe that this problem is equivalent with the delay integral equation

$$x(t) = \begin{cases} \varphi(t) & \text{for } t \in [-r, 0] \\ \varphi(0) + \int_0^t Lx_s\, ds & \text{for } t \in (0, +\infty). \end{cases} \tag{6.5.2}$$

Therefore, to complete the proof, it suffices to show that, for each $\varphi \in X$ and each $T > 0$, the equation (6.5.2) has a unique solution $x : [-r, T] \to \mathbb{R}^n$. In order to show this, let $Y = C([-r, T]; \mathbb{R}^n)$ which, endowed with the sup-norm $\|\cdot\|_Y$, is a Banach space and let $Q : Y \to Y$ be defined by

$$(Qy)(t) = \begin{cases} \varphi(t) & \text{for } t \in [-r, 0] \\ \varphi(0) + \int_0^t Ly_s\, ds & \text{for } t \in (0, T]. \end{cases}$$

Let us observe that $x \in Y$ is a solution of the delay integral equation (6.5.2) if and only if x is a fixed point of Q. So, in that follows, we shall prove that Q has a unique fixed point. A simple inductive argument shows that, for each $k \in \mathbb{N}^*$, and each $t \in [-r, T]$, we have

$$\|(Q^k y)(t) - (Q^k z)(t)\| \leq \frac{\|L\|^k t^k}{k!} \|x - y\|_Y.$$

Clearly, this inequality implies

$$\|Q^k y - Q^k z\|_Y \leq \frac{\|L\|^k T^k}{k!} \|x - y\|_Y.$$

Therefore, for sufficiently large k, Q^k is a strict contraction, which, by the Banach fixed point theorem, has a unique fixed point $x \in Y$. Inasmuch as

$$\|Qx - x\| = \|Q^k Qx - Q^k x\| \leq q\|Qx - x\|$$

with $q \in (0, 1)$, it follows that x is a fixed point of Q. In addition, this is the unique because each one of its fixed points is necessarily a fixed point of Q^k. The proof is complete. □

Let $t \geq 0$, and let $S(t) : X \to X$ be defined by $S(t)\varphi = x_t$, where x is the unique solution of the problem (6.5.1) corresponding to the "initial datum" φ. By Proposition 6.5.1, from the linearity and continuity of the operator L, it follows that, for each $t \geq 0$, $S(t)$ is linear continuous from X to X. More than this, we have:

Theorem 6.5.1. *The family $\{S(t)\,;\ t \geq 0\}$, defined as above, is a semigroup of class C_0 in X whose generator, $A : D(A) \subseteq X \to X$, is given by $D(A) = \{\varphi \in C^1([-r,0];\,\mathbb{R}^n)\,;\ \varphi'(0) = L\varphi\}$, and*

$$(A\varphi)(\theta) = \begin{cases} \varphi'(\theta) & \text{for } \theta \in [-r,0) \\ L\varphi & \text{for } \theta = 0. \end{cases}$$

In addition, for each $t \geq r$, $S(t)$ is a compact operator and the semigroup is differentiable on $[\,r,+\infty)$.

Proof. Let us observe that, for each $t \geq 0$, and each $\theta \in [-r,0\,]$, we have

$$[S(t)\varphi](\theta) = \begin{cases} \varphi(t+\theta) & \text{for } t+\theta \leq 0 \\ \varphi(0) + \displaystyle\int_0^{t+\theta} L(S(\tau)\varphi)\,d\tau & \text{for } t+\theta > 0. \end{cases}$$

From this representation formula, after a simple calculation, we deduce the expression of A. From this, and Theorem A.2.1, it follows that $S(t)$ is compact for each $t \geq r$ and the semigroup is differentiable on $[\,r,+\infty)$. □

Problems

Problem 6.1. Let $p \in [\,1,+\infty)$, $X = l_p$, $(a_n)_{n\in\mathbb{N}^*}$ a sequence of positive real numbers, and let $\{S(t)\,;\ t \geq 0\}$ the semigroup in Problem 2.1. Prove that this semigroup is equicontinuous. Using this, and Problem 2.1 show that, if $a_n = n^2$ for each $n \in \mathbb{N}^*$, then $\{S(t)\,;\ t \geq 0\}$, although equicontinuous, is not uniformly continuous.

Problem 6.2. Let $X = c_0$, $(a_n)_{n\in\mathbb{N}^*}$ a sequence of real positive numbers, and $\{S(t)\,;\ t \geq 0\}$ the semigroup in Problem 2.2. Prove that this semigroup is equicontinuous. Using this, and Problem 2.2, show that, if $a_n = n^2$ for each $n \in \mathbb{N}^*$, then $\{S(t)\,;\ t \geq 0\}$, although equicontinuous, is not uniformly continuous.

Problem 6.3. Let $p \in [\,1,+\infty)$, $X = l_p$, $(a_n)_{n\in\mathbb{N}^*}$ a sequence of positive real numbers and let $\{S(t)\,;\ t \geq 0\}$ the semigroup in Problem 2.1. Prove that this semigroup is compact if and only if $\lim_{n\to\infty} a_n = \infty$.

Problem 6.4. Let $X = c_0$, $(a_n)_{n\in\mathbb{N}^*}$ a sequence of real positive numbers and $\{S(t)\,;\ t \geq 0\}$ the semigroup in Problem 2.2. Prove that this semigroup is compact if and only if $\lim_{n\to\infty} a_n = \infty$.

Problem 6.5. Let H be the complexification of the space l_2 (we notice that H is considered as a real Hilbert space– see Section 1.6), $t \geq 0$, and let

$S(t) : l_2 \to l_2$ be defined by $(S(t)(x_n)_{n\in\mathbb{N}^*})_{k\in\mathbb{N}^*} = \left(e^{-k^2t+ie^{k^4}t}x_k\right)_{k\in\mathbb{N}^*}$ for each $(x_n)_{n\in\mathbb{N}^*} \in H$. Prove that $\{S(t)\,;\ t \geq 0\}$ is a compact C_0-semigroup of contractions. Find its infinitesimal generator, and show that this semigroup is not differentiable. This is Example 4.4.2, p. 181 in Balakrishnan [**12**].

Let H be a separable Hilbert space. An operator $T \in \mathcal{L}(H)$ is called *a Hilbert-Schmidt operator*, if there exists an orthonormal, and complete system $\{e_n\,;\ n \in \mathbb{N}^*\}$ in H, such that $\sum_{n=1}^{\infty} \|Te_n\|^2 < +\infty$. A C_0-semigroup of contractions on H, $\{S(t)\,;\ t \geq 0\}$, is *Hilbert-Schmidt*, if, for each $t > 0$, $S(t)$ is a Hilbert-Schmidt operator.

Problem 6.6. Prove that each Hilbert-Schmidt C_0-semigroup is compact.

Problem 6.7. Prove that the semigroup in the Problem 2.1 with $p = 2$, and $a_n = n^2$ for each $n \in \mathbb{N}^*$, is Hilbert-Schmidt, and therefore compact.

Problem 6.8. Let $X = L^2(0, \pi)$ and let $\{S(t)\,;\ t \geq 0\}$ be the semigroup in Problem 4.2. Prove that this semigroup is Hilbert-Schmidt and therefore compact.

Problem 6.9. Show that a C_0-group is differentiable if and only if is uniformly continuous.

Notes. The results in Section 6.1 are due to Vrabie [**126**]. Theorem 6.2.2 extends a remark in Cârjă and Vrabie [**34**]. A similar result was proved by Balakrishnan [**12**] in the case in which A is self-adjoint. The necessity of Theorem 6.3.2 is due to Lax (see Hille and Phillips [**70**], Theorem 10.2.2, p. 304), while the sufficiency, which is by far the most interesting, to Pazy [**98**]. Theorem 6.2.3 was established in Vrabie [**126**]; Theorems 6.3.1 and 6.3.2 were proved in Vrabie [**130**]. We recall also

Theorem (Pazy) *The operator $A : D(A) \subseteq X \to X$ is the infinitesimal generator of an eventually differentiable C_0-semigroup of type (M, ω) if and only if there exist $a \in \mathbb{R}$ and $b, C \in \mathbb{R}_+^*$ such that*

$$\{\lambda \in \mathbb{C};\ \operatorname{Re}\lambda \geq a - b\log|\operatorname{Im}\lambda\} = \Sigma \subset \rho(A) \quad \text{and}$$
$$\|R(\lambda\,; A)\|_{\mathcal{L}(X)} \leq C|\operatorname{Im}\lambda| \ \text{ for } \ \lambda \in \Sigma,\ \operatorname{Re}\lambda \leq \omega.$$

See Pazy [**101**], Theorem 4.7, p. 54. Theorem 6.4.1 is from Brezis and Cazenave, while Example 6.5.1 is due to Hale [**67**]. For other significant examples of functional equations treated by semigroup methods we refer to Corduneanu [**40**], pp. 245-261. For applications to control theory see Ahmed [**3**] and Cârjă [**33**]. The problems are from Balakrishnan [**12**] and Pazy [**101**].

CHAPTER 7

Analytic Semigroups

In this chapter we introduce a class of differentiable semigroups on a complex Banach space X which can be extended as analytic functions acting from a certain sector in the complex plane including the resolvent set of the generator to $\mathcal{L}(X)$. After analyzing several remarkable examples as the heat equation, the Stokes equation, some parabolic and elliptic problems with dynamic boundary conditions, we define and study the fractional powers of some closed operators, with main emphasis on the negative generators of analytic semigroups.

7.1. Definition and Characterizations

Let X be a complex Banach space, and let $A : D(A) \subseteq X \to X$ be a $\mathbb{C}$-linear operator generating a C_0-semigroup of contractions $\{S(t)\,;\ t \geq 0\}$. For $0 < \theta \leq \pi$, we define the sector

$$\mathbb{C}_\theta = \{z \in \mathbb{C};\ -\theta < \arg z < \theta\}.$$

Clearly,

$$\overline{\mathbb{C}}_\theta = \{z \in \mathbb{C};\ -\theta \leq \arg z \leq \theta\} \cup \{0\}.$$

Definition 7.1.1. We say that the C_0-semigroup $\{S(t)\,;\ t \geq 0\}$ is *analytic*, if there exists $0 < \theta \leq \pi$, and a mapping $\tilde{S} : \overline{\mathbb{C}}_\theta \to \mathcal{L}(X)$ such that:

(i) $S(t) = \tilde{S}(t)$ for each $t \geq 0$;

(ii) $\tilde{S}(z + w) = \tilde{S}(z)\tilde{S}(w)$ for each $z, w \in \overline{\mathbb{C}}_\theta$;

(iii) $\lim_{z \in \overline{\mathbb{C}}_\theta,\, z \to 0} \tilde{S}(z)x = x$ for each $x \in X$;

(iv) the mapping $z \mapsto \tilde{S}(z)$ is analytic from $\mathbb{C}_\theta$ to $\mathcal{L}(X)$.

If, in addition, for each $0 < \delta < \theta$, the mapping $z \mapsto \tilde{S}(z)$ is bounded from $\mathbb{C}_\delta$ to $\mathcal{L}(X)$, the C_0-semigroup $\{S(t);\ t \geq 0\}$ is called *analytic and uniformly bounded.*

Theorem 7.1.1. *Let $A : D(A) \subseteq X \to X$ be a $\mathbb{C}$-linear operator which generates a C_0-semigroup of contractions $\{S(t)\,;\ t \geq 0\}$. If $0 \in \rho(A)$, then, the following conditions are equivalent:*

(i) *the semigroup* $\{S(t)\,;\ t \geq 0\}$ *is analytic and uniformly bounded;*

(ii) $\{\lambda \in \mathbb{C};\ \operatorname{Re}\lambda > 0\} \subseteq \rho(A)$, *and there exists* $C > 0$ *such that, for each* $\lambda \in \mathbb{C}$ *with* $\operatorname{Re}\lambda > 0$ *and* $\operatorname{Im}\lambda \neq 0$, *we have*

$$\|R(\lambda; A)\|_{\mathcal{L}(X)} \leq \frac{C}{|\operatorname{Im}\lambda|};$$

(iii) *there exist* $\delta \in (0, \frac{\pi}{2})$ *and* $M > 0$, *such that* $\mathbb{C}_{\frac{\pi}{2}+\delta} \subseteq \rho(A)$ *and, for each* $\lambda \in \mathbb{C}_{\frac{\pi}{2}+\delta}$, *we have*

$$\|R(\lambda; A)\|_{\mathcal{L}(X)} \leq \frac{M}{|\lambda|};$$

(iv) *the semigroup* $\{S(t)\,;\ t \geq 0\}$ *is uniformly differentiable for* $t > 0$ *and there exists* $C > 0$ *so that*

$$\|S'(t)\|_{\mathcal{L}(X)} \leq \frac{C}{t}$$

for each $t > 0$.

Proof. We begin by showing that (i) implies (ii). Let $\delta > 0$ for which there exists $C_1 > 0$ such that

$$\|\tilde{S}(z)\|_{\mathcal{L}(X)} \leq C_1$$

for each $z \in \mathbb{C}$ with $|\arg(z)| < \delta$. By virtue of Remark 3.1.2, for each $x \in X$, $\sigma > 0$, and $\tau \in \mathbb{R}$, we have

$$R(\sigma + i\tau; A)x = \int_0^\infty e^{-(\sigma+i\tau)t} S(t)x\,dt.$$

Let $\theta \in (0, \frac{\pi}{2})$. Since the semigroup is analytic, for $\tau > 0$, we can shift the path of integration from $(0, +\infty)$ to the ray $\{\rho e^{-i\theta};\ 0 < \rho < +\infty\}$ oriented from 0 to $+\infty$. We obtain

$$\|R(\sigma + i\tau; A)x\| \leq \int_0^\infty e^{-\rho(\sigma\cos\theta + \tau\sin\theta)} C_1\|x\|\,d\rho$$

$$= \frac{C_1}{\sigma\cos\theta + \tau\sin\theta}\|x\| \leq \frac{C}{|\tau|}\|x\|.$$

Analogously, for $\tau < 0$, shifting the path of integration from $(0, +\infty)$ to the ray $\{\rho e^{i\theta};\ 0 < \rho < +\infty\}$, oriented from 0 to $+\infty$, we obtain

$$\|R(\sigma + i\tau; A)x\| \leq \int_0^\infty e^{-\rho(\sigma\cos\theta - \tau\sin\theta)} C_1\|x\|\,d\rho$$

$$= \frac{C_1}{\sigma\cos\theta - \tau\sin\theta}\|x\| \leq \frac{C}{|\tau|}\|x\|,$$

which proves (ii).

To prove that (ii) implies (iii), let us observe that, thanks to Remark 3.1.2, for each $\lambda \in \mathbb{C}$ with $\operatorname{Re}\lambda > 0$, we have

$$\|R(\lambda; A)\|_{\mathcal{L}(X)} \leq \frac{1}{\operatorname{Re}\lambda}.$$

On the other hand, by (ii), we know that, for each $\lambda \in \mathbb{C}$ satisfying $\operatorname{Re}\lambda > 0$, and $\operatorname{Im}\lambda \neq 0$, we have

$$\|R(\lambda; A)\|_{\mathcal{L}(X)} \leq \frac{C}{|\operatorname{Im}\lambda|}.$$

But, from these two inequalities, it follows that there exists $M > 0$ such that

$$\|R(\lambda; A)\|_{\mathcal{L}(X)} \leq \frac{M}{|\lambda|}$$

for each $\lambda \in \mathbb{C}$ with $\operatorname{Re}\lambda > 0$. Let $\sigma > 0$ and $\tau \in \mathbb{R}$. The Taylor expansion of the resolvent function around $\sigma + i\tau$ is

$$R(\lambda; A) = \sum_{n=0}^{\infty}(-1)^n R(\sigma + i\tau; A)^{n+1}(\sigma + i\tau - \lambda)^n.$$

This series is convergent in $\mathcal{L}(X)$ for each $\lambda \in \mathbb{C}$ satisfying

$$\|R(\sigma + i\tau; A)\|_{\mathcal{L}(X)}|\sigma + i\tau - \lambda| \leq k < 1.$$

Taking $\lambda = \operatorname{Re}\lambda + i\tau$ in the above series and using the inequality in (ii), we observe that it is convergent in $\mathcal{L}(X)$ for $|\sigma - \operatorname{Re}\lambda| \leq k|\tau|/C$. Since both $\sigma > 0$ and $k \in (0, 1)$ are arbitrary, it follows that $\rho(A)$ includes all complex numbers λ with $\operatorname{Re}\lambda \leq 0$ satisfying $|\operatorname{Re}\lambda|/|\operatorname{Im}\lambda| < 1/C$. In particular, we have

$$C_{\frac{\pi}{2}+\delta} = \left\{\lambda \in \mathbb{C};\ |\arg\lambda| < \frac{\pi}{2} + \delta\right\} \subseteq \rho(A),$$

where $\delta = k\arctan(1/C)$, $k \in (0, 1)$. Moreover, on $C_{\frac{\pi}{2}+\delta}$, we have

$$\|R(\lambda; A)\|_{\mathcal{L}(X)} \leq \frac{C}{1-k} \leq \frac{\sqrt{C^2+1}}{(1-k)}\frac{1}{|\lambda|} = \frac{M}{|\lambda|},$$

from where it follows that A satisfies (iii).

To prove that (iii) implies (iv) first let us remark that, if (iii) holds, for each $t > 0$, we have $\sigma(tA) \subseteq \{\lambda \in \mathbb{C};\ \operatorname{Re}\lambda < 0\}$. Then, for a fixed $\frac{\pi}{2} < \theta < \frac{\pi}{2} + \delta$, the Dunford integral of the analytic function e^z calculated at tA is well-defined, and

$$e^{tA} = \frac{1}{2\pi i}\int_{\Gamma} e^{\mu t}R(\mu; A)\,d\mu, \tag{7.1.1}$$

where Γ is the path consisting of the two rays $\{\rho e^{-i\theta};\ 0 < \rho < +\infty\}$, and $\{\rho e^{i\theta};\ 0 < \rho < +\infty\}$, oriented in the sense of the increase of the imaginary part of λ. We emphasize that, from (iii) and from the condition $0 \in \rho(A)$, it follows that Γ is entirely contained in the domain of analyticity of the resolvent function. We shall prove next that, for each $t \geq 0$,

$$e^{tA} = S(t), \tag{7.1.2}$$

where e^{tA} is defined by (7.1.1). First, let us observe that, from (i) in Theorem 1.8.1, it follows that $\{e^{tA}\,;\ t > 0\} \cup \{I\}$ is a semigroup. Therefore, to check (7.1.2), it suffices to show that, for each $\lambda > 0$, we have

$$R(\lambda; A) = \int_0^{+\infty} e^{-\lambda t} e^{tA}\, dt.$$

Let $\lambda > 0$. Let us multiply (7.1.1) by $e^{-\lambda t}$, and let us integrate from 0 to b. From Fubini theorem and the Residues theorem, we deduce

$$\int_0^b e^{-\lambda t} e^{tA}\, dt = \frac{1}{2\pi i} \int_\Gamma \frac{1}{\mu - \lambda}(e^{(\mu-\lambda)b} - 1) R(\mu; A)\, d\mu$$

$$= R(\lambda; A) + \frac{1}{2\pi i} \int_\Gamma \frac{e^{(\mu-\lambda)b}}{\mu - \lambda} R(\mu; A)\, d\mu.$$

As

$$\lim_{b \to +\infty} \int_\Gamma \frac{e^{(\mu-\lambda)b}}{\mu - \lambda} R(\mu; A)\, d\mu = 0,$$

from the above equality, we get (7.1.2). At this point, let us observe that the integral on the right-hand side in (7.1.1) can be differentiated with respect to the parameter $t > 0$, and we can interchange the integration with the differentiation, because the integral $\frac{1}{2\pi i} \int_\Gamma \lambda e^{\lambda t} R(\lambda; A)\, d\lambda$ is convergent for $t > 0$ in the uniform operator norm of the space $\mathcal{L}(X)$. This last assertion follows from the simple observation that, for each $t > 0$, we have

$$\left\| \frac{1}{2\pi i} \int_\Gamma \mu e^{\mu t} R(\mu; A)\, d\mu \right\| \leq \frac{1}{\pi} \int_0^{+\infty} e^{-\rho \cos\theta t} d\rho = \frac{1}{\pi \cos\theta} \frac{1}{t}.$$

Differentiating both sides in (7.1.1) with respect to $t > 0$, using (7.1.2) and the above inequality, we deduce (iv) with $C = \frac{1}{\pi \cos\theta}$.
To prove that (iv) implies (i), let us observe that, for each $t > 0$ and each $n \in \mathbb{N}^*$, we have

$$S^{(n)}(t) = \left[S'\left(\frac{t}{n}\right) \right]^n. \tag{7.1.3}$$

This follows by observing that, if the semigroup $\{S(t);\ t \geq 0\}$ is uniformly differentiable, then, for each $x \in X$, each $t > 0$ and each $n \in \mathbb{N}^*$, we have

$S(t)x \in D(A^n)$. From Corollary 2.4.1 and the fact that the semigroup commutes with its infinitesimal generator, we have

$$S^{(n)}(t) = A^n S(t) = A^n S\left(\frac{t}{n}\right)^n = \left[AS\left(\frac{t}{n}\right)\right]^n = \left[S'\left(\frac{t}{n}\right)\right]^n,$$

which proves (7.1.3). From (7.1.3) and the inequality $n^n \leq n!e^n$, we deduce

$$\frac{1}{n!}\|S^{(n)}(t)\|_{\mathcal{L}(X)} \leq \left(\frac{Ce}{t}\right)^n, \tag{7.1.4}$$

for each $t > 0$, and each $n \in \mathbb{N}^*$.
Let us consider now the power series

$$\tilde{S}(z) = S(t) + \sum_{n=1}^{\infty} \frac{(z-t)^n}{n!} S^{(n)}(t) \tag{7.1.5}$$

which, by virtue of (7.1.4), is uniformly convergent in the norm of $\mathcal{L}(X)$ for $|z-t| \leq k(\frac{t}{Ce})$, for each $k \in (0,1)$. Obviously, the family of linear operators $\{\tilde{S}(z)\,;\ z \in \mathbb{C}_\theta\}$, with $\theta = \arctan(\frac{1}{Ce})$, extends $\{S(t)\,;\ t \geq 0\}$ to $\mathbb{C}_\theta$ and therefore it satisfies (i) in Definition 1.5.1. By (i) and (ii) in Theorem 1.8.1, it follows that $\{\tilde{S}(z)\,;\ z \in \overline{\mathbb{C}}_\theta\}$ satisfies (ii) in Definition 7.1.1. By (7.1.5), we conclude that $\{\tilde{S}(z)\,;\ z \in \overline{\mathbb{C}}_\theta\}$ satisfies (iii) and (iv) in Definition 7.1.1. Since, by virtue of (7.1.4), the uniform boundedness condition is obviously satisfied, this achieves the proof. □

From Theorem 7.1.1 and Theorem 6.3.2 it follows:

Theorem 7.1.2. *Let $A : D(A) \subseteq X \to X$ be a $\mathbb{C}$-linear operator generating a C_0-semigroup of contractions $\{S(t)\,;\ t \geq 0\}$. Then $\{S(t)\,;\ t \geq 0\}$ is analytic if and only if for each $\alpha \in (0,1)$ there exists*

$$\lim_{n\to\infty} A\left(I - \frac{t}{n}A\right)^{-n} = T(t)$$

in the usual sup-norm topology of $C([\alpha, 1/\alpha]; \mathcal{L}(X))$, and there exists $C > 0$ such that

$$\|T(t)\|_{\mathcal{L}(X)} \leq \frac{C}{t}$$

for each $t > 0$.

If H is a complex Hilbert space, we have:

Corollary 7.1.1. *If $A : D(A) \subseteq H \to H$ is self-adjoint and generates a C_0-semigroup of contractions $\{S(t)\,;\ t \geq 0\}$, then $\{S(t)\,;\ t \geq 0\}$ is analytic.*

Proof. First, let us observe that A has a $\mathbb{C}$-linear self-adjoint canonical extension. On the other hand, by (ii) in Theorem 6.4.1, it follows that (iv) in Theorem 7.1.1 holds. The proof is complete. □

7.2. The Heat Equation

Using the results established in Chapter 6 and in the preceding section, we can obtain additional information concerning the regularity of solutions of many partial differential equations of parabolic nature. We begin with the heat equation. More precisely, let Ω a nonempty and open subset in $\mathbb{R}^n$ whose boundary is denoted by Γ, let $Q_\infty = \mathbb{R}_+ \times \Omega$, $\Sigma_\infty = \mathbb{R}_+ \times \Gamma$, and let us consider the equation

$$\begin{cases} u_t = \Delta u & (t,x) \in Q_\infty \\ u = 0 & (t,x) \in \Sigma_\infty \\ u(0,x) = \xi(x) & x \in \Omega, \end{cases} \tag{7.2.1}$$

known as *the heat, or diffusion equation.*

Theorem 7.2.1. *For each $\xi \in H^{-1}(\Omega)$, the problem* (7.2.1) *has a unique continuous solution $u \in C(\mathbb{R}_+; H^{-1}(\Omega))$ which satisfies:*

(i) $u \in C(\mathbb{R}_+^*; H_0^1(\Omega)) \cap C^1(\mathbb{R}_+^*; H^{-1}(\Omega))$;
(ii) *for each $m \in \mathbb{N}$, we have $\Delta^m u \in C^\infty(\mathbb{R}_+^*; H_0^1(\Omega))$;*
(iii) $u \in C^\infty(\mathbb{R}_+^* \times \Omega)$;
(iv) *if $\xi \in L^2(\Omega)$, then $u \in C(\mathbb{R}_+; L^2(\Omega))$;*
(v) *if $\xi \in H_0^1(\Omega)$, then $u \in C(\mathbb{R}_+; H_0^1(\Omega)) \cap C^1(\mathbb{R}_+; H^{-1}(\Omega))$;*
(vi) *if $\xi \in H_0^1(\Omega)$ and $\Delta\xi \in L^2(\Omega)$, then $\Delta u \in C(\mathbb{R}_+; L^2(\Omega))$ and $u \in C^1(\mathbb{R}_+; L^2(\Omega))$.*

Proof. The item (i) follows from both Theorem 4.1.1 combined with (i) in Theorem 6.4.1. To check (ii), let us observe that, in view of Theorem 4.1.1 and Corollary 7.1.1, the operator Δ with the Dirichlet boundary condition is self-adjoint, and generates a C_0-semigroup in $H^{-1}(\Omega)$. So, we are in the hypotheses of Corollary 6.4.1, from where it follows that, for each $m \in \mathbb{N}$, we have $u \in C^\infty((0,+\infty); X_m)$. Since $X_m = D(\Delta^m)$, and $u^{(m)} = \Delta^m u$ (see Corollary 2.4.1), it follows that (ii) holds. As $\cap_{m\geq 0} X_m \subseteq C^\infty(\Omega)$, from (ii) we get (iii). The items (iv), (v) and (vi) follow from (vi) in Corollary 3.5.1, observing that the C_0-semigroup generated by the Laplace operator with the Dirichlet boundary condition in $L^2(\Omega)$ coincides with the $L^2(\Omega)$-restriction of the C_0-semigroup generated by the homonymous operator in $H^{-1}(\Omega)$. The proof is complete. □

In the case in which Ω is bounded and Γ sufficiently smooth, the unique solution u of the problem (7.2.1) enjoys additional regularity properties. More precisely we have:

Theorem 7.2.2. *If Ω is bounded and $\xi \in H_0^1(\Omega) \cap H^2(\Omega)$, then:*

(i) *if Γ is C^2, then $u \in C([0,+\infty); H^2(\Omega)) \cap C^1([0,+\infty); L^2(\Omega))$;*

(ii) *if Γ is C^{2m} and $m \in \mathbb{N}^*$, then $u \in C^\infty((0,+\infty); H^{2m}(\Omega))$;*

(iii) *if Γ is C^∞, then $u \in C^\infty([\varepsilon,+\infty) \times \overline{\Omega})$ for each $\varepsilon > 0$;*

(iv) *if Γ is C^∞, $\xi \in C^\infty(\overline{\Omega})$ and, in addition, ξ satisfies the usual compatibility conditions $\xi = \Delta\xi = \cdots = \Delta^m\xi = \cdots = 0$ on Γ, then $u \in C^\infty([0,+\infty) \times \overline{\Omega})$.*

Proof. The conclusion follows from Theorem 7.2.1 and Remark 4.1.1. □

Theorem 7.2.3. *Let $\xi \in L^2(\Omega)$, and let u be the unique continuous solution of the problem (7.2.1). Then:*

(i) $\|\Delta u(t)\|_{L^2(\Omega)} \leq \dfrac{1}{t\sqrt{2}}\|\xi\|_{L^2(\Omega)}$ *for each* $t > 0$;

(ii) $\displaystyle\int_0^\infty s\|\Delta u(s)\|_{L^2(\Omega)}^2 ds \leq \frac{1}{4}\|\xi\|_{L^2(\Omega)}^2$;

(iii) $\|\nabla u(t)\|_{L^2(\Omega)} \leq \dfrac{1}{\sqrt{2t}}\|\xi\|_{L^2(\Omega)}$ *for each* $t > 0$;

(iv) $\displaystyle\int_0^\infty \|\nabla u(s)\|_{L^2(\Omega)}^2 ds \leq \frac{1}{2}\|\xi\|_{L^2(\Omega)}^2$;

(v) *if* $\xi \in H_0^1(\Omega)$, $\|\Delta u(t)\|_{L^2(\Omega)} \leq \dfrac{1}{\sqrt{2t}}\|\nabla\xi\|_{L^2(\Omega)}$ *for* $t > 0$;

(vi) *if* $\xi \in H_0^1(\Omega)$, $\displaystyle\int_0^\infty s\|\Delta u(s)\|_{L^2(\Omega)}^2 ds \leq \frac{1}{2}\|\nabla\xi\|_{L^2(\Omega)}^2$.

Proof. The conclusion follows from Theorem 6.4.1, taking into account both Theorem 4.1.1 and Remark 6.4.1. □

Theorem 7.2.4. *If Ω is bounded, and Γ is of class C^1, then the Laplace operator with the Dirichlet boundary condition generates a compact C_0-semigroup of contractions in $H^{-1}(\Omega)$.*

Proof. We apply Theorem 4.1.1 and Corollary 6.3.2, observing that, by the Sobolev-Rellich-Kondrachov Theorem 1.5.4, the operator $(\lambda I - \Delta)^{-1}$ is compact in $\mathcal{L}(H^{-1}(\Omega))$. □

Theorem 7.2.5. *If Ω is bounded and Γ is of class C^1, then the Laplace operator with the Dirichlet boundary condition generates a compact C_0-semigroup of contractions in $L^2(\Omega)$.*

Proof. We apply Theorem 4.1.2 and Corollary 6.3.2 observing that, from the Sobolev-Rellich-Kondrachov Theorem 1.5.4, $(\lambda I-\Delta)^{-1}$ is compact from $L^2(\Omega)$ to $L^2(\Omega)$. □

We conclude this section with some useful regularity results concerning the semigroup generated by the Laplace operator on $L^q(\Omega)$ for $q \geq 1$. First we need the following simple lemma.

Lemma 7.2.1. *Let Ω be a nonempty, bounded and open subset in $\mathbb{R}^n$ whose boundary is of class C^1 and let $p \in [1,+\infty]$. Let A_p be the Laplace operator subjected to the Dirichlet boundary condition on $L^p(\Omega)$, with the convention that, for $p=\infty$, $L^p(\Omega)$ is replaced with $C_0(\overline{\Omega})$. Let $\{S_p(t)\,;\ t \geq 0\}$ be the C_0-semigroup generated by A_p on $L^p(\Omega)$. Then, for each $p,q \in [1,+\infty]$, each $\xi \in C_0(\overline{\Omega})$, and each $t \geq 0$, we have $S_p(t)\xi = S_q(t)\xi$.*

Proof. One may easily see that, for each $p \in [1,+\infty]$, $C_0(\overline{\Omega}) \subset L^p(\Omega)$, $\mathrm{graph}(A_\infty) \subseteq \mathrm{graph}(A_p)$ and $D(A_\infty) \subset D(A_p)$, all the inclusions being dense. This completes the proof. □

Lemma 7.2.1 allows us to denote the C_0-semigroup generated by the Laplace operator subjected to the Dirichlet boundary condition on any of the spaces $L^p(\Omega)$, by the very same symbol $\{S(t)\,;\ t \geq 0\}$.

Theorem 7.2.6. *Let Ω be a nonempty, bounded and open subset in $\mathbb{R}^n$ whose boundary is of class C^1. Let $q \geq 1$ and let $\{S(t)\,;\ t \geq 0\}$ be the C_0-semigroup generated by the Laplace operator subjected to the Dirichlet boundary condition on $L^q(\Omega)$. Then, for each $1 \leq q \leq p \leq +\infty$, each $\xi \in L^q(\Omega)$ and each $t > 0$, we have*

$$\|S(t)\xi\|_{L^p(\Omega)} \leq (4\pi t)^{-\frac{n}{2}\left(\frac{1}{q}-\frac{1}{p}\right)}\|\xi\|_{L^q(\Omega)}. \tag{7.2.2}$$

We shall prove Theorem 7.2.6 with the help of the next lemmas.

Lemma 7.2.2. *Let $K : (0,+\infty) \times \mathbb{R}^n \to \mathbb{R}$ be defined by*

$$K(t)(x) = (4\pi t)^{-\frac{n}{2}} e^{-\frac{\|x\|^2}{4t}}.$$

If $\psi \in C(\mathbb{R}^n)$ has compact support, then $v(t,x) = K(t)(x) * \psi(x)$ satisfies

(i) $v \in C([0,+\infty); C_b(\mathbb{R}^n)) \cap C((0,+\infty); C_b^2(\mathbb{R}^n))$[1]

(ii) *if* $1 \leq p \leq +\infty$, $v \in C([0,+\infty); L^p(\mathbb{R}^n)) \cap C^\infty((0,+\infty); L^p(\mathbb{R}^n))$

(iii) $v_t = \Delta v$ *for* $t > 0$ *and* $v(0) = \psi$

---

[1] $C_b^k(\mathbb{R}^n)$ denotes the space of all real-valued functions defined on $\mathbb{R}^n$ which are bounded and of class C^k.

(iv) $\|v(t)\|_{L^p(\Omega)} \le (4\pi t)^{-\frac{n}{2}\left(\frac{1}{q}-\frac{1}{p}\right)}\|\psi\|_{L^q(\Omega)}$ *for each* $1 \le q \le p \le +\infty$ *and* $t > 0$.

Proof. The regularity properties (i) and (ii), as well as (iii) follow from direct calculations. See also Problems 4.3 and 4.4. As concerns (iv), it follows from Hausdorff-Young inequality, by observing that

$$\|K(t)\|_{L^p(\mathbb{R}^n)} \le p^{-\frac{n}{2p}}(4\pi t)^{-\frac{n}{2}\left(1-\frac{1}{p}\right)} \le (4\pi t)^{-\frac{n}{2}\left(1-\frac{1}{p}\right)}$$

for $1 \le p \le +\infty$ and $t > 0$. This completes the proof. □

Lemma 7.2.3. *Let* $\varphi \in L^2(\Omega)$ *be nonnegative, i.e.* $\varphi \ge 0$ *a.e. in* Ω *and let* $\{S(t)\,;\ t \ge 0\}$ *be the* C_0*-semigroup generated by the Laplace operator subjected to Dirichlet boundary condition in* $L^2(\Omega)$*. Then, for each* $t > 0$*, we have* $S(t)\varphi \ge 0$ *a.e. in* Ω.

Proof. The conclusion follows by a simple density argument observing that, by virtue of (v) in Theorem 7.2.1, for each $\varphi \in C_0(\overline{\Omega}) \cap H_0^1(\Omega)$ with $\Delta u \in C_0(\overline{\Omega})$, we have $u \in C([0,+\infty); H_0^1(\Omega))$. Therefore $u^- = \min\{u, 0\}$ satisfies $u^- \in C([0,+\infty); H_0^1(\Omega))$, and thus

$$\begin{aligned}\frac{d}{dt}\int_\Omega (u^-)^2\,dx &= -\int_\Omega u_t u^-\,dx = -\int_\Omega u^-\Delta u\,dx\\ &= \int_\Omega \nabla u\cdot\nabla u^-\,dx = -\int_\Omega \|\nabla u^-\|^2\,dx \le 0.\end{aligned}$$

Since $u^-(0) = 0$, this inequality shows that $\int_\Omega (u^-)^2\,dx \le 0$ and hence $u(t) \ge 0$ for each $t \ge 0$ and a.e. in Ω. The proof is complete. □

We may now pass to the proof of Theorem 7.2.6.

Proof. By a density argument, we may confine ourselves only to the case in which $\varphi \in C_0(\overline{\Omega}) \cap H_0^1(\Omega)$. Denote by $\eta = |\varphi|$ and let us observe that, by virtue of Lemma 7.2.3, we have

$$-S(t)\eta \le S(t)\varphi \le S(t)\eta,$$

a.e. in Ω. Thus

$$\|S(t)\varphi\|_{L^p(\Omega)} \le \|S(t)\eta\|_{L^p(\Omega)}. \tag{7.2.3}$$

Let us define $\psi : \mathbb{R}^n \to \mathbb{R}$ by

$$\psi(x) = \begin{cases} \eta(x) & \text{for } x \in \Omega\\ 0 & \text{for } x \in \mathbb{R}^n \setminus \Omega.\end{cases}$$

Clearly $\psi \in C(\mathbb{R}^n)$ and has compact support. Define both

$$\begin{cases} v(t,x) = K(t)(x) * \eta(x)\\ u(t) = v(t)_{|\Omega} - S(t)\eta.\end{cases}$$

By Lemma 7.2.3 and Theorem 7.2.1 we deduce that

$$u \in C([0,+\infty); C_0(\overline{\Omega})) \cap C((0,+\infty); H_0^1(\Omega)) \cap C^1((0,+\infty); L^2(\Omega))$$

and

$$\Delta u \in C((0,+\infty); L^2(\Omega)).$$

In addition, $u(t) = v(t) \geq 0$ a.e. on Γ, $u_t = \Delta u$ for $t > 0$ and $u(0) = 0$. Hence

$$\begin{aligned}\frac{d}{dt}\int_\Omega (u^-)^2\,dx &= -\int_\Omega u_t u^-\,dx = -\int_\Omega u^-\Delta u\,dx\\ &= \int_\Omega \nabla u\cdot\nabla u^-\,dx = -\int_\Omega \|\nabla u^-\|^2\,dx \leq 0\end{aligned}$$

and thus $u(t) \leq 0$ a.e. in Ω for each $t \geq 0$. From (7.2.3) and the definition of ψ, we get

$$\|S(t)\varphi\|_{L^p(\Omega)} \leq \|v(t)\|_{L^p(\Omega)},$$

and the conclusion follows from Lemma 7.2.2. □

Theorem 7.2.7. *If Ω is a nonempty, bounded and open subset in $\mathbb{R}^n$ with C^1 boundary Γ, then the Laplace operator subjected to Dirichlet boundary condition generates an analytic and compact C_0-semigroup of contractions in $L^1(\Omega)$.*

Proof. Taking $q = 1$ and $p = 2$ in (7.2.2) we get

$$\|S(t)\xi\|_{L^2(\Omega)} \leq (4\pi t)^{-\frac{n}{4}}\|\xi\|_{L^1(\Omega)}$$

for each $\xi \in L^1(\Omega)$ and $t > 0$. So, $S(t)\xi \in L^2(\Omega)$.
As the semigroup generated by the Laplace operator subjected to the Dirichlet boundary condition in $L^2(\Omega)$ is analytic and compact and, by Lemma 7.2.1, it coincides with the homonymous semigroup in $L^1(\Omega)$, the conclusion follows from Theorems 7.2.3 and 7.2.5. □

The next theorem exhibits some asymptotic properties of the semigroup $\{S(t)\,;\ t \geq 0\}$ on each of the spaces $L^p(\Omega)$, for $p \in [1,+\infty]$. We recall that the first eigenvalue λ of the $-$Laplace operator in $H^{-1}(\Omega)$ is given by

$$\lambda = \inf\{\|\nabla u\|^2_{L^2(\Omega)}\,;\ u \in H_0^1(\Omega),\ \|u\|_{L^2(\Omega)} = 1\}, \tag{7.2.4}$$

and that $\lambda > 0$.

Theorem 7.2.8. *Let Ω be a nonempty, bounded and open subset in $\mathbb{R}^n$ and let $\lambda > 0$ be given by (7.2.4). Then*

$$\|S(t)\|_{\mathcal{L}(L^2(\Omega))} \leq e^{-\lambda t}$$

for each $t \geq 0$.

Proof. Let $\xi \in \mathcal{D}(\Omega)$, let $u(t) = S(t)\xi$, and let us define the function $f : \mathbb{R}_+ \to \mathbb{R}_+$ by $f(t) = \left(e^{\lambda t}\|S(t)\xi\|_{L^2(\Omega)}\right)^2$. A simple calculation involving Green's formula and the definition of λ, shows that

$$\begin{aligned} e^{-2\lambda t} f'(t) &= 2\lambda \int_\Omega u(t)^2 dx + 2 \int_\Omega u(t)u'(t)\, dx \\ &= 2\lambda \int_\Omega u(t)^2 dx + 2 \int_\Omega u(t)\Delta u(t)\, dx \\ &= 2\lambda \int_\Omega u(t)^2 dx - 2 \int_\Omega \|\nabla u(t)\|^2 dx \leq 0. \end{aligned}$$

Hence $\|S(t)\xi\|_{L^2(\Omega)} \leq e^{-\lambda t}\|\xi\|_{L^2(\Omega)}$ for each $\xi \in \mathcal{D}(\Omega)$ and $t \geq 0$. The conclusion follows by a simple density argument. □

Remark 7.2.1. Using mainly the same arguments one may prove that, if A is symmetric and generates a C_0-semigroup of contractions in a Hilbert space H, $\{S(t)\, ;\ t \geq 0\}$, and

$$\lambda = \inf\{\langle A\xi, \xi\rangle\, ; \xi \in D(A),\ \|\xi\| = 1\} > 0,$$

then

$$\|S(t)\|_{\mathcal{L}(H)} \leq e^{-\lambda t}$$

for each $t \geq 0$.

Theorem 7.2.9. *Let Ω be a nonempty, bounded and open subset in $\mathbb{R}^n$ whose Lebesgue measure is $|\Omega|$ and let $M = e^{\lambda|\Omega|^{2/n}}/(4\pi)$, where $\lambda > 0$ is given by (7.2.4). Then, for each $t \geq 0$, we have*

$$\|S(t)\|_{\mathcal{L}(L^\infty(\Omega))} \leq Me^{-\lambda t}. \tag{7.2.5}$$

In addition, for each $p \in [1, +\infty)$, there exists $M_p > 1$ such that

$$\|S(t)\|_{\mathcal{L}(L^p(\Omega))} \leq M_p e^{-\lambda t} \tag{7.2.6}$$

for each $t \geq 0$.

Proof. Let $\xi \in \mathcal{D}(\Omega)$ and let $T > 0$. If $0 \leq t \leq T$, we clearly have

$$\|S(t)\xi\|_{L^\infty(\Omega)} \leq \|\xi\|_{L^\infty(\Omega)} \leq e^{-\lambda t}e^{\lambda T}\|\xi\|_{L^\infty(\Omega)}.$$

If $T \leq t$, from (7.2.2) and Theorem 7.2.8, we get

$$\begin{aligned} \|S(t)\xi\|_{L^\infty(\Omega)} &\leq (4\pi T)^{-\frac{n}{4}}\|S(t-T)\xi\|_{L^\infty(\Omega)} \\ \leq (4\pi T)^{-\frac{n}{2}}e^{-\lambda t}e^{-\lambda T}\|\xi\|_{L^2(\Omega)} &\leq |\Omega|^{\frac{1}{2}}(4\pi T)^{-\frac{n}{2}}e^{-\lambda t}e^{-\lambda T}\|\xi\|_{L^\infty(\Omega)}. \end{aligned}$$

Taking $T = |\Omega|^{\frac{2}{n}}/(4\pi)$ in this inequality, we deduce (7.2.5). To prove (7.2.6), let $\xi \in \mathcal{D}(\Omega)$, let $1 \leq 2 \leq p \leq +\infty$ and let us observe that, by virtue of (7.2.2) and Theorem 7.2.8, for each $t \geq 1$, we have

$$\begin{aligned}\|S(t)\xi\|_{L^p(\Omega)} &\leq (4\pi)^{-\frac{n}{2}\left(\frac{1}{q}-\frac{1}{p}\right)}\|S(t-1)\xi\|_{L^2(\Omega)} \\ &\leq (4\pi)^{-\frac{n}{2}\left(\frac{1}{2}-\frac{1}{p}\right)}e^{-\lambda t}e^{\lambda}\|\xi\|_{L^2(\Omega)}.\end{aligned}$$

Since for $t \in [0,1]$, we clearly have

$$\|S(t)\xi\|_{L^2(\Omega)} \leq e^{-\lambda t}e^{\lambda}\|\xi\|_{L^2(\Omega)},$$

this completes the proof in the case of $2 \leq p \leq +\infty$. The case $1 \leq p < 2$ follows from the preceding one with the help of (7.2.2) and this achieves the proof. □

Remark 7.2.2. One may prove that in the inequalities (7.2.5) and (7.2.6) the constants M and M_p cannot be 1. On the other hand, if $p \in [2,+\infty)$, then

$$\|S(t)\|_{\mathcal{L}(L^p(\Omega))} \leq e^{-\frac{4\lambda t}{p^2}}$$

and this inequality is the best in the sense that

$$\frac{d}{dt}\left\{\|S(t)\|_{\mathcal{L}(L^p(\Omega))}\right\}_{t=0} = -\frac{4\lambda}{p^2}.$$

See Cazenave and Haraux [**36**], Remark, p. 46.

7.3. The Stokes Equation

In this section, by using the abstract results proved in this chapter, we shall study the existence and regularity properties of one of the most important partial differential equation in Fluid Mechanics, the Stokes equation. This equation describes the flow of "moderate speed" of a viscous incompressible fluid within a domain Ω in $\mathbb{R}^3$. We begin with the simplest case $\Omega = \mathbb{R}^3$.

Example 7.3.1. (*The Stokes operator in* $\mathbb{R}^3$). Let

$$H = \{u \in \left[L^2(\mathbb{R}^3)\right]^3 ; \nabla \cdot u = 0\},$$

where the condition $\nabla \cdot u = 0$ is understood in the sense of distributions over $\mathbb{R}^3$. One can easily see that, endowed with the standard inner product of the space $\left[L^2(\mathbb{R}^3)\right]^3$, defined by

$$\langle u, v\rangle = \sum_{i=1}^{3}\langle u_i, v_i\rangle_{L^2(\mathbb{R}^3)},$$

H is a real Hilbert space. We define $A : D(A) \subseteq H \to H$, called *the Stokes operator on* $\mathbb{R}^3$, by

$$\begin{cases} D(A) = \{u \in \left[H^2(\mathbb{R}^3)\right]^3 \cap H\,;\ \Delta u \in H\} \\ Au = \Delta u \quad \text{for} \quad u \in D(A), \end{cases}$$

where $\Delta u = (\Delta u_1, \Delta u_2, \Delta u_3)$ in the sense of distributions over $\mathbb{R}^3$.

Theorem 7.3.1. *The operator A, defined as above, is the infinitesimal generator of an analytic C_0-semigroup of contractions in H.*

Proof. Since $C_\sigma^\infty(\mathbb{R}^3; \mathbb{R}^3) = \{u \in \left[\mathcal{D}(\mathbb{R}^3)\right]^3\,;\ \nabla \cdot u = 0\}$ is both dense in H and included in $D(A)$, it follows that A is densely defined. In addition, from Theorem 4.1.2, we deduce that A is closed and symmetric. On the other hand, for each $\lambda > 0$ and each $f \in H$, the equation

$$\lambda u - \Delta u = f$$

has a unique solution $u \in \left[H^2(\mathbb{R}^3)\right]^3$. Let us denote by $v = \nabla \cdot u$, which clearly belongs to $H^1(\mathbb{R}^3)$. Since $\nabla \cdot f = 0$ in $H^{-1}(\mathbb{R}^3)$, it follows that $\lambda v - \Delta v = 0$ in $H^{-1}(\mathbb{R}^3)$, which implies $v = 0$. So, $u \in D(A)$ and therefore $(0, +\infty) \subseteq \rho(A)$. More than this, for each $\lambda > 0$, we have

$$\|R(\lambda; A)\|_{\mathcal{L}(X)} \leq \frac{1}{\lambda}.$$

In view of Theorem 3.1.1, we know that A is the generator of a C_0-semigroup of contractions on H. From Lemma 1.6.1, we have that A is self-adjoint, while by Corollary 7.1.1, we deduce that the semigroup generated by A is analytic. The proof is complete. □

Let us consider next *the Stokes equation in* $\mathbb{R}^3$,

$$\begin{cases} u_t = \Delta u & (t, x) \in R_\infty \\ \nabla \cdot u = 0 & (t, x) \in R_\infty \\ u(0, x) = \xi(x) & x \in \mathbb{R}^3, \end{cases} \tag{7.3.1}$$

where $R_\infty = \mathbb{R}_+ \times \mathbb{R}^3$ and $u(t, x)$ represents the instantaneous velocity of the particle occupying the point of coordinates x at the time t.

Theorem 7.3.2. *For each $\xi \in H$ the problem* (7.3.1) *has a unique solution $u \in C([0, +\infty); H)$ which satisfies:*

(i) $u \in C((0, +\infty); \left[H^2(\mathbb{R}^3)\right]^3) \cap C^1((0, +\infty); H)$;

(ii) *for each* $m \in \mathbb{N}$, $\Delta^m u \in C^\infty((0, +\infty); \left[H^2(\mathbb{R}^3)\right]^3) \cap H)$;

(iii) *if* $\xi \in D(A)$, *then* $u \in C([0, +\infty); \left[H^2(\mathbb{R}^3)\right]^3) \cap C^1([0, +\infty); H)$.

Proof. We apply Theorem 7.3.1, and Corollary 2.4.1. □

Theorem 7.3.3. *Let $\xi \in H$ and let u the unique continuous solution of the problem* (7.3.1). *Then:*

(i) $\|\Delta u(t)\|_H \leq \dfrac{1}{t\sqrt{2}}\|\xi\|_H$ *for each* $t > 0$;

(ii) $\displaystyle\int_0^\infty s\|\Delta u(s)\|_H^2 ds \leq \frac{1}{4}\|\xi\|_H^2$;

(iii) $\displaystyle\sum_{i=1}^3 \|\nabla u_i(t)\|_{L^2(\mathbb{R}^3)} \leq \frac{1}{\sqrt{2t}}\|\xi\|_H$ *for each* $t > 0$;

(iv) $\displaystyle\int_0^\infty \sum_{i=1}^3 \|\nabla u_i(s)\|_{L^2(\mathbb{R}^3)}^2 ds \leq \frac{1}{2}\|\xi\|_H^2$;

(v) *if* $\xi \in D(A)$, *then* $\displaystyle\|\Delta u(t)\|_H \leq \frac{1}{\sqrt{2t}}\sum_{i=1}^3 \|\nabla \xi_i\|_{L^2(\mathbb{R}^3)}$ *for each* $t > 0$,

and

(vi) $\displaystyle\int_0^\infty s\|\Delta u(s)\|_H^2 ds \leq \frac{1}{2}\sum_{i=1}^3 \|\nabla \xi_i\|_{L^2(\mathbb{R}^3)}^2$.

Proof. The conclusion follows from Theorems 6.4.1 and 7.3.1. □

Example 7.3.2. (*The Stokes operator in* Ω). Let Ω be a bounded domain in $\mathbb{R}^3$ with boundary Γ of class C^2 and let $\nu(x)$ be the outward normal to Γ at the point $x \in \Gamma$. Let

$$Y = \{u \in [\mathcal{D}(\Omega)]^3 \; ; \; \nabla \cdot u = 0 \text{ in } \Omega, \; u \cdot \nu = 0 \text{ on } \Gamma\}$$

and let H be the closure of Y in $\left[L^2(\Omega)\right]^3$. Let us observe that, endowed with the usual inner product of $\left[L^2(\Omega)\right]^3$, H is a Hilbert space. Let us denote by $P : \left[L^2(\Omega)\right]^3 \to H$ the orthogonal projection on H and let us define *the Stokes operator on* Ω, $A : D(A) \subseteq H \to H$, by

$$\begin{cases} D(A) = \left[H^2(\Omega) \cap H_0^1(\Omega)\right]^3 \cap H \\ Au = P(\Delta u) \text{ for } u \in D(A). \end{cases}$$

Theorem 7.3.4. (Fujita-Kato) *The operator A, defined as above, is the generator of a compact and analytic C_0-semigroup of contractions in H.*

Proof. Since $C_\sigma^\infty(\Omega; \mathbb{R}^3) = \{u \in [\mathcal{D}(\Omega)]^3 \; ; \; \nabla \cdot u = 0\}$, which is dense in H, is included in $D(A)$, it follows that A is densely defined. From Theorem 4.1.2, we have that A closed and symmetric and thus self-adjoint. See Lemma 1.6.1. On the other hand, for each $\lambda > 0$ and each $f \in H$, the equation

$$\lambda u - P(\Delta u) = f$$

has a unique solution $u \in H$ which, in addition, satisfies

$$\|u\| \leq \frac{1}{\lambda}\|f\|.$$

Therefore $(0,+\infty) \subseteq \rho(A)$ and $\|R(\lambda;A)\|_{\mathcal{L}(H)} \leq \frac{1}{\lambda}$. By virtue of Hille-Yosida Theorem 3.1.1, it follows that A generates a C_0-semigroup of contractions on H. Since A is self-adjoint, by Corollary 7.1.1, the semigroup is analytic. Finally, thanks to Sobolev-Rellich-Kondrachov Theorem 1.5.4, for each $\lambda > 0$, $R(\lambda;A)$ is compact. An appeal to Corollary 6.3.2 shows that the semigroup is compact and this achieves the proof. □

Let us consider next *the Stokes equation in* Ω

$$\begin{cases} u_t = P(\Delta)u & (t,x) \in Q_\infty \\ \nabla \cdot u = 0 & (t,x) \in Q_\infty \\ u = 0 & (t,x) \in \Sigma_\infty \\ u(0,x) = \xi(x) & x \in \Omega, \end{cases} \tag{7.3.2}$$

where $Q_\infty = \mathbb{R}_+ \times \Omega$ and $\Sigma_\infty = \mathbb{R}_+ \times \Gamma$.

Theorem 7.3.5. *For each* $\xi \in H$ *the problem* (7.3.2) *has a unique solution* $u \in C([0,+\infty);H)$ *which satisfies*:

(i) $u \in C((0,+\infty);[H^2(\Omega)]^3) \cap C^1((0,+\infty);H)$;

(ii) *for each* $m \in \mathbb{N}$, $[P(\Delta)]^m u \in C^\infty((0,+\infty);[H^2(\Omega)]^3) \cap H)$;

(iii) *if* $\xi \in D(A)$, *then* $u \in C([0,+\infty);[H^2(\Omega)]^3) \cap C^1([0,+\infty);H)$.

Proof. We apply Theorem 7.3.4 and Corollary 2.4.1. □

Theorem 7.3.6. *Let* $\xi \in H$, *and let* u *be the unique continuous solution of the problem* (7.3.2). *Then*:

(i) $\|P(\Delta)u(t)\|_H \leq \dfrac{1}{t\sqrt{2}}\|\xi\|_H$ *for each* $t > 0$;

(ii) $\displaystyle\int_0^\infty s\|P(\Delta)u(s)\|_H^2 ds \leq \frac{1}{4}\|\xi\|_H^2$;

(iii) $\langle -P(\Delta)u(t), u(t)\rangle_{L^2(\Omega)} \leq \dfrac{1}{2t}\|\xi\|_H^2$ *for each* $t > 0$;

(iv) $\displaystyle\int_0^\infty \langle -P(\Delta)u(s), u(s)\rangle_{L^2(\Omega)} ds \leq \frac{1}{2}\|\xi\|_H^2$;

(v) *if* $\xi \in D(A)$ *then* $\|P(\Delta)u(t)\|_H^2 \leq \dfrac{1}{2t}\langle -P(\Delta)\xi, \xi\rangle_{L^2(\Omega)}$ *for each* $t > 0$, *and*

(iv) $\displaystyle\int_0^\infty s\|P(\Delta)u(s)\|_H^2 ds \leq \frac{1}{2}\langle -P(\Delta)\xi, \xi\rangle_{L^2(\Omega)}$.

Proof. The conclusion follows from Theorems 6.4.1 and 7.3.4. □

7.4. A Parabolic Problem with Dynamic Boundary Conditions

Let Ω be a bounded domain in $\mathbb{R}^n$ whose boundary Γ is of class C^2. Let $Q_T = [0,T] \times \Omega$, $\Sigma_T = [0,T] \times \Gamma$, $u_0^\Omega \in L^2(\Omega)$, $u_0^\Gamma \in L^2(\Gamma)$, and let us consider the parabolic problem subjected to dynamic boundary conditions

$$\begin{cases} u_t = \Delta & (t,x) \in Q_T \\ u_t + u_\nu = 0 & (t,x) \in \Sigma_T \\ u(0,x) = u_0^\Omega(x) & x \in \Omega \\ u(0,x) = u_0^\Gamma(x) & x \in \Gamma, \end{cases} \tag{7.4.1}$$

where u_ν is the outward normal derivative of u on Γ. For $n = 3$, this problem represents the mathematical model which describes the evolution of the heat distribution $u(t,x)$ in a homogenous and isotropic solid body (represented by Ω), imbedded in a moving fluid. The body has at the initial time a heat distribution on the boundary Γ, distribution which, at that time, might be different from that of the corresponding boundary of the moving fluid. We begin with a simple lemma which we need in the sequel.

Lemma 7.4.1. *Let Ω be a bounded domain in $\mathbb{R}^n$ with C^2 boundary Γ and let $\mu \geq 0$, $\lambda > 0$. Then, for each $f \in L^2(\Omega)$ and $g \in L^2(\Gamma)$, the elliptic problem*

$$\begin{cases} \mu u - \Delta u = f \\ \lambda v + u_\nu = g \\ u_{|\Gamma} = v \end{cases} \tag{7.4.2}$$

has a unique solution $u \in H^{3/2}(\Omega)$ with $\Delta u \in L^2(\Omega)$, $u_{|\Gamma} \in H^1(\Gamma)$ and $u_\nu \in L^2(\Gamma)$.

Proof. By the classical theory of variational problems based on the Lax-Milgram theorem (see for instance Brezis [**29**], IX.5., pp. 175-181), (7.4.2) has a unique solution $u \in H^1(\Omega)$ with $u_{|\Gamma} \in L^2(\Gamma)$ and $\Delta u \in L^2(\Omega)$. Then it follows that $u_\nu \in H^{-1/2}(\Gamma)$ and so, since $u_{|\Gamma}, g \in L^2(\Gamma)$, it follows that $u_\nu \in L^2(\Gamma)$. But then, $u \in H^{3/2}(\Omega)$ and $u_\Gamma \in H^1(\Gamma)$. The proof is complete. □

Theorem 7.4.1. *The operator*

$$A : D(A) \subseteq L^2(\Omega) \times L^2(\Gamma) \to L^2(\Omega) \times L^2(\Gamma),$$

defined by

$$D(A) = \{(u,v) \in L^2(\Omega) \times L^2(\Gamma)\,;\ \Delta u \in L^2(\Omega),\ u_\nu \in L^2(\Gamma),\ u_{|\Gamma} = v\}$$

$$A(u,v) = (\Delta u, -u_\nu) \text{ for each } (u,v) \in D(A),$$

is the infinitesimal generator of a compact and analytic C_0-semigroup of contractions.

Proof. Clearly $D(A)$ is dense in $L^2(\Omega) \times L^2(\Gamma)$, because each function in $L^2(\Gamma)$ can be approximated with functions of class C^2 on Γ, and each function in $L^2(\Omega)$ can be approximated with functions of class $C^2(\overline{\Omega})$ whose restrictions to the boundary are preassigned C^2 functions on Γ. Moreover, let us observe that, for each $(u,v),(\varphi,\psi) \in D(A)$, we have

$$\begin{aligned}\langle A(u,v),(\varphi,\psi)\rangle &= \langle \Delta u, \varphi\rangle - \langle u_\nu, \psi\rangle_{L^2(\Gamma)} = -\langle \nabla u, \nabla \varphi\rangle \\ &= \langle u, \Delta\varphi\rangle - \langle u, \varphi_\nu\rangle_{L^2(\Gamma)} = \langle (u,v), A(\varphi,\psi)\rangle.\end{aligned}$$

Consequently A is symmetric. In addition, for each $\lambda > 0$, the equation $(\lambda I - A)(u,v) = (f,g)$ rewrites equivalently under the form

$$\begin{cases} \lambda u - \Delta u = f \\ \lambda v + u_\nu = g \\ u_{|\Gamma} = v. \end{cases}$$

From Lemma 7.4.1, we know that, for each $(f,g) \in L^2(\Omega) \times L^2(\Gamma)$, the problem above has a unique solution u satisfying $u \in H^{3/2}(\Omega)$, $\Delta u \in L^2(\Omega)$, $u_{|\Gamma} \in H^1(\Gamma)$ and $u_\nu \in L^2(\Gamma)$. Taking the $L^2(\Omega)$-inner product both sides of the first two equations in the system above by u and respectively by v and taking into account the third equality, we deduce

$$\begin{cases} \lambda\|u\|^2_{L^2(\Omega)} + \|\nabla u\|^2_{L^2(\Omega)} - \langle u_\nu, u\rangle_{L^2(\Gamma)} = \langle f, u\rangle_{L^2(\Omega)} \\ \lambda\|v\|^2_{L^2(\Gamma)} + \langle u_\nu, u\rangle_{L^2(\Gamma)} = \langle g, v\rangle_{L^2(\Gamma)}. \end{cases}$$

Adding side by side the two equalities and using Cauchy-Schwarz inequality, we obtain

$$\lambda\|(u,v)\|^2_{L^2(\Omega)\times L^2(\Gamma)} \leq \|(f,g)\|_{L^2(\Omega)\times L^2(\Gamma)}\|(u,v)\|_{L^2(\Omega)\times L^2(\Gamma)}.$$

From this, we deduce, on one hand that $(0,+\infty) \subseteq \rho(A)$ and, on the other hand, that, for each $\lambda > 0$, we have $\|(\lambda I - A)^{-1}\|_{\mathcal{L}(X)} \leq \frac{1}{\lambda}$. By virtue of Theorem 3.1.1, it follows that A generates a C_0-semigroup of contractions, and from Lemma 1.6.1, we conclude that A is self-adjoint. Since $D(A)$, endowed with the graph-norm, is compactly imbedded in $L^2(\Omega) \times L^2(\Gamma)$, by Corollary 6.3.2, it follows that the semigroup generated by A is compact. The proof is complete. □

Theorem 7.4.2. *For each $u_0^\Omega \in L^2(\Omega)$ and each $u_0^\Gamma \in L^2(\Gamma)$, the problem (7.4.1) has a unique continuous solution $u : \mathbb{R}_+ \to L^2(\Omega)$ satisfying*

(i) $u \in C([0,+\infty); L^2(\Omega)) \cap C^1((0,+\infty); L^2(\Omega))$;
(ii) $u_{|\Gamma} \in C([0,+\infty); L^2(\Gamma)) \cap C^1((0,+\infty); L^2(\Gamma))$;

(iii) $\Delta u \in C((0,+\infty);L^2(\Omega))$, *and* $u_\nu \in C((0,+\infty);L^2(\Gamma))$.

If, in addition, $\Delta u_0^\Omega \in L^2(\Omega)$, $u_{0\,|\Gamma}^\Omega \in L^2(\Gamma)$ *and* $u_{0\,|\Gamma}^\Omega = u_0^\Gamma$, *then*

(iv) $u \in C^1([0,+\infty);L^2(\Omega))$;

(v) $u_{|\Gamma} \in C^1([0,+\infty);L^2(\Gamma))$;

(vi) $\Delta u \in C([0,+\infty);L^2(\Omega))$, *and* $u_{|\Gamma} \in C([0,+\infty);L^2(\Gamma))$.

Proof. The problem (7.4.1) can be rewritten as an ordinary differential equation, in the space $H = L^2(\Omega) \times L^2(\Gamma)$, of the form

$$\begin{cases} u' = Au \\ u(0) = u_0, \end{cases}$$

where A is defined as in Theorem 7.4.1 and $u_0 = (u_0^\Omega, u_0^\Gamma)$. The conclusion follows from Theorem 7.4.1 combined with Theorem 6.4.1, and this achieves the proof. □

7.5. An Elliptic Problem with Dynamic Boundary Conditions

Let Ω be a bounded domain in $\mathbb{R}^n$ whose boundary Γ is of class C^2. Let $Q_T = [0,T] \times \Omega$, $\Sigma_T = [0,T] \times \Gamma$, $u_0^\Omega \in L^2(\Omega)$, $u_0^\Gamma \in L^2(\Gamma)$, and let us consider the elliptic problem subjected to dynamic boundary conditions

$$\begin{cases} -\Delta u = 0 & (t,x) \in Q_T \\ u_t + u_\nu = 0 & (t,x) \in \Sigma_T \\ u(0,x) = u_0(x) & x \in \Gamma, \end{cases} \tag{7.5.1}$$

where u_ν is outward normal derivative of u on Γ.

Let $y \in H^{1/2}(\Gamma)$ and let us consider the nonhomogeneous elliptic problem

$$\begin{cases} -\Delta u = 0 & \text{in } \Omega \\ u = y & \text{on } \Gamma. \end{cases} \tag{7.5.2}$$

It is known that this problem has a unique solution $u \in H^1(\Omega)$. See Brezis [**29**], Exemple 2, p. 176. Let us denote this solution by $u = \mathcal{S}(y)$.

Theorem 7.5.1. *The operator* $A : D(A) \subseteq L^2(\Gamma) \to L^2(\Gamma)$, *defined by*

$$\begin{cases} D(A) = \{y \in H^{1/2}(\Gamma); \text{for which } u = \mathcal{S}(y) \text{ satisfies } u_\nu \in L^2(\Gamma)\} \\ Ay = -u_\nu \text{ for } y \in D(A), \end{cases}$$

is the infinitesimal generator of an analytic and compact C_0*-semigroup of contractions in* $L^2(\Gamma)$.

Proof. Let us observe that, for each $y, z \in D(A)$, we have

$$\langle Ay, z\rangle_{L^2(\Gamma)} = -\langle \nabla u, \nabla v\rangle_{L^2(\Omega)} - \langle \Delta u, v\rangle_{[H^1(\Omega)]^*, H^1(\Omega)}$$

and

$$\langle y, Az\rangle_{L^2(\Gamma)} = -\langle \nabla u, \nabla v\rangle_{L^2(\Omega)} - \langle \Delta v, u\rangle_{[H^1(\Omega)]^*, H^1(\Omega)},$$

where u satisfies (7.5.2), while v is the solution of a similar equation with y substituted by z. Here $\langle\cdot,\cdot\rangle_{[H^1(\Omega)]^*,H^1(\Omega)}$ is the usual duality between $H^1(\Omega)$ and its topological dual $[H^1(\Omega)]^*$. Since $\Delta u = \Delta v = 0$, it follows that

$$\langle Ay, z\rangle_{L^2(\Gamma)} = -\langle \nabla u, \nabla v\rangle_{L^2(\Omega)} = \langle y, Az\rangle_{L^2(\Gamma)}$$

for each $y, z \in D(A)$. So, A is symmetric. We shall prove next that A is self-adjoint in $H = L^2(\Gamma)$ and satisfies the conditions in Theorem 3.1.1. To this aim, we shall show that, for each $\lambda > 0$, $(\lambda I - A)^{-1} \in \mathcal{L}(H)$, and $\|(\lambda I - A)^{-1}\|_{\mathcal{L}(H)} \leq \frac{1}{\lambda}$. One may easily see that $\lambda y - Ay = f$ if and only if $y = u_{|\Gamma}$, where u is the unique $H^1(\Omega)$ solution of the elliptic problem

$$\begin{cases} -\Delta u = 0 & \text{in } \Omega \\ \lambda u + u_\nu = f & \text{on } \Gamma. \end{cases} \tag{7.5.3}$$

So, $(\lambda I - A)$ is surjective if and only if for each $f \in L^2(\Gamma)$, the unique solution in $H^1(\Omega)$ of (7.5.3) satisfies

$$u \in L^2(\Gamma) \text{ and } u_\nu \in L^2(\Gamma). \tag{7.5.4}$$

But this follows from Lemma 7.4.1.

Finally, we have merely to prove that, for each $\lambda > 0$, we have

$$\|(\lambda I - A)^{-1}\|_{\mathcal{L}(H)} \leq \frac{1}{\lambda}. \tag{7.5.5}$$

To this aim, let us multiply both sides of the equation $-\Delta u = 0$ in (7.5.3) by u. Integrating over Ω yields

$$\|\nabla u\|^2_{L^2(\Omega)} - \langle u_\nu, u\rangle_{L^2(\Gamma)} = 0,$$

from where, taking into account the boundary condition in (7.5.3) to substitute u_ν by $f - \lambda u$, we deduce

$$\|\nabla u\|^2_{L^2(\Omega)} + \lambda\|u\|^2_{L^2(\Gamma)} \leq \|f\|_{L^2(\Gamma)}\|u\|_{L^2(\Gamma)}.$$

From this inequality, observing that $(\lambda I - A)y = f$ if and only if $y = u_{|\Gamma}$, we obtain (7.5.5). So, A satisfies all the conditions in Theorem 3.1.1, and Lemma 1.6.1. Moreover, from Corollary 7.1.1, it follows that A generates an analytic semigroup. Finally, the fact that this semigroup is compact follows from the last assertion by observing that the imbedding $D(A) \subseteq L^2(\Gamma)$ is compact. See Corollary 6.3.2. The proof is complete. □

Theorem 7.5.2. *For each* $u_0 \in L^2(\Gamma)$, *the problem* (7.5.1) *has a unique solution* $u : \mathbb{R}_+ \to H^1(\Omega)$, *satisfying*:

(i) $u \in C((0,+\infty); H^1(\Omega)) \cap L^2(0,+\infty; H^1(\Omega))$, *and* $\|\nabla u\|^2_{L^2((0,+\infty);L^2(\Omega))} \leq \frac{1}{2}\|\xi\|^2_{L^2(\Gamma)}$;

(ii) $u_{|\Gamma} \in C([0,+\infty); L^2(\Gamma)) \cap C^1((0,+\infty); L^2(\Gamma))$;
(iii) $u_\nu \in C((0,+\infty); L^2(\Gamma))$.

If, in addition, $u_0 \in H^{1/2}(\Gamma)$, *and the unique solution* v *of the problem* $-\Delta v = 0$ *on* Ω *and* $v = u_0$ *on* Γ, *satisfies* $v_\nu \in L^2(\Gamma)$, *then*

(iv) $u \in C^1([0,+\infty); H^1(\Omega))$;
(v) $u_{|\Gamma} \in C^1([0,+\infty); L^2(\Gamma))$;
(vi) $u_\nu \in C([0,+\infty); L^2(\Gamma))$.

Proof. The problem (7.5.1) can be rewritten as an ordinary differential equation in $H = L^2(\Omega) \times L^2(\Gamma)$, of the form

$$\begin{cases} u' = Au \\ u(0) = \xi, \end{cases}$$

where A is defined as in Theorem 7.5.1 and $\xi = u_0$. Except for (i) and (iv), the conclusion follows from Theorem 7.5.1 combined with Theorem 6.4.1. In order to check (i) and (iv), let us observe that, for each $t, s \in [0,+\infty)$, we have $-\Delta(u(t) - u(s)) = 0$. Taking the $L^2(\Omega)$ inner product both sides by $u(t)-u(s)$, and taking into account the boundary conditions, we deduce

$$\|\nabla(u(t) - u(s))\|^2_{L^2(\Omega)} \leq \|u_\nu(t) - u_\nu(s)\|_{L^2(\Gamma)} \|u(t) - u(s)\|_{L^2(\Gamma)}.$$

From this inequality, from (ii) and (iii), we get the first assertion in (i). From the same inequality, from (iv) and (v), we deduce (vi). To prove the second assertion in (i), it suffices to observe that

$$\|\nabla u(t)\|^2_{L^2(\Omega)} - \langle u_\nu(t), u(t)\rangle_{L^2(\Gamma)} = 0,$$

or equivalently

$$\|\nabla u(t)\|^2_{L^2(\Omega)} - \frac{1}{2}\frac{d}{dt}\|u(t)\|^2_{L^2(\Gamma)} = 0.$$

Integrating this equality from 0 to T, we obtain

$$\int_0^T \|\nabla u(t)\|^2_{L^2(\Omega)}\,dt \leq \frac{1}{2}\|\xi\|^2_{L^2(\Gamma)}$$

for each $T > 0$, which achieves the proof of (i). The proof is complete. □

7.6. Fractional Powers of Closed Operators

In this section we shall define the fractional power of the negative of an infinitesimal generator of a C_0-semigroup. First, let us recall that, for each $\alpha \in (0,+\infty)$, the improper integral

$$\Gamma(\alpha) = \int_0^\infty t^{\alpha-1} e^{-t}\,dt$$

is convergent, and its values is the so-called Euler Γ-function. Let $a > 0$ and $\alpha > 0$. The change of variable $ta = \tau$ shows that

$$a^{-\alpha} = \frac{1}{\Gamma(\alpha)} \int_0^\infty t^{\alpha-1} e^{-ta}\, dt.$$

Then, if $A : D(A) \subseteq X \to X$ is a linear operator with $-A$ the infinitesimal generator of a semigroup of class C_0, $\{S(t)\,;\ t \geq 0\}$, it is quite natural to consider that, substituting a with A and e^{-ta} with $S(t)$ in the relation above, we obtain an operator playing the role of $A^{-\alpha}$. We recall that, at least formally, $S(t)$ can be interpreted as $S(t) = e^{-tA}$, and this suggests and explains the above analogy. More precisely, we have

Definition 7.6.1. The operator $A^{-\alpha} : D(A^{-\alpha}) \subseteq X \to X$ defined by

$$A^{-\alpha}x = \frac{1}{\Gamma(\alpha)} \int_0^\infty t^{\alpha-1} S(t)x\, dt, \tag{7.6.1}$$

for each $x \in D(A^{-\alpha})$, where

$$D(A^{-\alpha}) = \left\{ x \in X\,;\ \int_0^\infty t^{\alpha-1} S(t)x\, dt \text{ is convergent} \right\},$$

is called *the power of exponent* $-\alpha$ *of the operator* A. By definition, $D(A^0) = X$ and $A^0 = I$.

Remark 7.6.1. Clearly $D(A^{-\alpha})$ is a linear subspace in X and $A^{-\alpha}$ is a linear, closed operator. Moreover, for each $0 \leq \alpha \leq \beta$, we have

$$D(A^{-\beta}) \subseteq D(A^{-\alpha}).$$

Remark 7.6.2. If the C_0-semigroup, $\{S(t)\,;\ t \geq 0\}$, generated by $-A$, is of type $(M, -\omega)$, with $M \geq 1$ and $\omega > 0$, i.e., for each $t \geq 0$, we have

$$\|S(t)\|_{\mathcal{L}(X)} \leq M e^{-\omega t},$$

then, for each $\alpha \geq 0$, $D(A^{-\alpha}) = X$. Indeed, this follows from the remark that, in this case

$$\int_0^\infty t^{\alpha-1} \|S(t)\|_{\mathcal{L}(X)}\, dt < +\infty.$$

In addition, since $\int_0^\infty t^{\alpha-1} S(t)\, dt$ is convergent in the uniform operator norm, it follows that, for each $\alpha \geq 0$, $A^{-\alpha}$ is a linear bounded operator.

Example 7.6.1. Let H be a real Hilbert space, whose inner product is denoted by $\langle\cdot,\cdot\rangle$, and let $A : D(A) \subseteq H \to H$ be such that $-A$ generates a C_0-semigroup of contractions on H. Let us assume that A is self-adjoint and invertible with compact inverse A^{-1}. Then, by a theorem of Hilbert, there exists a sequence of positive numbers $\mu_k > 0$, $\mu_{k+1} \leq \mu_k$, for $k \in \mathbb{N}^*$ and an orthonormal basis $\{e_k\,;\ k \in \mathbb{N}^*\}$ of H such that

$$A^{-1}e_k = \mu_k e_k$$

for $k \in \mathbb{N}^*$. Let us denote $\lambda_k = {\mu_k}^{-1}$, and let us observe that $e_k \in D(A)$ and

$$Ae_k = \lambda_k e_k$$

for $k \in \mathbb{N}^*$. We also have that $\lim_{k\to\infty} \lambda_k = +\infty$. Let $\alpha > 0$ and let us define $A_\alpha : D(A_\alpha) \subseteq H \to H$ by

$$D(A_\alpha) = \left\{ u \in H\,;\ u = \sum_{k=1}^{\infty} u_k e_k,\ \sum_{k=1}^{\infty} \lambda_k^{2\alpha} |u_k|^2 < +\infty \right\}$$

$$A_\alpha u = \sum_{k=1}^{\infty} {\lambda_k}^{\alpha} u_k e_k \quad \text{for} \quad u \in D(A_\alpha).$$

A simple calculation shows that $A_\alpha = A^\alpha$. We notice that $D(A^\alpha)$, endowed with the natural inner product $\langle \cdot, \cdot \rangle_\alpha : D(A^\alpha) \times D(A^\alpha) \to \mathbb{R}$, defined by

$$\langle u, v \rangle_\alpha = \sum_{k=1}^{\infty} \lambda_k^{2\alpha} \langle u_k, v_k \rangle$$

for each $u, v \in D(A^\alpha)$, $u = \sum_{k=1}^{\infty} u_k e_k$ and $v = \sum_{k=1}^{\infty} v_k e_k$, is a real Hilbert space. In addition, with respect to this inner product, the family $\{\lambda_k^{-\alpha} e_k\,;\ k \in \mathbb{N}^*\}$ is an orthonormal basis in $D(A^\alpha)$.

Lemma 7.6.1. *If the C_0-semigroup, $\{S(t)\,;\ t \geq 0\}$, generated by $-A$, is of type $(M, -\omega)$, with $M \geq 1$ and $\omega > 0$, then, for each $\alpha, \beta \in [0, +\infty)$, we have*

$$A^{-(\alpha+\beta)} = A^{-\alpha} A^{-\beta}.$$

Proof. Let us observe that

$$\begin{aligned} A^{-\alpha}A^{-\beta} &= \frac{1}{\Gamma(\alpha)\Gamma(\beta)} \int_0^\infty \int_0^\infty t^{\alpha-1} s^{\beta-1} S(t)S(s)\,dt\,ds \\ &= \frac{1}{\Gamma(\alpha)\Gamma(\beta)} \int_0^\infty t^{\alpha-1} \int_t^\infty (u-t)^{\beta-1} dt S(u)\,du \\ &= \frac{1}{\Gamma(\alpha)\Gamma(\beta)} \int_0^\infty \left(\int_0^u t^{\alpha-1}(u-t)^{\beta-1} dt \right) S(u)\,du \\ &= \frac{1}{\Gamma(\alpha)\Gamma(\beta)} \int_0^1 v^{\alpha-1}(1-v)^{\beta-1} dv \int_0^\infty u^{\alpha+\beta-1} S(u)\,du. \end{aligned}$$

Inasmuch as

$$\int_0^1 v^{\alpha-1}(1-v)^{\beta-1} dv = \frac{\Gamma(\alpha)\Gamma(\beta)}{\Gamma(\alpha+\beta)},$$

from the equalities in above, we deduce

$$A^{-\alpha}A^{-\beta} = \frac{1}{\Gamma(\alpha+\beta)}\int_0^\infty u^{\alpha+\beta-1}S(u)\,du = A^{-(\alpha+\beta)}$$

and this achieves the proof. □

Lemma 7.6.2. *If the C_0-semigroup, $\{S(t)\,;\ t \geq 0\}$, generated by $-A$, is of type $(M, -\omega)$, with $M \geq 1$ and $\omega > 0$, then, for each $\alpha \in (0,1)$, we have*

$$A^{-\alpha} = \frac{\sin \pi\alpha}{\pi}\int_0^\infty \lambda^{-\alpha}(\lambda I + A)^{-1}\,d\lambda. \tag{7.6.2}$$

Proof. Since $-A$ is the infinitesimal generator of a C_0-semigroup of type (M, ω), by Theorem 3.3.1, we have that

$$\|(\lambda I + A)^{-1}\| \leq \frac{M}{\lambda+\omega}$$

for each $\lambda > -\omega$, and therefore the integral on the right-hand side in (7.6.2) is convergent. In addition, as

$$(\lambda I + A)^{-1} = \int_0^\infty e^{-\lambda t}S(t)\,dt,$$

we have

$$\frac{\sin \pi\alpha}{\pi}\int_0^\infty \lambda^{-\alpha}(\lambda I + A)^{-1}d\lambda = \frac{\sin \pi\alpha}{\pi}\int_0^\infty \lambda^{-\alpha}e^{-\lambda t}d\lambda\int_0^\infty S(t)\,dt$$

$$= \frac{\sin \pi\alpha}{\pi}\int_0^\infty v^{-\alpha}e^{-v}\,dv\int_0^\infty t^{\alpha-1}S(t)\,dt = \frac{\sin \pi\alpha}{\pi}\Gamma(\alpha)\Gamma(1-\alpha)A^{-\alpha}.$$

Since

$$\Gamma(\alpha)\Gamma(1-\alpha) = \frac{\pi}{\sin \pi\alpha},$$

this completes the proof. □

Lemma 7.6.3. *If the C_0-semigroup $\{S(t)\,;\ t \geq 0\}$, generated by $-A$, is of type $(M, -\omega)$, with $M \geq 1$ and $\omega > 0$, then there exists $C > 0$ such that, for each $\alpha \in [0,1]$, we have*

$$\|A^{-\alpha}\|_{\mathcal{L}(X)} \leq C.$$

Proof. It suffices to prove the inequality only for $\alpha \in (0,1)$. From Lemma 7.6.2, we have

$$\|A^{-\alpha}\|_{\mathcal{L}(X)} \leq \left\|\frac{\sin \pi\alpha}{\pi}\int_0^1 \lambda^{-\alpha}(\lambda I + A)^{-1}\,d\lambda\right\|_{\mathcal{L}(X)}$$

$$+\left\|\frac{\sin \pi\alpha}{\pi}\int_1^\infty \lambda^{-\alpha}(\lambda I + A)^{-1}\,d\lambda\right\|_{\mathcal{L}(X)}.$$

Inasmuch as $\|(\lambda I + A)^{-1}\|_{\mathcal{L}(X)} \leq \frac{M}{\lambda+\omega}$ with $\omega > 0$, there exist $C_0 > 0$ and $C_1 > 0$ such that

$$\|(\lambda I + A)^{-1}\|_{\mathcal{L}(X)} \leq C_0$$

for each $\lambda \in [0, 1]$, and

$$\|\lambda(\lambda I + A)^{-1}\|_{\mathcal{L}(X)} \leq C_1$$

for each $\lambda \geq 1$. Accordingly, we have

$$\|A^{-\alpha}\|_{\mathcal{L}(X)} \leq C_0 \left|\frac{\sin \pi(1-\alpha)}{\pi(1-\alpha)}\right| + C_1 \left|\frac{\sin \pi\alpha}{\pi\alpha}\right| \leq C,$$

where $C = C_0 + C_1$. The proof is complete. □

Lemma 7.6.4. *If the C_0-semigroup, $\{S(t)\,;\; t \geq 0\}$, generated by $-A$, is of type $(M, -\omega)$, with $M \geq 1$ and $\omega > 0$, then, for each $x \in X$, we have*

$$\lim_{\alpha \downarrow 0} A^{-\alpha}x = x.$$

Proof. Let $x \in D(A)$. Since $(-\omega, +\infty) \subset \rho(-A)$, we have $0 \in \rho(A)$ and therefore there exists $y \in X$ with $x = A^{-1}y$. We then have

$$A^{-\alpha}x - x = A^{-(\alpha+1)}y - A^{-1}y = \int_0^\infty \left(\frac{t^\alpha}{\Gamma(1+\alpha)} - 1\right) S(t)y\,dt.$$

From the growth condition $\|S(t)\|_{\mathcal{L}(X)} \leq Me^{-\omega t}$, we deduce

$$\|A^{-\alpha}x - x\| \leq M\|y\| \int_0^\infty \left|\frac{t^\alpha}{\Gamma(1+\alpha)} - 1\right| e^{-\omega t} dt.$$

On the other hand, there exists $C > 0$, such that, for each $\alpha \in [0, 1]$ and each $t \geq 1$, we have

$$\left|\frac{t^\alpha}{\Gamma(1+\alpha)} - 1\right| \leq Ct.$$

It then follows that, for each $k \geq 1$, we have

$$\|A^{-\alpha}x - x\|$$

$$\leq M\|y\| \int_0^k \left|\frac{t^\alpha}{\Gamma(1+\alpha)} - 1\right| e^{-\omega t} dt + CM\|y\| \int_k^\infty te^{-\omega t} dt.$$

Let $\varepsilon > 0$. Let us fix a sufficiently large $k \geq 1$, such that the second term on the right-hand side in the above inequality be less than $\varepsilon/2$. We observe that there exists $\delta(\varepsilon) > 0$ such that, for each $\alpha \in (0, \delta(\varepsilon))$, the first term on the right-hand side in the above inequality is less than $\varepsilon/2$. Then, for each $x \in D(A)$, we have $\lim_{\alpha\downarrow 0} A^{-\alpha}x = x$. Since $D(A)$ is dense in X and, by virtue of Lemma 7.6.3, the family of operators $\{A^{-\alpha};\; \alpha \in [0, 1]\}$ is bounded in $\mathcal{L}(X)$, the proof is complete. □

From Lemmas 7.6.2 and 7.6.4, it follows:

Corollary 7.6.1. *If the C_0-semigroup, $\{S(t)\,;\ t \geq 0\}$, generated by $-A$, is of type $(M, -\omega)$, with $M \geq 1$ and $\omega > 0$, then $\{A^{-t};\ t \geq 0\}$ is a semigroup of class C_0.*

Lemma 7.6.5. *If the C_0-semigroup, $\{S(t)\,;\ t \geq 0\}$, generated by $-A$, is of type $(M, -\omega)$, with $M \geq 1$ and $\omega > 0$, then, for each $\alpha \geq 0$, $A^{-\alpha}$ is injective.*

Proof. Obviously A^{-1} is injective and thus, for each $n \in \mathbb{N}^*$, A^{-n} has the same property. Let $\alpha \geq 0$ and $x \in X$ with $A^{-\alpha}x = 0$. Let us fix $n \in \mathbb{N}^*$ with $n \geq \alpha$. We have $A^{-n}x = A^{-n+\alpha}A^{-\alpha}x = 0$. Accordingly $x = 0$, and therefore $A^{-\alpha}$ is injective. The proof is complete. □

Lemma 7.6.5 allows us to define the power of an arbitrary real exponent, of any operator A of the form $A = -B$, with B the infinitesimal generator of a C_0-semigroup of type $(M, -\omega)$. More precisely, we have:

Definition 7.6.2. Let $A : D(A) \subseteq X \to X$ be a linear operator, with $-A$ the infinitesimal generator of a C_0-semigroup of type $(M, -\omega)$, with $M \geq 1$ and $\omega > 0$, and let $\alpha > 0$. By definition

$$A^{\alpha} = (A^{-\alpha})^{-1}. \tag{7.6.3}$$

In all that follows, A^{α} is defined by $A^{\alpha} = A^{-\beta}$ with $\beta = -\alpha$, where $A^{-\beta}$ is given by (7.6.1) if $\alpha \leq 0$, and respectively by (7.6.3) if $\alpha > 0$. According to the previous convention, we have $A^0 = I$.

Theorem 7.6.1. *If the C_0-semigroup $\{S(t)\,;\ t \geq 0\}$, generated by $-A$, is of type $(M, -\omega)$, with $M \geq 1$ and $\omega > 0$, then:*

(i) *for each $\alpha \in \mathbb{R}$, A^{α} is a closed operator with $D(A^{\alpha}) = R(A^{-\alpha})$;*
(ii) *for each $0 \leq \alpha \leq \beta$, we have $D(A^{\alpha}) \subseteq D(A^{\beta})$;*
(iii) *for each $\alpha \geq 0$, $D(A^{\alpha})$ is dense in X;*
(iv) *for each $\alpha, \beta \in \mathbb{R}$ and each $x \in D(A^{\gamma})$ ($\gamma = \max\{\alpha, \beta, \alpha + \beta\}$), we have*

$$A^{\alpha+\beta}x = A^{\alpha}A^{\beta}x.$$

Proof. The items (i) and (ii) follow from Remark 7.6.1, even in a more general setting when $\{S(t)\,;\ t \geq 0\}$ is only of class C_0. The item (iii) follows from the observation that $D(A^n)$ is dense in X and, for each $\alpha \geq n$, we have $D(A^n) \subseteq D(A^{\alpha})$. See Theorem 2.4.3, and (ii). Finally, (iv) is a direct consequence of both Lemma 7.6.1 and Definition 7.6.2, and this achieves the proof. □

For $0 < \alpha < 1$ and $x \in D(A) \subseteq D(A^\alpha)$, we can define A^α explicitly. More precisely, we have:

Theorem 7.6.2. *If the C_0-semigroup $\{S(t)\,;\ t \geq 0\}$, generated by $-A$ is of type $(M, -\omega)$, with $M \geq 1$ and $\omega > 0$, $\alpha \in (0,1)$, and $x \in D(A)$, then:*

$$A^\alpha x = \frac{\sin \pi\alpha}{\pi} \int_0^\infty \lambda^{\alpha-1}(\lambda I + A)^{-1} Ax\, d\lambda.$$

Proof. By (i) and (ii) in Theorem 7.6.1, we have $R(A^{\alpha-1}) = D(A^{1-\alpha})$ and $D(A^{1-\alpha}) \subseteq D(A)$. Therefore, $A^{\alpha-1}x \in D(A)$, for each $x \in X$. On the other hand, by Lemma 7.6.2, we get

$$A^{\alpha-1}x = \frac{\sin \pi\alpha}{\pi} \int_0^\infty \lambda^{\alpha-1}(\lambda I + A)^{-1} x\, d\lambda, \tag{7.6.4}$$

while from the preceding observation, both sides of the equality above are in $D(A)$. Applying A to both sides in (7.6.4), and using Theorem 1.2.2 and (iv) in Theorem 7.6.1, we obtain the conclusion. The proof is complete. □

Theorem 7.6.3. *Let $A : D(A) \subseteq X \to X$, where $-A$ is the generator of a C_0-semigroup of type $(M, -\omega)$, with $M \geq 1$ and $\omega > 0$, and let $\alpha \in (0,1)$. Then there exists $C > 0$ such that, for each $x \in D(A)$ and each $\rho > 0$, we have*

$$\|A^\alpha x\| \leq C(\rho^\alpha \|x\| + \rho^{\alpha-1}\|Ax\|) \tag{7.6.5}$$

and

$$\|A^\alpha x\| \leq 2C\|x\|^{1-\alpha}\|Ax\|^\alpha. \tag{7.6.6}$$

Proof. By Theorem 7.6.3, since $(\lambda I + A)^{-1}Ax = x - \lambda(\lambda I + A)^{-1}x$, we deduce

$$\begin{aligned}
\|A^\alpha x\| &\leq \left|\frac{\sin \pi\alpha}{\pi}\right| \int_0^\rho \lambda^{\alpha-1}\|A(\lambda I + A)^{-1}\|_{\mathcal{L}(X)}\|x\|\, d\lambda \\
&\quad + \left|\frac{\sin \pi\alpha}{\pi}\right| \int_\rho^\infty \lambda^{\alpha-1}\|(\lambda I + A)^{-1}\|_{\mathcal{L}(X)}\|Ax\|\, d\lambda \\
&\leq \left|\frac{\sin \pi\alpha}{\pi}\right| (1+M)\rho^\alpha\|x\| + \left|\frac{\sin \pi(1-\alpha)}{\pi(1-\alpha)}\right| M\rho^{\alpha-1}\|Ax\| \\
&\leq C(\rho^\alpha\|x\| + \rho^{\alpha-1}\|Ax\|),
\end{aligned}$$

where

$$C = (1+M) \sup_{\alpha\in(0,1)} \left\{\left|\frac{\sin \pi\alpha}{\pi}\right| + \left|\frac{\sin \pi(1-\alpha)}{\pi(1-\alpha)}\right|\right\}.$$

So, (7.6.5) holds. Since (7.6.6) is obviously true for $x = 0$, and follows from (7.6.5), putting $\rho = \|Ax\|/\|x\|$, for $x \neq 0$, the proof is complete. □

Corollary 7.6.2. *Let $A : D(A) \subseteq X \to X$, where $-A$ is the generator of a C_0-semigroup of type $(M, -\omega)$, with $M \geq 1$ and $\omega > 0$, and let $\alpha \in (0, 1]$. Let $B : D(B) \subseteq X \to X$ be a closed operator with $D(A^\alpha) \subseteq D(B)$. Then there exists $C > 0$ such that, for each $x \in D(A^\alpha)$, we have*

$$\|Bx\| \leq C\|A^\alpha x\|. \tag{7.6.7}$$

Moreover, there exists $C_1 > 0$ such that, for each $\rho > 0$ and each $x \in D(A)$, we have

$$\|Bx\| \leq C_1(\rho^\alpha\|x\| + \rho^{\alpha-1}\|Ax\|). \tag{7.6.8}$$

Proof. Since $D(A^\alpha) \subseteq D(B)$, the operator $BA^{-\alpha}$ is defined on the whole X and is closed. From the closed graph theorem (see Dunford and Schwartz, Theorem 4, p. 57), it is in $\mathcal{L}(X)$, which proves (7.6.7). Since (7.6.8) follows from (7.6.5) and (7.6.7), the proof is complete. □

7.7. Further Investigations in the Analytic Case

Lemma 7.7.1. *Let $A : D(A) \subseteq X \to X$, where $-A$ is the infinitesimal generator of an analytic C_0-semigroup of type $(M, -\omega)$, with $M \geq 1$ and $\omega > 0$. Then:*

(i) *there exists $C > 0$ such that, for each $t > 0$, we have*

$$\|AS(t)\|_{\mathcal{L}(X)} \leq Ct^{-1};$$

(ii) *there exist $C_0 > 0$ and $a \in (0, \omega)$ such that, for each $t > 0$, we have*

$$\|AS(t)\|_{\mathcal{L}(X)} \leq C_0 t^{-1} e^{-at};$$

(iii) *for each $m \in \mathbb{N}^*$, we have*

$$\|A^m S(t)\| \leq (C_0 m)^m t^{-m} e^{-at},$$

where $C_0 > 0$ and $a \in (0, \omega)$ are the constants given by (ii).

Proof. The item (i) follows from Theorem 7.1.1. On the other hand, since the semigroup is of type $(M, -\omega)$, we have

$$\begin{aligned}\|AS(t)\|_{\mathcal{L}(X)} &= \|S(t/2)AS(t/2)\|_{\mathcal{L}(X)}\\ &\leq \|S(t/2)\|_{\mathcal{L}(X)}\|AS(t/2)\|_{\mathcal{L}(X)} \leq 2MCt^{-1}e^{-(\omega/2)t},\end{aligned}$$

which proves (ii) with $C_0 = 2MC$ and $a = \omega/2$. To check (iii), let us observe that

$$\begin{aligned}\|A^m S(t)\| = \|(AS(t/m))^m\| \leq \|AS(t/m)\|^m &\leq \left(C_0 m t^{-1} e^{-a(t/m)}\right)^m\\ &= (C_0 m)^m t^{-m} e^{-at},\end{aligned}$$

and this achieves the proof. □

Theorem 7.7.1. *Let $A : D(A) \subseteq X \to X$, where $-A$ is the infinitesimal generator of an analytic C_0-semigroup of type $(M, -\omega)$, with $M \geq 1$ and $\omega > 0$, let $\alpha \in (0,1)$, and let $B : D(B) \subseteq X \to X$ be a closed operator with $D(A) \subseteq D(B)$. If there exist $C > 0$, $\gamma \in (0,1)$, and $\rho_0 > 0$, such that, for each $x \in D(A)$ and each $\rho \geq \rho_0$,*

$$\|Bx\| \leq C(\rho^{\gamma}\|x\| + \rho^{\gamma-1}\|Ax\|), \tag{7.7.1}$$

then, for each $\gamma < \alpha \leq 1$, we have $D(A^{\alpha}) \subseteq D(B)$.

Proof. Let $x \in D(A^{1-\alpha})$. Then $A^{-\alpha}x \in D(A) \subseteq D(B)$. As B is closed, from Theorem 1.2.2, it follows that

$$BA^{-\alpha}x = \frac{1}{\Gamma(\alpha)} \int_0^{\infty} t^{\alpha-1} BS(t)x\, dt$$

if the integral on the right-hand side is convergent. Let us observe that

$$\|BA^{-\alpha}x\| \leq \frac{1}{\Gamma(\alpha)} \left(\int_0^{\rho} t^{\alpha-1}\|BS(t)x\|\, dt + \int_{\rho}^{\infty} t^{\alpha-1}\|BS(t)x\|\, dt \right).$$

Taking $\delta = {\rho_0}^{-1}$, and using (7.7.1), with $\rho = t^{-1}$ in the first integral on the right-hand side, and respectively, with $\rho = \rho_0$, in the second integral, and using (i) and (ii) in Lemma 7.7.1, we deduce that there exists $C_1 > 0$ such that $\|BA^{-\alpha}x\| \leq C_1\|x\|$ for each $x \in D(A^{1-\alpha})$. Since $BA^{-\alpha}$ is closed and $D(A^{1-\alpha})$ is dense in X, it follows that the preceding inequality holds true for each $x \in X$. Hence $D(A^{\alpha}) \subseteq D(B)$, and this achieves the proof. □

Theorem 7.7.2. *Let $A : D(A) \subseteq X \to X$, where $-A$ is the infinitesimal generator of an analytic C_0-semigroup of type $(M, -\omega)$, with $M \geq 1$ and $\omega > 0$. Then:*

(i) *for each $t > 0$ and $\alpha \geq 0$, we have $S(t)X \subseteq D(A^{\alpha})$;*
(ii) *for each $\alpha \geq 0$ and $x \in D(A^{\alpha})$, we have $S(t)A^{\alpha}x = A^{\alpha}S(t)x$;*
(iii) *for each $t > 0$, and each $\alpha \geq 0$, the operator $A^{\alpha}S(t)$ is in $\mathcal{L}(X)$. In addition, there exist $M_{\alpha} > 0$ and $a \in (0, \omega)$ such that, for each $t > 0$, we have*

$$\|A^{\alpha}S(t)\|_{\mathcal{L}(X)} \leq M_{\alpha}t^{-\alpha}e^{-at};$$

(iv) *for each $\alpha \in (0,1]$, there exists $C_{\alpha} > 0$ such that, for each $t > 0$ and each $x \in D(A^{\alpha})$, we have*

$$\|S(t)x - x\| \leq C_{\alpha}t^{\alpha}\|A^{\alpha}x\|.$$

Proof. Since the semigroup generated by $-A$ is of type $(M, -\omega)$, with $M \geq 1$ and $\omega > 0$, from Lemma 7.6.5, we deduce that, for each $\alpha \geq 0$, A^{α}

is well-defined. Inasmuch as $\{S(t)\,;\ t \geq 0\}$ is analytic, for each $\alpha \geq 0$ and each $t > 0$, we have

$$S(t)X \subseteq \bigcap_{n=0}^{\infty} D(A^n) \subseteq D(A^\alpha),$$

which proves (i).
Let $\alpha \geq 0$ and $x \in D(A^\alpha)$. By virtue of Lemma 7.6.4, there exists $y \in X$ such that $x = A^{-\alpha}y$. We then have

$$S(t)x = S(t)A^{-\alpha}y = \frac{1}{\Gamma(\alpha)}\int_0^\infty s^{\alpha-1}S(s)S(t)y\,ds$$

$$= A^{-\alpha}S(t)y = A^{-\alpha}S(t)A^\alpha x,$$

and thus we get (ii). Clearly (iii) holds for $\alpha = 0$. So, let us assume that $\alpha > 0$. Since A^α is closed, for each $t > 0$, $A^\alpha S(t)$ enjoys the very same property. Moreover, in view of (i), $A^\alpha S(t)$ is defined on the whole X. By virtue of the closed graph theorem, it follows that it is bounded. Let $n \in \mathbb{N}^*$ with $n - 1 < \alpha \leq n$. From (iii) in Lemma 7.7.1, we deduce

$$\|A^\alpha S(t)\|_{\mathcal{L}(X)} = \|A^{\alpha-n}A^nS(t)\|_{\mathcal{L}(X)}$$

$$\leq \frac{1}{\Gamma(n-\alpha)}\int_0^\infty s^{n-\alpha-1}\|A^nS(t+s)\|_{\mathcal{L}(X)}\,ds$$

$$\leq \frac{(Cn)^n}{\Gamma(n-\alpha)}\int_0^\infty s^{n-\alpha-1}(t+s)^{-n}e^{-a(t+s)}ds$$

$$\leq \frac{(Cn)^ne^{-at}}{\Gamma(n-\alpha)t^\alpha}\int_0^\infty u^{n-\alpha-1}(1+u)^{-n}du = \frac{M_\alpha}{t^\alpha}e^{-at},$$

which proves (iii).
Finally, (iv) follows from the fact that, by virtue of (iii), for each $\alpha \in (0,1)$, we have

$$\|S(t)x - x\| = \left\|\int_0^t AS(s)x\,ds\right\| = \left\|\int_0^t A^{1-\alpha}S(s)A^\alpha x\,ds\right\|$$

$$\leq \int_0^t \|A^{1-\alpha}S(s)\|_{\mathcal{L}(X)}\|A^\alpha x\|\,ds \leq M_\alpha\int_0^t s^{\alpha-1}e^{-as}\|A^\alpha x\|\,ds$$

$$\leq M_\alpha\int_0^t s^{\alpha-1}\|A^\alpha x\|\,ds = C_\alpha t^\alpha\|A^\alpha x\|.$$

The proof is complete. □

Finally, we notice that we can define the fractional powers of the negative of an infinitesimal generator of an analytic C_0-semigroup by using the functional calculus for unbounded operators presented in Section 1.9. More precisely, if $-A$ is the infinitesimal generator of an analytic C_0-semigroup with $0 \in \rho(-A)$ and $\alpha > 0$, we can define

$$A^{-\alpha} = \frac{1}{2\pi i} \int_\Gamma \lambda^{-\alpha}(A - \lambda I)^{-1}\, d\lambda,$$

where Γ is a simple rectifiable curve included in $\rho(A)$, which does not intersect the negative axis, curve consisting of two rays connected by an arc of circle centered at the origin and oriented from $\infty e^{-i\theta}$ to $\infty e^{i\theta}$, with θ fixed in $(0, \pi)$. In addition, $\lambda^{-\alpha}$ is taken to be positive for λ real and positive. For more details with regards to this manner of defining $A^{-\alpha}$, see Pazy [**101**], Section 2.6, p. 69.

Problems

Problem 7.1. Show that if a C_0-group has an analytic extension in a sector of the complex plane, then its infinitesimal generator is bounded. See Goldstein [**61**], Exercise 14, p. 75.

Problem 7.2. Let $H = L^2(0,\pi)$ and let $A : D(A) \subseteq H \to H$ be defined by

$$\begin{cases} D(A) = H_0^1(0,\pi) \cap H_0^1(0,\pi) \\ Au = -u'' \ \text{ for } \ u \in D(A) \end{cases}$$

and let $\alpha \in [\,\frac{1}{2}, 1\,]$. Find $D(A^\alpha)$.

Problem 7.3. Let $X = L^2(\mathbb{R}_+)$, and $A : D(A) \subseteq X \to X$ with $-A$ the generator of the C_0-semigroup of translations $S(t) : X \to X$, defined by

$$[S(t)f](s) = f(t+s)$$

for each $t \geq 0$, each $f \in X$, and a.e. for $s \in \mathbb{R}_+$. Let $\alpha \in (0,1)$. Show that $C_0^\infty(0,+\infty) \subseteq D(A^{-\alpha})$, and that, for each $f \in D(A^{-\alpha})$, and a.e. for $s \in (0,+\infty)$, $A^{-\alpha} f$ is *the Weyl fractional integral of order α of f*, i.e.

$$(A^{-\alpha} f)(s) = \frac{1}{\Gamma(\alpha)} \int_s^\infty (u-s)^{\alpha-1} f(u)\, du = (K^\alpha f)(s).$$

Prove that, for each $f, g \in C_0^\infty(0,+\infty)$, we have

$$\int_0^\infty f(s)(K^\alpha g)(s)\, ds = \int_0^\infty (J^\alpha f)(u) g(u)\, du,$$

where $J^\alpha f$ is the Riemann-Liouville integral, of exponent α, of f, i.e.

$$(J^\alpha f)(u) = \frac{1}{\Gamma(u)} \int_0^u (u-s)^{\alpha-1} f(s)\, ds.$$

See also Problem 4.7. See McBride [**89**], p. 103.

Problem 7.4. Let $\{S(t)\,;\ t \geq 0\}$ be the C_0-semigroup in Problem 4.3, and let A be its generator. Prove that $A_{1/2} = -\,(-A)^{1/2}$ is defined by means of the singular integral operator

$$[A_{1/2}f](x) = \lim_{h\to 0} \frac{1}{\pi} \int_{-\infty}^{+\infty} \frac{f(x-y) - f(x)}{y^2 + h^2}\, dy$$

for each $f \in D(A_{1/2})$. Prove that the C_0-semigroup generated by $A_{1/2}$ is the Poisson semigroup in Problem 4.5. See Goldstein [**61**], pp. 62-63.

Notes. Theorem 7.1.1 was proved by Hille in 1942, and Theorem 7.1.2 is due to Vrabie [**130**] and is somehow related to the following characterization of generators of analytic semigroups due to Crandall, Pazy and Tartar [**43**].

Theorem (Crandall-Pazy-Tartar) *Let $A : D(A) \subseteq X \to X$ be the generator of a C_0-semigroup of type (M, ω), $\{S(t)\,;\ t \geq 0\}$. Then $\{S(t)\,;\ t \geq 0\}$ is analytic if and only if there exist $C > 0$ and $\Lambda \geq 0$ such that*

$$\|AR(\lambda; A)^{n+1}\|_{\mathcal{L}(X)} \leq \frac{C}{n\lambda^n}$$

for $\lambda > n\Lambda$, $n = 1, 2, \ldots$.

See Pazy [**101**], Theorem 5.5, p. 65. The results in Section 7.1, 7.2 and 7.3 are from Brezis and Cazenave [**31**] and Cazenave and Haraux [**36**]. For similar decay estimates, as those in Theorem 7.2.6, in the case of the C_0-group generated by the linear Schrödinger operator in the whole space, see Sulem and Sulem [**118**], pp. 41-51. Sharper time-space estimates, for the Schrödinger C_0-group, known as *Strichartz inequalities*, may be found also in Sulem and Sulem [**118**] *loc. cit.*, as well as in Brezis and Cazenave [**31**], pp. 88-94. Sections 4 and 5 are from Bejenaru, Diaz and Vrabie [**23**]. Several interesting examples of analytic semigroups, arising from thermoelastic plate systems subjected to various boundary conditions and requiring a much more delicate analysis, can be found in Lasiecka and Triggiani [**84**], Chapter 3, pp. 311-413. Sections 7.6 and 7.7 present, in a slightly different manner, the main concepts and results in Pazy [**101**] concerning fractional powers of closed operators, introduced by Bochner [**25**], Phillips [**104**], redefined and studied within a more general frame by Balakrishnan [**10**], and Komatsu [**75**], [**76**], [**77**], [**78**], [**79**]. For more details on this subject see Amann [**5**] and Pazy [**101**]. We note that whenever A generates an analytic semigroup, then $-(-A)^{1/2}$ generates an analytic semigroup too, and the unique solution of the boundary-value problem $u'' = -Au$, $\lim_{t\downarrow 0} u(t) = \xi$ and u bounded is the solution of $u' = -(-A)^{1/2}u$, $t > 0$, $\lim_{t\downarrow 0} u(t) = \xi$. This is in fact the definition given by Balakrishnan [**11**] for the square root of $-A$. We note that Barbu, in [**15**] and [**16**], extended this definition to the nonlinear case as well, i.e. when A is maximal monotone operator acting in a real Hilbert space H. Further details on the fractional powers of the Stokes operator may be found in Fujita and Kato [**58**] and Constantin and Foiaş [**38**]. The problems in this chapter are from Goldstein [**61**] and McBride [**89**].

CHAPTER 8

The Nonhomogeneous Cauchy Problem

In this chapter we introduce several concepts of solution for the nonhomogeneous Cauchy problem

$$\begin{cases} u' = Au + f \\ u(a) = \xi, \end{cases}$$

where $A : D(A) \subseteq X \to X$ is the infinitesimal generator of a C_0-semigroup of contractions, $\xi \in X$, and $f \in L^1(a, b; X)$. Namely, we introduce the concepts of C^1, strong and respectively C^0-solution and we analyze the relationship between these types of solutions. We then study the compactness properties of the C^0-solution operator $f \mapsto u$ which will prove useful later in order to obtain existence results for some classes of semilinear Cauchy problems.

8.1. The Cauchy Problem $u' = Au + f$, $u(a) = \xi$

As we already have seen in the preceding sections, if $A : D(A) \subseteq X \to X$ is the infinitesimal generator of a C_0-semigroup $\{S(t)\,;\ t \geq 0\}$, then, for each $a \geq 0$, and $\xi \in D(A)$, the function $u : [a, +\infty) \to X$, defined by $u(t) = S(t - a)\xi$ for each $t \geq 0$, is the unique solution of the homogeneous Cauchy problem

$$\begin{cases} u' = Au \\ u(a) = \xi. \end{cases} \tag{8.1.1}$$

From this reason, it is quite natural to consider that, for each $\xi \in X$, the function u, defined as above, is a solution for (8.1.1), in a generalized sense. The aim of this section is to consider the nonhomogeneous problem

$$\begin{cases} u' = Au + f \\ u(a) = \xi, \end{cases} \tag{8.1.2}$$

where A is as before, $\xi \in X$, and $f \in L^1(a, b; X)$.

Definition 8.1.1. The function $u : [a, b] \to X$ is called *classical, or C^1-solution* of the problem (8.1.2), if u is continuous on $[a, b]$, continuously differentiable on $(a,b]$, $u(t) \in D(A)$ for each $t \in (a,b]$ and it satisfies $u'(t) = Au(t) + f(t)$ for each $t \in [a,b]$ and $u(a) = \xi$.

Definition 8.1.2. The function $u : [a,b] \to X$ is called *absolutely continuous, or strong solution*, of the problem (8.1.2), if u is absolutely continuous on $[a,b]$, $u' \in \mathcal{L}^1(a,b;X)$, $u(t) \in D(A)$ a.e. for $t \in (a,b)$, and it satisfies $u'(t) = Au(t) + f(t)$ a.e. for $t \in (a,b)$ and $u(a) = \xi$.

Remark 8.1.1. Each classical solution of (8.1.2) is a strong solution of the same problem, but not conversely.

Remark 8.1.2. As we already have seen, if A generates a uniformly continuous semigroup and f is continuous from $[a,b]$ to $\mathbb{R}^n$, then a function $u : [a,b] \to X$ is a classical solution of the nonhomogeneous problem (8.1.2) if and only if it is given by the so-called *variation of constants formula*

$$u(t) = S(t-a)\xi + \int_a^t S(t-s)f(s)\,ds \tag{8.1.3}$$

for each $t \in [a,b]$. See Theorem 2.1.2. Simple examples show that, in the case in which X is infinite-dimensional and A is unbounded, i.e. it generates only a C_0-semigroup, the problem (8.1.2) may fail to have any classical solution and this, no matter as regular is the datum f. See Example 8.1.1 below.

The next simple result is fundamental in the study of the nonhomogeneous problem (8.1.2).

Theorem 8.1.1. (Duhamel Principle) *Each strong solution of* (8.1.2) *is given by* (8.1.3). *In particular, each classical solution of the problem* (8.1.2) *is given by* (8.1.3).

Proof. Let u be a strong solution of (8.1.2), $t \in (a,b]$ and let us define $g : [a,t] \to X$ by $g(s) = S(t-s)u(s)$ for each $s \in [a,t]$. Then g is a.e. differentiable on (a,t), its derivative belongs to $L^1(a,t;X)$, and

$$g'(s) = -AS(t-s)u(s) + S(t-s)u'(s)$$

$$= -AS(t-s)u(s) + S(t-s)Au(s) + S(t-s)f(s) = S(t-s)f(s)$$

a.e. for $s \in (a,t)$. Since $f \in L^1(a,b;X)$, $s \mapsto S(t-s)f(s)$ is integrable on $[a,t]$. Integrating the above equality from a to t, we obtain (8.1.3). $\square$

From Theorem 8.1.1 we deduce the following uniqueness result.

Corollary 8.1.1. *For each* $\xi \in X$ *and each* $f \in L^1(a,b;X)$, *the problem* (8.1.2) *may have at most one classical (strong) solution.*

The previous considerations justify why, in the case of infinite-dimensional spaces X, the variation of constants formula is promoted to the rank of definition. More precisely, we introduce:

Definition 8.1.3. The function $u : [a, b] \to X$, defined by (8.1.3) is called C^0, *or mild, solution* of the problem (8.1.2).

In all that follow, we focus our attention on some sufficient conditions in order that a C^0-solution of (8.1.2) be a C^1, or strong solution of the same problem. We begin with an example which shows that, the continuity of f alone is not sufficient for the C^0-solution given by (8.1.3) to be a strong one and, even less, a C^1-solution.

Example 8.1.1. Let $A : D(A) \subseteq X \to X$ be the infinitesimal generator of a C_0-semigroup $\{S(t);\ t \geq 0\}$, for which there exists $\eta \in X$ such that $S(t)\eta \notin D(A)$ for each $t \geq 0$. (Such semigroup is that in Example 2.4.1 in which η is a nowhere differentiable function). Let us define $f(s) = S(s)\eta$ for each $s \in [0, T]$ and let us observe that f is continuous. On the other hand, the problem (8.1.2) with $\xi = 0$ has no strong solution, and so it has no classical solution. Indeed, if u is a strong solution of the problem (8.1.2), in view of Remark 8.1.2, it is given by

$$u(t) = \int_0^t S(t-s)S(s)\eta\, ds = tS(t)\eta$$

for each $t \in [0, T]$, function which clearly is not a.e. differentiable on $[0, T]$.

Theorem 8.1.2. *Let* $A : D(A) \subseteq X \to X$ *be the infinitesimal generator of a* C_0*-semigroup* $\{S(t);\ t \geq 0\}$, *let* $f \in L^1(a, b; X)$ *be continuous on* (a, b), *and let*

$$v(t) = \int_a^t S(t-s)f(s)\, ds$$

for $t \in [a, b]$. *If at least one of the conditions below is satisfied*

(i) v *is continuously differentiable on* (a, b);
(ii) $v(t) \in D(A)$ *for each* $t \in (a, b)$, *and* $t \mapsto Av(t)$ *is continuous on* (a, b)

then, for each $\xi \in D(A)$, (8.1.2) *has a unique classical solution. If there exists* $\xi \in D(A)$ *such that* (8.1.2) *has a classical solution, then* v *satisfies both* (i) *and* (ii).

Proof. Let us observe that, for each $t \in (a, b)$ and $h > 0$, we have

$$\frac{1}{h}(S(h) - I)v(t) = \frac{v(t+h) - v(t)}{h} - \frac{1}{h}\int_t^{t+h} S(t+h-s)f(s)\, ds. \quad (8.1.4)$$

Let us assume that (i) holds. Since f is continuous, and v is continuously differentiable on (a,b), it follows that the right-hand side of the equality above converges for h tending to 0. Hence, the left-hand side converges too, and consequently $v(t) \in D(A)$, and

$$v'(t) = Av(t) + f(t) \tag{8.1.5}$$

for each $t \in (a,b)$. If (ii) holds, then v is differentiable from the right on (a,b), and its right derivative is continuous on (a,b). Since v is clearly continuous, it follows that v is continuously differentiable on (a,b), and satisfies (8.1.5). Hence, in both cases (i) and (ii), v satisfies (8.1.5). Since $v(a) = 0$, it follows that, for each $\xi \in D(A)$, $t \mapsto u(t) = S(t-a)\xi + v(t)$ for $t \in [\,a,b\,]$ is a classical solution of (8.1.2).
Let us assume now that there exists $\xi \in D(A)$ such that (8.1.2) has a classical solution u which, in view of Remark 8.1.2, is given by (8.1.3). So, the function $t \mapsto v(t) = u(t) - S(t-a)\xi$ is differentiable on (a,b), and in addition $v'(t) = u'(t) - S(t-a)A\xi$ is continuous on (a,b). Thus v satisfies (i). If $\xi \in D(A)$, we have $S(t-a)\xi \in D(A)$ for each $t \in [\,a,b\,]$, and therefore it follows that $v(t) = u(t) - S(t-a)\xi \in D(A)$ for each $t \in (a,b)$, and $t \mapsto Av(t) = Au(t) - AS(t-a)\xi = u'(t) - f(t) - S(t-a)A\xi$ is continuous on (a,b). So, v satisfies (ii), and this achieves the proof. □

Corollary 8.1.2. *If $A : D(A) \subseteq X \to X$ is the infinitesimal generator of a C_0-semigroup $\{S(t);\ t \geq 0\}$, and f is of class C^1 on $[\,a,b\,]$, then, for each $\xi \in D(A)$, the problem* (8.1.2) *has a unique classical solution.*

Proof. Let us observe that

$$t \mapsto v(t) = \int_a^t S(t-s)f(s)\,ds = \int_0^{t-a} S(s)f(t-s)\,ds$$

is continuously differentiable on (a,b). Indeed, a simple calculation shows that

$$\begin{aligned} v'(t) &= S(t-a)f(a) + \int_0^{t-a} S(s)f'(t-s)\,ds \\ &= S(t-a)f(a) + \int_a^t S(t-s)f'(s)\,ds \end{aligned}$$

for each $t \in (a,b)$. The conclusion follows from (i) in Theorem 8.1.2. □

Corollary 8.1.3. *If $A : D(A) \subseteq X \to X$ is the infinitesimal generator of a C_0-semigroup $\{S(t);\ t \geq 0\}$, $f \in L^1(a,b;X)$ is continuous on (a,b), $f(s) \in D(A)$ for each $s \in (a,b)$, and $Af(\cdot) \in L^1(a,b;X)$, then, for each $\xi \in D(A)$, the problem* (8.1.2) *has a unique classical solution.*

Proof. For each $t \in (a, b)$ and $s \in (a, t]$, we have $S(t-s)f(s) \in D(A)$. Moreover, the function $t \mapsto AS(t-s)f(s) = S(t-s)Af(s)$ is integrable, and thus v, defined as in Theorem 8.1.2, satisfies $v(t) \in D(A)$ for each $t \in (a, b)$. In addition, the function

$$t \mapsto Av(t) = A\int_a^t S(t-s)f(s)\,ds = \int_a^t S(t-s)Af(s)\,ds$$

is continuous on $[a, b]$. We conclude the proof with the help of (ii) in Theorem 8.1.2. □

Theorem 8.1.3. *Let $A : D(A) \subseteq X \to X$ be the infinitesimal generator of a C_0-semigroup $\{S(t);\ t \geq 0\}$, let $f \in L^1(a, b; X)$, and let*

$$v(t) = \int_a^t S(t-s)f(s)\,ds$$

for $t \in [a, b]$. If at least one of the conditions below is satisfied

(i) *v is a.e. differentiable on (a, b), and $v' \in L^1(a, b; X)$;*
(ii) *$v(t) \in D(A)$ a.e. for $t \in (a, b)$, and $Av(\cdot) \in L^1(a, b; X)$,*

then, for each $\xi \in D(A)$, the problem (8.1.2) *has a unique strong solution. If there exists $\xi \in D(A)$ such that the problem* (8.1.2) *has a strong solution, then v satisfies both* (i) *and* (ii).

As the proof of Theorem 8.1.3 is a copy of that of Theorem 8.1.2, we do not enter into details. We state, also without proof, two consequences of Theorem 8.1.3 which are variants of Corollaries 8.1.2 and 8.1.3.

Corollary 8.1.4. *If $A : D(A) \subseteq X \to X$ is the infinitesimal generator of a C_0-semigroup $\{S(t);\ t \geq 0\}$, f is a.e. differentiable on (a, b), and $f' \in L^1(a, b; X)$, then, for each $\xi \in D(A)$, the problem* (8.1.2) *has a unique strong solution.*

Corollary 8.1.5. *If $A : D(A) \subseteq X \to X$ is the infinitesimal generator of a C_0-semigroup $\{S(t);\ t \geq 0\}$, $f \in L^1(a, b; X)$ is continuous on (a, b), $f(s) \in D(A)$ a.e. for $s \in (a, b)$, and $Af(\cdot) \in L^1(a, b; X)$, then, for each $\xi \in D(A)$, the problem* (8.1.2) *has a unique strong solution.*

Theorem 8.1.4. *Let $A : D(A) \subseteq X \to X$ be the infinitesimal generator of a C_0-semigroup $\{S(t);\ t \geq 0\}$ and let $f \in C([a, b]; X)$. If at least one of the two conditions below is satisfied*

(i) *$f \in L^1(a, b; D(A))$;*
(ii) *$f \in W^{1,1}(a, b; X)$,*

then, for each $\xi \in D(A)$, the problem (8.1.2) *has a unique classical solution.*

Proof. It suffices to consider only the case $x = 0$. So, we will prove first that, whenever f satisfies either (i), or (ii), the function

$$v(t) = \int_a^t S(t-s)f(s)\,ds$$

belongs to $C^1([\,a,b\,];X)$. Let $t \in [\,a,b)$ and $h \in (0,b-a)$. We assume first that f satisfies (i). We then have

$$\frac{v(t+h)-v(t)}{h} = \int_a^t S(t-s)\frac{S(h)-I}{h}f(s)\,ds + \frac{1}{h}\int_t^{t+h} S(t+h-s)f(s)\,ds.$$

Since $\|S(h)x - x\| \leq \|Ax\|h$ for each $x \in D(A)$ and $h > 0$, by the Lebesgue dominated convergence theorem, it follows that

$$\lim_{h\downarrow 0}\frac{S(h)-I}{h}f = Af$$

in $L^1(a,b;X)$. So

$$\frac{d^+v}{dt}(t) = \int_a^t S(t-s)Af(s)\,ds + f(t)$$

for each $t \in [\,a,b)$. Now, if we assume that $f \in W^{1,1}(a,b;X)$, for each $t \in [\,a,b)$ and $h \in (0,b-a)$, we have

$$\frac{v(t+h)-v(t)}{h} = \int_a^t S(s)\frac{f(t+h-s)-f(t-s)}{h}\,ds$$
$$+S(h)\left(\frac{1}{h}\int_a^{a+h} S(t-s)f(s)\,ds\right).$$

Since

$$\lim_{h\downarrow 0}\frac{f(t+h-\cdot)-f(t-\cdot)}{h} = f'(\cdot)$$

in $L^1(a,t;X)$ and

$$\lim_{h\downarrow 0} S(h)\left(\frac{1}{h}\int_a^{a+h} S(t-s)f(s)\,ds\right) = S(t-a)f(a),$$

we have

$$\frac{d^+v}{dt} = \int_a^t S(s)f'(t-s)\,ds + S(t-a)f(a)$$

for each $t \in [\,a,b)$. So in both cases (i) and (ii), $\frac{d^+v}{dt} \in C([\,a,b);X)$. Similar arguments show that $\frac{d^-v}{dt} \in C((a,b\,];X)$ and thus $v \in C^1([\,a,b\,];X)$. The proof is complete. $\square$

A Banach space X has the *Radon-Nicodým property*, if every absolutely continuous function $f : [a,b] \to X$ is almost everywhere differentiable on (a,b), $f' \in L^1(a,b;X)$ and

$$f(t) = f(a) + \int_a^t f'(s)\,ds$$

for each $t \in [a,b]$. We notice that all reflexive Banach spaces, as well as all separable duals have the Radon-Nicodým property. See Diestel and Uhl [**46**], Theorem 1, p. 79 and Corollary 4, p. 82.

Corollary 8.1.6. *Let $A : D(A) \subseteq X \to X$ be the infinitesimal generator of a C_0-semigroup $\{S(t);\ t \geq 0\}$ and let $f : [a,b] \to X$ be an absolutely continuous functions. If X has the Radon-Nicodým property then, for each $\xi \in D(A)$, the problem* (8.1.2) *has a unique classical solution.*

Proof. Since $f \in W^{1,1}(a,b;X)$, we can apply Theorem 8.1.4 and this achieves the proof. □

A specific form of Corollary 8.1.6, very useful in applications, is:

Corollary 8.1.7. *Let $A : D(A) \subseteq X \to X$ be the infinitesimal generator of a C_0-semigroup $\{S(t);\ t \geq 0\}$ and let $f : [a,b] \to X$ be a Lipschitz continuous functions. If X is reflexive, for each $\xi \in D(A)$, the problem* (8.1.2) *has a unique classical solution.*

8.2. Smoothing Effect. The Hilbert Space Case

Let H be a Hilbert space.

Lemma 8.2.1. *Let $A : D(A) \subseteq H \to H$ be a linear self-adjoint operator which generates a C_0-semigroup of contractions. Let $u \in W^{1,2}(0,T;H)$ with $u(t) \in D(A)$ a.e. for $t \in [a,b]$, and $Au \in L^2(a,b;H)$. Then the function $t \mapsto \frac{1}{2}\langle Au(t), u(t)\rangle$ is absolutely continuous on $[a,b]$ and*

$$\frac{d}{dt}\left(\frac{1}{2}\langle Au(t), u(t)\rangle\right) = \langle Au(t), u'(t)\rangle. \tag{8.2.1}$$

Proof. Let $\lambda > 0$ and let $A_\lambda \in \mathcal{L}(H)$ be the Yosida approximation of A. Obviously A_λ is a self-adjoint operator. By Theorem 1.3.4, we know that u is absolutely continuous on $[0,T]$, a.e. differentiable on $(0,T)$, its derivative belongs to $L^2(a,b;H)$, and u is given by

$$u(t) = u(0) + \int_0^t u'(s)\,ds.$$

We have

$$\frac{d}{dt}\left(\frac{1}{2}\langle A_\lambda u(t), u(t)\rangle\right) = \frac{1}{2}\langle A_\lambda u(t), u'(t)\rangle + \frac{1}{2}\langle A_\lambda u'(t), u(t)\rangle$$
$$= \langle A_\lambda u(t), u'(t)\rangle.$$

Integrating this equality from s to t, we obtain

$$\frac{1}{2}\langle A_\lambda u(t), u(t)\rangle - \frac{1}{2}\langle A_\lambda u(s), u(s)\rangle = \int_s^t \langle A_\lambda u(\tau), u'(\tau)\rangle \, d\tau.$$

By virtue of Lemma 3.2.1 and Lebesgue dominated convergence theorem, we can pass to the limit in this equality. We get

$$\frac{1}{2}\langle Au(t), u(t)\rangle - \frac{1}{2}\langle Au(s), u(s)\rangle = \int_s^t \langle Au(\tau), u'(\tau)\rangle \, d\tau$$

for each $a \leq s \leq t \leq b$. Therefore, the function $t \mapsto \frac{1}{2}\langle Au(t), u(t)\rangle$ is absolutely continuous on $[a, b]$, and (8.2.1) holds. The proof is complete. □

The next theorem gives a sufficient condition in order that, for each $\xi \in X$ and each $f \in L^2(a, b; X)$, the unique C^0-solution of the problem (8.1.2) be a strong solution.

Theorem 8.2.1. *Let $A : D(A) \subseteq H \to H$ be the infinitesimal generator of a C_0-semigroup of contractions $\{S(t);\ t \geq 0\}$. If A is self-adjoint then, for each $\xi \in X$ and $f \in L^2(a, b; X)$, the unique C^0-solution of* (8.1.2) *is strong. Moreover, the function $t \mapsto (t-a)^{1/2}u'(t)$ belongs to $L^2(a, b; H)$, the function $t \mapsto \frac{1}{2}\langle Au(t), u(t)\rangle$ belongs to $L^1(a, b)$, and, for each $c \in (a, b)$, is absolutely continuous on $[c, b]$. If $\xi \in D(A)$, then, for each $f \in L^2(a, b; X)$, the unique C^0-solution of* (8.1.2) *is strong and satisfies $u' \in L^2(a, b; H)$, and the function $t \mapsto \frac{1}{2}\langle Au(t), u(t)\rangle$ is absolutely continuous on $[a, b]$.*

Proof. From Theorem 6.4.1 we know that, for each $\xi \in X$ and $t > 0$, we have $S(t)\xi \in D(A)$. This means that the function

$$u(t) = S(t-a)\xi + \int_a^t S(t-s)f(s)\, ds$$

is a.e. differentiable on $[a, b]$ and satisfies

$$u'(t) = Au(t) + f(t)$$

a.e. for $t \in [a, b]$. Let us take the inner product both sides in this equality by $(t-a)u'(t)$. Inasmuch as A is self-adjoint and generates a C_0-semigroup, by virtue of Lemma 8.2.1, we obtain

$$(t-a)\|u'(t)\|^2 = (t-a)\frac{1}{2}\frac{d}{dt}\left(\langle Au(t), u(t)\rangle\right) + (t-a)\langle f(t), u'(t)\rangle.$$

Integrating from a to b this inequality we get

$$\int_a^b (t-a)\|u'(t)\|^2\,dt = \frac{1}{2}(t-a)\langle Au(t),u(t)\rangle\big|_a^b$$
$$-\frac{1}{2}\int_a^b \langle Au(t),u(t)\rangle\,dt + \int_a^b (t-a)\langle f(t),u'(t)\rangle\,dt.$$

Recalling that, by virtue of Lemma 6.4.2, $\langle Ax,x\rangle \le 0$ for all $x \in D(A)$, and using the inequality $\langle f(t),u'(t)\rangle \le \frac{1}{2}\|f(t)\|^2 + \frac{1}{2}\|u'(t)\|^2$, we deduce

$$\int_a^b (t-a)\|u'(t)\|^2\,dt \le \int_a^b (t-a)\|f(t)\|^2\,dt - \int_a^b \langle Au(t),u(t)\rangle\,dt.$$

But $\langle Au(t),u(t)\rangle = \langle u'(t)-f(t),u(t)\rangle = \frac{1}{2}\frac{d}{dt}\left(\|u(t)\|^2\right) - \langle f(t),u(t)\rangle$, and therefore we have

$$\int_a^b |\langle Au(t),u(t)\rangle|\,dt = -\int_a^b \langle Au(t),u(t)\rangle\,dt$$
$$\le \frac{1}{2}\|\xi\|^2 + \int_a^b \|f(t)\|\|u(t)\|\,dt,$$

and

$$\int_a^b (t-a)\|u'(t)\|^2\,dt \le \int_a^b (t-a)\|f(t)\|^2\,dt + \frac{1}{2}\|\xi\|^2 + \int_a^b \|f(t)\|\|u(t)\|\,dt.$$

So, $t \mapsto \langle Au(t),u(t)\rangle$ is in $L^1(a,b)$. Inasmuch as the semigroup generated by A is of contractions, from (8.1.3), we have

$$\|u(t)\| \le \|\xi\| + \int_a^b \|f(s)\|\,ds,$$

and accordingly

$$\int_a^b (t-a)\|u'(t)\|^2\,dt \le \int_a^b (t-a)\|f(t)\|^2\,dt + \frac{1}{2}\left(\|\xi\| + \int_a^b \|f(t)\|\,dt\right)^2.$$

From this inequality we deduce the first part of the conclusion. In order to prove the second one, let us consider $\xi \in D(A)$, and let us observe that, in this case, $u' \in L^2(a,b;H)$. Indeed, taking the inner product both sides in (8.1.1) by u', and integrating over $[a,b]$, we obtain

$$\int_a^b \|u'(t)\|^2 dt = \frac{1}{2}\langle Au(b),u(b)\rangle - \frac{1}{2}\langle A\xi,\xi\rangle + \int_a^b \langle f(t),u'(t)\rangle\,dt.$$

From this equality, the fact A is dissipative, and from the inequality

$$\int_a^b \langle f(t),u'(t)\rangle\,dt \le \frac{1}{2}\left(\int_a^b \|f(t)\|^2 dt + \int_a^b \|u'(t)\|^2 dt\right),$$

we deduce

$$\int_a^b \|u'(t)\|^2\,dt \le -\langle A\xi,\xi\rangle + \int_a^b \|f(t)\|^2\,dt.$$

Then it follows that $u' \in L^2(a,b;H)$. Consequently, $Au \in L^2(a,b;H)$ too, and the conclusion follows from Lemma 8.2.1. The proof is complete. □

8.3. An Approximation Result

The next result shows that each C^0-solution of the problem (8.1.2) can be approximated uniformly on $[a,b]$ by classical solutions of certain suitably defined approximating Cauchy problems.

Theorem 8.3.1. *Let $A : D(A) \subseteq X \to X$ be the infinitesimal generator of a C_0-semigroup of type (M,ω), let $f \in L^1(a,b;X)$ and $\xi \in X$. Then there exist a sequence $(\xi_n)_{n\in\mathbb{N}}$ in $D(A)$ and a sequence $(f_n)_{n\in\mathbb{N}}$ in $C^1([a,b];X)$ such that*

$$\begin{cases} \lim_{n\to\infty} \xi_n = \xi & \text{strongly in } X \\ \lim_{n\to\infty} f_n = f & \text{strongly in } L^1(a,b;X) \\ \lim_{n\to\infty} u_n(t) = u(t) & \text{uniformly on } [a,b], \end{cases}$$

where u_n is the unique classical solution of (8.1.2) *corresponding to ξ_n and f_n, and u is the unique C^0-solution of the problem* (8.1.2).

Proof. As, by Theorem 2.4.1, $D(A)$ is dense in X, for each $\xi \in X$ there exists $(\xi_n)_{n\in\mathbb{N}}$ such that $\lim_{n\to\infty} \xi_n = \xi$. Moreover, since $C^1([a,b];X)$ is dense in $L^1(a,b;X)$, for each $f \in L^1(a,b;X)$, there exists $(f_n)_{n\in\mathbb{N}}$ such that $\lim_{n\to\infty} f_n = f$. By virtue of Corollary 8.1.2, for each $n \in \mathbb{N}$, the Cauchy problem

$$\begin{cases} u_n' = Au_n + f_n \\ u_n(a) = \xi_n \end{cases}$$

has a unique classical solution given by the variation of constants formula

$$u_n(t) = S(t-a)\xi_n + \int_a^t S(t-s)f_n(s)ds.$$

See Remark 8.1.2. But the semigroup is of type (M,ω) and thus

$$\|S(t)\|_{\mathcal{L}(X)} \le Me^{\omega t}$$

for each $t \in \mathbb{R}_+$. We then have

$$\|u(t)-u_n(t)\| \le \|S(t-a)(\xi-\xi_n)\| + \int_a^t \|S(t-s)(f(s)-f_n(s))\|\,ds$$

$$\le \|S(t-a)\|_{\mathcal{L}(X)}\|\xi-\xi_n\| + \int_a^t \|S(t-s)\|_{\mathcal{L}(X)}\|f(s)-f_n(s)\|\,ds$$

$$\leq Me^{\omega(t-a)}\|\xi-\xi_n\|+M\int_a^t e^{\omega(t-s)}\|f(s)-f_n(s)\|\,ds$$

$$\leq Me^{|\omega|(b-a)}\left(\|\xi-\xi_n\|+\int_a^b\|f(s)-f_n(s)\|\,ds\right),$$

inequality which completes the proof. □

8.4. Compactness of the Solution Operator from $L^p(a,b;X)$

We start with a necessary and sufficient condition in order that a given set of C^0-solutions be relatively compact in $C([a,b];X)$. Let us consider the nonhomogeneous Cauchy problem

$$\begin{cases} u' = Au + f \\ u(a) = \xi, \end{cases} \tag{8.4.1}$$

where $A : D(A) \subseteq X \to X$ generates a C_0-semigroup of contractions, $\xi \in X$ and $f \in L^1(a,b;X)$. We notice that, although we confine ourselves only to the case of C_0-semigroups of contractions, all the results here included hold true in general, i.e. for C_0-semigroups of type (M,ω).

Definition 8.4.1. The operator

$$Q : X \times L^1(a,b;X) \to C([a,b];X),$$

defined by $Q(\xi,f) = u$, where u is the unique C^0-solution of the problem (8.4.1) corresponding to $\xi \in X$ and $f \in L^1(a,b;X)$, i.e.

$$u(t) = S(t-a)\xi + \int_a^t S(t-s)f(s)\,ds$$

for each $t \in [0,T]$, is called *the solution operator* attached to the problem (8.4.1).

Remark 8.4.1. The operator Q is nonexpansive and so, it maps each bounded subset in $X \times L^1(a,b;X)$ into a bounded subset in $C([a,b];X)$. Indeed, we have

$$\|(Q(\xi,f)(t)-(Q(\eta,g)(t)\|$$

$$\leq \|S(t-a)\xi - S(t-a)\eta\| + \int_a^t \|S(t-s)(f(s)-g(s))\|\,ds$$

$$\leq \|\xi-\eta\| + \int_a^t \|S(t-s)\|_{\mathcal{L}(X)}\|f(s)-g(s)\|\,ds$$

$$\leq \|\xi-\eta\| + \int_a^b \|f(s)-g(s)\|\,ds$$

for each $(\xi, f), (\eta, g) \in X \times L^1(a, b; X)$ and each $t \in [a, b]$. Taking the supremum for $t \in [a, b]$ in this inequality, we deduce

$$\|Q(\xi, f) - Q(\eta, g)\|_\infty \le \|\xi - \eta\| + \|f - g\|_{L^1(a,b;X)},$$

which proves the assertion.

Next, we shall concentrate our attention on some compactness properties of the operator Q.

Theorem 8.4.1. *Let $A : D(A) \subseteq X \to X$ be the generator of a C_0-semigroup of contractions $\{S(t);\ t \ge 0\}$, let $\mathcal{D}$ be a bounded subset in X, and $\mathcal{F}$ a uniformly integrable subset in $L^1(a, b; X)$. Then $Q(\mathcal{D}, \mathcal{F})$ is relatively compact in $C([c, b]; X)$ for each $c \in (a, b)$, if and only if there exists a dense subset D in $[a, b]$ such that, for each $t \in D$, the t-section of $Q(\mathcal{D}, \mathcal{F})$, i.e. $Q(\mathcal{D}, \mathcal{F})(t) = \{Q(\xi, f)(t);\ (\xi, f) \in \mathcal{D} \times \mathcal{F}\}$, is relatively compact in X. Moreover, if the latter condition is satisfied, and $a \in D$, then $Q(\mathcal{D}, \mathcal{F})$ is relatively compact in $C([a, b]; X)$.*

Proof. By virtue of Theorem A.2.1, the necessity is obvious. In order to prove the sufficiency, we shall make use again of Theorem A.2.1. To this aim, we shall show that $Q(\mathcal{D}, \mathcal{F})$ is equicontinuous on $(a, b]$. Let $\varepsilon > 0$. Since $\mathcal{F}$ is uniformly integrable, there exists $\gamma(\varepsilon) > 0$ such that

$$\int_E \|f(s)\|\, ds \le \varepsilon$$

for each measurable subset E in $[a, b]$ with $\mu(E) \le \gamma(\varepsilon)$, and uniformly for $f \in \mathcal{F}$. We recall that μ is the Lebesgue measure on $\mathbb{R}$. Let $t \in (a, b]$ and let us fix $\lambda = \lambda(\varepsilon) > 0$ such that $t - \lambda > a$, $2\lambda \le \gamma(\varepsilon)$ and $t - \lambda \in D$. This is always possible because D is dense in $[a, b]$. Inasmuch as $Q(\mathcal{D}, \mathcal{F})(t - \lambda)$ is relatively compact in X, for each $\varepsilon > 0$, there exists a finite family $\{(\xi_1, f_1), (\xi_2, f_2), \dots (\xi_{k(\varepsilon)}, f_{k(\varepsilon)})\}$ in $\mathcal{D} \times \mathcal{F}$ such that, for each $(\xi, f) \in \mathcal{D} \times \mathcal{F}$, there exists $i \in \{1, 2, \dots k(\varepsilon)\}$ with the property

$$\|Q(\xi, f)(t - \lambda) - Q(\xi_i, f_i)(t - \lambda)\| \le \varepsilon.$$

On the other hand, the family $\{Q(\xi_1, f_1), Q(\xi_2, f_2), \dots Q(\xi_{k(\varepsilon)}, f_{k(\varepsilon)})\}$ is equicontinuous at t, being a finite family of continuous functions on $[a, b]$. Therefore, for the very same $\varepsilon > 0$, there exists $\delta(\varepsilon) \in (0, \lambda]$, such that

$$\|Q(\xi_i, f_i)(t + h) - Q(\xi_i, f_i)(t)\| \le \varepsilon$$

for each $i = 1, 2, \dots k(\varepsilon)$, and each $h \in \mathbb{R}$ with $|h| \le \delta(\varepsilon)$. We then have

$$\|Q(\xi, f)(t + h) - Q(\xi, f)(t)\| \le \|Q(\xi, f)(t + h) - Q(\xi_i, f_i)(t + h)\|$$
$$+\|Q(\xi_i, f_i)(t + h) - Q(\xi_i, f_i)(t)\| + \|Q(\xi_i, f_i)(t) - Q(\xi, f)(t)\|.$$

Since for each $(\eta, g) \in \mathcal{D} \times \mathcal{F}$ and each $h \in (-\delta(\varepsilon), \delta(\varepsilon))$, we have

$$Q(\eta, g)(t+h) = S(\lambda+h)Q(\eta, g)(t-\lambda) + \int_{t-\lambda}^{t+h} S(t+h-s)g(s)\,ds,$$

from the preceding inequality, we get

$$\begin{aligned}
&\|Q(\xi, f)(t+h) - Q(\xi, f)(t)\| \\
&\leq \|S(\lambda+h)(Q(\xi, f)(t-\lambda) - Q(\xi_i, f_i)(t-\lambda))\| \\
&\quad + \int_{t-\lambda}^{t+h} \|S(t+h-s)\|_{\mathcal{L}(X)} \|f(s) - f_i(s)\|\,ds \\
&\quad + \|Q(\xi_i, f_i)(t+h) - Q(\xi_i, f_i)(t)\| \\
&\quad + \|S(\lambda)(Q(\xi, f)(t-\lambda) - Q(\xi_i, f_i)(t-\lambda))\| \\
&\quad + \int_{t-\lambda}^{t} \|S(t-s)\|_{\mathcal{L}(X)} \|f(s) - f_i(s)\|\,ds \\
&\leq \|S(\lambda+h)\|_{\mathcal{L}(X)} \|Q(\xi, f)(t-\lambda) - Q(\xi_i, f_i)(t-\lambda)\| \\
&\quad + \int_{t-\lambda}^{t+h} \|S(t+h-s)\|_{\mathcal{L}(X)} (\|f(s)\| + \|f_i(s)\|)\,ds \\
&\quad + \|Q(\xi_i, f_i)(t+h) - Q(\xi_i, f_i)(t)\| \\
&\quad + \|S(\lambda)\|_{\mathcal{L}(X)} \|Q(\xi, f)(t-\lambda) - Q(\xi_i, f_i)(t-\lambda)\| \\
&\quad + \int_{t-\lambda}^{t} \|S(t-s)\|_{\mathcal{L}(X)} (\|f(s)\| + \|f_i(s)\|)\,ds \\
&\leq 2\|Q(\xi, f)(t-\lambda) - Q(\xi_i, f_i)(t-\lambda)\| \\
&\quad + 2\int_{t-\lambda}^{t+\lambda} (\|f(s)\| + \|f_i(s)\|)\,ds \\
&\quad + \|Q(\xi_i, f_i)(t+h) - Q(\xi_i, f_i)(t)\|
\end{aligned}$$

for each $(\xi, f) \in \mathcal{D} \times \mathcal{F}$ and each $h \in (-\delta(\varepsilon), \delta(\varepsilon))$. From this inequality and from the manner of choice of both $\lambda > 0$ and $\delta(\varepsilon) > 0$, it follows

$$\|Q(\xi, f)(t+h) - Q(\xi, f)(t)\| \leq 7\varepsilon$$

for each $(\xi, f) \in \mathcal{D} \times \mathcal{F}$ and each $h \in \mathbb{R}$ with $|h| < \delta(\varepsilon)$. Therefore $Q(\mathcal{D}, \mathcal{F})$ is equicontinuous on $(a, b]$, and, by virtue of Theorem A.2.1, it is relatively compact in $C([c, b]; X)$ for each $c \in (a, b)$. If $a \in D$, then $Q(\mathcal{D}, \mathcal{F})$ is equicontinuous at a as well, and this achieves the proof. □

Remark 8.4.2. The condition of uniform integrability which circumvents the general frame of Theorem 8.4.1 cannot be relaxed as we can see from the simple example below. Let $A : \mathbb{R} \to \mathbb{R}$ be defined by $Au = 0$ for each $u \in \mathbb{R}$, let $\mathcal{D} = \{0\}$ and let $\mathcal{F} = \{f_n;\ n \in \mathbb{N}^*\}$, where, for each $n \in \mathbb{N}^*$, $f_n : [\,0,1\,] \to \mathbb{R}$ is defined by

$$f_n(t) = nt^{n-1}$$

for $t \in (0,1)$. A simple calculation shows that $Q(\mathcal{D},\mathcal{F}) = \{u_n;\ n \in \mathbb{N}^*\}$, where for each $n \in \mathbb{N}^*$, $u_n : [\,0,1\,] \to \mathbb{R}$ is defined by

$$u_n(t) = t^n$$

for each $t \in [\,0,1\,]$. So, $Q(\mathcal{D},\mathcal{F})$ is not relatively compact in $C([\,0,1\,])$ even though it satisfies (ii) in Theorem 8.4.1 with $D = [\,0,1\,]$. We emphasize that this is a consequence of the fact that $\mathcal{F}$ lacks uniform integrability "near" $t = 1$". On the other hand, $Q(\mathcal{D},\mathcal{F})$ is relatively compact in $L^p(a,b)$ for each $p \in [\,1,+\infty)$. This observation suggests the possibility to get a variant of Theorem 8.4.1 concerning the relative compactness of $Q(\mathcal{D},\mathcal{F})$ in $L^p(a,b;X)$, by assuming only the boundedness of $\mathcal{F}$ in $L^1(a,b;X)$. We notice that this is the topic of Section 8.6.

An extremely useful consequence of Theorem 8.4.1 is the following sufficient condition of relative compactness of the set $Q(\mathcal{D},\mathcal{F})$.

Theorem 8.4.2. *Let $A : D(A) \subseteq X \to X$ be the generator of a compact C_0-semigroup of contractions, let $\xi \in X$, $\mathcal{D} = \{\xi\}$ and let $\mathcal{F}$ be a uniformly integrable subset in $L^1(a,b;X)$. Then $Q(\mathcal{D},\mathcal{F})$ is relatively compact in $C([\,a,b\,];X)$.*

Proof. By virtue of Theorem 8.4.1, it suffices to check that, for each $t \in (a,b)$, $Q(\mathcal{D},\mathcal{F})(t)$ is relatively compact in X. Let $t \in (a,b)$ and let $\lambda > 0$ with $t - \lambda \geq a$. We have

$$Q(\xi,f)(t) = S(t-a)\xi + S(\lambda)\int_a^{t-\lambda} S(t-\lambda-s)f(s)\,ds + \int_{t-\lambda}^{t} S(t-s)f(s)\,ds.$$

Since $S(\lambda)$ is compact, from Remark 8.4.1, we deduce that the operator $P_\lambda : Q(\mathcal{D},\mathcal{F})(t) \to X$, defined by

$$P_\lambda Q(\xi,f)(t) = S(t-a)\xi + S(\lambda)\int_a^{t-\lambda} S(t-\lambda-s)f(s)\,ds,$$

maps the set $Q(\mathcal{D},\mathcal{F})(t)$ into a relatively compact subset in X. In addition, from the preceding relation and from the uniform integrability of $\mathcal{F}$, it follows that

$$\lim_{\lambda\downarrow 0} \|P_\lambda Q(\xi,f)(t) - Q(\xi,f)(t)\| = 0$$

uniformly for $f \in \mathcal{F}$. From Lemma A.1.2 it follows that $Q(\mathcal{D}, \mathcal{F})(t)$ is relatively compact in X for each $t \in [a, b]$. An appeal to Theorem 8.4.1 completes the proof. □

8.5. The Case when $(\lambda I - A)^{-1}$ is Compact

Let $A : D(A) \subseteq X \to X$ be the infinitesimal generator of a C_0-semigroup of contractions $\{S(t)\,;\ t \geq 0\}$, $\xi \in X$, $f \in L^1(a, b\,; X)$ and let us consider the Cauchy problem

$$\begin{cases} u' = Au + f \\ u(a) = \xi. \end{cases} \tag{8.5.1}$$

Let $Q : X \times L^1(a, b\,; X) \to C([a, b]; X)$ be the solution operator attached to (8.5.1), i.e.

$$Q(\xi, f)(t) = S(t - a)\xi + \int_a^t S(t - s)f(s)\, ds.$$

See Definition 8.4.1. Our aim here is to prove a necessary and sufficient condition in order that $Q(\mathcal{D}, \mathcal{F})$ be relatively compact in $C([a, b]; X)$. The main result in this section refers to the case in which, for each $\lambda > 0$, $(\lambda I - A)^{-1}$ is a compact operator.

Theorem 8.5.1. *Let $A : D(A) \subseteq X \to X$ be the infinitesimal generator of a C_0-semigroup of contractions with $(\lambda I - A)^{-1}$ compact for each $\lambda > 0$, let $\mathcal{D}$ be a bounded subset in X, and $\mathcal{F}$ a uniformly integrable subset in $L^1(a, b\,; X)$. Then $Q(\mathcal{D}, \mathcal{F})$ is relatively compact in $C([a, b]; X)$ if and only if it is equicontinuous from the right at each $t \in [a, b)$*

Proof. *Necessity.* By virtue of Theorem A.2.1 the necessity is evident. *Sufficiency.* Let $\mathcal{D}$ be a bounded subset in X, $\mathcal{F}$ a uniformly integrable subset in $L^1(a, b\,; X)$, and let us assume that $Q(\mathcal{D}, \mathcal{F})$ is equicontinuous from the right on $[a, b)$. According to Remarks A.5.1 and 6.3.1, $Q(\mathcal{D}, \mathcal{F})$ is bounded in $C([a, b]; X)$. More precisely, there exists $M > 0$ such that

$$\|Q(\xi, f)(t)\| \leq M \tag{8.5.2}$$

for each $(\xi, f) \in \mathcal{D} \times \mathcal{F}$ and $t \in [a, b]$. Let $t \in [a, b)$, and let $\lambda > 0$ with $t + \lambda \leq b$. Let us define $I_\lambda : Q(\mathcal{D}, \mathcal{F})(t) \to X$ by

$$I_\lambda Q(\xi, f)(t) = \lambda R(\lambda; A) Q(\xi, f)(t)$$

for each $(\xi, f) \in \mathcal{D} \times \mathcal{F}$. In view of (8.5.2) and of the compactness of the operator $R(\lambda; A)$, it follows that $I_\lambda Q(\mathcal{D}, \mathcal{F})(t))$ is relatively compact. We shall show that

$$\lim_{\lambda \to \infty} \|I_\lambda Q(\xi, f)(t) - Q(\xi, f)(t)\| = 0, \tag{8.5.3}$$

uniformly for $f \in \mathcal{F}$. Indeed,

$$
\begin{aligned}
&\|I_\lambda Q(\xi, f)(t) - Q(\xi, f)(t)\| \\
&= \left\| \lambda \int_0^\infty e^{-\lambda\tau}(S(\tau)Q(\xi, f)(t) - Q(\xi, f)(t))\, d\tau \right\| \\
&\leq \lambda \int_0^\infty e^{-\lambda\tau}\|S(\tau)Q(\xi, f)(t) - Q(\xi, f)(t)\|\, d\tau. \qquad (8.5.4)
\end{aligned}
$$

At this point, let us observe that

$$
\begin{aligned}
\|S(\tau)Q(\xi, f)(t) - Q(\xi, f)(t)\| &\leq \|S(\tau)Q(\xi, f)(t) - Q(\xi, f)(t + \tau)\| \\
&+\|Q(\xi, f)(t + \tau) - Q(\xi, f)(t)\| \qquad (8.5.5)
\end{aligned}
$$

for each $(\xi, f) \in \mathcal{D} \times \mathcal{F}$, and $\tau > 0$ with $t + \tau \leq b$. Since the family $Q(\mathcal{D}, \mathcal{F})$ is equicontinuous from the right at t, for each $\varepsilon > 0$ there exists $\delta = \delta(\varepsilon) > 0$ such that, for each $(\xi, f) \in \mathcal{D} \times \mathcal{F}$, and each $\tau \in (0, \delta)$, we have $\|Q(\xi, f)(t + \tau) - Q(\xi, f)(t)\| \leq \varepsilon$. As

$$
\begin{aligned}
\|S(\tau)Q(\xi, f)(t) - Q(\xi, f)(t + \tau)\| &\leq \int_t^{t+\tau} \|S(t + \tau - \theta)\|_{\mathcal{L}(X)}\|f(\theta)\|\, d\theta \\
&\leq \int_t^{t+\tau} \|f(\theta)\|\, d\theta
\end{aligned}
$$

for each $(\xi, f) \in \mathcal{D} \times \mathcal{F}$, and each $\tau > 0$ with $t + \tau \leq b$, by virtue of the uniform integrability of $\mathcal{F}$, for the very same $\varepsilon > 0$, there exists $\gamma(\varepsilon) > 0$, with $t + \gamma(\varepsilon) \leq b$, and such that

$$
\|S(\tau)Q(\xi, f)(t) - Q(\xi, f)(t + \tau)\| \leq \varepsilon
$$

for each $f \in \mathcal{F}$ and each $\tau > 0$ with $\tau \leq \gamma(\varepsilon)$. Let us fix $\mu > 0$ with $\mu \leq \max\{\delta(\varepsilon), \gamma(\varepsilon)\}$. From (8.5.2), (8.5.4), (8.5.5), it follows

$$
\begin{aligned}
&\|I_\lambda Q(\xi, f)(t) - Q(\xi, f)(t)\| \\
&\leq \lambda \int_0^\mu e^{-\lambda\tau}\|S(\tau)Q(\xi, f)(t) - Q(\xi, f)(t + \tau)\|\, d\tau \\
&+\lambda \int_0^\mu e^{-\lambda\tau}\|Q(\xi, f)(t + \tau) - Q(\xi, f)(t)\|\, d\tau \\
&+\lambda \int_\mu^\infty e^{-\lambda\tau}\|S(\tau)Q(\xi, f)(t) - Q(\xi, f)(t)\|\, d\tau \\
&\leq 2(1 - e^{-\lambda\mu})\varepsilon + 2e^{-\lambda\mu}M.
\end{aligned}
$$

Passing to the lim sup for λ tending to $+\infty$ in the inequality above, we obtain

$$
\limsup_{\lambda\to\infty} \|I_\lambda Q(\xi, f)(t) - Q(\xi, f)(t)\| \leq 2\varepsilon,
$$

uniformly for $(\xi, f) \in \mathcal{D} \times \mathcal{F}$. Since $\varepsilon > 0$ is arbitrary, this implies (8.5.3). But (8.5.3), along with the relative compactness of the sets $I_\lambda Q(\mathcal{D}, \mathcal{F})(t)$, for each $\lambda > 0$ with $t - \lambda > a$, shows that we are in the hypotheses of Lemma A.1.2. Therefore, for each $t \in [a, b)$, $Q(\mathcal{D}, \mathcal{F})(t)$ is relatively compact. An appeal to Theorem 8.4.1 completes the proof. □

8.6. Compactness of the Solution Operator from $L^1(a,b;X)$

The aim of this section is to prove a necessary and sufficient condition in order that a family of C^0-solutions be relatively compact in $L^p(a,b;X)$ for each $p \in [1, +\infty)$. More precisely, let $A : D(A) \subseteq X \to X$ be the infinitesimal generator of a C_0-semigroup of contractions $\{S(t)\,;\ t \geq 0\}$, let $\xi \in X$ and $f \in L^1(a,b;X)$. We denote by $u = Q(\xi, f)$ the unique C^0-solution of the nonhomogeneous Cauchy problem

$$\begin{cases} u' = Au + f \\ u(a) = \xi. \end{cases} \tag{8.6.1}$$

The main result in this section is:

Theorem 8.6.1. *Let $A : D(A) \subseteq X \to X$ be the infinitesimal generator of a C_0-semigroup of contractions $\{S(t)\,;\ t \geq 0\}$ and let $\mathcal{D} \times \mathcal{F}$ be a bounded subset in $X \times L^1(a,b;X)$. Then $Q(\mathcal{D}, \mathcal{F})$ is relatively compact in $L^p(a,b;X)$ for each $p \in [1, +\infty)$ if and only if for each $\varepsilon > 0$ there exists a relatively compact subset C_ε in X such that, for each $(\xi, f) \in \mathcal{D} \times \mathcal{F}$ there exists a subset $E_{\varepsilon,\xi,f}$ in $[a, b]$ whose Lebesgue measure is less that ε, and such that $Q(\xi, f)(t) \in C_\varepsilon$ for each $(\xi, f) \in \mathcal{D} \times \mathcal{F}$ and $t \in [a, b] \setminus E_{\varepsilon,\xi,f}$.*

Proof. *Necessity.* By Theorem A.5.1 the necessity is obvious.
Sufficiency. We shall use again Theorem A.5.1. First, let us observe that, by virtue of Lebesgue dominated convergence theorem, it suffices to show that $Q(\mathcal{D}, \mathcal{F})$ is relatively compact in $L^1(a,b;X)$ and bounded in $L^\infty(a,b;X)$. To this aim, let us recall that there exist $m_\mathcal{D} > 0$ and $m_\mathcal{F} > 0$ such that

$$\|\xi\| \leq m_\mathcal{D} \quad \text{and} \quad \|f\|_{L^1(a,b;X)} \leq m_\mathcal{F}, \tag{8.6.2}$$

for each $(\xi, f) \in \mathcal{D} \times \mathcal{F}$. Then, by virtue of Remark 6.3.1, we have

$$\|Q(\xi, f)(t)\| \leq m_\mathcal{D} + m_\mathcal{F} \tag{8.6.3}$$

for each $(\xi, f) \in \mathcal{D} \times \mathcal{F}$ and $t \in [a, b]$.
We will prove next that $Q(\mathcal{D}, \mathcal{F})$ is 1-equiintegrable. To this aim let $\varepsilon > 0$, let $(\xi, f) \in \mathcal{D} \times \mathcal{F}$ and let C_ε and $E_{\varepsilon,\xi,f}$ be the sets with the properties mentioned in the hypotheses. We have

$$\int_a^{b-h} \|Q(\xi, f)(t+h) - Q(\xi, f)(t)\|\, dt$$

$$\leq \int_a^{b-h} \|Q(\xi,f)(t+h) - S(h)Q(\xi,f)(t)\| \, dt$$

$$+ \int_a^{b-h} \|S(h)Q(\xi,f)(t) - Q(\xi,f)(t)\| \, dt \leq \int_a^{b-h} \int_t^{t+h} \|f(s)\| \, ds \, dt$$

$$+ \int_{[a,b]\setminus E_{\varepsilon,\xi,f}} \|S(h)Q(\xi,f)(t) - Q(\xi,f)(t)\| \, dt$$

$$+ \int_{E_{\varepsilon,\xi,f}} \|S(h)Q(\xi,f)(t) - Q(\xi,f)(t)\| \, dt$$

$$\leq h\|f\|_{L^1([a,b];X)} + \int_{[a,b]\setminus E_{\varepsilon,\xi,f}} \|S(h)Q(\xi,f)(t) - Q(\xi,f)(t)\| \, dt + 2M_{\mathfrak{F}}\varepsilon,$$

for each $f \in \mathfrak{F}$ and $h \in (0, b-a]$, where $M_{\mathfrak{F}} = m_{\mathfrak{D}} + m_{\mathfrak{F}}$. As $Q(\xi,f)(t) \in C_\varepsilon$ for each $(\xi,f) \in \mathfrak{D} \times \mathfrak{F}$, and each $t \in [a,b] \setminus E_{\varepsilon,\xi,f}$, and C_ε is relatively compact in X, there exists $\delta(\varepsilon) \in (0, b-a]$ such that, for each $h \in (0, \delta(\varepsilon)]$

$$\|S(h)Q(\xi,f)(t) - Q(\xi,f)(t)\| \leq \varepsilon,$$

uniformly for $f \in \mathfrak{F}$ and $t \in [a,b] \setminus E_{\varepsilon,\xi,f}$. Then, taking into account (8.6.2), (8.6.3) and the preceding inequalities, we obtain

$$\int_a^{b-h} \|Q(\xi,f)(t+h) - Q(\xi,f)(t)\| \, dt \leq (b - a + m_{\mathfrak{F}} + 2M_{\mathfrak{F}})\varepsilon$$

for each $f \in \mathfrak{F}$ and $h \in (0, \delta(\varepsilon)] \cap (0, \varepsilon]$. Obviously, this relation shows that $Q(\mathfrak{D}, \mathfrak{F})$ is 1-equiintegrable. From Theorem A.5.1 it follows that it is relatively compact in $L^1(a,b;X)$. An appeal to Lebesgue dominated convergence theorem completes the proof. □

Two remarkable consequences of Theorem 8.6.1 are listed below.

Corollary 8.6.1. *Let $A : D(A) \subseteq X \to X$ be the infinitesimal generator of a C_0-semigroup of contractions and let $\mathfrak{D} \times \mathfrak{F}$ be a bounded subset in $X \times L^1(a,b;X)$. If $\{Q(\xi,f)(t);\ (\xi,f) \in \mathfrak{D} \times \mathfrak{F},\ t \in [a,b]\}$ is relatively compact in X, then $Q(\mathfrak{D},\mathfrak{F})$ is relatively compact in $L^p(a,b;X)$, for each $p \in [1,+\infty)$.*

Corollary 8.6.2. *Let X be a finite-dimensional Banach space, $A \in \mathcal{L}(X)$ and let $\mathfrak{D} \times \mathfrak{F}$ be a bounded subset in $X \times L^1(a,b;X)$. Then $Q(\mathfrak{D},\mathfrak{F})$ is relatively compact in $L^p(a,b;X)$, for each $p \in [1,+\infty)$.*

The next lemma, which is closely related to Egoroff uniform asymptotic convergence theorem (see Dunford and Schwartz [**49**], Theorem 12, p. 149), will prove useful later.

Lemma 8.6.1. *Let $U \subset L^1(a,b;X)$ be nonempty and bounded, and let*

$$\{Q_h : U \to L^1(a,b;X), h \in (0, b-a]\}$$

be a family of (possible nonlinear) operators such that

$$\lim_{h \downarrow 0} \int_a^b \|Q_h u(t)\| \, dt = 0,$$

uniformly for $u \in U$. Then, for each $\varepsilon > 0$, there exists a sequence $(h_n)_{n \in \mathbb{N}}$, decreasing to 0, with the property that, for each $u \in U$, there exists a subset $E_{\varepsilon,u}$ in $[a,b]$ whose Lebesgue measure $\mu(E_{\varepsilon,u})$ is less than ε, and

$$\lim_{n \to \infty} Q_{h_n} u(t) = 0,$$

uniformly for $u \in U$ and $t \in [a,b] \setminus E_{\varepsilon,u}$.

Proof. Let $\varepsilon > 0$ and let us choose two sequences $(h_n)_{n \in \mathbb{N}}$ and $(a_n)_{n \in \mathbb{N}}$, both decreasing to 0, such that

$$\sum_{n=0}^{\infty} a_n \leq \varepsilon \quad \text{and} \quad \int_a^b \|Q_{h_n} u(t)\| \, dt < a_n^2 \tag{8.6.4}$$

for each $n \in \mathbb{N}$ and $u \in U$. For $u \in U$ and $n \in \mathbb{N}$ let us define

$$E_u^n = \{t \in [a,b];\ \|Q_{h_n} u(t)\| \geq a_n\} \quad \text{and} \quad E_u = \bigcup_{n \in \mathbb{N}} E_u^n.$$

By virtue of the second relation in (8.6.4), it follows that $\mu(E_u^n) < a_n$ for each $n \in \mathbb{N}$, and therefore $\mu(E_{\varepsilon,u}) < \varepsilon$. From the definition of $E_{\varepsilon,u}$, we deduce that

$$\|Q_{h_n} u(t)\| < a_n,$$

for each $n \in \mathbb{N}$, $u \in U$ and $t \in [a,b] \setminus E_{\varepsilon,u}$ and this completes the proof. □

Theorem 8.6.2. *Let $A : D(A) \subseteq X \to X$ be the infinitesimal generator of a compact C_0-semigroup of contractions. Then, for each bounded subset $\mathcal{D} \times \mathcal{F}$ in $X \times L^1(a,b;X)$, and for each $p \in [1, +\infty)$, the set $Q(\mathcal{D}, \mathcal{F})$ is relatively compact in $L^p(a,b;X)$.*

Proof. We will show that the set $Q(\mathcal{D}, \mathcal{F})$ satisfies the hypotheses of Theorem 8.6.1. To this aim, let us observe that

$$\int_{a+h}^b \|Q(\xi, f)(t) - S(h)Q(\xi, f)(t-h)\| \, dt$$

$$\leq \int_{a+h}^b \int_{t-h}^t \|f(s)\| \, ds \, dt \leq h \, \|f\|_{L^1(a,b;X)},$$

for each $(\xi, f) \in \mathcal{F}$ and $h \in (0, b-a]$. Since $\mathcal{F}$ is bounded in $L^1(a,b;X)$, we have

$$\lim_{h \downarrow 0} \int_{a+h}^{b} \|Q(\xi,f)(t) - S(h)Q(\xi,f)(t-h)\| \, dt = 0, \tag{8.6.5}$$

uniformly for $(\xi, f) \in \mathcal{D} \times \mathcal{F}$.
Let us define $U = Q(\mathcal{D}, \mathcal{F})$ and $Q_h : U \to L^1(a,b;X)$ by

$$(Q_h u)(t) = \begin{cases} 0 & \text{if } t \in [a, a+h] \\ u(t) - S(h)u(t-h), & \text{if } t \in (a+h, b] \end{cases}$$

for $u \in U$. From (8.6.5) it follows that Lemma 8.6.1 applies, and therefore for each $\varepsilon > 0$ there exists $E_{\varepsilon,u} \subseteq [a,b]$ with $\mu(E_{\varepsilon,u}) < \varepsilon$ and $(h_n)_{n \in \mathbb{N}}$ decreasing to 0 such that

$$\lim_{n \to \infty} \|u(t) - S(h_n)u(t-h_n)\| = 0,$$

uniformly for $u \in Q(\mathcal{D}, \mathcal{F})$ and $t \in [a,b] \setminus E_{\varepsilon,u}$. Inasmuch as $S(h_n)$ is compact for each $n \in \mathbb{N}$ and $\{u(s); u \in Q(\mathcal{D}, \mathcal{F}), s \in [a,b]\}$ is bounded in X, it follows that the set

$$C_\varepsilon = \{u(t);\ u \in Q(\mathcal{D}, \mathcal{F}),\ t \in [a,b] \setminus E_{\varepsilon,u}\}$$

is relatively compact in X. Consequently we are in the hypotheses of Theorem 8.6.1. So $Q(\mathcal{D}, \mathcal{F})$ is relatively compact in $L^p(a,b;X)$ for each $p \in [1, +\infty)$, and this achieves the proof. □

Problems

Problem 8.1. Let $A : D(A) \subseteq X \to X$ be the infinitesimal generator of a C_0-semigroup of contractions in X, $\xi \in X$, and $f \in L^1(0, +\infty; X)$. Prove that the nonhomogeneous equation

$$\begin{cases} u' = Au + f(t) \\ u(0) = \xi \end{cases}$$

can be equivalently rewritten as a homogeneous equation of the form

$$\begin{cases} U' = \mathcal{A}U \\ U(0) = U_0 \end{cases}$$

in the Banach space $\mathcal{X} = X \times L^1(0, +\infty; X)$, endowed with the norm

$$\|(u,f)\|_{\mathcal{X}} = \|u\| + \|f\|_{L^1(0,+\infty;X)}$$

where $U_0 = (\xi, f)$, $U(t) = (u(t), f(t+\cdot))$, and $\mathcal{A} : D(\mathcal{A}) \subseteq \mathfrak{X} \to \mathfrak{X}$ is defined by

$$\begin{cases} D(\mathcal{A}) = \{(\xi, f) \in \mathfrak{X};\ \xi \in D(A),\ f \in W^{1,1}(0, +\infty; X) \\ \mathcal{A}(\xi, f) = (A\xi + f(0), f') \text{ for each } (\xi, f) \in D(\mathcal{A}). \end{cases}$$

Prove that $\mathcal{A}$ is the infinitesimal generator of a C_0-semigroup of contractions in $\mathfrak{X}$.

Problem 8.2. Let $a \in \mathbb{R}^3$, $u_0 \in L^p(\mathbb{R}^3)$, and $f \in L^1(\mathbb{R}_+; L^p(\mathbb{R}^3))$. Find the solution of the Cauchy problem for *the nonhomogeneous transport equation*

$$\begin{cases} u_t + a \cdot \nabla u = f(t, x) & (t, x) \in \mathbb{R}_+ \times \mathbb{R}^3 \\ u(0, x) = u_0(x) & x \in \mathbb{R}^3. \end{cases}$$

Problem 8.3. Let $\mathcal{A}$ be an $n \times n$ matrix with real entries, $u_0 \in L^p(\mathbb{R}^n)$, and let $f \in L^1(\mathbb{R}_+; L^p(\mathbb{R}^n))$. Solve the Cauchy problem

$$\begin{cases} u_t + \langle \mathcal{A}x, \nabla u \rangle = f(t, x) & (t, x) \in \mathbb{R}_+ \times \mathbb{R}^n \\ u(0, x) = u_0(x) & x \in \mathbb{R}^n. \end{cases}$$

Let $X = L^2_{2\pi}(\mathbb{R})$ be the space of a.e. equivalence classes of measurable functions from $\mathbb{R}$ to $\mathbb{R}$, periodic of principal period 2π, and square integrable on $[\,0,\ 2\pi\,]$. Endowed with the inner product and with the norm of $L^2(0, 2\pi)$, this is a real Hilbert space isometric to $L^2(0, 2\pi)$.

Problem 8.4. Let $X = L^2_{2\pi}(\mathbb{R})$, and let $A : D(A) \subseteq X \to X$ be defined by

$$\begin{cases} D(A) = \{u \in L^2_{2\pi}(\mathbb{R})\,;\, u' \in L^2_{2\pi}(\mathbb{R})\} \\ Au = u' \text{ for each } u \in D(A). \end{cases}$$

Prove that, for each $\lambda \in \mathbb{R}_+$, $(\lambda I - A)^{-1}$ is a compact operator.
Let $\mathfrak{F} = \{\sin(n(\cdot + x))\,;\ n \in \mathbb{N}^*\}$, and let $Q(\mathfrak{F})$ the set of all C_0 solutions of the problem

$$\begin{cases} u' = Au + f \\ u(0) = 0, \end{cases}$$

corresponding to $f \in \mathfrak{F}$. Show that $Q(\mathfrak{F})$ is not relatively compact in $C([\,0, 1\,]; X)$, although $\mathfrak{F}$ is uniformly integrable in $L^1(0, 1; X)$. This shows that, in Theorem 8.4.2, the compactness of the C_0-semigroup is essential.

Problem 8.5. Let $A : D(A) \subseteq X \to X$ be the infinitesimal generator of a C_0-semigroup of contractions. If $(\lambda I - A)^{-1}$ is compact for some $\lambda > 0$, then, for each $\xi \in X$ and $f \in L^1(\mathbb{R}_+; X)$, the orbit $\{u(t)\,;\ t \geq 0\}$, of the C^0-solution u of the Cauchy problem

$$\begin{cases} u' = Au + f(t) \\ u(0) = \xi, \end{cases}$$

is relatively compact in X.

Notes. The results in Section 8.1 are classical and were adapted mainly from Pazy [**101**] and Brezis and Cazenave [**31**]. Theorem 8.1.4 is from Cazenave and Haraux [**36**]. The concept of C^0-solution, known mainly under the name of "mild solution" proved very useful in the study of many problems of practical interest. For details, see Barbu [**20**] and Pavel [**97**]. Theorem 8.4.1 was proved by Vrabie [**125**] in the general case in which A generates a contraction semigroup, possible nonlinear, and Theorem 8.4.2 is in Baras, Hassan and Veron [**14**]. We notice that an L^∞ variant of this result was established independently by Pazy [**99**] in 1975, by using a proof very similar to that in [**14**]. Nonlinear versions of this compactness result were obtained by Baras [**13**] and Vrabie [**124**]. Theorem 8.5.1 is a "linear" consequence of a general result in Mitidieri and Vrabie [**90**], Theorem 8.6.1 is also a specific case of a result in Vrabie [**128**], and Theorem 8.6.2 is from Baras, Hassan and Veron [**14**]. As concerns the problems here included, the first one is a particular case of a classical trick due to Dafermos and Slemrod, while the other ones are simple applications of the abstract results in this chapter.

CHAPTER 9

Linear Evolution Problems with Measures as Data

In this chapter we consider the abstract nonhomogeneous Cauchy problem

$$\begin{cases} du = \{Au\}dt + dg \\ u(a) = \xi, \end{cases}$$

where $A : D(A) \subseteq X \to X$ is the infinitesimal generator of a C_0-semigroup of contractions, $\xi \in X_A$, and $g \in BV([a,b]; X_A)$, while X_A is a space of abstract measures including X^{**}. Namely, we introduce the concept of $\mathfrak{L}^\infty$-solution and we study the compactness properties of the $\mathfrak{L}^\infty$-solution operator $(\xi, g) \mapsto u$, which will prove useful in obtaining existence results for some classes of semilinear Cauchy problems involving measures.

9.1. The Problem $du = \{Au\}dt + dg$, $u(a) = \xi$

Let X be a Banach space, $A : D(A) \subseteq X \to X$ the infinitesimal generator of a C_0-semigroup of contractions on X, $\xi \in X$, $g \in BV([a,b]; X)$, and let us consider the Cauchy problem

$$\begin{cases} du = \{Au\}dt + dg \\ u(a) = \xi. \end{cases} \tag{9.1.1}$$

In order to give a precise meaning of (9.1.1), we begin with the concept of the so-called $\mathfrak{L}^\infty$-solution (see Definition 9.1.1 below), which is inspired by the variation of constants formula. To this aim, let $t \in (a,b]$, $\mathcal{P} \in \mathcal{P}[a,t]$, $\mathcal{P} : a = t_0 < t_1 < \cdots < t_k = t$, let $\tau_i \in [t_{i-1}, t_i]$, $i = 1, 2, \ldots, k$, and let us consider the Riemann-Stieltjes sum of $\tau \mapsto S(t-\tau)$ over $[a,t]$ with respect to g, i.e.

$$\sigma_{[a,t]}(\mathcal{P}, S, g, \tau_i) = \sum_{i=1}^{k} S(t-\tau_i)(g(t_i) - g(t_{i-1})).$$

Let $\{S(t)^\odot\,;\ t \geq 0\}$ be the sun dual semigroup (see Definition 3.8.2), and let us observe that, for each $x^\odot \in X^\odot$ and each $\varepsilon > 0$, there exists $\delta(x^\odot, \varepsilon) > 0$

such that, for each $\mathcal{P} \in \mathcal{P}[a,t]$ with

$$\lambda(\mathcal{P}) = \max\{t_i - t_{i-1}\,;\; i = 1,2,\ldots,k\} \leq \delta(x^{\odot},\varepsilon)$$

we have

$$\left|\left(\sigma_{[a,t]}(\mathcal{P},S,g,\tau_i) - \sigma_{[a,t]}(\mathcal{P},S,g,\theta_i), x^{\odot}\right)\right| \leq \varepsilon, \tag{9.1.2}$$

for each $\tau_i, \theta_i \in [t_{i-1}, t_i]$, $i = 1,2,\ldots,k$. Indeed, let us remark that

$$\left|\left(\sigma_{[a,t]}(\mathcal{P},S,g,\tau_i) - \sigma_{[a,t]}(\mathcal{P},S,g,\theta_i), x^{\odot}\right)\right|$$

$$= \left|\sum_{i=1}^{k} \left(g(t_i) - g(t_{i-1}), (S(t-\tau_i)^{\odot} - S(t-\theta_i)^{\odot})x^{\odot}\right)\right|$$

$$\leq \sup_{i=1,2,\ldots,k} \{\|(S(t-\tau_i)^{\odot} - S(t-\theta_i)^{\odot})x^{\odot}\|\} \mathrm{Var}\,(g,[a,t]).$$

Since the sun dual semigroup is a C_0-semigroup of contractions, it readily follows that, for each $x^{\odot} \in X^{\odot}$ and $\varepsilon > 0$, there exists $\delta(x^{\odot},\varepsilon) > 0$ such that, for each $\tau_i, \theta_i \in [a,t]$ with $|\tau_i - \theta_i| \leq \delta(x^{\odot},\varepsilon)$, we have

$$\|(S(t-\tau_i)^{\odot} - S(t-\theta_i)^{\odot})x^{\odot}\| \leq \frac{\varepsilon}{\mathrm{Var}\,(g,[a,t])},$$

and this completes the proof of (9.1.2). But (9.1.2) shows that the mapping $\tau \mapsto S(t-\tau)$ satisfies the usual Cauchy condition of dg-integrability on $[a,t]$ in $\sigma(X, X^{\odot})$, called the weak-$\odot$ topology on X. Let $X_A = (X^{\odot})^*$. We obviously have $X^{**} \subseteq X_A$ and the arguments above show that there exists a unique element $\int_a^t S(t-s)\,dg(s) \in X_A$ such that

$$\int_a^t S(t-s)\,dg(s) = \lim_{\lambda(\mathcal{P})\downarrow 0} \sum_{i=1}^{k} S(t-\tau_i)(g(t_i) - g(t_{i-1}))$$

weakly-$\odot$ and it is called *the Riemann-Stieltjes integral* on $[a,t]$ of the operator-valued function $\tau \mapsto S(t-\tau)$ with respect to the vector-valued function g. If $\alpha : [a,b] \to \mathbb{R}$ is a given function, by a similar procedure, we can define

$$\int_a^t \alpha(s)S(t-s)\,dg(s) = \lim_{\lambda(\mathcal{P})\downarrow 0} \sum_{i=1}^{k} \alpha(\tau_i)S(t-\tau_i)(g(t_i) - g(t_{i-1})), \tag{9.1.3}$$

of course, whenever the limit on the right-hand side exists in the weak-$\odot$ topology on X. It it easy to see that this happens, for instance, if α is the characteristic function of a proper subinterval of $[a,t]$.

Remark 9.1.1. An elementary computational argument shows that, for each $c \in [a, b)$ and each $\delta > 0$, such that $c+\delta \in [a, b]$, and each $t \in [c+\delta, b]$, we have

$$\int_c^{c+\delta} S(t-s)\,dg(s) = \int_c^{c+\delta} \chi_{(c,c+\delta]}(s)S(t-s)\,dg(s) \\ +S(t-c)(g(c+0)-g(c)), \tag{9.1.4}$$

where $\chi_{(c,c+\delta]}$ denotes the characteristic function of $(c, c+\delta]$. Similarly, for each $c \in (a, b]$, and each $\delta > 0$, such that $c - \delta \in [a, b]$, and each $t \in [c, b]$, we have

$$\int_{c-\delta}^{c} S(t-s)\,dg(s) = \int_{c-\delta}^{c} \chi_{[c-\delta,c)}(s)S(t-s)\,dg(s) \\ +S(t-c)(g(c)-g(c-0)). \tag{9.1.5}$$

Remark 9.1.2. If X is reflexive the weak-$\odot$ topology on X is nothing else than the weak topology on X, and therefore $X_A = X$. In general this is not the case as the following simple example shows.

Example 9.1.1. Let $X = L^1(\mathbb{R})$ and let $\{S(t)\,;\ t \geq 0\}$ be the C_0-group of translations, i.e.

$$(S(t)f)(x) = f(x+t)$$

for each $f \in X$, each $t \in \mathbb{R}$, and a.e. for $x \in \mathbb{R}$. It is well-known that, in this case, $X^{\odot} = C_{ub}(\mathbb{R})$, i.e. that space of all bounded, uniformly continuous functions on $\mathbb{R}$, space endowed with the usual sup-norm. See Example 3.8.1. At this point let us observe that the weak topology on X does not coincide with the weak-$\odot$ topology. More than this, X is not sequentially weakly-$\odot$ complete, and its sequential completion is a space of measures strictly larger that X. One may easily verify that all Dirac measures belong to X_A.

Remark 9.1.3. If g is defined by a density, i.e. there exists a function $f \in L^1(a, b; X)$ such that $dg(s) = f(s)\,ds$, then, for each $t \in [a, b]$, we have

$$\int_a^t S(t-s)\,dg(s) = \int_a^t S(t-s)f(s)\,ds \in X.$$

From Remarks 9.1.2 and 9.1.3, it became clear that, in order for the Riemann-Stieltjes integral defined above to belong to X, we must impose some extra-conditions on X, on g, or on A. Since in applications X is a space with quite bad geometric properties, as $L^1(\Omega)$ for instance, and thus nonreflexive, and g is not a priori known being the "weak limit" of a certain sequence of L^1-functions which excludes the situation in Remark 9.1.3, in that follows, we will mainly focus our attention only on the properties of A which may ensure the existence of a "good integral". The next theorem gives a useful sufficient condition in this respect. The simple remark below will prove useful in that follows.

Remark 9.1.4. If $g \in BV([a,b];X)$, then $g([a,b])$ is strongly relatively compact in X. This is an easy consequence of the fact that g is piecewise continuous on $[a,b]$ (see Proposition 1.4.2).

Theorem 9.1.1. *If $\{S(t);\ t \geq 0\}$ is continuous from $(0,+\infty)$ to $\mathcal{L}(X)$ in the uniform operator topology, then, for each $g \in BV([a,b];X)$ and each $t \in (a,b)$, the limit in (9.1.3) exists in the norm topology of X and consequently we have*

$$\int_a^t S(t-s)\,dg(s) \in X. \tag{9.1.6}$$

Proof. It suffices to show that, for some fixed $\ell > 0$, for each $t \in (a,b]$, and each $\varepsilon > 0$, there exists $\eta(\varepsilon) > 0$, such that

$$\|\sigma_{[a,t]}(\mathcal{P},S,g,\tau_i) - \sigma_{[a,t]}(\mathcal{P}',S,g,\tau_j')\| \leq \ell\varepsilon,$$

provided $\lambda(\mathcal{P}) \leq \eta(\varepsilon)$ and $\mathcal{P}'$ is finer than $\mathcal{P}$, i.e. contains all the points of $\mathcal{P}$. To fix the notation, let $\mathcal{P} \in \mathcal{P}[a,t]$, $\mathcal{P} : a = t_0 < t_1 < \cdots < t_k = t$ be a partition of $[a,t]$, and let τ_i be arbitrary in $[t_i, t_{i+1}]$, $i = 0,1,\ldots,k-1$. If $\mathcal{P}'$ is another partition of $[a,t]$ which is finer that $\mathcal{P}$, relabelling if necessary, we have $\mathcal{P}' : t_i = t_{i,0} \leq \tau_{i,0} \leq t_{i,1} \leq \tau_{i,1} \leq, \ldots, \leq \tau_{i,m_i-1} \leq t_{i,m_i} = t_{i+1}$, $i = 0,1,\ldots,k-1$. Let $\varepsilon > 0$, and fix $\delta > 0$ with $t - 2\delta \in [a,b]$, and such that

$$\|(S(s) - I)(g(\tau) - g(\tau'))\| \leq \varepsilon \tag{9.1.7}$$

for each $s \in [0,\delta]$ and each $\tau, \tau' \in [a,t)$, and

$$\operatorname{Var}(g,[t-2\delta,t)) \leq \varepsilon. \tag{9.1.8}$$

This is possible because the semigroup is strongly continuous, $g([a,t))$ is strongly relatively compact in X (see Remark 9.1.4) while, by virtue of Lemma 1.4.1, $\lim_{\delta\downarrow 0} \operatorname{Var}(g,[t-\delta,t)) = 0$. Since $\tau \mapsto S(\tau)$ is continuous at $\tau = \delta > 0$ in the uniform operator topology, there exists $\eta \in (0,\delta]$ such that

$$\|S(s)S(\delta) - S(\delta)\| \leq \varepsilon \tag{9.1.9}$$

for each $s \in [0,\eta]$.
Let us assume now that $\lambda(\mathcal{P}) \leq \eta$ and let us fix $p = p(\delta) \in \{1,2,\ldots,k-1\}$ such that $t_p < t - \delta$ and $t_{p+1} \geq t - \delta$. Let us denote by

$$S_n^h = \sum_{i=n}^{h-1}\sum_{j=0}^{m_i-1}(S(t-\tau_i) - S(t-\tau_{i,j}))(g(t_{i,j+1}) - g(t_{i,j})),$$

and let us observe that

$$\sigma_{[a,t]}(\mathcal{P}, S, g, \tau_i) - \sigma_{[a,t]}(\mathcal{P}', S, g, \tau_{i,j}) = S_0^k.$$

Set $S_0^k = T_1 + T_2 + T_3$, where $T_1 = S_0^p$, $T_2 = S_p^k - T_3$, and T_3 is the last term in S_0^k. By (9.1.9), and the fact that $\delta < t - t_p$, we deduce

$$\|T_1\| \leq \sum_{i=0}^{p-1} \sum_{j=0}^{m_i-1} \|[S(t_p - \tau_i) - S(t_p - \tau_{i,j})]S(t - t_p)(g(t_{i,j+1}) - g(t_{i,j}))\|$$

$$\leq \sum_{i=0}^{p-1} \sum_{j=0}^{m_i-1} \|S(|\tau_i - \tau_{i,j}|)S(\delta) - S(\delta)\| \|g(t_{i,j+1}) - g(t_{i,j})\|$$

$$\leq \operatorname{Var}(g, [a, b])\varepsilon.$$

Since $t_{p+1} - t_p \leq \eta$, $t_{k-1,m_{k-1}-1} - t_{p+1} \leq t - t_{p+1} \leq \delta$, and $\eta \in (0, \delta]$, in view of (9.1.8), we get

$$\|T_2\| \leq 2\operatorname{Var}(g, [t - \eta - \delta, t)) \leq 2\operatorname{Var}(g, [t - 2\delta, t)) \leq 2\varepsilon.$$

Finally, by virtue of (9.1.7), we have

$$\|T_3\| \leq \|(S(|\tau_k - \tau_{k-1,m_{k-1}-1}|) - I)(g(t) - g(t_{k-1,m_{k-1}-1}))\| \leq \varepsilon.$$

Consequently

$$\left\|\sigma_{[a,t]}(\mathcal{P}, S, g, \tau_i) - \sigma_{[a,t]}(\mathcal{P}', S, g, \tau_{i,j})\right\| \leq (\operatorname{Var}(g, [a, b]) + 3)\varepsilon,$$

and this completes the proof. □

Let $\xi \in X$ and $g \in BV([a, b]; X)$.

Definition 9.1.1. The function $u : [a, b] \to X_A$, defined by

$$u(t) = S(t - a)\xi + \int_a^t S(t - s)\, dg(s)$$

for each $t \in [a, b]$, is called *the $\mathcal{L}^\infty$-solution* of the problem (9.1.1) on the interval $[a, b]$ with the initial condition $u(a) = \xi$ if it satisfies (9.1.6).

Remark 9.1.5. Clearly, whenever dg is defined by means of a density f, i.e. $dg = f dt$ with $f \in L^1(a, b; X)$, the $\mathcal{L}^\infty$-solution of (9.1.1) corresponding to (ξ, g) is nothing else than the mild, or C^0-solution of the Cauchy problem

$$\begin{cases} u' = Au + f \\ u(a) = \xi. \end{cases}$$

This happens, for example, whenever X has the Radon-Nicodým property (see Diestel and Uhl [**46**], Definition 3, p. 61), and g is absolutely continuous on $[a, b]$, case in which $f = g'$ a.e. on $[a, b]$. Some specific but important such instances are those in which X is either reflexive, or a separable dual. See also Diestel and Uhl [**46**], Theorem 1, p. 79, and Corollary 4, p. 82.

The proofs of the next two propositions follow from a simple computational argument, and therefore we do not give details.

Proposition 9.1.1. *Let $v : [a,c] \to X$ be the $\mathcal{L}^\infty$-solution of the problem* (9.1.1) *on $[a,c]$ with the initial condition $v(a) = \xi$, and $w : [c,b] \to X$ the $\mathcal{L}^\infty$-solution of the very same problem on $[c,b]$ with the initial condition $w(c) = v(c)$. Then the function $u : [a,b] \to X$, defined by*

$$u(t) = \begin{cases} v(t) & \text{if } t \in [a,c] \\ w(t) & \text{if } t \in (c,b], \end{cases}$$

is the $\mathcal{L}^\infty$-solution of the problem (9.1.1) *on the interval $[a,b]$ with the initial datum $u(a) = \xi$.*

Proposition 9.1.1, to which we refer as to the concatenation principle, is equivalent to the fact that the family $\{U(t,s);\ a \leq s \leq t \leq b\}$ in $\mathcal{L}(X)$, defined by $U(t,s)\xi = S(t-s)\xi + \int_s^t S(t-\tau)dg(\tau)$ for each $\xi \in X$, is an evolution system, i.e. $U(s,s) = I$ and $U(t,s) = U(t,\tau)U(\tau,s)$ for each $a \leq s \leq \tau \leq t \leq b$, and this follows from the simple proposition below.

Proposition 9.1.2. *If $u : [a,b] \to X$ is the $\mathcal{L}^\infty$-solution of the problem* (9.1.1), *then, for each $c \in (a,b)$, we have*

$$u(t) = S(t-c)u(c) + \int_c^t S(t-s)\,dg(s).$$

9.2. Regularity of $\mathcal{L}^\infty$-Solutions

We begin with the following fundamental regularity result.

Theorem 9.2.1. *Let $\xi \in X$, $g \in BV([a,b];X)$, and let u be the $\mathcal{L}^\infty$-solution of* (9.1.1) *corresponding to ξ and g. Then, for each $t \in [a,b)$ and each $s \in (a,b]$, there exist $u(t+0)$ and $u^*(t-0) = \lim_{h\downarrow 0} S(h)u(s-h)$, and*

$$u(t+0) - u(t) = g(t+0) - g(t), \tag{9.2.1}$$

and

$$u(s) - u^*(s-0) = g(s) - g(s-0). \tag{9.2.2}$$

If, in addition, either the semigroup generated by A is continuous from $(0,+\infty)$ to $\mathcal{L}(X)$ in the uniform operator topology, or it can be imbedded into a group, then there exists $u(s-0) = u^*(s-0)$ and accordingly

$$u(s) - u(s-0) = g(s) - g(s-0). \tag{9.2.3}$$

So, in this case, u is continuous from the right (left) at $t \in [a,b]$ if and only if g is continuous from the right (left) at t. Also in this specific case, u is continuous at any point at which g is continuous and thus u is piecewise continuous on $[a,b]$.

Proof. Since, by Proposition 9.1.2, u satisfies the evolution property and by (9.1.6), for each $t \in [a,b]$ and $h > 0$ with $t+h \leq b$, we have

$$u(t+h) - u(t) = S(h)u(t) + \int_t^{t+h} S(t+h-s)\,dg(s) - u(t)$$

$$= S(h)u(t) - u(t) + \int_t^{t+h} \chi_{(t,t+h]}(s)S(t+h-s)\,dg(s)$$
$$+S(h)(g(t+0) - g(t)),$$

we get

$$\|u(t+h) - u(t) - g(t+0) + g(t)\|$$
$$\leq \|S(h)(u(t) + g(t+0) - g(t)) - u(t) - g(t+0) + g(t)\| + \operatorname{Var}(g,(t,t+h]).$$

From the strong continuity of the semigroup and Lemma 1.4.1, we deduce (9.2.1). To check (9.2.2), let $s \in (a,b]$ and $h > 0$ with $s - h \geq a$ and let us observe that, by using similar arguments, we obtain

$$u(s) = S(h)u(s-h) + \int_{s-h}^{s} S(s-\tau)\,dg(\tau)$$

$$= S(h)u(s-h) + \int_{s-h}^{s} \chi_{[s-h,s)}(\tau)S(s-\tau)\,dg(\tau) + g(s) - g(s-0),$$

and therefore

$$\|u(s) - S(h)u(s-h) - g(s) + g(s-0)\| \leq \operatorname{Var}(g,[s-h,s)).$$

An appeal to Lemma 1.4.1 completes the proof of (9.2.2). Let us assume next that the semigroup $\{S(t)\,;\, t \geq 0\}$ is continuous in the uniform operator topology from $(0,+\infty)$ to $\mathcal{L}(X)$. Let $s \in (a,b]$ and $\delta > 0$ with $s - \delta \geq a$ and let $h \in (0,\delta]$. We have

$$\|u(s-h) - S(h)u(s-h)\|$$

$$\leq \|S(h)\xi - \xi\| + \left\| \int_a^{s-h} (S(s-\tau) - S(s-h-\tau))\,dg(\tau) \right\|.$$

Since the semigroup is strongly continuous at 0, to complete the proof, it suffices to show that the second term on the right-hand side tends to 0 when h tends to 0. Let $\varepsilon > 0$. By virtue of Lemma 9.4.1, there exists $\delta > 0$ such that

$$\operatorname{Var}(g,[s-\delta,s)) \leq \varepsilon.$$

So, in view of (9.1.5), for each $h \in (0, \delta)$, we have

$$\left\| \int_a^{s-h} (S(s-\tau) - S(s-h-\tau))\, dg(\tau) \right\|$$

$$\leq \left\| \int_a^{s-\delta} (S(s-\tau) - S(s-h-\tau))\, dg(\tau) \right\|$$

$$+ \left\| \int_{s-\delta}^{s-h} (S(s-\tau) - S(s-h-\tau))\, dg(\tau) \right\|$$

$$\leq \|S(\delta) - S(\delta - h)\| \mathrm{Var}\,(g, [\,a, b\,]) + 2\mathrm{Var}\,(g, [\,s-\delta, s-h))$$

$$+ \|(S(h) - I)(g(s-h) - g(s-h-0))\| \leq \|S(\delta) - S(\delta - h)\| \mathrm{Var}\,(g, [\,a, b\,])$$

$$+ 2\mathrm{Var}\,(g, [\,s-\delta, s)) + \|(S(h) - I)(g(s-h) - g(s-h-0))\|.$$

Since the semigroup is continuous in the uniform operator topology from $(0, +\infty)$ to $\mathcal{L}(X)$ and, by Remark 9.1.4, $g([\,a, b\,])$ is relatively compact in X, for the very same $\varepsilon > 0$, there exists $\eta \in (0, \delta)$ such that, for each $h \in (0, \eta)$, we have

$$\sup_{\theta \in (0,\eta)} \|S(\delta) - S(\delta - \theta)\| \leq \varepsilon$$

and

$$\|(S(h) - I)(g(s-h) - g(s-h-0))\| \leq \varepsilon.$$

Summing up, we get

$$\left\| \int_a^{s-h} (S(s-\tau) - S(s-h-\tau))\, dg(\tau) \right\| \leq (\mathrm{Var}\,(g, [\,a, b\,]) + 2)\varepsilon$$

which shows that $\lim_{h \downarrow 0} \|u(s-h) - S(h)u(s-h)\| = 0$. Recalling that, as we already have shown, the limit $\lim_{h \downarrow 0} S(h)u(s-h) = u^*(s-0)$ exists, we conclude that there exists $u(s-0)$ and $u(s-0) = u^*(s-0)$. This achieves the proof of (9.2.3) in the case in which the semigroup is continuous in the uniform operator topology on $(0, +\infty)$. Finally, if $\{S(t)\,;\ t \geq 0\}$ can be imbedded into a group, for each $t \geq 0$, $S(t)$ is invertible and so

$$u(t-h) = S(h)^{-1} S(h) u(t-h).$$

Thus, by virtue of (9.2.2) and the strong continuity of both $S(h)$ and $S(h)^{-1}$, we get (9.2.3). The proof is complete. □

Remark 9.2.1. It is a simple exercise to show that, even for general C_0-semigroups, there exists $\lim_{h \downarrow 0} u(s-h)$ in the weak-$\odot$ topology on X, and equals $u^*(s-0)$. So, we have

$$u(s) - u(s-0) = g(s) - g(s-0) \tag{9.2.4}$$

for each $s \in (a,b]$, where the one-sided limit $u(s-0)$ is considered in the weak-$\odot$ topology on X.

Corollary 9.2.1. *Let $g \in BV([a,b];X)$. If g is right continuous on $[a,b]$, then, for each $\xi \in X$, the $\mathfrak{L}^\infty$-solution of* (9.1.1), *corresponding to ξ and g, is right continuous on $[a,b]$. If, either the semigroup generated by A is continuous from $(0,+\infty)$ to $\mathfrak{L}(X)$ in the uniform operator topology, or it can be imbedded into a group and g is continuous, then for each $\xi \in X$, the $\mathfrak{L}^\infty$-solution of* (9.1.1), *corresponding to ξ and g, is continuous on $[a,b]$.*

Remark 9.2.2. A quite natural question we may raise is whether or not any $\mathfrak{L}^\infty$-solution is of bounded variation. The answer to this question is in the negative, as we can see from the next simple example. Let X be a reflexive Banach space and let us assume that the semigroup generated by A is not differentiable, i.e. there exists $x \in X$ such that $S(t)x \notin D(A)$ for each $t \geq 0$[1]. Now, let $g : [0,2] \to X$ be defined by

$$g(t) = \begin{cases} 0 & \text{for } t \in [0,1) \\ x & \text{for } t \in [1,2]. \end{cases}$$

Clearly, g is of bounded variation on $[1,2]$. Moreover, one may easily see that the unique $\mathfrak{L}^\infty$-solution u of the problem (9.1.1), corresponding to $\xi = 0$ and to g, is defined by

$$u(t) = \begin{cases} 0 & \text{for } t \in [0,1) \\ S(t-1)x & \text{for } t \in [1,2]. \end{cases}$$

At this point, let us recall that, whenever X is reflexive, and $u : [a,b] \to X$ is of bounded variation, then u is a.e. differentiable on $[a,b]$. See Bochner and Taylor **[26]**, Theorem 5.2. From this remark, it is clear that u cannot be of bounded variation on $[0,2]$ since it is nowhere differentiable on $[1,2]$. We conclude by noticing that it should be of great interest to know, under what circumstances on X, ξ, A and g, the corresponding $\mathfrak{L}^\infty$-solution of (9.1.1) is of bounded variation. One may prove that this is the case whenever $A \in \mathfrak{L}(X)$, i.e. when the generated semigroup is uniformly continuous, but we don't know any other relevant situations.

9.3. A Characterization of $\mathfrak{L}^\infty$-Solutions

In this section we prove a characterization of $\mathfrak{L}^\infty$-solutions in the terms of the duality between X and X_A. Namely, let $A^\odot : D(A^\odot) \subseteq X^\odot \to X^\odot$

[1] A typical instance of this sort is: $X = L^2(\mathbb{R}_+;\mathbb{R})$ and $A : D(A) \subseteq X \to X$ defined by $Af = f'$, for each $f \in D(A) = \{f \in L^2(\mathbb{R}_+;\mathbb{R});\ f' \in L^2(\mathbb{R}_+;\mathbb{R})\}$, which generates the well-known translation semigroup.

be the infinitesimal generator of the sun dual semigroup $\{S(t)^{\odot};\ t \geq 0\}$. Throughout this section we denote by

$$C([a,b]; D(A^{\odot})) = \{f \in C([a,b]; X^{\odot});\ f(t) \in D(A^{\odot}) \text{ for } t \in [a,b]\}.$$

If $g \in BV([a,b]; X)$ and $\varphi^{\odot} \in C([a,b]; X^{\odot})$, we denote by

$$\int_a^b (dg(t), \varphi^{\odot}(t)) = \lim_{\lambda(\mathcal{P})\downarrow 0} \sum_{i=1}^{k} (g(t_i) - g(t_{i-1}), \varphi^{\odot}(\tau_i)),$$

i.e. the Riemann-Stieltjes integral of the function $\varphi^{\odot}$ with respect to g. We introduce:

Definition 9.3.1. A function $u : [a,b] \to X$ is called a *variational*, or *$\mathcal{V}$-solution* of the problem (9.1.1) if u satisfies (9.2.1) and (9.2.4) and, for each $f^{\odot} \in C([a,b]; D(A^{\odot}))$, we have

$$\int_a^b (u(t), f^{\odot}(t))\, dt + \int_a^b (dg(t), \varphi^{\odot}(t)) + (\xi, \varphi^{\odot}(a)) = 0, \tag{9.3.1}$$

where $\varphi^{\odot} \in C([a,b]; D(A^{\odot})) \cap C^1([a,b]; X^{\odot})$ is the unique strong solution of the $\odot$-adjoint problem

$$\begin{cases} \varphi^{\odot\prime} = -A^{\odot}\varphi^{\odot} + f^{\odot} \\ \varphi^{\odot}(b) = 0. \end{cases} \tag{9.3.2}$$

Clearly, each $\mathcal{V}$-solution of (9.1.1) is weakly-$\odot$ piecewise continuous on $[a,b]$. Since $D(A^{\odot})$ is dense in $X^{\odot}$, it follows that $C([a,b]; D(A^{\odot}))$ is dense in $C([a,b]; X^{\odot})$ too and, accordingly, we have:

Theorem 9.3.1. *For each $\xi \in X$ and each $g \in BV([a,b]; X)$, the problem (9.1.1) has at most one $\mathcal{V}$-solution defined on $[a,b]$.*

Proof. The conclusion follows from the simple remark that, whenever u and v are two $\mathcal{V}$-solutions of (9.1.1), $u - v$ is weakly-$\odot$ continuous on $[a,b]$ (see (9.2.1) and (9.2.4)) and

$$\int_a^b (u(t) - v(t), f^{\odot}(t))\, dt = 0$$

for each $f \in C([a,b]; X^{\odot})$. □

As concern $\mathcal{V}$-solution, we have the following characterization theorem.

Theorem 9.3.2. *Let $\xi \in X$, and $g \in BV([a,b]; X)$. Then $u \in \mathcal{L}^{\infty}(a,b; X)$ is the $\mathcal{V}$-solution of the problem (9.1.1) on $[a,b]$ if and only if u is the $\mathcal{L}^{\infty}$-solution of the same problem on the same interval.*

Proof. Let u be an $\mathcal{L}^\infty$-solution of (9.1.1). Since, by Remark 9.2.1, u satisfies (9.2.4), by virtue of the uniqueness Theorem 9.3.1, it suffices to verify (9.3.1). We have

$$\int_a^b (u(t), f^\odot(t))\,dt + \int_a^b (dg(t), \varphi^\odot(t)) + (\xi, \varphi^\odot(a))$$
$$= \int_a^b (S(t-a)\xi, f^\odot(t))\,dt + \int_a^b \left(\int_a^t S(t-s)\,dg(s), f^\odot(t)\right)\,dt$$
$$+ \int_a^b (dg(t), \varphi^\odot(t)) + (\xi, \varphi^\odot(a)). \tag{9.3.3}$$

Let us observe that, for each $t, s \in [a,b]$, we have $\chi_{[a,t]}(s) = \chi_{[s,b]}(t)$, where χ_E is the characteristic function of the subset $E \subseteq [a,b]$, and therefore

$$\int_a^b \left(\int_a^t S(t-s)\,dg(s), f^\odot(t)\right)\,dt$$
$$= \int_a^b \left(\int_a^b \chi_{[a,t]}(s) S(t-s)\,dg(s), f^\odot(t)\right)\,dt$$
$$= \int_a^b \left(\int_a^b \chi_{[s,b]}(t) S(t-s) dg(s), f^\odot(t)\right)\,dt$$
$$= \int_a^b \left(dg(s), \int_a^b \chi_{[s,b]}(t) S(t-s)^\odot f^\odot(t) dt\right)$$
$$= \int_a^b \left(dg(s), \int_s^b S(t-s)^\odot f^\odot(t)\,dt\right).$$

We also have

$$\int_a^b (S(t-a)\xi, f^\odot(t))\,dt = \int_a^b (\xi, S(t-a)^\odot f^\odot(t))\,dt,$$

and

$$\begin{cases} \varphi^\odot(s) = -\int_s^b S(t-s)^\odot f^\odot(t)\,dt \\ \varphi^\odot(a) = -\int_a^b S(t-a)^\odot f^\odot(t)\,dt. \end{cases} \tag{9.3.4}$$

From these relations and (9.3.3), we deduce (9.3.1).
Conversely, if u is a $\mathcal{V}$-solution of (9.1.1) on $[a,b]$ and $f^\odot \in C([a,b]; X^\odot)$, then $\varphi^\odot$ satisfies (9.3.4) and a simple backward calculation shows that (9.3.1) implies

$$\int_a^b \left(u(t) - S(t-a)\xi - \int_a^t S(t-s)\,dg(s), f^\odot(t)\right)\,dt = 0.$$

Now the conclusion follows from the arbitrariness of $f^{\odot} \in C([a,b];X^{\odot})$ combined with the fact that, thanks to (9.2.1), and (9.2.4), the first factor under the integral above is weakly-$\odot$ continuous on $[a,b]$. The proof is complete. □

9.4. Compactness of the $\mathfrak{L}^{\infty}$-Solution Operator

From now on we shall assume that X and A are fixed, and such that, for each $(\xi,g) \in X \times BV([a,b];X)$, the Cauchy problem (9.1.1) has a unique $\mathfrak{L}^{\infty}$-solution. See Remark 9.1.2 and Theorem 9.1.1. Our goal here is to prove a necessary and sufficient condition in order that the family of all $\mathfrak{L}^{\infty}$-solutions of the semilinear Cauchy problem (9.1.1), when ξ ranges in a bounded subset in X, and g ranges in a subset in $BV([a,b];X)$ of equibounded variation, be relatively compact in $L^p(a,b;X)$ for each p in $[1,+\infty)$. This condition is in fact an extension of Theorem 8.6.1 from mild, or C^0-solutions, to $\mathfrak{L}^{\infty}$-solutions, allowing of course the right-hand side in (9.1.1) to be a measure generated by a function of bounded variation. Let $\xi \in X$ and $g \in BV([a,b];X)$, and let us denote by $u = Q(\xi,g)$ the unique $\mathfrak{L}^{\infty}$-solution of the Cauchy problem (9.1.1) corresponding to ξ and g.

Remark 9.4.1. Since $\{S(t)\,;\ t \geq 0\}$ is a semigroup of contractions, we have

$$\|u(t)\| \leq \|\xi\| + \operatorname{Var}(g,[a,b]).$$

The next simple lemma will prove useful in what follows.

Lemma 9.4.1. *For each $g \in BV([a,b];X)$ and $h \in (0,b-a)$, we have*

$$\int_a^{b-h} \left\| \int_t^{t+h} S(t+h-s)\,dg(s) \right\| dt \leq h \operatorname{Var}(g,[a,b]),$$

and

$$\int_{a+h}^{b} \left\| \int_{t-h}^{t} S(t-s)\,dg(s) \right\| dt \leq h \operatorname{Var}(g,[a,b]).$$

Proof. We shall prove only the first equality, the second one being obtained via very similar arguments. Since $t < s \leq t+h$ if and only if $s-h \leq t < s$, we have $\chi_{(t,t+h]}(s) = \chi_{[s-h,s)}(t)$ for each $t \in [a,b-h]$ and each $s \in [a+h,b]$. Let us denote by $V_g(s) = \operatorname{Var}(g,[a,s])$. Since

$$\int_t^{t+h} S(t+h-s)\,dg(s)$$

$$= \int_a^b \chi_{(t,t+h]} S(t+h-s)\,dg(s) + S(h)[g(t+0)-g(t)]$$

and $\|g(t+0)-g(t)\|=0$ a.e. for $t\in[a,b]$, we have

$$\int_a^{b-h}\left\|\int_t^{t+h}S(t+h-s)\,dg(s)\right\|dt$$
$$=\int_a^{b-h}\left\|\int_a^{b}\chi_{(t,t+h]}(s)S(t+h-s)\,dg(s)\right\|dt$$
$$\leq\int_a^{b}\int_{a+h}^{b}\chi_{[s-h,s)}(t)\,dV_g(s)\,dt$$
$$=\int_{a+h}^{b}\left(\int_a^{b}\chi_{[s-h,s)}(t)\,dt\right)dV_g(s)\leq h\,\mathrm{Var}\,(g,[a,b]).$$

The proof is complete. □

We recall that a subset $\mathcal{G}$ in $BV([a,b];X)$ is of *equibounded variation* if there exists $m_{\mathcal{G}}>0$ such that

$$\mathrm{Var}\,(g,[a,b])\leq m_{\mathcal{G}}$$

for each $g\in\mathcal{G}$.
We may now proceed to the statement of the main result in this section.

Theorem 9.4.1. *Let $A:D(A)\subseteq X\to X$ be the infinitesimal generator of a C_0-semigroup of contractions $\{S(t)\,;\ t\geq 0\}$, let $\mathcal{D}$ be a bounded subset in X and $\mathcal{G}$ a subset in $BV([a,b];X)$ of equibounded variation. Then $Q(\mathcal{D},\mathcal{G})$ is relatively compact in $L^p(a,b;X)$ for each $p\in[1,+\infty)$ if and only if for each $\varepsilon>0$ there exists a relatively compact subset C_ε in X, such that, for each $(\xi,g)\in\mathcal{D}\times\mathcal{G}$, there exists a subset $E_{\varepsilon,\xi,g}$ in $[a,b]$ whose Lebesgue measure is less than ε, and such that $Q(\xi,g)(t)\in C_\varepsilon$ for each $(\xi,g)\in\mathcal{D}\times\mathcal{G}$, and each $t\in[a,b]\setminus E_{\varepsilon,\xi,g}$.*

Proof. By virtue of Theorem A.5.1, the necessity is obvious. To prove the sufficiency, we also make use of the same Theorem A.5.1. First, let us observe that, by virtue of Lebesgue dominated convergence theorem, it suffices to show that $Q(\mathcal{D},\mathcal{F})$ is relatively compact in $L^1(a,b;X)$, and bounded in $\mathcal{L}^\infty(a,b;X)$. To this aim, let us recall that there exist $m_{\mathcal{D}}>0$, and $m_{\mathcal{G}}>0$, such that

$$\|\xi\|\leq m_{\mathcal{D}}\ \text{ and }\ \mathrm{Var}\,(g,[a,b])\leq m_{\mathcal{G}},\tag{9.4.1}$$

for each $(\xi,g)\in\mathcal{D}\times\mathcal{G}$. Then, by virtue of Remark 9.4.1, we have

$$\|Q(\xi,g)(t)\|\leq m_{\mathcal{D}}+m_{\mathcal{G}}\tag{9.4.2}$$

for each $(\xi,g)\in\mathcal{D}\times\mathcal{G}$ and $t\in[a,b]$. In order to prove that $Q(\mathcal{D},\mathcal{G})$ is 1-equiintegrable, let $\varepsilon>0$, $(\xi,g)\in\mathcal{D}\times\mathcal{G}$, and let C_ε and $E_{\varepsilon,\xi,g}$ be the sets having the properties mentioned by hypotheses. A simple computational argument, along with Lemma 9.4.1, shows that

$$\int_a^{b-h} \|Q(\xi,g)(t+h) - Q(\xi,g)(t)\|\,dt$$

$$\leq \int_a^{b-h} \|Q(\xi,g)(t+h) - S(h)Q(\xi,g)(t)\|\,dt$$

$$+ \int_a^{b-h} \|S(h)Q(\xi,g)(t) - Q(\xi,g)(t)\|\,dt$$

$$\leq \int_a^{b-h} \left\| \int_t^{t+h} S(t+h-s)\,dg(s) \right\| dt$$

$$+ \int_{[a,b]\setminus E_{\varepsilon,\xi,g}} \|S(h)Q(\xi,g)(t) - Q(\xi,g)(t)\|\,dt$$

$$+ \int_{E_{\varepsilon,\xi,g}} \|S(h)Q(\xi,g)(t) - Q(\xi,g)(t)\|\,dt$$

$$\leq h\,\mathrm{Var}\,(g,[\,a,b\,]) + \int_{[\,a,b\,]\setminus E_{\varepsilon,\xi,g}} \|S(h)Q(\xi,g)(t) - Q(\xi,g)(t)\|\,dt + 2M_{\mathcal{G}}\varepsilon,$$

for each $(\xi,g) \in \mathcal{D} \times \mathcal{G}$, and $h \in (0, b-a\,]$, where $M_{\mathcal{G}} = m_{\mathcal{D}} + m_{\mathcal{G}}$. As $Q(\xi,g)(t) \in C_\varepsilon$ for each $(\xi,g) \in \mathcal{D} \times \mathcal{G}$, and each $t \in [\,a,b\,] \setminus E_{\varepsilon,\xi,g}$, while C_ε is relatively compact in X, there exists $\delta(\varepsilon) \in (0, b-a\,]$ such that, for each $h \in (0, \delta(\varepsilon)\,]$,

$$\|S(h)Q(\xi,g)(t) - Q(\xi,g)(t)\| \leq \varepsilon,$$

uniformly for $(\xi,g) \in \mathcal{D} \times \mathcal{G}$, and $t \in [\,a,b\,] \setminus E_{\varepsilon,\xi,g}$. Then, taking into account (9.4.1), (9.4.2) and the preceding inequalities, we obtain

$$\int_a^{b-h} \|Q(\xi,g)(t+h) - Q(\xi,g)(t)\|\,dt \leq (b-a+m_{\mathcal{G}}+2M_{\mathcal{G}})\varepsilon$$

for each $(\xi,g) \in \mathcal{D} \times \mathcal{G}$, and $h \in (0, \delta(\varepsilon)\,] \cap (0, \varepsilon\,]$. Obviously, this relation shows that $Q(\mathcal{D},\mathcal{G})$ is 1-equiintegrable. From Theorem A.5.1, it follows that it is relatively compact in $L^1(a,b;X)$, and this completes the proof. □

Remark 9.4.2. If X is finite dimensional then, for each bounded subset $\mathcal{D}$ in X, and each subset $\mathcal{G}$ in $BV([\,a,b\,];X)$ of equibounded variation, $Q(\mathcal{D},\mathcal{G})$ is relatively compact in $L^p(a,b;X)$, and thus p-equiintegrable. This follows from the observation that, by virtue of Remark 9.4.1, the set $\{Q(\xi,g)(t);\ (\xi,g) \in \mathcal{D} \times \mathcal{G},\ t \in [\,a,b\,]\}$ is bounded and, inasmuch as X is finite dimensional, the set above is relatively compact. So, we are in the hypotheses of Theorem 9.4.1 and the conclusion follows.

To see that in infinite dimensional spaces the p-equiintegrability condition is not an intrinsic property of the set $Q(\mathcal{D}, \mathcal{G})$ with $\mathcal{D}$ and $\mathcal{G}$ bounded, and respectively of equibounded variation, let us analyze the following example.

Example 9.4.1. Let $X = L^2_{2\pi}(\mathbb{R})$ be the space of all equivalence classes, with respect to the almost everywhere equality on $\mathbb{R}$, of measurable and 2π-periodic functions from $\mathbb{R}$ to $\mathbb{R}$. Endowed with the $L^2(0, 2\pi)$-norm this is a real Hilbert space. Let $A : D(A) \subseteq H \to H$ be the operator defined by $D(A) = \{u \in H\,;\ u' \in H\}$ and $Au = u'$ for each $u \in D(A)$. Obviously, A generates a C_0-group of isometries on H, i.e. the translation group. Let $\mathcal{D} = \{0\}$, and $\mathcal{G} = \{t \mapsto -\frac{1}{n}\cos\{n(t + \cdot)\}\,;\ n \in \mathbb{N}^*\}$. It is easy to see that $\mathcal{G}$ is of equibounded variation on $[0, 1]$. On the other hand, in this case, $Q(\mathcal{D}, \mathcal{G}) = \{t \mapsto t\sin\{n(t + \cdot)\}\,;\ n \in \mathbb{N}^*\}$, which is not L^p-equicontinuous on $[0, 1]$, because it is not relatively compact in $L^p(0, 1\,; L^2(0, 2\pi))$.

Corollary 9.4.1. *Let $A : D(A) \subseteq X \to X$ be the generator of a C_0-semigroup of contractions $\{S(t)\,;\ t \geq 0\}$, let $\mathcal{D}$ be a bounded subset in X, and $\mathcal{G}$ a subset in $BV([a, b]; X)$ of equibounded variation. If*

$$\{Q(\xi, g)(t);\ (\xi, g) \in \mathcal{D} \times \mathcal{G},\ t \in [a, b]\}$$

is relatively compact in X, then, for each $p \in [1, +\infty)$, $Q(\mathcal{D}, \mathcal{G})$ is relatively compact in $L^p(a, b\,; X)$.

We conclude this section with a significant extension of Theorem 8.6.2.

Theorem 9.4.2. *If $A : D(A) \subseteq X \to X$ is the infinitesimal generator of a compact C_0-semigroup of contractions, $\mathcal{D}$ is a bounded subset in X, and $\mathcal{G}$ is a subset in $BV([a, b]\,; X)$ of equibounded variation, then, for each $p \in [1, +\infty)$, the set $Q(\mathcal{D}, \mathcal{G})$ is relatively compact in $L^p(a, b\,; X)$.*

Proof. We show that $Q(\mathcal{D}, \mathcal{G})$ satisfies the hypotheses of Theorem 9.4.1. To this aim let us observe that, by virtue of Lemma 9.4.1, we have

$$\int_{a+h}^{b} \|u(t) - S(h)u(t-h)\|\,dt \leq \int_{a+h}^{b} \left\| \int_{t-h}^{t} S(t-s)\,dg(s) \right\| dt$$
$$\leq h \operatorname{Var}(g, [a, b]),$$

for each $(\xi, g) \in \mathcal{D} \times \mathcal{G}$ and $h \in (0, b-a]$. Inasmuch as $\mathcal{G}$ is of equibounded variation, we get

$$\lim_{h \downarrow 0} \int_{a+h}^{b} \|u(t) - S(h)u(t-h)\|\,dt = 0, \tag{9.4.3}$$

uniformly for $f \in \mathcal{G}$.

Let us define $U = \mathcal{Q}(\mathcal{D}, \mathcal{G})$ and $\mathcal{Q}_h : U \to L^1(a, b; X)$ by

$$(\mathcal{Q}_h u)(t) = \begin{cases} 0 & \text{if } t \in [\,a, a+h\,] \\ u(t) - S(h)u(t-h) & \text{if } t \in (a+h, b\,]. \end{cases}$$

From (9.4.3), it follows that Lemma 8.6.1 applies, and therefore, for each $\varepsilon > 0$, there exists $E_{\varepsilon,u} \subseteq [\,a, b\,]$ with $\mu(E_{\varepsilon,u}) < \varepsilon$, and $(h_n)_{n\in\mathbb{N}}$ decreasing to 0, such that

$$\lim_{n\to\infty} \|u(t) - S(h_n)u(t-h_n)\| = 0,$$

uniformly for $u \in \mathcal{Q}(\mathcal{D}, \mathcal{G})$, and $t \in [\,a, b\,] \setminus E_{\varepsilon,u}$. Since $S(h_n)$ is compact for each $n \in \mathbb{N}$ and $\{u(s); u \in \mathcal{Q}(\mathcal{D}, \mathcal{G}), s \in [\,a, b\,]\}$ is bounded in X, it follows that the set

$$C_\varepsilon = \{u(t);\ u \in \mathcal{Q}(\mathcal{D}, \mathcal{G}),\ t \in [\,a, b\,] \setminus E_{\varepsilon,u}\}$$

is relatively compact in X. So, we are in the hypotheses of Theorem 9.4.1, and thus $\mathcal{Q}(\mathcal{D}, \mathcal{G})$ is relatively compact in $L^p(a, b; X)$ for each $p \in [1, +\infty)$. This completes the proof. □

9.5. Evolution Equations with "Spatial" Measures as Data

Let X be a real Banach space and let us consider the Cauchy problem (9.1.1), where $A : D(A) \subseteq X \to X$ generates a compact C_0-semigroup of contractions, $\xi \in X_A$ and $g \in BV([\,a, b\,]; X_A)$, where $X_A = (X^{\odot})^*$. See Section 9.1. By the Hahn-Banach theorem (see Hille and Phillips [**70**], Theorem 2.1.2, p. 29) it follows that X_A is a closed subspace of X^{**}. We notice that whenever X is reflexive, and thus $X^{\odot} = X^*$, we have $X_A = X$, and therefore the problem (9.1.1) can be easily treated by the previously developed theory. This is no longer true in the nonreflexive case when $X_A \neq X$, and this explains why, throughout this section, we assume that X is nonreflexive, although all the abstract results hold (trivially) true in general. Another reason, much more subtle for doing this, is that the analysis of partial differential equations involving measures with respect to the spatial argument relies heavily on nonreflexive settings and techniques, as L^1 spaces, and vague topologies. See Example 9.5.1 below.
In order to give a precise sense of (9.1.1) in this more general setting, we need the following convergence result.

Theorem 9.5.1. *Let $A : D(A) \subseteq X \to X$ be the generator of a compact C_0-semigroup of contractions, let $\xi \in X_A$ and $g \in BV([\,a, b\,]; X_A)$, and let $(\xi_k)_{k\in\mathbb{N}}$ and $(g_k)_{k\in\mathbb{N}}$ be two sequences in X, and respectively in*

$BV([a,b];X)$, such that $(g_k)_{k\in\mathbb{N}}$ has equibounded variation, and

$$\begin{cases} \lim_{k\to\infty} \xi_k = \xi & \text{in } \sigma(X_A, X^{\odot}) \\ \lim_{k\to\infty} g_k(t) = g(t) & \text{for each } t \in [a,b] \text{ in } \sigma(X_A, X^{\odot}). \end{cases}$$

Then there exists $u \in \mathcal{L}^\infty(a,b;X)$, such that, for each $p \in [1,+\infty)$,

$$\lim_{k\to\infty} Q(\xi_k, g_k) = u \tag{9.5.1}$$

strongly in $L^p(a,b;X)$, and pointwise in $\sigma(X_A, X^{\odot})$. In addition, for each $x^{\odot} \in X^{\odot}$, we have

$$(u(t), x^{\odot}) = (\xi, S(t-a)^{\odot}x^{\odot}) + \int_a^t (dg(s), S(t-s)^{\odot}x^{\odot}), \tag{9.5.2}$$

$$u(t+0)-u(t) = g(t+0)-g(t) \text{ and } u(s)-u(s-0) = g(s)-g(s-0), \tag{9.5.3}$$

where the one sided limits on the left-hand sides of (9.5.3) *are considered in the weak-$\odot$ topology on X.*

Proof. Let us observe that, for each $k,p \in \mathbb{N}$, each $t \in [a,b]$, and each $x^{\odot} \in X^{\odot}$, we have

$$(Q(\xi_k, g_k)(t) - Q(\xi_p, g_p)(t), x^{\odot}) = \int_a^t (d(g_k - g_p)(s), S(t-s)^{\odot}x^{\odot}),$$

where $\{S(t)^{\odot};\ t \geq 0\}$ is the sun dual semigroup. See Section 3.8. Since $\{S(t);\ t \geq 0\}$ is compact, by virtue of Schauder's theorem, p. 282 in Yosida [**136**], it follows that its sun dual is a compact semigroup too and therefore both X and $X^{\odot}$ are separable. See Theorem 6.2.2. So, we are in the hypotheses of Lemma 1.4.7, and accordingly

$$\lim_{k,p\to\infty} \int_a^t (d(g_k - g_p)(s), S(t-s)^{\odot}x^{\odot}) = 0$$

for each $t \in [a,b]$, and $x^{\odot} \in X^{\odot}$. Furthermore, we have

$$\lim_{k,p\to\infty} (\xi_k - \xi_p, S(t-a)^{\odot}x^{\odot}) = 0$$

for each $t \in [a,b]$ and $x^{\odot} \in X^{\odot}$. Summing up, we conclude that, for each $t \in [a,b]$, $(Q(\xi_k, g_k)(t))_{k\in\mathbb{N}}$ is a Cauchy sequence in the weak-$\odot$ topology on X_A. Therefore there exists $u : [a,b] \to X_A$ such that

$$\lim_{k\to\infty} Q(\xi_k, g_k)(t) = u(t)$$

weakly-$\odot$ in X_A. Fix $p \in [1,+\infty)$, and let us observe that, by virtue of Theorem 9.4.1, on a subsequence at least, we have

$$\lim_{k\to\infty} Q(\xi_k, g_k) = v$$

strongly in $L^p(a,b;X)$. So, v coincides with u. As $\{Q(\xi_k, g_k)\,;\ k \in \mathbb{N}\}$ is relatively compact in $L^p(a,b;X)$, we get that $(Q(\xi_k, g_k))_{k\in\mathbb{N}}$ itself converges in $L^p(a,b;X)$ to u, and this proves (9.5.1). Finally, let us observe that (9.5.2) is a direct consequence of Lemma 1.4.7, while (9.5.3) follows by using the very same arguments as in Theorem 9.2.1, and this achieves the proof. □

Remark 9.5.1. By Lemma 1.4.7, one may easily verify that the limit in (9.5.1) does not depend on the choice of the sequences $(\xi_k)_{k\in\mathbb{N}}$ and $(g_k)_{k\in\mathbb{N}}$ which approximate ξ and respectively g. Therefore, Theorem 9.5.1 allows us to extend the concept of the $\mathcal{L}^\infty$-solution to the case in which $\xi \in X_A$ and $g \in BV([\,a,b];X_A)$, whenever the latter can be approximated in the pointwise convergence weak-$\odot$ topology by a sequence of functions $(g_k)_{k\in\mathbb{N}}$ with equibounded variation and, of course, the semigroup generated by A is compact. More precisely, let $\xi \in X_A$ and $g \in BV([\,a,b];X_A)$.

Definition 9.5.1. A function $u \in \mathcal{L}^\infty(a,b;X)$ satisfying (9.5.2) is called an $\mathcal{L}^\infty$-*generalized solution* of the problem (9.1.1) on $[\,a,b\,]$.

We also notice that, under these circumstances, the operator A has a smoothing effect on the data in the sense that, for each $\xi \in X_A$ and $g \in BV([\,a,b];X_A)$, as in Theorem 9.5.1, the $\mathcal{L}^\infty$-generalized solution u is an X-valued function and not an X_A-valued one, as we might expect at a first glance.

Since X_A is obviously dependent of A, in all that follows, we call it the *space of admissible measures for A*. A prototype of the situation described in Theorem 9.5.1 is illustrated by the following suggestive example.

Example 9.5.1. Let Ω be a bounded domain in $\mathbb{R}^n$ with sufficiently smooth boundary Γ, and let us consider the linear parabolic problem

$$\begin{cases} u_t = \Delta u + \delta(t-t_0)\otimes\delta(x-x_0) & \text{on } (0,T)\times\Omega \\ u = 0 & \text{in } (0,T)\times\Gamma \\ u(0) = \delta(x-y_0) & \text{on } \Omega, \end{cases} \tag{9.5.4}$$

where $t_0 \in (0,T)$ and $x_0, y_0 \in \Omega$. To get an $\mathcal{L}^\infty$-generalized solution of (9.5.4) we proceed as follows. Let $X = L^1(\Omega)$, and let $A : D(A) \subseteq X \to X$ be defined by

$$D(A) = \{u \in W_0^{1,1}(\Omega);\ \Delta u \in L^1(\Omega)\} \ \text{ and } \ Au = \Delta u$$

for each $u \in D(A)$. We recall that A generates a compact C_0-semigroup of contractions $\{S(t)\,;\ t \geq 0\}$ on X. See Theorem 7.2.7. Let $\{S(t)^\odot\,;\ t \geq 0\}$ be the sun dual semigroup on $X^\odot$. In our case, one may easily verify that $X^\odot = C_0(\overline{\Omega})$, i.e. the space of all continuous functions from $\overline{\Omega}$ to

$\mathbb{R}$ vanishing on Γ. At this point, let us take two sequences $(\xi_k)_{k\in\mathbb{N}}$ and $(f_k)_{k\in\mathbb{N}}$ in $L^1(\Omega)$, and respectively in $L^1(0,T;L^1(\Omega))$, satisfying

$$\|\xi_k\|_{L^1(\Omega)} = \|f_k\|_{L^1(0,T;L^1(\Omega))} = 1$$

for each $k \in \mathbb{N}$, and

$$\begin{cases} \lim_{k\to\infty} \int_\Omega \int_0^T f_k(s,x)\varphi(s,x)\,ds\,dx = \varphi(t_0,x_0), \\ \\ \lim_{k\to\infty} \int_\Omega \xi_k(x)\psi(x)\,dx = \psi(y_0) \end{cases}$$

for each $\varphi \in C([0,T];C_0(\overline{\Omega}))$ and $\psi \in C_0(\overline{\Omega})$. Now, let us consider the sequence of problems

$$\begin{cases} u_{kt} = \Delta u_k + f_k & \text{on } (0,T)\times\Omega \\ u_k = 0 & \text{in } (0,T)\times\Gamma \\ u_k(0) = \xi_k & \text{on } \Omega, \end{cases}$$

and let us observe that we are in the hypotheses of Theorem 9.5.1. Thus, the problem (9.5.3) has a unique $\mathcal{L}^\infty$-generalized solution.

Remark 9.5.2. Many other problems can be approached by using a similar scheme. We mention here only one of the simplest, i.e. the heat equation with point control

$$\begin{cases} u_t = \Delta u + f\delta(x - x_0) & \text{on } (0,T)\times\Omega \\ u = 0 & \text{in } (0,T)\times\Gamma \\ u(0) = \xi & \text{on } \Omega, \end{cases}$$

where the control $f \in \mathbb{R}$.

Problems.

Problem 9.1. Let $X = C_{ub}(\mathbb{R})$ and let $\{G(t);\ t \in \mathbb{R}\}$ the translation group, i.e. $[G(t)f](s) = f(s-t)$ for $f \in X$ and $t, s \in \mathbb{R}$. Find the sun dual group $\{S(t)^\odot;\ t \geq 0\}$.

Problem 9.2. Let X be a real Banach space, $\{S(t)\,;\ t \geq 0\}$ a C_0-semigroup of contractions on X, $x \in X$ with $x \neq 0$, and let $g : \mathbb{R} \to X$ be defined by

$$g(t) = \begin{cases} -x & \text{if } t < 0 \\ 0 & \text{if } t = 0 \\ x & \text{if } t > 0. \end{cases}$$

Prove that $g \in BV(\mathbb{R};X)$ and, for every $a,t \in \mathbb{R}$ with $a < t$, the Riemann-Stieltjes integral $\int_a^t S(t-s)\,dg(s)$ exists in the norm topology of X and

$$\int_a^t S(t-s)\,dg(s) = \begin{cases} 0 & \text{if } a \le t < 0 \\ x & \text{if } a < 0 = t \\ 2x & \text{if } a < 0 < t \\ x & \text{if } 0 = a < t \\ 0 & \text{if } 0 < a < t. \end{cases}$$

This example shows that, even for nonreflexive Banach spaces X, for semigroups which are not continuous in the uniform operator topology from $(0,+\infty)$ to $\mathcal{L}(X)$ and for functions g which are not defined by densities, it may happen that the Riemann-Stieltjes integral above not only to belong to X, but to exist even in the norm topology of X. See Remarks 9.1.2, 9.1.3 and 9.1.5.

Problem 9.3. Let X be a real Banach space, $\{S(t)\,;\ t \ge 0\}$ a C_0-semigroup of contractions on X, $x \in X$ with $x \neq 0$, $h \in BV(\mathbb{R})$ and let $g : \mathbb{R} \to X$ be defined by $g(t) = h(t)x$ for every $t \in \mathbb{R}$. Show that $g \in BV(\mathbb{R};X)$ and, for each $a,t \in \mathbb{R}$ with $a < t$, the Riemann-Stieltjes integral $\int_a^t S(t-s)\,dg(s)$ exists in the norm topology of X.

Problem 9.4. Let X be a real Banach space, $\{S(t)\,;\ t \ge 0\}$ a C_0-semigroup of contractions on X, let $x_i \in X$ satisfying $x_i \neq 0$ and $h_i \in BV(\mathbb{R})$ for $i = 1,2,\dots n$ and let $g : \mathbb{R} \to X$ be defined by $g(t) = \sum_{i=1}^n h_i(t)x_i$ for every $t \in \mathbb{R}$. Show that $g \in BV(\mathbb{R};X)$ and, for each $a,t \in \mathbb{R}$ with $a < t$, the Riemann-Stieltjes integral $\int_a^t S(t-s)\,dg(s)$ exists in the norm topology of X.

Problem 9.5. Let X be a real Banach space, $A : D(A) \subseteq X \to X$ the generator of a C_0-semigroup of contractions, and let $BV_A([\,a,b\,];X)$ be the space of all $g \in BV([\,a,b\,];X)$ enjoying the property that, for every $t \in [\,a,b\,]$, there exists the integral $\int_0^t S(t-s)\,dg(s)$ in the norm topology of X (and belongs to X, of course). Show that $BV_A([\,a,b\,];X)$ is a closed subspace in $BV([\,a,b\,];X)$, the later being endowed with the seminorm $\|g\| = \mathrm{Var}\,(g,[\,a,b\,])$.

A subset $\mathcal{G}$ in $BV([\,a,b];X)$ is *equi-absolutely continuous* if there exists a function $\ell : \mathbb{R}_+ \to \mathbb{R}_+$, nondecreasing, continuous at 0, with $\ell(0) = 0$ and such that $\mathrm{Var}\,(g,[\,c,d\,]) \le \ell(d-c)$ for all $g \in \mathcal{G}$ and $[\,c,d\,] \subseteq [\,a,b\,]$.

Problem 9.6. Let $A : D(A) \subseteq X \to X$ be the generator of a C_0-semigroup of contractions $\{S(t)\,;\ t \ge 0\}$, let $\mathcal{D}$ be a bounded subset in X, and $\mathcal{F}$ an equi-absolutely continuous subset in $BV([\,a,b\,];X)$. Show that $Q(\mathcal{D},\mathcal{F})$ is in $C([a,b];X)$ and is relatively compact in $C([c,b];X)$ for each $c \in (a,b)$, if and only if there exists a dense subset D in $[a,b]$ such that, for each $t \in D$, the t-section of $Q(\mathcal{D},\mathcal{F})$, i.e. $Q(\mathcal{D},\mathcal{F})(t) = \{Q(\xi,f)(t);\ (\xi,f) \in \mathcal{D} \times \mathcal{F}\}$, is relatively compact in X. Moreover, if the latter condition is satisfied, and $a \in D$, then $Q(\mathcal{D},\mathcal{F})$ is relatively compact in $C([a,b];X)$.

Problem 9.7. Prove that if $A : D(A) \subseteq X \to X$ is the generator of a compact C_0-semigroup of contractions, $\mathcal{D}$ is a bounded subset in X, and $\mathcal{G}$ is an equi-absolutely continuous subset in $BV([a,b];X)$, then, for each $c \in (a,b)$, the set

$$Q(\mathcal{D},\mathcal{G}) = \{Q(\xi,g);\ (\xi,g) \in \mathcal{D} \times \mathcal{G}\}$$

is relatively compact in $C([c,b];X)$.

Notes. The results in this chapter were proved by Vrabie [**131**]. Related results referring to linear evolution equations with measures as data were obtained recently by Amann [**6**] by assuming that A generates an analytic semigroup and dg is a Radon measure. See also Ahmed [**2**]. We notice that the first-order necessary conditions of optimality for optimal control problems with state constraints are of the form (9.1.1). See Barbu [**18**], Casas [**35**], Fattorini [**52**], Lasiecka [**83**], [**82**], Raymond [**107**], Raymond-Tröltzsch [**108**], Raymond-Zidani [**109**] and the references therein. The main results in Section 9.4, Theorems 9.4.1 and 9.4.2, extend Theorem 8.6.1 and Baras-Hassan-Veron's Theorem 8.6.2 to the general case of distributed measures, while Theorem 9.5.1 in Section 9.5 exhibits a closedness property of the $\mathcal{L}^\infty$-solution operator. Concerning the problems here included we notice that all of them are new. There are many other interesting open problems from which I list only a few.

1. It would be of great interest to prove a relative compactness sufficient condition of the same form as that in Theorem 9.4.2 not only in $L^p(a,b;X)$, but also in the pointwise convergence topology.

2. A second problem is to develop an existence and regularity theory for another type of solution, more general than that of $\mathcal{L}^\infty$-solution. Namely, eliminating the condition (9.1.6) imposed in Definition 9.1.1, we get another concept of solution, weaker than that of $\mathcal{L}^\infty$-solution. More precisely, we say that $u : [a,b] \to X_A$ is an X_A*-valued solution* of (9.1.1) on $[a,b]$ if u is given by the variation of constants formula, i.e.

$$u(t) = S(t-s)\xi + \int_a^t S(t-s)\,dg(s)$$

for each $t \in [a,b]$, where the integral is considered in the $\sigma(X_A, X^{\odot})$-topology. See Section 9.1.

3. It would be of great interest to use the results here established in order to get information on control problems of the form:

$$\text{minimize} \quad \left\{\int_0^T C(u_f(t))\,dt\,;\ f \in \mathcal{F}\right\}$$

subjected to

$$\begin{cases} u_f' = Au_f + f(t) \\ u(a) = \xi, \end{cases}$$

where $A : D(A) \subseteq X \to X$ is the infinitesimal generator of a compact C_0-semigroup of contractions, and the set $\mathcal{F}$ of admissible controls is the unit closed ball centered at zero in X. Here X is assumed to be a nonreflexive real Banach space.

4. Another problem is to find necessary and/or sufficient conditions in order for a locally closed subset[2] D in X to be invariant with respect to the evolution equation

$$du = \{Au\}dt + dg, \tag{$\mathcal{E}$}$$

where $A : D(A) \subseteq X \to X$ is the infinitesimal generator of a compact C_0-semigroup and $g \in BV([\,a,b\,];X)$. We recall that D is *invariant with respect to* $(\mathcal{E})$ if for each $c \in [\,a,b)$ and each $\xi \in X$ with $g(c+0)-g(c)+\xi \in D$, the unique $\mathcal{L}^\infty$-solution $u : [\,a,c\,] \to X$ satisfies $u(t+0) \in D$ for each $t \in [\,c,b)$. We notice that, in the case in which $g(t) = \int_a^t f(s)\,ds$ with $f \in C([\,a,b\,];X)$, D is invariant with respect to $(\mathcal{E})$ if and only if

$$\liminf_{h\downarrow 0} \frac{1}{h} d(S(h)\xi + hf(t); D) = 0 \tag{$\mathcal{T}$}$$

for each $t \in [\,a,b)$ and each $\xi \in D$. Here $d(x;D)$ denotes the distance between $x \in X$ and D. For details on this interesting subject see Cârjă and Vrabie [**34**].

---

[2] We recall that D is locally closed if for each $\xi \in D$ there exists $r > 0$ such that $B(\xi,r) \cap D$ is closed. Both closed subsets and open subsets in X are locally closed, but there are examples of sets which, although neither closed, nor open, are locally closed. An example of such set is an open 2-dimensional disk in $\mathbb{R}^3$.

CHAPTER 10

Some Nonlinear Cauchy Problems

In order to illustrate the power of the abstract methods already developed in the preceding chapters, we shall present here some remarkable results concerning the local existence problem for several classes of nonlinear differential equations in general Banach spaces. We begin with an extension of Peano's local existence theorem to infinite dimensions. Further, we prove another generalization of the later and present some basic facts about saturated and global solutions. As a first application, we consider the Klein-Gordon semilinear equation which, surprisingly, can be rewritten and studied as an ordinary first-order differential equation driven by a continuous right-hand side. We conclude with a partial differential equation from mechanics which also can be handled as an ordinary differential equation with continuous right-hand side

---

10.1. Peano's Local Existence Theorem

Let X be a real Banach space, $\mathcal{D}$ a nonempty and open subset in $\mathbb{R} \times X$, and $f : \mathcal{D} \to X$. Let us consider the Cauchy problem

$$\begin{cases} u' = f(t, u) \\ u(a) = \xi, \end{cases} \tag{10.1.1}$$

and let us recall that, if X is finite-dimensional, and f is continuous, then, for each $(a, \xi) \in \mathcal{D}$, (10.1.1) has at least one local solution. This result, due to Peano, fundamental in the theory of ordinary differential equations, does not extend exactly in this form to the infinite-dimensional case, as we shall see in the counterexample below. To fix the ideas, we recall first that by a *solution* of (10.1.1) we mean a C^1 function $u : [a, b) \to X$ such that $(t, u(t)) \in \mathcal{D}$ for each $t \in [a, b)$ and which satisfies $u'(t) = f(t, u(t))$ for each $t \in [a, b)$ and $u(a) = \xi$.

Counterexample 10.1.1. (Dieudonné) Let $X = c_0$ be the space of all real sequences $(x_n)_{n \in \mathbb{N}^*}$ with $\lim_{n \to \infty} x_n = 0$, which, endowed with the sup-norm defined by $\|(x_n)_{n \in \mathbb{N}^*}\|_\infty = \sup\{|x_n| ;\ n \in \mathbb{N}^*\}$ for each $(x_n)_{n \in \mathbb{N}^*} \in X$, is a real Banach space, which is infinite-dimensional. Let $f : X \to X$ be defined by

$$(f((x_n)_{n\in\mathbb{N}^*})_k)_{k\in\mathbb{N}^*} = \left(2\sqrt{|x_k|}\right)_{k\in\mathbb{N}^*}$$

for each $(x_n)_{n\in\mathbb{N}^*} \in c_0$, let $\xi = (1/n)_{n\in\mathbb{N}^*}$, and let us consider the Cauchy problem

$$\begin{cases} u' = f(u) \\ u(0) = \xi. \end{cases}$$

Let us observe that $u : [\,0,\delta) \to X$ is a solution of the problem above if and only if $(u_k)_{k\in\mathbb{N}^*} : [\,0,\delta) \to X$ is a solution of the system of infinitely many differential equations

$$\begin{cases} u'_k = 2\sqrt{|u_k|} \\ u_k(0) = 1/k \qquad k = 1, 2, \ldots. \end{cases}$$

Let us assume that the above Cauchy problem, or equivalently the above system, has at least one solution $(u_k)_{k\in\mathbb{N}^*} : [\,0,\delta) \to X$, with $\delta > 0$. As this system contains infinitely many uncoupled differential equations with separate variables, whose solutions, if there exist, are necessarily of the form

$$u_k(t) = (t + 1/k)^2$$

for each $k \in \mathbb{N}^*$ and each $t \in [\,0,\delta)$, it follows that $(u_k)_{k\in\mathbb{N}^*}$ is defined by the above equalities. But, for all $t > 0$, we have $\lim_{k\to\infty} u_k(t) = t^2$, relation in contradiction with the fact that $(u_k(t))_{k\in\mathbb{N}^*}$ belongs to c_0, i.e. $\lim_{k\to\infty} u_k(t) = 0$. This contradiction can be eliminated only if the Cauchy problem in question has no local solution. Analyzing the proof of Peano's local existence theorem (see Corduneanu [**39**], Theorem 2.4, p. 33), we easily conclude that this nonexistence phenomenon is due to the lack of relative compactness of bounded subsets in infinite-dimensional Banach spaces.

This counterexample shows that, in order to obtain an existence result applying to (10.1.1), besides the continuity condition on f, we need an extra-condition to compensate the lack of relative compactness of bounded subsets in a an infinite-dimensional Banach space. Roughly speaking, such an extra-condition would require f to carry bounded subsets in $\mathcal{D}$ into relatively compact subsets in X. In fact, we will use a slightly weaker condition. Namely, we introduce:

Definition 10.1.1. A function $f : \mathcal{D} \to X$ is called *b-compact* if for each $[\,a,b\,] \subset \mathbb{R}$, $\xi \in X$ and $r > 0$ with $[\,a,b\,] \times B(\xi,r) \subset \mathcal{D}$, $f([\,a,b\,] \times B(\xi,r))$ is relatively compact in X. The function f is called *compact* if it carries bounded subsets in $\mathcal{D}$ into relatively compact subsets in X.

Clearly each compact function is b-compact, and whenever $\mathcal{D} = \mathbb{R} \times X$, the converse statement is also true. However, we notice that, if $\mathcal{D}$ does not coincide with $\mathbb{R} \times X$, there exists b-compact functions which are not compact. A simple example in the case of $X = \mathbb{R}$ is furnished by the function $f : (0, +\infty) \times (0, +\infty) \to \mathbb{R}$, $f(t, u) = \frac{1}{u}$, which is b-compact but not compact.

In this section, we shall prove that, whenever $f : \mathcal{D} \to X$ is b-compact, for each $(a, \xi) \in \mathcal{D}$, the Cauchy problem (10.1.1) has at least one local solution. This result, fundamental in the theory of ordinary differential equation in infinite dimensions, is an extension of the well-known Peano's local existence theorem.

We analyze first the simplest case when $\mathcal{D} = \mathbb{I} \times X$, with $\mathbb{I}$ a nonempty and open interval, and f is continuous on $\mathbb{I} \times X$ with $f(\mathbb{I} \times X)$ relatively compact in X, and then we shall show how we can pass to the general case. So, let $f : \mathbb{I} \times X \to X$, let $(a, \xi) \in \mathbb{I} \times X$, let $\lambda > 0$ and let $\delta > 0$ be such that $[\,a, a + \delta\,] \subset \mathbb{I}$. Let us consider the integral equation with delay

$$u_\lambda(t) = \begin{cases} \xi & \text{for } t \in [\,a - \lambda, a\,] \\ \xi + \displaystyle\int_a^t f(\tau, u_\lambda(\tau - \lambda))\, d\tau & \text{for } t \in (a, a + \delta\,]. \end{cases} \tag{10.1.2}$$

The following result will prove useful in the sequel.

Lemma 10.1.1. *If $f : \mathbb{I} \times X \to X$ is continuous, then, for each $\lambda > 0$, each $(a, \xi) \in \mathbb{I} \times X$ and each $\delta > 0$ with $[\,a, a+\delta\,] \subset \mathbb{I}$,* (10.1.2) *has one and only one solution defined on $[\,a - \lambda, a + \delta\,]$.*

Proof. Obviously, u_λ is uniquely determined on $[\,a - \lambda, a\,]$ by (10.1.2). Let then $t \in [\,a, a + \lambda\,]$. Let us remark that, for each $\tau \in [\,a, t\,]$, we have $\tau - \lambda \in [\,a - \lambda, a\,]$, and therefore $u_\lambda(\tau - \lambda) = \xi$. Accordingly,

$$u_\lambda(t) = \xi + \int_a^t f(\tau, \xi)\, d\tau$$

and u_λ is uniquely determined on $[\,a, a + \lambda\,]$. Proceeding analogously, we can successively determine u_λ on $[\,a + \lambda, a + 2\lambda\,]$, $[\,a + 2\lambda, a + 3\lambda\,]$, and so on. Obviously, after m steps, with $m\lambda \geq a + \delta$, we can define u_λ on the whole interval $[\,a, a + \delta\,]$. We complete the proof by observing that u_λ is continuous and uniquely determined. □

As we already have noticed, we shall prove first the following existence result which, although auxiliary, is interesting by itself.

Lemma 10.1.2. *If $f : \mathbb{I} \times X$ is continuous and $f(\mathbb{I} \times X)$ is relatively compact, then, for each $(a, \xi) \in \mathbb{I} \times X$ and $\delta > 0$ with $[\,a, a + \delta\,] \subset \mathbb{I}$,* (10.1.1) *has at least one solution defined on $[\,a, a + \delta\,]$.*

Proof. Let $(a, \xi) \in \mathbb{I} \times X$, and $\delta > 0$ such that $[a, a+\delta] \subset \mathbb{I}$, let $m \in \mathbb{N}^*$, and let us consider the integral equation with the delay $\delta_m = \delta/m$

$$u_m(t) = \begin{cases} \xi & \text{for } t \in [a - \delta_m, a] \\ \xi + \displaystyle\int_a^t f(\tau, u_m(\tau - \delta_m))\, d\tau & \text{for } t \in (a, a+\delta]. \end{cases} \tag{10.1.3}$$

Let us remark that, by virtue of Lemma 10.1.1, for each $m \in \mathbb{N}^*$, (10.1.3) has a unique continuous solution $u_m : [a - \delta_m, a + \delta] \to X$. We shall prove next that $\{u_m;\ m \in \mathbb{N}^*\}$ satisfies the hypotheses of Theorem A.2.1. Namely, we shall show that, for each $t \in [a, a+\delta]$, $\{u_m(t)\,;\ m \in \mathbb{N}^*\}$ is relatively compact in X, and $\{u_m;\ m \in \mathbb{N}^*\}$ is equicontinuous on $[a, a+\delta]$. Let us recall that $f(\mathbb{I} \times X)$ is relatively compact in X and, therefore, from Lemma A.1.3, we conclude that

$$\{u_m(t)\,; m \in \mathbb{N}^*\} = \left\{\xi + \int_a^t f(s, u_m(s - \delta_m))\, ds\,; m \in \mathbb{N}^*\right\}$$

is relatively compact in X for each $t \in (a, a+\delta]$.
Next, let us observe that, also from (10.1.3), we have

$$\|u_m(t) - u_m(s)\| \leq \left| \int_s^t \|f(\tau, u_m(\tau - \delta_m))\|\, d\tau \right| \leq M|t - s|,$$

for each $m \in \mathbb{N}^*$ and $t, s \in [a, a+\delta]$. Consequently, $\{u_m;\ m \in \mathbb{N}^*\}$ is equicontinuous on $[a, a+\delta]$. From Theorem A.2.1, it follows that $(u_m)_{m \in \mathbb{N}^*}$ has at least one subsequence, denoted for simplicity again by $(u_m)_{m \in \mathbb{N}^*}$, which is uniformly convergent on $[a, a+\delta]$ to a continuous function u. Obviously, we have

$$\lim_{m \to \infty} u_m(\tau - \delta_m) = u(\tau),$$

uniformly for $\tau \in [a, a+\delta]$. Since f is continuous on $\mathbb{I} \times X$, the relation above and Corollary A.2.2 show that we may pass to the limit in (10.1.3) for $m \to \infty$. Moreover, u satisfies

$$u(t) = \xi + \int_a^t f(\tau, u(\tau))\, d\tau, \quad \text{for each } \ t \in [a, a+\delta],$$

which shows that $u : [a, a+\delta] \to X$ is a solution for (10.1.1), and this achieves the proof. □

Remark 10.1.1. In the hypotheses of Lemma 10.1.1, we can prove that, for each initial datum $(a, \xi) \in \mathbb{I} \times X$, (10.1.3) has at least one global solution.

We may now proceed to the statement of the main result in this section. To this aim, let $\mathcal{D}$ be a nonempty and open subset in $\mathbb{R} \times X$ and $f : \mathcal{D} \to X$ a given function.

Theorem 10.1.1. *If* $f : \mathcal{D} \to X$ *is b-compact, then for each* $(a, \xi) \in \mathcal{D}$, (10.1.1) *has at least one local solution.*

Proof. Let $(a, \xi) \in \mathcal{D}$. As $\mathcal{D}$ is open, there exist $d > 0$, and $r > 0$, such that $[a-d, a+d] \times B(\xi, r) \subset \mathcal{D}$, where

$$B(\xi, r) = \{\eta \in X;\ \|\eta - \xi\| \le r\}.$$

Let us define $\rho : X \to X$ by

$$\rho(y) = \begin{cases} y & \text{for } y \in B(\xi, r) \\ \dfrac{r}{\|y-\xi\|}(y-\xi) + \xi & \text{for } y \in X \setminus B(\xi, r). \end{cases}$$

We may easily see that ρ maps X to $B(\xi, r)$ and is continuous on X. Next, let us define $g : (a-d, a+d) \times X \to X$ by

$$g(t, y) = f(t, \rho(y)), \text{ for each } (t, y) \in (a-d, a+d) \times X.$$

As f is b-compact, $f([a-d, a+d] \times B(\xi, r))$ is relatively compact. Hence g is continuous and $g((a-d, a+d) \times X)$ is relatively compact. Thanks to Lemma 10.1.1, we know that, for each $d' \in (0, d)$, the Cauchy problem

$$\begin{cases} u' = g(t, u) \\ u(a) = \xi \end{cases}$$

has at least one solution $u : [a, a+d'] \to X$. Inasmuch as u is continuous at $t = a$ and $u(a) = \xi$, for $r > 0$ there exists $\delta \in (0, d']$ such that, for each $t \in [a, a+\delta]$, $\|u(t) - \xi\| \le r$. But in this case $g(t, u(t)) = f(t, u(t))$, and therefore $u : [a, a+\delta] \to X$ is a solution of the problem (10.1.1). The proof is complete. $\square$

10.2. The Problem $u' = f(t,u) + g(t,u)$

Let X be a real Banach space and $\mathcal{D}$ a nonempty and open subset in $\mathbb{R} \times X$. In this section we shall prove a local existence result referring to a class of Cauchy problem of the type

$$\begin{cases} u' = f(t, u) + g(t, u) \\ u(a) = \xi, \end{cases} \tag{10.2.1}$$

where $f : \mathcal{D} \to X$ is a b-compact function, and $g : \mathcal{D} \to X$ is continuous on $\mathcal{D}$, and locally Lipschitz with respect of its last argument. We recall that $g : \mathcal{D} \to X$ is *locally Lipschitz with respect to its last argument* if for each $(a, \xi) \in \mathcal{B} \subset \mathcal{D}$ there exists $b > a$, $r > 0$ and $L = L_{a,\xi} > 0$ such that $[a, b] \times B(\xi, r) \subset \mathcal{D}$ and

$$\|g(t, u) - g(t, v)\| \le L\|u - v\|$$

for each $(t,u),(t,v) \in [a,b] \times B(\xi,r)$.
As expected, by a solution of (10.2.1) we mean a C^1-function $u : [a,b] \to X$ satisfying $(t,u(t)) \in \mathcal{D}$, $u'(t) = f(t,u(t)) + g(t,u(t))$ for each $t \in [a,b]$, and $u(a) = \xi$.

Theorem 10.2.1. *Let $f : \mathcal{D} \to X$ be b-compact, and let $g : \mathcal{D} \to X$ be continuous on $\mathcal{D}$, and locally Lipschitz with respect to its last argument. Then, for each $(a,\xi) \in \mathcal{D}$, there exists $b > a$ such that* (10.2.1) *has at least one solution defined on $[a,b]$.*

We shall prove this theorem with the help of three lemmas. First, we prove a variant of the well-known Gronwall's lemma. Having in mind our later purposes, we shall formulate this lemma in a more general framework than that needed here.

Lemma 10.2.1. (Gronwall) *Let $x,k : [a,b) \to \mathbb{R}_+$ be measurable with $s \mapsto k(s)x(s)$ locally integrable on $[a,b)$, let $m \geq 0$ and assume that*

$$x(t) \leq m + \int_a^t k(s)x(s)\,ds$$

a.e. for $t \in [a,b)$. Then

$$x(t) \leq m e^{\int_a^t k(s)\,ds}$$

a.e. for $t \in [a,b)$.

Proof. Let $y(t) = m + \int_a^t k(s)x(s)\,ds$. Then y is a.e. differentiable on $[a,b)$ and we have

$$y'(t) = k(t)x(t) \leq k(t)y(t)$$

a.e. for $t \in [a,b)$. Hence

$$\frac{d}{dt}\left(y(t)e^{-\int_a^t k(s)\,ds}\right) \leq 0$$

a.e. for $t \in [a,b)$. Integrating from a to t, after some rearrangements, we get

$$y(t) \leq m e^{\int_a^t k(s)\,ds}$$

for each $t \in [a,b)$. Since $x(t) \leq y(t)$ a.e. for $t \in [a,b)$, this achieves the proof. □

Throughout the next two lemmas, $\mathbb{I}$ denotes an interval with nonempty interior but which is not necessarily open.

Lemma 10.2.2. *Let $g : \mathbb{I} \times X \to X$ be continuous on $\mathbb{I} \times X$ and Lipschitz on X. Then, for each $(a,\xi) \in \mathbb{I} \times X$, each $b > 0$ with $[a,b] \subset \mathbb{I}$, and each $h \in L^1(a,b;X)$, the problem*

$$\begin{cases} u' = g(t,u) + h(t) \\ u(a) = \xi \end{cases} \tag{10.2.2}$$

has a unique solution defined on $[a,b]$. In addition, for each (a,ξ), and b, fixed as above, the mapping $h \mapsto \mathcal{S}(h)$, i.e. the unique solution of the problem (10.2.2) corresponding to h, satisfies

$$\|\mathcal{S}(h_1) - \mathcal{S}(h_2)\|_{C([a,b];X)} \le e^{\delta L}\|H_1 - H_2\|_{C([a,b];X)}, \tag{10.2.3}$$

where $L > 0$ is the Lipschitz constant of g, and

$$H_i(t) = \int_a^t h_i(s)\,ds$$

for $i = 1,2$ and $t \in [a,b]$.

Proof. Let $a \in \mathbb{I}$, $b > a$ with $[a,b] \subset \mathbb{I}$, $\xi \in X$ and $h \in L^1(a,b;X)$. Let us define $Q : C([a,b];X) \to C([a,b];X)$ by

$$(Qu)(t) = \xi + \int_a^t g(s,u(s))\,ds + \int_a^t h(s)\,ds$$

for each $u \in C([a,b];X)$, and each $t \in [a,b]$. Let us observe that, for each $u, v \in C([a,b];X)$ and $t \in [a,b]$, we have

$$\|(Qu)(t) - (Qv)(t)\| \le L(t-a)\|u - v\|_{C([a,b];X)}.$$

From this inequality, by a simple inductive argument, we deduce that

$$\|(Q^n u)(t) - (Q^n v)(t)\| \le \frac{L^n(t-a)^n}{n!}\|u - v\|_{C([a,b];X)}$$

for each $n \in \mathbb{N}^*$, each $u, v \in C([a,b];X)$, and each $t \in [a,b]$. Consequently

$$\|Q^n u - Q^n v\|_{C([a,b];X)} \le \frac{L^n(b-a)^n}{n!}\|u - v\|_{C([a,b];X)}$$

for each $n \in \mathbb{N}^*$, and each $u, v \in C([a,b];X)$. But, this shows that, for $n \in \mathbb{N}^*$ large enough, Q^n is a strict contraction. By virtue of Banach fixed point theorem, Q^n has a unique fixed point $u \in C([a,b];X)$. But

$$\|Qu - u\| = \|Q^n Qu - Q^n u\| \le q\|Qu - u\|,$$

where $q \in (0,1)$. Hence $Qu = u$. But u is the only fixed point of Q, because each fixed point of Q is a fixed point for Q^n, which is a strict contraction. As each fixed point of Q is a solution of the problem (10.2.2) and conversely, this proves the first part of lemma.

Let now h_1 and h_2 be two elements in $L^1(a,b;X)$. Let us observe that

$$\|(\mathcal{S}(h_1))(t)-(\mathcal{S}(h_2))(t)\| \le \left\|\int_a^t (h_1(s)-h_2(s))\,ds\right\|$$

$$+\int_a^t \|g(s,(\mathcal{S}(h_1))(s))-g(s,(\mathcal{S}(h_2))(s))\|\,ds$$

$$\le \|H_1-H_2\|_{C([\,a,b\,];X)}+L\int_a^t \|(\mathcal{S}(h_1))(s)-(\mathcal{S}(h_2))(s)\|\,ds$$

for each $t\in[\,a,b\,]$. From this, and Gronwall Lemma 10.2.1 we deduce

$$\|(\mathcal{S}(h_1))(t)-(\mathcal{S}(h_2))(t)\| \le e^{\delta L}\|H_1-H_2\|_{C([\,a,b\,];X)}.$$

Passing to the supremum for $t\in[\,a,b\,]$, we obtain (10.2.3), and this achieves the proof. □

Lemma 10.2.3. *Let $f:\mathbb{I}\times X\to X$ be a continuous function, with $f(\mathbb{I}\times X)$ relatively compact, and let $g:\mathbb{I}\times X\to X$ be continuous on $\mathbb{I}\times X$, and Lipschitz on X. Then, for each $(a,\xi)\in\mathbb{I}\times X$, and each $b>a$ with $[\,a,b\,]\subset\mathbb{I}$, the problem* (10.2.1) *has at least one solution defined on $[\,a,b\,]$.*

Proof. Let $\lambda>0$. In view of Lemma 10.2.2, it follows that the problem

$$\begin{cases} u'_\lambda = g(t,u_\lambda)+f(t,u_\lambda(\cdot-\lambda)) \\ u_\lambda(t)=\xi & \text{for } t\in[\,a-\lambda,a\,], \end{cases}$$

which is equivalent with the integral equation with the delay λ

$$u_\lambda(t)=\begin{cases} \xi & \text{in } [a-\lambda,a] \\ \xi+\displaystyle\int_a^t g(s,u_\lambda(s))\,ds+\int_a^t f(s,u_\lambda(s-\lambda))\,ds & \text{in } [a,b], \end{cases} \tag{10.2.4}$$

has a unique continuous solution u_λ. Let now $\lambda=1/n$, and let us denote by u_n the corresponding solution of (10.2.4). Let us denote by $F_n:[\,a,b\,]\to X$ the function defined by

$$F_n(t)=\int_a^t f(s,u_n(s-1/n))\,ds$$

for each $n\in\mathbb{N}^*$ and each $t\in[\,a,b\,]$. As $f(\mathbb{I}\times X)$ is relatively compact, from Lemma A.1.3, it follows that, for each $t\in[\,a,b\,]$, the set $\{F_n(t)\,;\ n\in\mathbb{N}^*\}$ is relatively compact in X. Moreover, $\{F_n\,;\ n\in\mathbb{N}^*\}$ is equicontinuous on $[\,a,b\,]$. Indeed, as f is bounded on $\mathbb{I}\times X$, there exists $M>0$ such that

$$\|f(t,u)\|\le M$$

for each $(t,u) \in \mathbb{I} \times X$. We then have

$$\|F_n(t) - F_n(s)\| \leq M|t-s|$$

for each $n \in \mathbb{N}^*$ and each $t, s \in [a,b]$. In view of Theorem A.2.1, there exists $F \in C([a,b];X)$ such that, on a subsequence at least,

$$\lim_{n\to\infty} F_n(t) = F(t),$$

uniformly for $t \in [a,b]$. Let us observe that, by virtue of (10.2.4), we have

$$u_n(t) = \xi + \int_a^t g(s, u_n(s))\,ds + \int_a^t f(s, u_n(s - 1/n))\,ds, \tag{10.2.5}$$

relation which, according to (10.2.3) and to the definition of the function F_n, leads to

$$\|u_n - u_m\|_{C([a,b];X)} \leq e^{\delta L}\|F_n - F_m\|_{C([a,b];X)}$$

for each $n, m \in \mathbb{N}^*$. Inasmuch as $(F_n)_{n\in\mathbb{N}^*}$ is a Cauchy sequence, it follows that $(u_n)_{n\in\mathbb{N}^*}$ has the same property, and therefore it is convergent in $C([a,b];X)$ to some function u. Passing to the limit in (10.2.5), and using Corollary A.2.2, we deduce that u is a solution of the problem (10.2.1) defined on $[a,b]$, and this completes the proof. □

We can now proceed to the proof of Theorem 10.2.1.

Proof. Let $(a,\xi) \in \mathcal{D}$. Since $\mathcal{D}$ is open, there exist c, b and $r > 0$ with $c < a < b$, such that

$$[c,b] \times B(\xi, r) \subset \mathcal{D}.$$

In addition, as g is continuous on $\mathcal{D}$ and locally Lipschitz with respect to its last argument, diminishing r if necessary, we may assume that there exist $M > 0$ and $L > 0$ such that

$$\|g(t,u)\| \leq M$$

for each $(t,u) \in [c,b] \times B(\xi,r)$, and

$$\|g(t,u) - g(t,v)\| \leq L\|u - v\|$$

for each $(t,u), (t,v) \in [c,b] \times B(\xi,r)$. Let us define $\rho : X \to X$ by

$$\rho(y) = \begin{cases} y & \text{for } y \in B(\xi,r) \\ \dfrac{r}{\|y-\xi\|}(y-\xi) + \xi & \text{for } y \in X \setminus B(\xi,r), \end{cases}$$

and let us observe that ρ maps X in $B(\xi,r)$ and is Lipschitz continuous on X with Lipschitz constant 2. Next, we define $f_r : [c,b] \times X \to X$ by

$$f_r(t,y) = f(t,\rho(y)), \text{ for each } (t,y) \in [c,b] \times X,$$

and $g_r : [c,b] \times X \to X$ by $g_r(t,y) = g(t,\rho(y))$ for each $(t,y) \in [c,b] \times X$. As f is b-compact, $f([c,b] \times B(\xi,r))$ is relatively compact. Accordingly, f_r is continuous and $f_r((c,b) \times X)$ is relatively compact. Analogously, from the choice of c, b and r, we deduce that g_r is continuous and bounded on $[c,b] \times X$, and Lipschitz continuous on X with Lipschitz constant $2L$. From Lemma 10.2.3, we know that the Cauchy problem

$$\begin{cases} u' = f_r(t,u) + g_r(t,u) \\ u(a) = \xi \end{cases}$$

has at least one solution $u : [a,b] \to X$. As u is continuous at $t = a$ and $u(a) = \xi$, diminishing $b > a$ if necessary, we have $\|u(t) - \xi\| \leq r$ for each $t \in [a,b]$. But this means that we have both $f_r(t,u(t)) = f(t,u(t))$, and $g_r(t,u(t)) = g(t,u(t))$. Consequently $u : [a,b] \to X$ is a solution of the problem (10.2.1), and this achieves the proof. □

10.3. Saturated Solutions

Let $\mathcal{D}$ be a nonempty and open subset in $\mathbb{R} \times X$, let $f : \mathcal{D} \to X$ be a given function, and $(a,\xi) \in \mathcal{D}$. Let us consider the Cauchy problem

$$\begin{cases} u' = f(t,u) \\ u(a) = \xi. \end{cases} \tag{10.3.1}$$

A solution $u : [a,b) \to X$ of (10.3.1) is called *continuable* if there exists another solution $v : [a,c) \to X$ with $b < c$ and such that $u(t) = v(t)$ for each $t \in [a,b)$. A solution $u : [a,b) \to X$ is called *saturated* if it is not continuable. If the projection of $\mathcal{D}$ on $\mathbb{R}$ contains $\mathbb{R}_+$, a solution u is called *global* if it is defined on $[a,+\infty)$.

Remark 10.3.1. We emphasize that all the results within this section hold true, with essentially the same proofs, also for the Cauchy problem

$$\begin{cases} u' = f(t,u) + g(t,u) \\ u(a) = \xi, \end{cases}$$

where f and g are as in Section 10.2, but, for the simplicity of writing, we consider here only this simpler case corresponding to $g \equiv 0$ and analyzed in Section 10.1.

Lemma 10.3.1. *Let $f : \mathcal{D} \to X$ be a b-compact function on $\mathcal{D}$. Then a solution $u : [a,b) \to X$ of* (10.3.1) *is continuable if and only if there exists*

$$u^* = \lim_{t \uparrow b} u(t) \tag{10.3.2}$$

and

$$(b,u^*) \in \mathcal{D}. \tag{10.3.3}$$

Proof. The necessity is obvious, while the sufficiency is a consequence of Theorem 10.1.1. □

Remark 10.3.1. From Lemma 10.3.1 it follows that each saturated solution of (10.3.1) is necessarily defined on an interval of the form $[a, b)$, i.e. on an interval which is open at the right.

A sufficient condition for the existence of the limit (10.3.2) is stated below.

Proposition 10.3.1. *Let $u : [a, b) \to X$ be a solution of* (10.3.1) *and let us assume that $b < +\infty$, and there exists $M > 0$ such that*

$$\|f(\tau, u(\tau))\| \leq M,$$

for each $\tau \in [a, b)$. Then there exists

$$u^* = \lim_{t \uparrow b} u(t).$$

Proof. Let us observe that, for each $t, s \in [a, b)$, we have

$$\|u(t) - u(s)\| \leq \left| \int_s^t \|f(\tau, u(\tau))\| d\tau \right| \leq M|t - s|,$$

and thus u satisfies the hypothesis of the Cauchy test on the existence of the limit at b. □

We may now proceed to the statement of a characterization of continuable solutions of (10.3.1).

Theorem 10.3.1. *Let $f : \mathcal{D} \to X$ be b-compact and let $u : [a, b) \to X$ be a solution of* (10.3.1). *A necessary and sufficient condition in order for u to be continuable is that the graph of u, i.e.*

$$\text{graph}\, u = \{(t, u(t)) \in \mathbb{R} \times X;\ t \in [a, b)\}$$

is included in a compact subset in $\mathcal{D}$.

Proof. *Necessity.* Assume that u is continuable. Then u may be extended by continuity to $[a, b]$. Denote this extension by v and let us observe that the mapping $t \mapsto (t, v(t))$ is continuous from $[a, b]$ to $\mathcal{D}$. Therefore, its range, which coincides with graph v is compact and included in $\mathcal{D}$. Since graph $u \subset$ graph v, this completes the proof of the necessity.
Sufficiency. Assume that graph u is included into a compact subset of $\mathcal{D}$. Then f is bounded on graph u, i.e. there exists $M > 0$ such that

$$\|f(\tau, u(\tau))\| \leq M,$$

for each $\tau \in [a, b)$. The conclusion is an easy consequence of Lemma 10.3.1 and Proposition 10.3.1. □

We continue with a fundamental existence result concerning saturated solutions for (10.3.1).

Theorem 10.3.2. *If* $f : \mathcal{D} \to X$ *is b-compact and* $u : [a, b) \to X$ *is a solution of* (10.3.1), *then either u is saturated, or u can be continued up to a saturated one.*

Proof. If u is saturated we have nothing to prove. Thus, let us assume that u is continuable and let us define $\mathcal{S}$ as the set of all solutions of (10.3.1) which extend u. Obviously, $u \in \mathcal{S}$ and thus $\mathcal{S}$ is nonempty. Moreover, since u is continuable, $\mathcal{S}$ contains at least two elements. On $\mathcal{S}$ let us define the relation " $\preceq$ " by $v \preceq w$ if w extends v. It is a simple exercise to show that $(\mathcal{S}, \preceq)$ is an inductively ordered set. So, from Zorn's Lemma, there exists at least one maximal element $u_m \in \mathcal{S}$ such that $u \preceq u_m$. From the definition of " $\preceq$ ", and from the maximality of u_m it follows that u_m is a saturated solution of (10.3.1) which coincides with u on the common part of the domains, thereby completing the proof. □

From Theorems 2.3.1 and 10.3.2 it follows:

Corollary 10.3.1. *If* $f : \mathcal{D} \to X$ *is b-compact, then for each initial data* $(a, \xi) \in \mathcal{D}$, (10.3.1) *has at least one saturated solution.*

We recall that a *limit point* of a function $u : [a, b) \to X$ as t tends to b is any element u^* in X for which there exists a sequence $(t_k)_{k \in \mathbb{N}}$ in $[a, b)$ tending to b and such that $\lim_{k \to \infty} u(t_k) = u^*$. We denote the set of all limit points of u as t tends to b by $\mathrm{Lim}_{t \uparrow b} u(t)$.
Concerning the behavior of saturated solutions at the right end point of their interval of definition, we have the following fundamental result.

Theorem 10.3.3. *Let* $f : \mathcal{D} \to X$ *be b-compact and let* $u : [a, b) \to X$ *be a saturated solution of* (10.3.1). *Then either*

(i) *u is unbounded or*
(ii) *u is bounded and, either u is global, or*
(iii) *u is bounded and non-global, and in this case, either* $\mathrm{Lim}_{t \uparrow b} u(t)$ *is empty, or for each* $u^* \in \mathrm{Lim}_{t \uparrow b} u(t)$, $(b, u^*) \in \partial\mathcal{D}$.

Proof. If u is unbounded on $[a, b)$, or if it is bounded and global, or bounded and $\mathrm{Lim}_{t \uparrow b} u(t)$ is empty, we have nothing to prove. Thus, let us assume that u is bounded on $[a, b)$ and not global, i.e. $b < +\infty$ and $\mathrm{Lim}_{t \uparrow b} u(t)$ is nonempty. To prove (iii) let us assume by contradiction that there exists at least one sequence $(t_k)_{k \in \mathbb{N}}$ in $[a, b)$ tending to b and such that $(u(t_k))_{k \in \mathbb{N}}$ is convergent to some $u^* \in X$, but $(b, u^*) \notin \partial\mathcal{D}$. So, we necessarily have $(b, u^*) \in \mathcal{D}$. In view of Lemma 10.3.1, to get a contradiction, it suffices to show that there exists $\lim_{t\uparrow b} u(t)$, which of course must coincide with u^*. To this aim, let us observe that, since $\mathcal{D}$ is open and $(b, u^*) \in \mathcal{D}$, there exists $c > b$ and $r > 0$ such that $[a, c) \times B(u^*, r) \subset \mathcal{D}$. Furthermore, inasmuch as $b < +\infty$, and f is continuous on $\mathcal{D}$, diminishing $r > 0$ if necessary, it follows that there exists $M > 0$ such that

$$\|f(\tau, v)\| \leq M, \tag{10.3.4}$$

for each $(\tau, v) \in [a, b] \times B(u^*, r)$. As $\lim_{k\to\infty} t_k = b$ and $\lim_{k\to\infty} u(t_k) = u^*$, we may choose $k \in \mathbb{N}$ such that

$$\begin{cases} b - t_k < \dfrac{r}{2M} \\ \|u(t_k) - u^*\| < \dfrac{r}{2}. \end{cases} \tag{10.3.5}$$

Fix k with the properties above. We shall show that for each $t \in [t_k, b)$, we have $u(t) \in B(u^*, r)$. Let

$$t^* = \sup\{t \in [t_k, b);\ u(s) \in B(u^*, r),\ \text{for } s \in [t_k, t]\}.$$

If $t^* = b$, the statement above is obviously true. Let us assume by contradiction that $t^* < b$. This means that $u(t) \in B(u^*, r)$ for each $t \in [t_k, t^*]$, $\|u(t^*) - u^*\| = r$ and there exist points $t > t^*$, as close to t^* as we wish, satisfying $\|u(t) - u^*\| > r$. In other words, t^* is the "first moment in (t_k, b) after which u leaves the set $B(u^*, r)$". The condition $\|u(t^*) - u^*\| = r$ signifies that, at t^*, u must cross the boundary of $B(u^*, r)$. Next, from the remark above combined with (10.3.4), and (10.3.5), we get

$$r = \|u(t^*) - u^*\| \leq \|u(t^*) - u(t_k)\| + \|u(t_k) - u^*\|$$

$$\leq \int_{t_k}^{t^*} \|f(\tau, u(\tau))\| d\tau + \|u(t_k) - u^*\| \leq (t^* - t_k)M + \|u(t_k) - u^*\|$$

$$\leq (b - t_k)M + \|u(t_k) - u^*\| < \frac{r}{2} + \frac{r}{2} = r.$$

This contradiction ($r < r$) is a consequence of our supposition that, for at least one $t \in [t_k, b)$, $u(t) \notin B(u^*, r)$. Then, for each $t \in [t_k, b)$, we must have $u(t) \in B(u^*, r)$. In view of (10.3.4) combined with Proposition 10.3.1 and Lemma 10.3.1, it follows that u is continuable thereby contradicting the hypothesis. This contradiction can be eliminated only if $u^* \in \partial\mathcal{D}$. The proof is complete. □

Under an additional hypothesis on f we shall prove a sharper result.

Theorem 10.3.4. *Let $f : \mathcal{D} \to X$ be b-compact and let us assume that it maps bounded subsets in $\mathcal{D}$ into bounded subsets in X. Let $u : [a, b) \to X$ be a saturated solution of* (10.3.1). *Then either*

(i') *u is unbounded and, if $b < +\infty$, $\lim_{t\uparrow b} \|u(t)\| = +\infty$, or*
(ii) *u is bounded and, either u is global, or*
(iii') *u is bounded and non-global, and in this latter case there exists $\lim_{t\uparrow b} u(t) = u^*$, and $(b, u^*) \in \partial\mathcal{D}$.*

Proof. In view of Theorem 10.3.3, the only fact we have to prove here is that whenever (ii) does not hold, then one of the two conditions (i') or (iii') must hold. So, let us assume for the beginning that (i') and (ii) do not hold. Then, from Lemma 10.3.1 and Proposition 10.3.1, it follows that there exists

$$\lim_{t\uparrow b} u(t) = u^*,$$

while from (iii) in Theorem 10.3.3 we know that $(b, u^*) \in \partial\mathcal{D}$. Thus (iii') holds.
Let us assume now that (ii) and (iii') do not hold and that $b < +\infty$. To show that $\lim_{t\uparrow b} \|u(t)\| = +\infty$, let us assume by contradiction that, under these circumstances, we have

$$\liminf_{t\uparrow b} \|u(t)\| < +\infty.$$

In other words, there exist at least one sequence $(t_k)_{k\in\mathbb{N}}$ in (a, b), convergent to b, and $r > 0$ such that $\|u(t_k)\| \leq r$, for each $k \in \mathbb{N}$.
Let $C = \{v \in \Omega;\ \|v\| \leq r + 1\}$. Since f maps bounded subsets in $\mathcal{D}$ into bounded subsets in X, $b < +\infty$ and C is bounded, there exists $M > 0$, such that

$$\|f(\tau, v)\| \leq M, \tag{10.3.6}$$

for each $(\tau, v) \in [a, b) \times C$. Now let us choose a number $d > 0$ satisfying

$$dM < 1, \tag{10.3.7}$$

and fix $k \in \mathbb{N}$ such that $b - d < t_k < b$. Since u is unbounded on $[a, b)$, it is necessarily unbounded on $[t_k, b)$. Then there exists $t^* \in (t_k, b)$ such that

$$\|u(\tau)\| < r + 1, \tag{10.3.8}$$

for each $\tau \in [t_k, t^*)$ and

$$\|u(t^*)\| = r + 1.$$

In fact, t^* is "the first moment in (t_k, b) at which u leaves $B(0, r + 1)$." Let us observe that

$$r + 1 = \|u(t^*)\| = \|u(t_k) + \int_{t_k}^{t^*} f(\tau, u(\tau))d\tau\|$$

$$\leq \|u(t_k)\| + \int_{t_k}^{t^*} \|f(\tau, u(\tau))\|d\tau.$$

Taking into account (10.3.6), (10.3.7), (10.3.8), and the obvious inequality $t^* - t_k < d$, we get

$$r + 1 \leq r + (t^* - t_k)M < r + dM < r + 1.$$

This contradiction can be eliminated only if

$$\lim_{t \uparrow b} \|u(t)\| = +\infty$$

and this completes the proof. □

Remark 10.3.2. If $\mathcal{D} = \mathbb{R}_+ \times X$, then, each compact function $f : \mathcal{D} \to X$ maps bounded subsets in $\mathcal{D}$ into bounded subsets in X.

Corollary 10.3.2. *Let $f : \mathbb{R}_+ \times X \to X$ be compact. Then a saturated solution $u : [\,a, b) \to X$ of* (10.3.1) *is either global, or it is not global, and in this case there exists*

$$\lim_{t \uparrow b} \|u(t)\| = +\infty.$$

Proof. Let us observe first that, if $b < +\infty$, u is necessarily unbounded on $[\,a, b)$. Indeed, if we assume the contrary, u has at least one limit point u^* as t tends to b. By Theorem 10.3.4, (b, u^*) must lie on the boundary of $\mathbb{R} \times X$ which is the empty set. Thus, the supposition that u is bounded on $[\,a, b)$ is false. The conclusion of Corollary 10.3.2 is then an easy consequence of Remark 10.3.2 combined with Theorem 10.3.4. □

If $b < +\infty$ and $\lim_{t \uparrow b} \|u(t)\| = +\infty$, we say that u *blows up in finite time.*

Corollary 10.3.3. *Let $f : \mathbb{R}_+ \times X \to X$ be compact. A necessary and sufficient condition in order for a solution $u : [\,a, b) \to X$ of* (10.3.1) *to be continuable is that $b < +\infty$ and u be bounded on $[\,a, b)$.*

Proof. The necessity is trivial, while the sufficiency is a reformulation of Corollary 10.3.2. □

We conclude this section with a sufficient conditions on f ensuring the existence of global solutions of (10.3.1).

Theorem 10.3.5. *Let $f : \mathbb{R}_+ \times X \to X$ be compact, and let us assume that there exist two continuous functions*[1] *$h, k : \mathbb{R}_+ \to \mathbb{R}_+$ such that*

$$\|f(\tau, v)\| \leq k(\tau)\|v\| + h(\tau), \tag{10.3.9}$$

for each $(\tau, v) \in \mathbb{R}_+ \times X$. Then, for each $(a, \xi) \in \mathbb{R}_+ \times X$, (10.3.1) *has at least one global solution.*

---

[1] In fact we can only ask that these functions belong to $\mathcal{L}^1_{\text{loc}}(\mathbb{R}_+)$.

Proof. Thanks to Corollary 10.3.1, it suffices to show that each saturated solution of (10.3.1) is global. To this aim, let $u : [a,b) \to X$ be a saturated solution of (10.3.1). From (10.3.9), we get

$$\|u(t)\| \leq \|\xi\| + \int_a^t h(\tau)d\tau + \int_a^t k(\tau)\|u(\tau)\|d\tau,$$

for each $t \in [a,b)$. We will show next that $b = +\infty$. Indeed, if we assume the contrary, inasmuch as $[a,b]$ is compact and h, k are continuous on $[a,b] \subset \mathbb{R}_+$, there exists $M > 0$ such that

$$h(t) \leq M \quad \text{and} \quad k(t) \leq M,$$

for each $t \in [a,b]$. The preceding inequalities and Gronwall's Lemma 10.2.1 show that

$$\|u(t)\| \leq [\|\xi\| + M(b-a)]e^{M(b-a)},$$

for each $t \in (a,b)$. Hence u is bounded on $[a,b)$ and therefore it has at least one limit point u^* as t tends to b. By Theorem 10.3.3, (b, u^*) must lie on the boundary of $\mathbb{R}_+ \times X$ which is the empty set. This contradiction can be eliminated only if $b = +\infty$, and this completes the proof. □

10.4. The Klein-Gordon Equation

In this section we shall present an interesting application of Theorem 10.1.1. Let X be a real Banach space and $A : D(A) \subset X \to X$ the generator of a C_0-group $\{G(t)\,;\ t \in \mathbb{R}\}$, let $\mathcal{D}$ be a nonempty and open subset in $\mathbb{R} \times X$, let $f : \mathcal{D} \to X$ be a continuous function and let us consider the problem

$$\begin{cases} u' = Au + f(t,u) \\ u(a) = \xi. \end{cases} \tag{10.4.1}$$

We begin with the following simple, but useful lemma.

Lemma 10.4.1. *A function $u : [a,b] \to X$ is a C^0-solution for the problem* (10.4.1) *if and only if the function $v : [a,b] \to X$, defined by*

$$v(t) = G(-t+a)u(t)$$

for each $t \in [a,b]$, is a C^1-solution of the problem

$$\begin{cases} v' = g(t,v) \\ v(a) = \xi, \end{cases} \tag{10.4.2}$$

where $g : \mathcal{E} \to X$ is defined by

$$g(t,v) = G(-t+a)f(t, G(t-a)v)$$

for each $(t,v) \in \mathcal{E}$, where $\mathcal{E} = \{(t, G(t-a)v)\,;\ (t,v) \in \mathcal{D}\}$.

Proof. First, let us recall that $u : [a,b] \to X$ is a C^0-solution of (10.4.1) if $(t, u(t)) \in \mathcal{D}$ and

$$u(t) = G(t-a)\xi + \int_a^t G(t-s)f(s, u(s))\, ds$$

for each $t \in [a,b]$. Consequently we get

$$G(-t+a)u(t) = \xi + \int_a^t G(-s+a)f(s, u(s))\, ds$$

for $t \in [a,b]$. Let $v : [a,b] \to X$ be defined by $v(t) = G(-t+a)u(t)$ for $t \in [a,b]$. Obviously it satisfies

$$v(t) = \xi + \int_a^t G(-s+a)f(s, G(s-a)v(s))\, ds$$

for each $t \in [a,b]$, and this achieves the proof. □

Theorem 10.4.1. *Let $A : D(A) \subset X \to X$ be the generator of a C_0-group $\{G(t);\ t \in \mathbb{R}\}$, $\mathcal{D}$ a nonempty and open subset in X, and let $f : \mathcal{D} \to X$ be a b-compact function. Then, for each $(a,\xi) \in \mathcal{D}$, there exists $b > a$ such that* (10.4.1) *has at least one C^0-saturated solution defined on $[a,b)$.*

Proof. As f is b-compact, we deduce that the function g, defined in Lemma 10.4.1, is b-compact too. An appeal to Theorems 10.1.1 and 10.3.2 completes the proof. □

Example 10.4.1. Let Ω be a nonempty, bounded and open subset in $\mathbb{R}^n$ whose boundary Γ is of class C^2, let $Q_T = (0,T) \times \Omega$ and $\Sigma_T = (0,T) \times \Gamma$. Let $\lambda > 0$ be the first eigenvalue of $-\Delta$ on $H^{-1}(\Omega)$, i.e.

$$\lambda = \inf\{\|\nabla u\|^2_{L^2(\Omega)}\ ;\ u \in H^1_0(\Omega),\ \|u\|_{L^2(\Omega)} = 1\}$$

and let $m > -\lambda$. We consider the nonlinear hyperbolic equation

$$\begin{cases} u_{tt} = \Delta u - mu + g(t,x,u) & (t,x) \in Q_T \\ u(t,x) = 0 & (t,x) \in \Sigma_T \\ u(0,x) = u_0(x) & x \in \Omega, \\ u_t(0,x) = v_0(x) & x \in \Omega \end{cases} \tag{10.4.3}$$

where $g : \mathbb{R}_+ \times \Omega \times \mathbb{R} \to \mathbb{R}$ is a continuous function. This equation is known as the *semilinear Klein-Gordon equation.*

Theorem 10.4.2. *Let $g : \mathbb{R}_+ \times \Omega \times \mathbb{R} \to \mathbb{R}$ be a continuous function for which there exist $c > 0$, $d > 0$, and $\alpha \in \mathbb{R}$, such that*

$$|g(t,x,u)| \le c|u|^\alpha + d \tag{10.4.4}$$

for $(t, x, u) \in \mathbb{R}_+ \times \Omega \times \mathbb{R}$, *where* $\alpha \geq 0$ *if* $n = 2$, *and* $\alpha < n/(n-2)$ *if* $n \geq 3$. *Then, for each* $u_0 \in H_0^1(\Omega)$ *and each* $v_0 \in L^2(\Omega)$, *there exists* $T > 0$ *such that the problem* (10.4.3) *has at least one saturated solution* u *satisfying*

(i) $u \in C([0, T); H_0^1(\Omega))$

(ii) $u_t \in C([0, T); L^2(\Omega))$.

In addition, if $T < +\infty$, *then*

$$\lim_{s \uparrow T} \left(\|u(s)\|_{H_0^1(\Omega)} + \|u_t(s)\|_{L^2(\Omega)} \right) = +\infty.$$

If $n = 1$, *the conclusion remains valid without condition* (10.4.4).

Proof. First, we observe that (10.4.3) can be rewritten as a first-order ordinary differential equation in an infinite-dimensional Hilbert space. Let

$$H = H_0^1(\Omega) \times L^2(\Omega)$$

which, endowed with the inner product $\langle \cdot, \cdot \rangle$, defined by

$$\langle (u, v), (\tilde{u}, \tilde{v}) \rangle = \int_\Omega u'(x)\tilde{u}'(x)\, dx + m \int_\Omega u(x)\tilde{u}(x)\, dx + \int_\Omega v(x)\tilde{v}(x)\, dx$$

for each $(u, v), (\tilde{u}, \tilde{v}) \in H$, is a real Hilbert space. We define the operator $A : D(A) \subset H \to H$ by

$$\begin{cases} D(A) = (H^2(\Omega) \cap H_0^1(\Omega)) \times H_0^1(\Omega) \\ A(u, v) = (v, \Delta u - mu) \ \text{ for each } (u, v) \in D(A). \end{cases}$$

Furthermore, let us define $f : \mathbb{R}_+ \times D(f) \to H$ by

$$\begin{cases} D_f = \{(u, v) \in H;\ g(t, \cdot, u(\cdot)) \in L^2(\Omega) \text{ for each } t \in [0, +\infty)\} \\ f(t, (u, v))(x) = (0, g(t, x, u(x))) \end{cases}$$

for each $t \in [0, +\infty)$, each $(u, v) \in D_f$ and a.e. for $x \in \Omega$. At this point, let us observe that the problem (10.4.3) can be rewritten under the equivalent form

$$\begin{cases} z' = Az + f(t, z) \\ z(a) = \xi, \end{cases}$$

where A and f are as above, $z(t)(x) = (u(t, x), v(t, x))$ a.e. for (t, x) in $(0, T) \times \Omega$ and $\xi = (u_0, v_0)$. In order to prove that A and f satisfy the general conditions in Theorem 10.4.1, let us remark first that, by virtue of Theorem 4.6.4, A generates a C_0-group of isometries.

As concerns f, from (10.4.4), Theorem 1.5.4 and Lemma A.6.1, it follows that $D_f = H$ and f is continuous on $\mathbb{R} \times H$. Again from Theorem 1.5.4, we know that $H_0^1(\Omega)$ is compactly imbedded in: $C(\overline{\Omega})$ if $n = 1$, in $L^q(\Omega)$ for each $q \geq 1$ if $n = 2$, and $q < 2n/(n-2)$ if $n \geq 3$. Thus f is compact, and we are in the hypotheses of Theorem 10.4.1, and Corollaries 10.3.1 and 10.3.2. The proof is complete. □

Remark 10.4.1. Using the same arguments and Theorem 10.3.4, one may prove a global existence result for the *Sine-Gordon equation*

$$\begin{cases} u_{tt} = \Delta u + \sin u & (t,x) \in Q_T \\ u(t,x) = 0 & (t,x) \in \Sigma_T \\ u(0,x) = u_0(x) & x \in \Omega, \\ u_t(0,x) = v_0(x) & x \in \Omega. \end{cases}$$

This equation is important in the study of the transformation of surfaces with constant negative curvature.

10.5. An Application to a Problem in Mechanics

The movement of a continuous medium, having the domain of reference a nonempty open and bounded subset Ω in $\mathbb{R}^n$, is described by the following pseudoparabolic partial differential equation

$$\begin{cases} u_t = \alpha \Delta u_t + \beta \Delta u + f(t,x) & (t,x) \in Q_\infty \\ u = 0 & (t,x) \in \Sigma_\infty \\ u(0,x) = g(x) & x \in \Omega, \end{cases} \tag{10.5.1}$$

where $\alpha > 0$, $\beta > 0$, $f \in C(\mathbb{R}_+; L^2(\Omega))$, $g \in H^2(\Omega) \cap H_0^1(\Omega)$, $Q_\infty = \mathbb{R}_+ \times \Omega$ and u is the field of velocities.

Theorem 10.5.1. *In the general hypotheses above, the problem* (10.5.1) *has a unique solution*

$$u \in C^1(\mathbb{R}_+; H^2(\Omega) \cap H_0^1(\Omega)).$$

Proof. Let us observe that (10.5.1) can be rewritten as the following implicit differential equation in the space $L^2(\Omega)$

$$\begin{cases} [(I - \alpha A)\, u]' = \beta A u + f(t) \\ u(0) = g, \end{cases} \tag{10.5.2}$$

where $A : D(A) \subseteq L^2(\Omega) \to L^2(\Omega)$ is defined by

$$\begin{cases} D(A) = H_0^1(\Omega) \cap H^2(\Omega) \\ Au = \Delta u \text{ for each } u \in D(A). \end{cases}$$

Let us denote by $v = (I - \alpha A)u$ and let us observe that

$$\beta A(I - \alpha A)^{-1} = \frac{\beta}{\alpha}\left[I - (I - \alpha A)\right](I - \alpha A)^{-1} = \frac{\beta}{\alpha}(I - \alpha A)^{-1} - \frac{\beta}{\alpha} I.$$

So, the problem (10.5.2) can be reformulated as

$$\begin{cases} v' = \omega(I - \alpha A)^{-1} v - \omega v + f(t) \\ v(0) = (I - \alpha A) g, \end{cases} \tag{10.5.3}$$

where $\omega = \beta\alpha^{-1} > 0$. As $\omega(I - \alpha A)^{-1} - \omega I$ is linear bounded, it is globally Lipschitz on $L^2(\Omega)$. By virtue of Lemma 10.2.2, the problem (10.5.3) has a unique solution $v \in C^1(\mathbb{R}_+; L^2(\Omega))$. The fact that $u = (I - \alpha A)^{-1}v$ satisfies

$$u \in C^1(\mathbb{R}_+; H^2(\Omega) \cap H_0^1(\Omega))$$

follows from the remark that $(I - \alpha A)^{-1}$ is linear continuous from $L^2(\Omega)$ to $H^2(\Omega) \cap H_0^1(\Omega)$, and this completes the proof. □

Let us consider now the semilinear version of the problem (10.5.1)

$$\begin{cases} u_t = \alpha\Delta u_t + \beta\Delta u + f(t,x,u) & (t,x) \in Q_\infty \\ u = 0 & (t,x) \in \Sigma_\infty \\ u(0,x) = g(x) & x \in \Omega, \end{cases} \tag{10.5.4}$$

where $\alpha > 0$, $\beta > 0$, $f : \mathbb{R}_+ \times \overline{\Omega} \times \mathbb{R}$ is continuous and $g \in H^2(\Omega) \cap H_0^1(\Omega)$.

Theorem 10.5.2. *In the general hypotheses above, let us assume that there exists $c > 0$, $d \geq 0$ such that*

$$|f(t,x,u)| \leq c|u| + d$$

for all $(t,x,u) \in \mathbb{R}_+ \times \overline{\Omega} \times \mathbb{R}$. Then there exists $T > 0$ such that problem (10.5.4) *has at least one solution*

$$u \in C^1([0,T]; H^2(0,\pi)).$$

If in addition, for each $(t,x) \in \mathbb{R}_+ \times \Omega$, $f(t,x,\cdot) : \mathbb{R} \to \mathbb{R}$ is decreasing, then u is the only solution of the problem (10.5.4) *on* $[0,T]$.

Proof. Repeating the arguments in the proof of Theorem 10.5.1, we deduce that (10.5.4) can be rewritten under the form

$$\begin{cases} v' = \omega(I - \alpha A)^{-1}v - \omega v + f(t, (I - \alpha A)^{-1}v) \\ v(0) = (I - \alpha A)g, \end{cases} \tag{10.5.5}$$

where $\omega = \beta\alpha^{-1} > 0$. Since $(I - \alpha A)^{-1}$ is continuous from $L^2(\Omega)$ to $H^2(\Omega) \cap H_0^1(\Omega)$, and $H^2(\Omega) \cap H_0^1(\Omega)$ is compactly imbedded in $C(\overline{\Omega})$, it follows that the function $F : \mathbb{R}_+ \times L^2(\Omega) \to L^2(\Omega)$, defined by

$$F(t,v)(x) = f(t,x,((I - \alpha A)^{-1}v)(x))$$

a.e. for $x \in \Omega$, is compact. Since $(I - \alpha A)^{-1} - \omega I$ is Lipschitz continuous, we are in the hypotheses of Theorem 8.5.1, and so there exists at least one local solution $u \in C^1([0,T]; H^2(\Omega) \cap H_0^1(\Omega))$.

Next, let u and v be two solutions of the problem (10.5.4). Subtracting side by side the corresponding two equalities, taking the inner product of both sides by $u - v$, recalling that $f(t, x, \cdot)$ is decreasing, and integrating over Ω, we obtain

$$\frac{1}{2}\frac{d}{dt}\left(\|u(t) - v(t)\|^2_{L^2(\Omega)} + \alpha\|u_x(t) - v_x(t)\|^2_{L^2(\Omega)}\right) = -\beta\|u_x(t) - v_x(t)\|^2_{L^2(\Omega)}.$$

Therefore,

$$\frac{\gamma}{2}\frac{d}{dt}\left(\|u(t) - v(t)\|^2_{L^2(\Omega)} + \|u_x(t) - v_x(t)\|^2_{L^2(\Omega)}\right) \leq 0$$

for each $t > 0$, where $\gamma = \min\{1, \alpha\} > 0$. Consequently, the function

$$t \mapsto \frac{\gamma}{2}\left(\|u(t) - v(t)\|^2_{L^2(\Omega)} + \|u_x(t) - v_x(t)\|^2_{L^2(\Omega)}\right)$$

is decreasing on $[0, T]$. As this function vanishes at $t = 0$, it is identically 0 on $[0, T]$, and this achieves the proof. □

Problems

Problem 10.1. Let X be a Banach space, $\mathbb{I}$ a nonempty and open interval, Ω nonempty and open subset in X, and $f : \mathbb{I}\times\Omega \to X$ a function satisfying:

(i) for a.e. for $t \in \mathbb{I}$, $f(t, \cdot) : \Omega \to X$ is continuous;
(ii) for each $x \in \Omega$, $f(\cdot, x) : \mathbb{I} \to X$ is measurable;
(iii) for each $\xi \in \Omega$, there exists $r > 0$, and a compact subset K in X such that $f(t, x) \in K$ for each $x \in B(\xi, r) \cap \Omega$, and a.e. for $t \in \mathbb{I}$.

Then, for each $a \in \mathbb{I}$, and $\xi \in \Omega$, there exists $b > a$ with $[a, b] \subset \mathbb{I}$, and such that the Cauchy problem

$$\begin{cases} u' = f(t, u) \\ u(a) = \xi \end{cases}$$

has at least one absolutely continuous solution $u : [a, b] \to \Omega$.

Problem 10.2. Let X be a Banach space, $\mathbb{I}$ a nonempty and open interval, and Ω a nonempty and open subset in X. Let $f, g : \mathbb{I} \times \Omega \to X$ be two functions satisfying (i) and (ii) in Problem 10.1. Assume, in addition, that f satisfies (iii) in the same problem, and g satisfies:

(iv) for each $\xi \in \Omega$ there exists $r > 0$, and a locally integrable function $\ell : \mathbb{I} \to \mathbb{R}_+$, such that

$$\|g(t, u) - g(t, v)\| \leq \ell(t)\|u - v\|$$

for each $u, v \in B(\xi, r) \cap \Omega$, and a.e. for $t \in \mathbb{I}$, and $g(\cdot, \xi) : \mathbb{I} \to X$ is locally integrable. Then, for each $a \in \mathbb{I}$, and each $\xi \in \Omega$, there exists $b > a$ with $[a, b] \subset \mathbb{I}$, and such that the Cauchy problem

$$\begin{cases} u' = f(t, u) + g(t, u) \\ u(a) = \xi \end{cases}$$

has at least one absolutely continuous solution $u : [a, b] \to \Omega$.

Notes. Theorem 10.1.1 in Section 10.1 is the most known extension to arbitrary infinite-dimensional Banach spaces of Peano's local existence theorem. For a survey of such extensions see Bogachev [**27**]. Our proof given here differs from the usual ones using either the Euler's polygonal lines method, or Schauder fixed point theorem in that is more direct and elementary. The proof of Theorem 10.2.1 in Section 10.2 is elementary, and as far as we know, new, avoiding the use of Krasnoselskii fixed point theorem in [**80**] which we recall below.

Theorem (Krasnoselskii) *Let K be a nonempty closed convex subset in a Banach space X and let $F, G : K \to X$ be two mappings satisfying*:

(i) $F(K) + G(K) \subset K$;
(ii) *F is continuous and compact*;
(iii) *G is a strict contraction, i.e. there exists $k \in (0, 1)$ such that $\|G(u) - G(v)\| \leq k\|u - v\|$ for all $u, v \in K$.*

Then $F + G$ has at least one fixed point in K.

For a proof using the latter method see Frigon and O'Reagan [**57**]. The results in Section 10.3 are simple extensions to the infinite-dimensional case of some qualitative results known for many decades within the theory of ordinary differential equations. As far as we know, in this form allowing non-uniqueness, Theorem 10.3.3 seems to be new. Regarding Section 10.4, the method of reducing a second order hyperbolic equation to an ordinary differential equation in an infinite-dimensional Hilbert space, equation which can be handled by Peano's local existence theorem, is due to Vrabie (1998) and described for the first time here. The application included in Section 10.5 is adapted from Showalter and Ting [**113**]. The two problems included are extensions to the Carathéodory case of the abstract existence results in Sections 10.1 and 10.2

CHAPTER 11

The Cauchy Problem for Semilinear Equations

Section 11.1 is mainly concerned with an existence and uniqueness result for the semilinear evolution equation $u' = Au + f(t, u)$ in the case in which A generates a C_0-semigroup of contractions and f is locally Lipschitz with respect to its last argument. In Section 11.2 we reconsider the same problem by assuming that A generates a C_0-compact semigroup of contractions and f is merely continuous. In Section 11.3 we present some basic results on the continuation of the solutions, while in Section 11.4 we prove a Poincaré-Liapunov-type theorem concerning the asymptotic behavior of solutions. In Sections 11.5 and 11.6 we include several examples illustrating the abstract theory.

11.1. The Problem $u' = Au + f(t, u)$ with f Lipschitz

Let X be a real Banach space, and $A : D(A) \subseteq X \to X$ the infinitesimal generator of a C_0-semigroup of contractions $\{S(t)\,;\ t \geq 0\}$. Let $\mathcal{D}$ be a nonempty and open subset in $\mathbb{R} \times X$, let $f : \mathcal{D} \to X$ a continuous function and $(a, \xi) \in \mathcal{D}$. We consider the Cauchy problem

$$\begin{cases} u' = Au + f(t, u) \\ u(a) = \xi. \end{cases} \tag{11.1.1}$$

Excepting for Section 11.4, here we confine ourselves to the simpler case of C_0-semigroups of contractions, although all the results in this chapter hold true in the general case, i.e. of the C_0-semigroups of type (M, ω). In analogy with the linear non homogeneous case, we introduce:

Definition 11.1.1. A C^0*-solution* of the problem (11.1.1) on $[\,a, b\,]$ is a continuous function $u : [\,a, b\,] \to X$ satisfying $(t, u(t)) \in \mathcal{D}$ and

$$u(t) = S(t - a)\xi + \int_a^t S(t - s) f(s, u(s))\, ds$$

for each $t \in [\,a, b\,]$.

Since the notions of *classical*, as well as of *strong*, solution for (11.1.1) are simply defined in analogy with the case $f(t,u) = f(t)$ (see Definitions 8.1.1 and 8.1.2), we do not give their explicit statements here.

Definition 11.1.2. The function $f : \mathcal{D} \to X$ is called:

(i) *locally Lipschitz with respect to its last argument* if for each (a,ξ) in $\mathcal{D}$ there exist $b > a$, $r > 0$ and $L_{a,\xi} = L > 0$ such that

$$[a,b] \times B(\xi,r) \subset \mathcal{D} \text{ and}$$
$$\|f(t,u) - f(t,v)\| \le L\|u - v\|$$

for each $(t,u),(t,v) \in [a,b] \times B(\xi,r)$;

(ii) *locally Lipschitz with respect to both arguments* if for each (a,ξ) in $\mathcal{D}$ there exist $b > a$, $r > 0$ and $L_{a,\xi} = L > 0$ such that

$$[a,b] \times B(\xi,r) \subset \mathcal{D} \text{ and}$$
$$\|f(t,u) - f(s,v)\| \le L(|t-s| + \|u-v\|)$$

for each $(t,u),(s,v) \in [a,b] \times B(\xi,r)$;

(iii) *globally Lipschitz with respect to its last argument* if there exists $L > 0$ such that

$$\|f(t,u) - f(t,v)\| \le L\|u - v\|$$

for each $(t,u),(t,v) \in \mathcal{D}$;

(iv) *globally Lipschitz with respect to both arguments* if there exists $L > 0$ such that

$$\|f(t,u) - f(s,v)\| \le L(|t-s| + \|u-v\|)$$

for each $(t,u),(s,v) \in \mathcal{D}$.

As we shall see later on, the next abstract local existence result concerns abstract semilinear problems of both "parabolic" and "hyperbolic" nature.

Theorem 11.1.1. *If $A : D(A) \subseteq X \to X$ generates a C_0-semigroup of contractions, $\{S(t)\,;\ t \ge 0\}$, and $f : \mathcal{D} \to X$ is continuous and locally Lipschitz with respect to its last argument, then, for each $(a,\xi) \in \mathcal{D}$, there exists $b > a$ such that the problem* (11.1.1) *has a unique C^0-solution defined on $[a,b]$.*

For the proof, we need the following lemma which is interesting by itself.

Lemma 11.1.1. *If $A : D(A) \subseteq X \to X$ generates a C_0-semigroup of contractions, $\{S(t)\,;\ t \ge 0\}$, and $f : [a,b] \times X \to X$ is continuous, bounded and globally Lipschitz with respect to its last argument then, for each $\xi \in X$, the problem* (11.1.1) *has a unique C^0-solution defined on $[a,b]$.*

Proof. Let $\xi \in X$ and let us define $Q : C([\,a,b\,];X) \to C([\,a,b\,];X)$ by

$$(Qu)(t) = S(t-a)\xi + \int_a^t S(t-s)f(s,u(s))\,ds$$

for each $u \in C([\,a,b\,];X)$, and each $t \in [\,a,b\,]$. Let us observe that, for each $u, v \in C([\,a,b\,];X)$ and $t \in [\,a,b\,]$, we have

$$\|(Qu)(t) - (Qv)(t)\| \leq L(t-a)\|u-v\|_{C([\,a,b\,];X)}.$$

From this inequality, by a simple inductive argument, we deduce that

$$\|(Q^n u)(t) - (Q^n v)(t)\| \leq \frac{L^n(t-a)^n}{n!}\|u-v\|_{C([\,a,b\,];X)}$$

for each $n \in \mathbb{N}^*$, each $u, v \in C([\,a,b\,];X)$, and each $t \in [\,a,b\,]$. Consequently

$$\|Q^n u - Q^n v\|_{C([\,a,b\,];X)} \leq \frac{L^n(b-a)^n}{n!}\|u-v\|_{C([\,a,b\,];X)}$$

for each $n \in \mathbb{N}^*$, and each $u, v \in C([\,a,b\,];X)$. But, this shows that, for $n \in \mathbb{N}^*$ large enough, Q^n is a strict contraction. By virtue of Banach fixed point theorem, Q^n has a unique fixed point $u \in C([\,a,b\,];X)$. But

$$\|Qu - u\| = \|Q^n Qu - Q^n u\| \leq q\|Qu - u\|,$$

where $q \in (0,1)$. Hence $Qu = u$. But u is the only fixed point of Q, because each fixed point of Q is a fixed point for Q^n, which is a strict contraction. As each fixed point of Q is a solution of the problem (11.1.1) and conversely, this achieves the proof. □

We may now proceed to the proof of Theorem 11.1.1.

Proof. Let $(a,\xi) \in \mathcal{D}$. As $\mathcal{D}$ is open, there exist $b > a$ and $r > 0$ with $[\,a,b\,] \times B(\xi,r) \subseteq \mathcal{D}$. In addition, as f is continuous and locally Lipschitz with respect to its last argument, diminishing $b > a$ and/or $r > 0$ if necessary, we may assume that there exists $M > 0$ such that

$$\|f(t,u)\| \leq M$$

for each $(t,u) \in [\,a,b\,] \times B(\xi,r)$ and

$$\|f(t,u) - f(t,v)\| \leq L\|u-v\|$$

for each $(t,u), (t,v) \in [\,a,b\,] \times B(\xi,r)$. Let $\rho : X \to X$ be defined by

$$\rho(y) = \begin{cases} y & \text{for } y \in B(\xi,r) \\[2mm] \dfrac{r}{\|y-\xi\|}(y-\xi) + \xi & \text{for } y \in X \setminus B(\xi,r). \end{cases}$$

We may easily see that ρ maps X to $B(\xi, r)$, and is Lipschitz continuous on X with Lipschitz constant 2.
Next, let us define $g : [a, b] \times X \to X$ by

$$g(t, y) = f(t, \rho(y))$$

for each $(t, y) \in [a, b] \times X$. Since f is continuous and globally Lipschitz with respect to its last argument on $[a, b] \times B(\xi, r)$, we conclude that g enjoys the very same properties. From Lemma 11.1.1, we know that the Cauchy problem

$$\begin{cases} u' = Au + g(t, u) \\ u(a) = \xi \end{cases}$$

has a unique C^0-solution $u : [a, b] \to X$. As $u(a) = \xi$ and u is continuous at $t = a$, diminishing $b > a$ if necessary, we get that $u(t) \in B(\xi, r)$ for each $t \in [a, b]$. Accordingly $g(t, u(t)) = f(t, u(t))$, and therefore $u : [a, b] \to X$ is a C^0-solution of the problem (11.1.1). The proof is complete. □

In the case in which X has good geometrical properties and f is locally Lipschitz with respect to both arguments, then, for each $\xi \in D(A)$, the C^0-solution of (11.1.1) is even a classical one. More precisely we have:

Theorem 11.1.2. *Let $A : D(A) \subseteq X \to X$ be the generator of a C_0-semigroup of contractions and $f : \mathcal{D} \to X$ locally Lipschitz with respect to both arguments. If X has the Radon-Nicodým property, then, for each $(a, \xi) \in \mathcal{D}$ with $\xi \in D(A)$, there exists $b > a$ such that the problem* (11.1.1) *has a unique classical solution defined on $[a, b]$.*

Proof. Let $u : [a, b] \to X$ be the unique C^0-solution whose existence is ensured by Theorem 11.1.1 and let $v : [a, b] \to X$ be defined by

$$v(t) = \int_a^t S(t - s) f(s, u(s))\, ds.$$

Let $t \in (a, b)$, let $h > 0$ and let us observe that

$$v(t + h) - v(t) = \int_a^{t+h} S(t + h - s) f(s, u(s))\, ds - \int_a^t S(t - s) f(s, u(s))\, ds$$

$$= \int_{a-h}^t S(t - s) f(s + h, u(s + h))\, ds - \int_a^t S(t - s) f(s, u(s))\, ds$$

$$= \int_{a-h}^a S(t - s) f(s + h, u(s + h))\, ds$$

$$+ \int_a^t S(t - s)[f(s + h, u(s + h)) - f(s, u(s))]\, ds.$$

Inasmuch as $t \mapsto f(t, u(t))$ is bounded on $[a, b]$, say by $M > 0$, we deduce

$$\|v(t+h) - v(t)\| \leq Mh + L \int_a^t (|h| + \|u(s+h) - u(s)\|)\, ds.$$

Since $u(t) = S(t-a)\xi + v(t)$ and

$$\|S(t+h-a)\xi - S(t-a)\xi\| \leq \|S(h)\xi - \xi\| \leq \|A\xi\| h,$$

from the preceding inequality, we get

$$\|u(t+h) - u(t)\| \leq [\|A\xi\| + M + L(b-a)]h + L \int_a^t \|u(s+h) - u(s)\|\, ds.$$

From Gronwall Lemma 10.2.1, it follows

$$\|u(t+h) - u(t)\| \leq e^{L(b-a)}[\|A\xi\| + M + L(b-a)]h$$

for each $t, t+h \in [a, b]$. Clearly this inequality shows that u is Lipschitz on $[a, b]$ and consequently $t \mapsto f(t, u(t))$ is Lipschitz on $[a, b]$. Since X has the Radon-Nicodým property, it follows that $t \mapsto f(t, u(t))$ is in $W^{1,1}(a, b; X)$ (in fact in $W^{1,+\infty}(a, b; X)$). Now the conclusion follows from Theorem 8.1.4 and this completes the proof. □

Since reflexive Banach spaces, as well as separable duals have the Radon-Nicodým property, we easily deduce:

Theorem 11.1.3. *Let $A : D(A) \subseteq X \to X$ be the generator of a C_0-semigroup of contractions and $f : \mathcal{D} \to X$ locally Lipschitz with respect to both arguments. If X is either reflexive, or a separable dual, then, for each $(a, \xi) \in \mathcal{D}$ with $\xi \in D(A)$, there exists $b > a$ such that the problem* (11.1.1) *has a unique classical solution defined on $[a, b]$.*

11.2. The Problem $u' = Au + f(t,u)$ with f Continuous

In this section we shall present a useful consequence of Theorem 6.3.2. Let X be a real Banach space, and $A : D(A) \subseteq X \to X$ the infinitesimal generator of a C_0-semigroup of contractions $\{S(t);\ t \geq 0\}$. Let $\mathcal{D}$ be a nonempty and open subset in $\mathbb{R} \times X$, and let $f : \mathcal{D} \to X$ a continuous function. Let $(a, \xi) \in \mathcal{D}$, and let us consider the Cauchy problem

$$\begin{cases} u' = Au + f(t,u) \\ u(a) = \xi \end{cases} \tag{11.2.1}$$

As we shall see in the sequel, the following abstract local existence result concerns abstract semilinear problems "of parabolic type".

Theorem 11.2.1. (Pazy) *If $A : D(A) \subseteq X \to X$ generates a compact C_0-semigroup of contractions, $\{S(t)\,;\ t \geq 0\}$, and $f : \mathcal{D} \to X$ is continuous then, for each $(a, \xi) \in \mathcal{D}$, there exists $b > a$ such that the problem* (11.2.1) *has at least one C^0-solution defined on $[\,a, b\,]$.*

For the proof, we need the following lemma which is interesting by itself.

Lemma 11.2.1. *If $A : D(A) \subseteq X \to X$ generates a compact C_0-semigroup of contractions, $\{S(t)\,;\ t \geq 0\}$, and $f : [\,a, b\,] \times X \to X$ is continuous and bounded then, for each $\xi \in X$, the problem* (11.2.1) *has at least one C^0-solution defined on $[\,a, b\,]$.*

Proof. Let $\xi \in X$, let $\lambda > 0$ and let us consider the integral equation with the delay λ

$$u_\lambda(t) = \begin{cases} \xi & t \in [\,a - \lambda, a\,] \\ S(t-a)\xi + \displaystyle\int_a^t S(t-s) f(s, u_\lambda(s-\lambda))\, ds & t \in [\,a, b\,]. \end{cases} \tag{11.2.2}$$

Let us observe that (11.2.2) has a unique continuous solution which is defined successively on $[\,a, a+\lambda\,]$, $[\,a+\lambda, a+2\lambda\,]$ and so on. For $n \in \mathbb{N}^*$, let us denote by u_n the unique solution of the problem (11.2.2) corresponding to $\lambda = 1/n$. As f is bounded, it follows that $\mathcal{F} = \{f(\cdot, u_n(\cdot - 1/n))\,;\ n \in \mathbb{N}^*\}$ is bounded, and therefore uniformly integrable. In view of Theorem 6.3.2, the set $\{u_n\,;\ n \in \mathbb{N}^*\}$ is relatively compact in $C([\,a, b\,]; X)$. Then we may assume without loss of generality that there exists

$$\lim_{n \to \infty} u_n(t) = u(t),$$

uniformly for $t \in [\,a, b\,]$. We also have

$$\lim_{n \to \infty} u_n(s - 1/n) = u(s),$$

uniformly for $s \in [\,a, b\,]$. From Corollary A.2.2, we deduce that

$$\lim_{n \to \infty} f(s, u_n(s - 1/n)) = f(s, u(s)),$$

uniformly for $s \in [\,a, b\,]$. Therefore, passing to the limit both sides in

$$u_n(t) = S(t-a)\xi + \int_a^t S(t-s) f(s, u_n(s - 1/n))\, ds,$$

we obtain

$$u(t) = S(t-a)\xi + \int_a^t S(t-s) f(s, u(s))\, ds$$

for each $t \in [\,a, b\,]$, which shows that u is a C^0-solution of the problem (11.2.1) on $[\,a, b\,]$. The proof is complete. □

We may now proceed to the proof of Theorem 11.2.1.

Proof. Let $(a,\xi) \in \mathcal{D}$. As $\mathcal{D}$ is open, there exist $b > a$ and $r > 0$ with $[a,b] \times B(\xi,r) \subseteq \mathcal{D}$. In addition, as f is continuous, diminishing r if necessary, we may assume that there exists $M > 0$ such that

$$\|f(t,u)\| \leq M$$

for each $(t,u) \in [a,b] \times B(\xi,r)$. Let $\rho : X \to X$ be defined by

$$\rho(y) = \begin{cases} y & \text{for } y \in B(\xi,r) \\ \dfrac{r}{\|y-\xi\|}(y-\xi)+\xi & \text{for } y \in X \setminus B(\xi,r). \end{cases}$$

We may easily see that ρ maps X to $B(\xi,r)$, and is continuous on X. Next, let us define $g : [a,b] \times X \to X$ by

$$g(t,y) = f(t,\rho(y))$$

for each $(t,y) \in [a,b] \times X$. Inasmuch as f is continuous and bounded on $[a,b] \times B(\xi,r)$, we conclude that g is continuous and bounded on $[a,b] \times X$. From Lemma 11.2.1, we know that the Cauchy problem

$$\begin{cases} u' = Au + g(t,u) \\ x(a) = \xi \end{cases}$$

has at least one C^0-solution $u : [a,b] \to X$. As $u(a) = \xi$ and u is continuous at $t = a$, diminishing $b > a$ if necessary, we get that $u(t) \in B(\xi,r)$ for each $t \in [a,b]$. Accordingly $g(t,u(t)) = f(t,u(t))$, and therefore $u : [a,b] \to X$ is a C^0-solution of the problem (11.2.1). The proof is complete. $\square$

11.3. Saturated Solutions

Let $\mathcal{D}$ be a nonempty and open subset in $\mathbb{R} \times X$, let $f : \mathcal{D} \to X$ be a given function, and $(a,\xi) \in \mathcal{D}$ and $A : D(A) \subseteq X \to X$. Let us consider the semilinear Cauchy problem

$$\begin{cases} u' = Au + f(t,u) \\ u(a) = \xi. \end{cases} \tag{11.3.1}$$

By a *C^0-solution of* (11.3.1) *on* $[a,b)$ we mean a function $u : [a,b) \to X$ whose restriction to any interval $[a,c] \subset [a,b)$ is a C^0-solution of (11.3.1) on $[a,c]$ in the sense of Definition 11.1.1. A C^0-solution $u : [a,b) \to X$ of (11.3.1) is *continuable* if there exists another C^0-solution $v : [a,c) \to X$ with $b < c$ and such that $u(t) = v(t)$ for each $t \in [a,b)$. A C^0-solution $u : [a,b) \to X$ is called *saturated* if it is not continuable. If the projection of $\mathcal{D}$ on $\mathbb{R}$ contains $\mathbb{R}_+$, a solution u is called *global* if it is defined on $[a,+\infty)$. We begin with a very simple but useful lemma.

Lemma 11.3.1. *Let $A : D(A) \subseteq X \to X$ be the generator of a C_0-semigroup of contractions and let $f : \mathcal{D} \to X$ be continuous. Assume that either the semigroup is compact, or f is locally Lipschitz with respect to its last argument. Then a C^0-solution $u : [a, b) \to X$ of (11.3.1) is continuable if and only if there exists*

$$u^* = \lim_{t \uparrow b} u(t) \tag{11.3.2}$$

and

$$(b, u^*) \in \mathcal{D} \tag{11.3.3}$$

Proof. The necessity is obvious, while the sufficiency is a consequence of Theorems 11.2.1 and 11.1.1. □

Remark 11.3.1. By virtue of Lemma 11.3.1, it follows that each saturated solution of (11.3.1) is necessarily defined on an interval of the form $[a, b)$, i.e. on an interval which is open at the right.

A sufficient condition for the existence of the limit (11.3.2) is stated below.

Proposition 11.3.1. *Let $A : D(A) \subseteq X \to X$ be the generator of a C_0-semigroup of contractions and let $f : \mathcal{D} \to X$ be continuous. Furthermore, let $u : [a, b) \to X$ be a C^0-solution of (11.3.1) and assume that $b < +\infty$, and $f(\cdot, u(\cdot)) \in \mathcal{L}^1(a, b; X)$. Then there exists*

$$u^* = \lim_{t \uparrow b} u(t).$$

Proof. Since $g = f(\cdot, u(\cdot))$ is integrable, it follows that

$$v(t) = S(t-a)\xi + \int_a^t S(t-s)g(s)\,ds$$

is continuous on $[a, b]$ and thus there exists $\lim_{t \uparrow b} v(t) = v(b)$. Inasmuch as $v(t) = u(t)$ for each $t \in [a, b)$, this achieves the proof. □

We proceed next to the statement of a useful characterization of continuable C^0-solutions of (11.3.1).

Theorem 11.3.1. *Let $A : D(A) \subseteq X \to X$ be the generator of a C_0-semigroup of contractions and let $f : \mathcal{D} \to X$ be continuous. Assume that either the semigroup is compact, or f is locally Lipschitz with respect to its last argument. A necessary and sufficient condition in order that a C_0-solution $u : [a, b) \to X$ of (11.3.1) be continuable is that the graph of u, i.e.*

$$\text{graph}\, u = \{(t, u(t)) \in \mathbb{R} \times X;\ t \in [a, b)\}$$

be included in a compact subset in $\mathcal{D}$.

The proof is a simple copy of that of Theorem 10.3.1, and so we do not give details. We continue with a fundamental existence result concerning saturated C^0-solutions for (11.3.1).

Theorem 11.3.2. *Let $A : D(A) \subseteq X \to X$ be the generator of a C_0-semigroup of contractions and let $f : \mathcal{D} \to X$ be continuous. Assume that either the semigroup is compact, or f is locally Lipschitz with respect to its last argument. Let $u : [a, b) \to X$ be a C^0-solution of* (11.3.1). *Then either u is saturated, or u can be continued up to a saturated one.*

The proof follows the same lines as those in the proof of Theorem 10.3.2 and therefore we omit it. From Theorems 2.3.1 and 11.3.2 it follows:

Corollary 11.3.1. *Let $A : D(A) \subseteq X \to X$ be the generator of a C_0-semigroup of contractions and let $f : \mathcal{D} \to X$ be continuous. Assume that either the semigroup is compact, or f is locally Lipschitz with respect to its last argument. Then for each initial data $(a, \xi) \in \mathcal{D}$,* (11.3.1) *has at least one saturated C^0-solution.*

We recall that a *limit point* of a function $u : [a, b) \to X$ as t tends to b is any element u^* in X for which there exists a sequence $(t_k)_{k\in\mathbb{N}}$ in $[a, b)$ tending to b and such that $\lim_{k\to\infty} u(t_k) = u^*$. We denote the set of all limit points of u as t tends to b by $\text{Lim}_{t\uparrow b}u(t)$.
The behavior of saturated C^0-solutions at the right end point of their interval of definition is described by the following fundamental result.

Theorem 11.3.3. *Let $A : D(A) \subseteq X \to X$ be the generator of a C_0-semigroup of contractions and let $f : \mathcal{D} \to X$ be continuous. Assume that either the semigroup is compact, or f is locally Lipschitz with respect to its last argument. Let $u : [a, b) \to X$ be a saturated C^0-solution of* (11.3.1). *Then either*

(i) *u is unbounded or*
(ii) *u is bounded and, either u is global, or*
(iii) *u is bounded and non-global, and in this case, either $\text{Lim}_{t\uparrow b}u(t)$ is empty, or for each $u^* \in \text{Lim}_{t\uparrow b}u(t)$, $(b, u^*) \in \partial\mathcal{D}$.*

Proof. Up to one point, the proof parallels that one of Theorem 10.3.3. So, if u is unbounded on $[a, b)$, or if it is bounded and global, or bounded and $\text{Lim}_{t\uparrow b}u(t)$ is empty, we have nothing to prove. Thus, let us assume that u is bounded on $[a, b)$ and non-global and $\text{Lim}_{t\uparrow b}u(t)$ is nonempty. To prove (iii) let us assume the contrary, i.e. that there exists at least one sequence $(t_k)_{k\in\mathbb{N}}$ in $[a, b)$ tending to b and such that $(u(t_k))_{k\in\mathbb{N}}$ is convergent to some $u^* \in X$, but $(b, u^*) \notin \partial\mathcal{D}$. So, a fortiori $(b, u^*) \in \mathcal{D}$. Thanks to Lemma 11.3.1, to get a contradiction, it suffices to show that there exists $\lim_{t\uparrow b} u(t)$, which clearly must coincide with u^*. To this goal, let us observe that there exists $c > b$ and $r > 0$ such that $[a, c) \times B(u^*, r) \subset \mathcal{D}$. Furthermore, inasmuch as $b < +\infty$, and f is continuous, diminishing $r > 0$ if necessary, we may assume that there exists $M > 0$ such that

$$\|f(\tau, v)\| \leq M, \tag{11.3.4}$$

for each $(\tau, v) \in [a, b] \times B(u^*, r)$. As $\lim_{k\to\infty} t_k = b$, the semigroup is strongly continuous at $t = 0$ and $\lim_{k\to\infty} u(t_k) = u^*$, we have

$$\lim_{h\downarrow 0} S(h)u(t_k) = u(t_k)$$

uniformly for $k \in \mathbb{N}$. So we may choose $k \in \mathbb{N}$ such that

$$\begin{cases} b - t_k < \dfrac{r}{2M} \\ \|S(t - t_k)u(t_k) - u(t_k)\| + \|u(t_k) - u^*\| < \dfrac{r}{2} \quad \text{for } t \in [t_k, b). \end{cases} \tag{11.3.5}$$

Fix k with the properties above. We shall show that for each $t \in [t_k, b)$, we have $u(t) \in B(u^*, r)$. Let

$$t^* = \sup\{t \in [t_k, b);\ u(s) \in B(u^*, r),\ \text{for } s \in [t_k, t]\}.$$

If $t^* = b$, we have nothing to prove. So, let us assume by contradiction that $t^* < b$. In other words, $u(t) \in B(u^*, r)$ for $t \in [t_k, t^*]$, $\|u(t^*) - u^*\| = r$ and there exist points $t > t^*$, as close to t^* as we like, such that $u(t)$ does not belong to $B(u^*, r)$. More precisely, t^* is the "first moment in (t_k, b) after which u leaves the set $B(u^*, r)$". The fact that $\|u(t^*) - u^*\| = r$ has a simple geometrical meaning, i.e. it signifies that, at t^*, u must cross the boundary of $B(u^*, r)$. From the considerations above combined with (11.3.4), and (11.3.5), we conclude

$$r = \|u(t^*) - u^*\| \leq \|u(t^*) - u(t_k)\| + \|u(t_k) - u^*\|$$

$$\leq \|S(t^* - t_k)u(t_k) - u(t_k)\| + \int_{t_k}^{t^*} \|S(t^* - \tau)f(\tau, u(\tau))\| d\tau + \|u(t_k) - u^*\|$$

$$\leq (b - t_k)M + \|S(t^* - t_k)u(t_k) - u(t_k)\| + \|u(t_k) - u^*\| < r.$$

This contradiction ($r < r$) is a consequence of our supposition that, for at least one $t \in [t_k, b)$, $u(t) \notin B(u^*, r)$. Then, for each $t \in [t_k, b)$, we must have $u(t) \in B(u^*, r)$. Thanks to (11.3.4) combined with Proposition 11.3.1 and Lemma 11.3.1, it follows that u is continuable, assertion which does not agree with the hypothesis. This contradiction can be eliminated only if $(b, u^*) \in \partial\mathcal{D}$ and this achieves the proof. □

Under an additional hypothesis on f we shall prove a sharper result.

Theorem 11.3.4. *Let $A : D(A) \subseteq X \to X$ be the generator of a C_0-semigroup of contractions and let $f : \mathcal{D} \to X$ be continuous. Assume that f maps bounded subsets in $\mathcal{D}$ into bounded subsets in X and, either the semigroup is compact, or f is locally Lipschitz with respect to its last argument. Let $u : [a, b) \to X$ be a saturated C^0-solution of* (11.3.1). *Then either*

(i') *u is unbounded and, if $b < +\infty$, $\lim_{t\uparrow b} \|u(t)\| = +\infty$, or*
(ii) *u is bounded and, either u is global, or*
(iii') *u is bounded and non-global, and in this later case there exists $\lim_{t\uparrow b} u(t) = u^*$, and $(b, u^*) \in \partial\mathcal{D}$.*

Proof. In view of Theorem 11.3.3, we have merely to prove that whenever (ii) does not hold, then one of the two conditions (i') or (iii') must hold. So, let us assume first that (i') and (ii) do not hold. Then, from Lemma 11.3.1 and Proposition 11.3.1, it follows that there exists

$$\lim_{t\uparrow b} u(t) = u^*,$$

while from (iii) in Theorem 11.3.3 we know that $(b, u^*) \in \partial\mathcal{D}$ and so (iii') holds.

Let us assume now that (ii) and (iii') do not hold and that $b < +\infty$. To show that $\lim_{t\uparrow b} \|u(t)\| = +\infty$, let us assume the contrary, i.e. that, under these circumstances, we have

$$\liminf_{t\uparrow b} \|u(t)\| < +\infty.$$

Equivalently, there exist at least one sequence $(t_k)_{k\in\mathbb{N}}$ in (a, b), convergent to b, and $r > 0$ such that $\|u(t_k)\| \le r$, for each $k \in \mathbb{N}$.

Let $C = \{v \in X;\ \|v\| \le r + 1\}$. Since f maps bounded subsets in $\mathcal{D}$ into bounded subsets in X, $b < +\infty$ and C is bounded, there exists $M > 0$, such that

$$\|f(\tau, v)\| \le M, \tag{11.3.6}$$

for each $(\tau, v) \in ([a, b) \times C) \cap \mathcal{D}$. Now let us choose $d > 0$ satisfying

$$dM < 1, \tag{11.3.7}$$

and fix $k \in \mathbb{N}$ such that $b - d < t_k < b$. Since u is unbounded on $[a, b)$, it is necessarily unbounded on $[t_k, b)$. Then there exists $t^* \in (t_k, b)$ such that

$$\|u(\tau)\| < r + 1, \tag{11.3.8}$$

for each $\tau \in [t_k, t^*)$ and

$$\|u(t^*)\| = r + 1.$$

Let us remark that

$$r+1=\|u(t^*)\|=\|S(t^*-t_k)u(t_k)+\int_{t_k}^{t^*}S(t^*-\tau)f(\tau,u(\tau))d\tau\|$$

$$\leq\|S(t^*-t_k)u(t_k)\|+\int_{t_k}^{t^*}\|S(t^*-\tau)|_{\mathcal{L}(X)}\|f(\tau,u(\tau))\|d\tau$$

$$\leq\|u(t_k)\|+\int_{t_k}^{t^*}\|f(\tau,u(\tau))\|d\tau.$$

Taking into account (11.3.6), (11.3.7), (11.3.8) and the obvious inequality $t^*-t_k<d$, we get

$$r+1\leq r+(t^*-t_k)M<r+dM<r+1.$$

This contradiction can be eliminated only if

$$\lim_{t\uparrow b}\|u(t)\|=+\infty$$

and this completes the proof. □

The proofs of the next results follow exactly the same lines as those of their counterparts in Section 10.3, and therefore we do not give details. We only notice that there, due to the fact that $f:\mathbb{R}\times X\to X$ is compact, it maps bounded subsets in $\mathbb{R}\times X$ into bounded subsets in X, and therefore it was no need to assume this explicitly, as we shall do here.

Corollary 11.3.2. *Let $A:D(A)\subseteq X\to X$ be the generator of a C_0-semigroup of contractions and let $f:\mathbb{R}\times X\to X$ be continuous. Assume that f maps bounded subsets in $\mathcal{D}$ into bounded subsets in X and, either the semigroup is compact, or f is locally Lipschitz with respect to its last argument. Then a saturated C^0-solution $u:[\,a,b)\to X$ of* (11.3.1) *is either global, or it is not global, and in this case there exists*

$$\lim_{t\uparrow b}\|u(t)\|=+\infty.$$

Corollary 11.3.3. *Let $A:D(A)\subseteq X\to X$ be the generator of a C_0-semigroup of contractions and let $f:\mathbb{R}\times X\to X$ be continuous. Assume that f maps bounded subsets in $\mathcal{D}$ into bounded subsets in X and, either the semigroup is compact, or f is locally Lipschitz with respect to its last argument. Then a necessary and sufficient condition in order that a C^0-solution $u:[\,a,b)\to X$ of* (11.3.1) *be continuable is that $b<+\infty$ and u be bounded on $[\,a,b)$.*

We conclude this section with a sufficient conditions on f ensuring the existence of global C^0-solutions of (11.3.1).

Theorem 11.3.5. *Let $A : D(A) \subseteq X \to X$ be the generator of a C_0-semigroup of contractions and let $f : \mathbb{R}_+ \times X \to X$ be continuous. Assume that either the semigroup is compact, or f is locally Lipschitz with respect to its last argument. Let us assume further that there exist two continuous functions $h, k : \mathbb{R}_+ \to \mathbb{R}_+$ such that*

$$\|f(\tau, v)\| \leq k(\tau)\|v\| + h(\tau), \tag{11.3.9}$$

for each $(\tau, v) \in \mathbb{R}_+ \times X$. Then, for each $(a, \xi) \in \mathbb{R}_+ \times X$, (11.3.1) *has at least one global C^0-solution.*

11.4. Asymptotic Behavior

Let Ω be an open neighborhood of $0 \in \Omega$, $f : \mathbb{R}_+ \times \Omega \to X$ a continuous function with $f(t, 0) = 0$ for each $t \in \mathbb{R}_+$, and let $A : D(A) \subseteq X \to X$ be the generator of a C_0 semigroup $\{S(t)\,;\ t \geq 0\}$. We consider the *perturbed problem*

$$u' = Au + f(t, u). \tag{11.4.1}$$

In this section we will show that the asymptotic behavior of the null solution of the linear system $u' = Au$ is inherited by the null solution of the system (11.4.1), provided that the perturbation f is dominated by the semigroup generated by A. First we recall:

Definition 11.4.1. The null solution $x \equiv 0$ of (11.4.1) is *asymptotically stable* if for each $a \geq 0$ and each $\varepsilon > 0$ there exists $\delta(a, \varepsilon) > 0$ and $\mu(a) > 0$ such that:

(i) for each $\xi \in \Omega$ with $\|\xi\| \leq \delta(a, \varepsilon)$, each saturated solution $u(\cdot, a, \xi)$ of (11.4.1) with $u(a, a, \xi) = \xi$ is global and $\|u(t, a, \xi)\| \leq \varepsilon$ for all $t \geq 0$;

(ii) for each $\xi \in \Omega$ with $\|\xi\| \leq \mu(a)$, each saturated solution $u(\cdot, a, \xi)$ of (11.4.1) with $u(a, a, \xi) = \xi$ is global and $\lim_{t \to \infty} \|u(t, a, \xi)\| = 0$.

We notice that, in our general setting, we have no uniqueness and this explains why, the conditions in Definition 11.4.1 refer to all global solutions. We begin with the following fundamental result.

Theorem 11.4.1. (Poincaré-Liapunov) *Let $A : D(A) \subseteq X \to X$ be the generator of a C_0-semigroup $\{S(t)\,;\ t \geq 0\}$ and let $f : \mathbb{R}_+ \times \Omega \to X$ be a continuous function. We assume that there exist $M \geq 1$, $\omega > 0$ and $L > 0$ such that*

$$\|S(t)\|_{\mathcal{L}(X)} \leq Me^{-\omega t} \tag{11.4.2}$$

for each $t \in \mathbb{R}_+$,

$$\|f(t, x)\| \leq L\|x\| \tag{11.4.3}$$

for any $(t,x) \in \mathbb{R}_+ \times \Omega$ *and*

$$LM - \omega < 0. \tag{11.4.4}$$

If either the semigroup is compact, or f is Lipschitz with respect to its last argument, then the null solution of (11.4.1) *is asymptotically stable.*

Proof. Let $\xi \in \Omega$, $a \geq 0$ and let $u(\cdot, a, \xi) : [a, T_m) \to \Omega$ be a given saturated solution of (11.4.1) satisfying the initial condition $u(a, a, \xi) = \xi$. We notice that, excepting the case in which f is Lipschitz with respect to its last argument, this solution might not be unique. We will prove first that, if $\|\xi\|$ is sufficiently small, then $u(\cdot, a, \xi)$ is global, i.e. it is defined on $[a, +\infty)$. To this aim, let us recall that for each $t \in [a, T_m)$, we have

$$u(t, a, \xi) = S(t-a)\xi + \int_a^t S(t-s) f(s, u(s, a, \xi))\, ds.$$

From this relation, we deduce

$$\|u(t, a, \xi)\| \leq \|S(t-a)\|_{\mathcal{L}(X)} \|\xi\| + \int_a^t \|S(t-s)\|_{\mathcal{L}(X)} \|f(s, (s, a, \xi))\|\, ds$$

from where, by virtue of (11.4.2) and (11.4.3), it follows that

$$\|u(t, a, \xi)\| \leq Me^{-\omega(t-a)} \|\xi\| + \int_a^t LMe^{-\omega(t-s)} \|u(s, a, \xi)\|\, ds$$

for any $t \in [a, T_m)$. Multiplying both sides of the inequality above by $e^{\omega t}$, we get

$$e^{\omega t} \|u(t, a, \xi)\| \leq Me^{\omega a} \|\xi\| + \int_a^t LMe^{\omega s} \|u(s, a, \xi)\|\, ds$$

for each $t \in [a, T_m)$. Let $v : [a, T_m) \to \mathbb{R}_+$ be the function defined by

$$v(t) = e^{\omega t} \|u(t, a, \xi)\|$$

for each $t \in [a, T_m)$. The preceding inequality takes the equivalent form

$$v(t) \leq Me^{\omega a} \|\xi\| + \int_a^t LMv(s)\, ds$$

for each $t \in [a, T_m)$. From Gronwall Lemma 10.2.1 it follows that

$$v(t) \leq Me^{\omega a} \|\xi\| e^{LM(t-a)}$$

from where, recalling the definition of the function v, we deduce

$$\|u(t, a, \xi)\| \leq M \|\xi\| e^{(LM-\omega)(t-a)} \tag{11.4.5}$$

for any $t \in [a, T_m)$.

Let now $\rho > 0$ be such that $B(0,\rho) \subset \Omega$ and let $\mu(a) > 0$ be defined by

$$\mu(a) = \frac{\rho}{2M}.$$

Then, by virtue of (11.4.4) and (11.4.5), for each $\xi \in \Omega$ with $\|\xi\| \leq \mu(a)$, we have

$$\|u(t,a,\xi)\| \leq \frac{\rho}{2}$$

for any $t \in [\,a, T_m)$. Assuming that $T_m < +\infty$, from this inequality and Proposition 10.3.1, it follows that there exists

$$\lim_{t \uparrow T_m} u(t,a,\xi) = u^*$$

and $u^* \in B(0, \frac{\rho}{2}) \subset \Omega$, relation which, by virtue of (iii) in Theorem 11.3.3, shows that $u(\cdot, a, \xi)$ is not saturated. This contradiction can be eliminated only if, for each $\xi \in \Omega$ satisfying $\|\xi\| \leq \mu(a)$, we have $T_m = +\infty$. Now, taking $\delta(a,\varepsilon) = \min\{\mu(a), \frac{\varepsilon}{M}\}$, we conclude that $\|u(t,a,\xi)\| \leq \varepsilon$ for each $t \geq a$. Finally, let us observe that, from the above proof and (11.4.5), it follows that, for each $\xi \in \Omega$ with $\|\xi\| \leq \mu(a)$, we have

$$\lim_{t \uparrow +\infty} u(t,a,\xi) = 0,$$

and this completes the proof. □

A useful consequence is:

Theorem 11.4.2. *Let $A : D(A) \subseteq X \to X$ be the generator of the C_0-semigroup of type $(M, -\omega)$, $\{S(t)\,;\ t \geq 0\}$, with $M \geq 1$ and $\omega > 0$ and let $f : \mathbb{R}_+ \times \Omega \to X$ be continuous. We assume that there exists $\alpha : \mathbb{R}_+ \to \mathbb{R}_+$ such that*

$$\|f(t,u)\| \leq \alpha(\|u\|)$$

for each $(t,u) \in \mathbb{R}_+ \times \Omega$, and

$$\lim_{r \downarrow 0} \frac{\alpha(r)}{r} = 0.$$

If either $\{S(t)\ \ t \geq 0\}$ is compact, or f is Lipschitz with respect to its last argument, then the null solution of (11.4.1) *is asymptotically stable.*

Proof. Let us fix $L > 0$ satisfying (11.4.4) and let us choose $\delta > 0$ such that

$$\alpha(r) \leq Lr$$

for $r \in [\,0,\delta)$. Let us consider the restriction of f to $\mathbb{R}_+ \times \{x \in \Omega\,;\, \|x\| < \delta\}$, and let us observe that we are in the hypotheses of Theorem 11.4.1. The proof is complete. □

11.5. The Klein-Gordon Equation Revisited

In this section we reconsider the Klein-Gordon equation in Example 10.4.1 in a different frame, i.e. in the case in which the domain Ω is not necessarily bounded and/or smooth, i.e. with boundary Γ of class C^2. So, let Ω be a nonempty and open subset in $\mathbb{R}^n$, let us denote $Q_T = (0,T) \times \Omega$, and $\Sigma_T = (0,T) \times \Gamma$. For the sake of simplicity, throughout this section, we consider only a perturbation g which does not depend on t and x, i.e. $g : \mathbb{R} \to \mathbb{R}$. Let us consider the Klein-Gordon semilinear equation

$$\begin{cases} u_{tt} = \Delta u + g(u) & (t,x) \in Q_T \\ u(t,x) = 0 & (t,x) \in \Sigma_T \\ u(0,x) = u_0(x) & x \in \Omega \\ u_t(0,x) = v_0(x) & x \in \Omega \end{cases} \tag{11.5.1}$$

where $g : \mathbb{R} \to \mathbb{R}$ is a continuous function.

Theorem 11.5.1. *Let $g : \mathbb{R} \to \mathbb{R}$ be a continuous function for which there exist $c > 0$ and $\alpha \in \mathbb{R}_+$ such that*

$$\begin{cases} g(0) = 0 \\ |g(u) - g(v)| \leq c(1 + |u|^\alpha + |v|^\alpha)|u - v| \end{cases} \tag{11.5.2}$$

for each $u, v \in \mathbb{R}$, where $\alpha \geq 0$, if $n = 2$, and $\alpha < 2/(n-2)$, if $n \geq 3$. Then, for each $u_0 \in H_0^1(\Omega)$, and each $v_0 \in L^2(\Omega)$, there exists $T > 0$, such that the problem (11.5.1) *has a unique saturated solution u satisfying*

(i) $u \in C([0,T); H_0^1(\Omega))$;
(ii) $u_t \in C([0,T); L^2(\Omega))$;
(iii) $u \in C^2([0,T]; H^{-1}(\Omega))$.

In addition, if $T < +\infty$, then

$$\lim_{s \uparrow T} \left(\|u(s)\|_{H_0^1(\Omega)} + \|u_t(s)\|_{L^2(\Omega)} \right) = +\infty.$$

Proof. We note that (11.5.1) can be rewritten as a differential equation of the form (11.1.1) in a suitably chosen infinite-dimensional space. So, let

$$H = H_0^1(\Omega) \times L^2(\Omega)$$

which, endowed with the inner product $\langle \cdot, \cdot \rangle$ defined by

$$\langle (u_1, v_1), (u_2, v_2) \rangle = \int_\Omega u_1'(x) u_2'(x)\, dx + \int_\Omega v_1(x) v_2(x)\, dx$$

for each $(u_1, v_1), (u_2, v_2) \in H$, is a real Hilbert space. We define the operator $A : D(A) \subset H \to H$ by

$$\begin{cases} D(A) = (H^2(\Omega) \cap H_0^1(\Omega)) \times (H_0^1(\Omega) \cap L^2(\Omega)) \\ A(u,v) = (v, \Delta u) \ \text{ for each } (u,v) \in D(A). \end{cases}$$

Next, let us define $f : D(f) \to H$ by

$$\begin{cases} D(f) = \{(u,v) \in H\,;\ g(u(\cdot)) \in L^2(\Omega) \text{ for each } t \in [0,+\infty)\} \\ f((u,v))(x) = (0, g(u(x))) \end{cases}$$

for each $(u,v) \in D(f)$. One may easily see that the problem (11.5.1) can be rewritten under the form

$$\begin{cases} z' = Az + f(z) \\ z(a) = \xi, \end{cases}$$

where $z(t)(x) = (u(t,x), v(t,x))$ a.e. for $(t,x) \in (0,T) \times (\Omega)$, $\xi = (u_0, v_0)$, and A and f are as above. In order to prove that A and f satisfy the general hypotheses in Theorem 11.1.1, let us remark first that, by virtue of Theorem 4.6.2, A generates a C_0-group of isometries, and therefore it generates a C_0-semigroup of contractions as well. On the other hand, let us observe that, from (11.5.2), Theorem 1.5.4 and Lemma A.6.1, it follows that $D(f) = H$ and f is continuous on H. Let us observe that in fact f is locally Lipschitz on bounded subsets in H. This easily follows again from the condition (11.5.2) and Lemma A.6.1. We are thus in the hypotheses of Section 4.6, of Theorem 11.1.1 and Corollaries 11.3.1 and 11.3.2 and this achieves the proof. □

11.6. A Parabolic Semilinear Equation

Our aim here is to illustrate the power of Theorem 11.2.1 by means of a simple, but non-trivial application. More precisely, as we shall show next, Theorem 11.2.1 is very appropriate in the study of some parabolic semilinear partial differential equations. Let Ω be a nonempty, bounded, and open subset in $\mathbb{R}^n$, whose boundary Γ is of class C^2, let $Q_T = (0,T) \times \Omega$, and $\Sigma_T = (0,T) \times \Gamma$. Let us consider the semilinear parabolic partial differential equation

$$\begin{cases} u_t = \Delta u + g(t,x,u) & (t,x) \in Q_T \\ u(t,x) = 0 & (t,x) \in \Sigma_T \\ u(0,x) = \xi(x) & x \in \Omega, \end{cases} \tag{11.6.1}$$

where $g : \mathbb{R}_+ \times \Omega \times \mathbb{R} \to \mathbb{R}$ is a continuous function.
Using Pazy's Theorem 11.2.1, we shall prove that the problem (11.6.1) has at least one local solution. More precisely, we have:

Theorem 11.6.1. *If $g : \mathbb{R}_+ \times \Omega \times \mathbb{R} \to \mathbb{R}$ is a continuous function for which there exist $c > 0$ and $d > 0$ such that*

$$|g(t,x,u)| \le c|u| + d \tag{11.6.2}$$

for each $(t, x, u) \in \mathbb{R}_+ \times \Omega \times \mathbb{R}$, *then, for each* $\xi \in L^2(\Omega)$, *there exists at least one solution* $u : [0, +\infty) \to L^2(\Omega)$ *of* (11.6.1) *satisfying for* $T > 0$:

(i) $u \in C([0, T]; L^2(\Omega))$;
(ii) $t \mapsto \sqrt{t}u'(t, \cdot) \in L^2(0, T; L^2(\Omega))$;
(iii) $t \mapsto \|\nabla u(t, \cdot)\|_{L^2(\Omega)} \in L^1(0, T) \cap AC(\delta, T)$ for each $\delta \in (0, T)$;
(iv) $u \in L^2(\delta, T; H^2(\Omega)) \cap W^{1,1}(\delta, T; H_0^1(\Omega))$.

If $\xi \in H^2(\Omega) \cap H_0^1(\Omega)$, *then*

(v) $u \in W^{1,2}(0, T; L^2(\Omega)) \cap L^2(0, T; H^2(\Omega)) \cap W^{1,1}(0, T; H_0^1(\Omega))$.

If, in addition, $d = 0$ *and* c *is small enough, then we have also*

(vi) $\lim_{t \to \infty} \|u(t)\|_{L^2(\Omega)} = 0$.

Proof. Let $A : D(A) \subseteq L^2(\Omega) \to L^2(\Omega)$ be the operator defined by

$$\begin{cases} D(A) = H_0^1(\Omega) \cap H^2(\Omega) \\ Au = \Delta u \quad \text{for } u \in D(A). \end{cases}$$

Then, by virtue of Theorems 4.1.1 and 7.2.5, A is the infinitesimal generator of a compact C_0 semigroup of contractions on $L^2(\Omega)$. Let us define the mapping $f : [0, +\infty) \times D(f) \subseteq L^2(\Omega) \to L^2(\Omega)$ by

$$\begin{cases} D(f) = \{u \in L^2(\Omega);\ g(t, \cdot, u(\cdot)) \in L^2(\Omega) \text{ for each } t \in \mathbb{R}_+\} \\ f(t, u)(x) = g(t, x, u(x)) \end{cases}$$

for each $u \in D(f)$ each $t \in [0, +\infty)$ and a.e. $x \in \Omega$. Let us observe that, since g is continuous and satisfies the sublinear growth condition (11.6.2), it follows that $D(f) = L^2(\Omega)$. Again, from (11.6.2) and Lemma A.6.1, we deduce that f is continuous on $\mathbb{R}_+ \times L^2(\Omega)$. Finally, let us observe that the problem (11.6.1) can be rewritten in the space $L^2(\Omega)$ under the form

$$\begin{cases} u' = Au + f(t, u) \\ u(0) = \xi, \end{cases} \tag{11.6.3}$$

with A and f defined as above. In view of Corollary 11.3.1, combined with Theorem 11.3.5, we know that, for each $\xi \in L^2(\Omega)$, there exists at least one C^0-solution of (11.6.3) defined on $[0, +\infty)$, solution which obviously satisfies (i). From Theorem 4.1.2, we have that A is self-adjoint, and thus we are in the hypotheses of Theorem 8.2.1, from where we deduce (ii), (iii), (iv) and (v). Since (vi) is a direct consequence of Theorem 11.4.1, the proof is complete. □

Remark 11.6.1. Using a truncation procedure for the function g and estimates of the gradient of the function $u(t, \cdot)$ obtained directly by (11.6.1), and making use of Theorem 1.5.4, one can show that Theorem 11.6.1 can be extended to the more general case in which, instead of (11.6.2), g satisfies

$$|g(t, x, u)| \leq c|u|^p + d$$

where $p > 1$ depends on n.

Problems

Problem 11.1. Let $A : D(A) \subseteq X \to X$ be the generator of a compact C_0-semigroup of contractions, let $\mathbb{I}$ be a nonempty and open interval, Ω a nonempty and open subset in X, and $f : \mathbb{I} \times \Omega \to X$ a function satisfying:

(i) for a.e. for $t \in \mathbb{I}$, $f(t,\cdot) : \Omega \to X$ is continuous;
(ii) for each $x \in \Omega$, $f(\cdot,x) : \mathbb{I} \to X$ is measurable;
(iii) for each $\xi \in \Omega$, there exists $r > 0$, and a compact subset K in X such that $f(t,x) \in K$ for each $x \in B(\xi,r) \cap \Omega$, and a.e. for $t \in \mathbb{I}$.

Then, for each $a \in \mathbb{I}$ and each $\xi \in \Omega$, there exists $b > a$ with $[\,a,b\,] \subset \mathbb{I}$, and such that the Cauchy problem

$$\begin{cases} u' = Au + f(t,u) \\ u(a) = \xi \end{cases}$$

has at least one C^0-solution $u : [\,a,b\,] \to \Omega$.

Problem 11.2. Let $A : D(A) \subseteq X \to X$ be the infinitesimal generator of a C_0-semigroup of contractions, let $\mathbb{I}$ be a nonempty and open interval, Ω a nonempty and open subset in X, and $f : \mathbb{I} \times \Omega \to X$ a function satisfying:

(i) for a.e. for $t \in \mathbb{I}$, $f(t,\cdot) : \Omega \to X$ is continuous;
(ii) for each $x \in \Omega$, $f(\cdot,x) : \mathbb{I} \to X$ is measurable;
(iv) for each $\xi \in \Omega$ there exists $r > 0$, and a locally integrable function $\ell : \mathbb{I} \to \mathbb{R}_+$, such that $\|g(t,u) - g(t,v)\| \leq \ell(t)\|u - v\|$ for each $u, v \in B(\xi,r) \cap \Omega$, and a.e. for $t \in \mathbb{I}$,

and $g(\cdot,\xi) : \mathbb{I} \to X$ is locally integrable. Then, for each $a \in \mathbb{I}$ and $\xi \in \Omega$, there exists $b > a$ with $[\,a,b\,] \subset \mathbb{I}$, and such that the Cauchy problem

$$\begin{cases} u' = Au + f(t,u) \\ u(a) = \xi \end{cases}$$

has a unique C^0-solution $u : [\,a,b\,] \to \Omega$. If, in addition, $\Omega = X$, and the function ℓ in (iv) is independent of $r > 0$, then this solution can be continued up to a global one, i.e. defined on $[\,a,b)$, where $b = \sup \mathbb{I}$.

Problem 11.3. Let $a \in \mathbb{R}^3$, and $f : \mathbb{R}_+ \times \mathbb{R}^3 \to \mathbb{R}^3$ a function satisfying (i), (ii) in Problem 11.1, and (iv) in Problem 11.2. If the function ℓ in (iv) is independent of $r > 0$, then, for each $u_0 \in L^p(\mathbb{R}^3)$, the semilinear Cauchy problem

$$\begin{cases} u_t + a \cdot \nabla u = f(t,u) & (t,x) \in \mathbb{R}_+ \times \mathbb{R}^3 \\ u(0,x) = u_0(x) & x \in \mathbb{R}^3 \end{cases}$$

has a unique solution defined on $\mathbb{R}_+$.

Problem 11.4. Let $\mathcal{A}$ be an $n \times n$ matrix with real elements, and let $f : \mathbb{R}_+ \times \mathbb{R}^n \to \mathbb{R}^n$ be a function satisfying (i), (ii) in Problem 11.1 and (iv) in Problem 11.2. If the function ℓ in (iv) is independent of $r > 0$, then, for each $u_0 \in L^p(\mathbb{R}^3)$, the semilinear Cauchy problem

$$\begin{cases} u_t + \langle \mathcal{A}x, \nabla u \rangle = f(t, u) & (t, x) \in \mathbb{R}_+ \times \mathbb{R}^3 \\ u(0, x) = u_0(x) & x \in \mathbb{R}^n \end{cases}$$

has a unique solution defined on $\mathbb{R}_+$.

Problem 11.5. Let $a > 0$, $c > 0$ and let $\varepsilon > 0$ be "very small". Let $\varphi : \mathbb{R} \to \mathbb{R}$ be a continuous function satisfying $|\varphi(x)| \leq c$ for each $x \in \mathbb{R}$ and

$$\varphi(x) = \begin{cases} c & \text{if } |x| \leq a \\ 0 & \text{if } |x| > a + \varepsilon \end{cases}$$

and let us consider the Cauchy problem for the neutron transport in a stab equation

$$\begin{cases} u_t(t, x, y) + y u_x(t, x, y) = \dfrac{\varphi(x)}{2} \displaystyle\int_{-1}^{1} u(t, x, z)\, dz \\ u(0, x, y) = \xi(x, y). \end{cases}$$

Prove that, for each $\xi \in C_{ub}(\mathbb{R} \times [-1, 1])$, the problem above has a unique solution $u : \mathbb{R}_+ \to C_{ub}(\mathbb{R} \times [-1, 1])$.

Notes. The main result in Section 11.1 is due to Segal [**111**] and the one in Section 11.2 was proved by Pazy [**99**]. A nonlinear version of the latter was obtained by Vrabie [**124**]. For other results of this type see Pavel [**97**] and Vrabie [**127**]. Both Sections 11.3 and 11.4 contain slight extensions of some classical results on the continuation of the solutions and on their asymptotic behavior. See Sections 2.4 and 5.3 in Vrabie [**129**], and Cazenave and Haraux [**36**], Theorem 10.2.2, p. 157. The example analyzed in Section 11.5 is a reformulation of Theorem 6.2.2 in Cazenave and Haraux [**36**], while that in Section 11.6 is inspired from Pazy [**101**]. The first two Problems 11.1 and 11.2 extend Theorems 11.2.1 and 11.1.1 respectively to the case of Carathéodory perturbations. Problems 11.3 and 11.4 are classic, while Problem 11.5 is inspired from Richtmyer [**110**].

CHAPTER 12

Semilinear Equations Involving Measures

Here, we consider the Cauchy problem for the semilinear differential equation with distributed and/or spatial measures

$$\begin{cases} du = \{Au\}dt + dg_u \\ u(a) = \xi. \end{cases}$$

In Section 12.1 we prove a local existence and uniqueness result in the case when $u \mapsto g_u$ is Lipschitz from $L^\infty(a,b;X)$ to $BV([a,b];X)$ and $A : D(A) \subseteq X \to X$ generates a C_0-semigroup of contractions, continuous from $(0,+\infty)$ to $\mathcal{L}(X)$ in the uniform operator topology. In Section 12.2 we study the problem of local existence assuming that A generates a compact C_0-semigroup of contractions and $u \mapsto g_u$ is continuous in some weak sense. In Section 12.3 we present the main results concerning saturated solutions and in Section 12.4 we reconsider the problem in Section 12.2 within a more general setting, i.e. allowing $\xi \in X_A$ and $g \in BV([a,b];X_A)$. Finally, in the last two Sections 12.5 and 12.6 we include some examples.

12.1. The Problem $du = \{Au\}dt + dg_u$ with $u \mapsto g_u$ Lipschitz

Let X be a real Banach space, let $A : D(A) \subseteq X \to X$ be the generator of a C_0-semigroup $\{S(t)\,;\ t \geq 0\}$, continuous in the uniform operator topology from $(0,+\infty)$ to $\mathcal{L}(X)$, and let $G : L^\infty(a,b;X) \to BV([a,b];X)$ be a given function. Let $\xi \in X$, and let us consider the Cauchy problem

$$\begin{cases} du = \{Au\}dt + dg_u \\ u(a) = \xi, \end{cases} \tag{12.1.1}$$

where $g_u = G(u)$.
As in the preceding chapters, here, we deal only with C_0-semigroups of contractions, but we emphasize that all the results hold true for general C_0-semigroups.

Definition 12.1.1. The mapping $G : L^\infty(a,b;X) \to BV([a,b];X)$ is called *hereditary* if for each $u, v \in L^\infty(a,b;X)$ satisfying $u = v$ a.e. on $[a,c]$, we have also $\mathrm{Var}\,(G(u)-G(v)),[a,c]) = 0$.

Definition 12.1.2. The function $G : L^\infty(a,b;X) \to BV([a,b];X)$ is called *locally Lipschitz* if there exists a function $\ell : (0,+\infty) \to (0,+\infty)$ with $\lim_{x\downarrow 0} \ell(x) = 0$ and such that

$$\mathrm{Var}\,(G(u)-G(v),[a,c]) \leq \ell(c-a)\|u-v\|_{L^\infty(a,c;X)}$$

for each $u, v \in L^\infty(a,b;X)$ and each $c \in [a,b]$.

Remark 12.1.1. It is easy to observe that each locally Lipschitz function is hereditary.

Definition 12.1.3. Let $G : L^\infty(a,b;X) \to BV([a,b];X)$ be hereditary. A function $u : [a,c] \to X$ is called an *$\mathfrak{L}^\infty$-solution* of (12.1.1) on $[a,c]$ if it is an $\mathfrak{L}^\infty$-solution on $[a,c]$ of the Cauchy problem below

$$\begin{cases} du = \{Au\}dt + dg_{\tilde{u}} \\ u(a) = \xi, \end{cases}$$

where $\tilde{u} \in L^\infty(a,b;X)$ satisfies $\tilde{u} = u$ a.e. for $t \in [a,c]$.

Remark 12.1.2. We notice that, since G is hereditary, for each $\tilde{u}$ and $\tilde{v}$ in $L^\infty(a,b;X)$ with $\tilde{u} = \tilde{v} = u$ a.e. for $t \in [a,c]$, it follows that $G(u) - G(v)$ is constant on $[a,c]$. Therefore, $dg_{\tilde{u}} = dg_{\tilde{v}}$ in the sense of vector measures on $[a,c]$. So, without any danger of confusion, we can denote both these measures by dg_u.

A very simple but useful local existence and uniqueness result is:

Theorem 12.1.1. *Let $A : D(A) \subseteq X \to X$ be the generator of a C_0-semigroup of contractions, $\{S(t)\,;\, t \geq 0\}$, continuous from $(0,+\infty)$ to $\mathfrak{L}(X)$ in the uniform operator topology. Let $G : L^\infty(a,b;X) \to BV([a,b];X)$ be a locally Lipschitz mapping. Then, for each $\xi \in X$, there exists $c \in (a,b]$ such that the problem (12.1.1) has a unique $\mathfrak{L}^\infty$-solution defined on $[a,c]$.*

Proof. Let $c \in (a,b]$ be such that

$$\ell(c-a) < 1.$$

We define the operator $Q : L^\infty(a,c;X) \to L^\infty(a,c;X)$ by

$$Q(u)(t) = S(t-a) + \int_a^t S(t-s)\,dg_u(s)$$

for each $u \in L^\infty(a,c;X)$ and each $t \in [a,c]$, where dg_u is defined as in Remark 12.1.2. At this point let us observe that u is an $\mathfrak{L}^\infty$-solution of

(12.1.1) on $[a,c]$ if and only if u is a fixed point of Q. So, to complete the proof we have merely to show that Q has a unique fixed point. We shall do this with help of Banach fixed point theorem. To this aim let us observe that

$$\|Q(u)(t) - Q(v)(t)\| \leq \left\| \int_a^t S(t-s)\, d(g_u - g_v)(s) \right\|$$

$$\leq \operatorname{Var}(g_u - g_v, [a,t]) \leq \operatorname{Var}(g_u - g_v, [a,c]) \leq \ell(c-a)\|u - v\|_{L^\infty(a,c;X)}$$

for each $u, v \in L^\infty(a,c;X)$. But this inequality clearly shows that

$$\|Q(u) - Q(v)\|_{L^\infty(a,c;X)} \leq \ell(c-a)\|u - v\|_{L^\infty(a,c;X)}$$

for each $u, v \in L^\infty(a,c;X)$. Since $\ell(c-a) < 1$, Q is a strict contraction and therefore, by virtue of Banach fixed point theorem, it has a unique fixed point as claimed. The proof is complete. □

Let us consider next a useful application of Theorem 12.1.1. Let $\mathfrak{D}$ be a nonempty and open subset in $\mathbb{R} \times X$, let $f : \mathfrak{D} \to X$ a continuous function, $g \in BV([a,b,];X)$ and $(a,\xi) \in \mathbb{R} \times X$. We consider the Cauchy problem

$$\begin{cases} du = \{Au + f(t,u)\}dt + dg \\ u(a) = \xi. \end{cases} \tag{12.1.2}$$

Definition 12.1.4. By an $\mathcal{L}^\infty$-*solution* of (12.1.2) on $[a,c]$ we mean a function $u : [a,c] \to X$ which satisfies:

(i) for each $t \in [a,c)$, there exists $u(t+0)$;
(ii) for each $t \in [a,c)$, $(t, u(t+0)) \in \mathfrak{D}$;
(iii) $t \mapsto f(t, u(t+0)) \in L^1(a,c;X)$

and u is an $\mathcal{L}^\infty$-solution on $[a,c]$ of

$$\begin{cases} du = \{Au\}dt + dh(t) \\ u(a) = \xi \end{cases}$$

in the sense of Definition 9.1.1, where, for each $t \in [a,c]$,

$$h(t) = \int_a^t f(s, u(s+0))\, ds + g(t).$$

Remark 12.1.3. If $u : [a,c] \to X$ is an $\mathcal{L}^\infty$-solution of (12.1.1) then u is piecewise continuous on $[a,c]$ and consequently $f(t,u(t+0)) = f(t,u(t))$ excepting for an at most countable subset in $[a,c]$. This shows that h in Definition 12.1.4 is given in fact by

$$h(t) = \int_a^t f(s, u(s))\, ds + g(t)$$

for each $t \in [a,c]$. So, we may question why, in Definition 12.1.4, instead of (ii), we do not ask $(t, u(t)) \in \mathfrak{D}$ for each $t \in [a,c]$, as in the case of

C^0-solutions. The answer rests in that, for $\mathcal{L}^\infty$-solutions, it may happen that, at some point of discontinuity $t \in (a, c]$, $(t, u(t-0)) \notin \mathcal{D}$, and thus, in order to enter again $\mathcal{D}$, $(t, u(t))$, which is uniquely determined by the "left jump" condition $u(t) - u(t-0) = g(t) - g(t-0)$, must be, not only outside $\mathcal{D}$, but even outside its closure.

Theorem 12.1.2. *If $A : D(A) \subseteq X \to X$ generates a C_0-semigroup of contractions, $\{S(t)\,;\ t \geq 0\}$, which is continuous in the uniform operator topology from $(0, +\infty)$ to $\mathcal{L}(X)$, $f : \mathcal{D} \to X$ is continuous and locally Lipschitz with respect to its last argument, and $g \in BV([a, b]; X)$ then, for each $(a, \xi) \in \mathbb{R} \times X$ satisfying*

$$(a, g(a+0) - g(a) + \xi) \in \mathcal{D},$$

there exists $c > a$ such that the problem (12.1.2) *has a unique $\mathcal{L}^\infty$-solution defined on $[a, c]$.*

Proof. Let $(a, \xi) \in \mathbb{R} \times X$ with $(a, g(a+0) - g(a) + \xi) \in \mathcal{D}$, and let us denote by $\eta = g(a+0) - g(a) + \xi$. Inasmuch as $\mathcal{D}$ is open, there exist $b > a$ and $r > 0$ such that $[a, b] \times B(\eta, r) \subset \mathcal{D}$. In addition, as f is continuous and locally Lipschitz with respect to its last argument, diminishing $b > a$ and/or $r > 0$ if necessary, we may assume that there exists $L > 0$ such that

$$\|f(t, u) - f(t, v)\| \leq L\|u - v\|$$

for each $(t, u), (t, v) \in [a, b] \times B(\eta, r)$. Let $\rho : X \to X$ be defined by

$$\rho(y) = \begin{cases} y & \text{for } y \in B(\eta, r) \\ \dfrac{r}{\|y - \eta\|}(y - \eta) + \eta & \text{for } y \in X \setminus B(\eta, r). \end{cases}$$

We may easily see that ρ maps X to $B(\eta, r)$, and is Lipschitz continuous on X with Lipschitz constant 2. Next, let us define $f_r : [a, b] \times X \to X$ by

$$f_r(t, y) = f(t, \rho(y))$$

for each $(t, y) \in [a, b] \times X$. Since f is continuous and globally Lipschitz with respect to its last argument on $[a, b] \times B(\eta, r)$, we conclude that f_r enjoys the same properties. Let us define $G : L^\infty(a, b; X) \to BV([a, b]; X)$ by

$$G(u)(t) = \int_a^t f_r(s, u(s))\, ds + g(t).$$

We note that G is well-defined because, for each $u \in L^\infty(a, b; X)$, the mapping $s \mapsto f_r(s, u(s))$ is strongly measurable and bounded. While the boundedness is obvious, the measurability follows from the observation that it can be approximated a.e. by a sequence of countably-valued functions simply because u enjoys this property and f_r is continuous. See (i) and (iii) in Definition 1.1.1. Now, let us remark that the problem (12.1.2) can be rewritten under the form (12.1.1), where $g_u = G(u)$, with G defined as above. Let us observe that

$$\begin{aligned}\operatorname{Var}(G(u) - G(v), [\,a, c\,]) &\leq \int_a^c \|f_r(s, u(s)) - f_r(s, v(s))\|\, ds \\ &\leq 2L(c-a)\|u - v\|_{L^\infty(a,c\,;X)}\end{aligned}$$

for each $c \in (a, b\,]$ and each $u, v \in L^\infty(a, c\,; X)$. This shows that G is locally Lipschitz (with $\ell(x) = 2Lx$). By virtue of Theorem 12.1.1, we know that the Cauchy problem

$$\begin{cases} du = \{Au\}dt + dg_u(t) \\ x(a) = \xi \end{cases}$$

has a unique $\mathfrak{L}^\infty$-solution $u : [\,a, c\,] \to X$ provided $c \in (a, b\,]$ is small enough. Recalling that, by virtue of (9.2.1) in Theorem 9.2.1,

$$u(a+0) = g(a+0) - g(a) + u(a) = \eta,$$

diminishing $c > a$ if necessary, we get that $u(t) \in B(\eta, r)$ for each $t \in [\,a, c\,]$. Hence $(t, u(t+0) \in [\,a, c\,] \times B(\eta, r) \subset \mathfrak{D}$ for each $t \in [\,a, c)$. Accordingly $f_r(t, u(t+0)) = f(t, u(t+0))$, and therefore $u : [\,a, c\,] \to X$ is in fact an $\mathfrak{L}^\infty$-solution of the problem (12.1.2). The proof is complete. □

12.2. The Problem $du = \{Au\}dt + dg_u$ with $u \mapsto g_u$ Continuous

Let X be a Banach space and let $A : D(A) \subseteq X \to X$ be the infinitesimal generator of a compact C_0-semigroup of contractions $\{S(t)\,;\ t \geq 0\}$, let $G : L^\infty(a, b\,; X) \to BV([\,a, b\,]; X)$ be a given function, $\xi \in X$ and let us consider the Cauchy problem

$$\begin{cases} du = \{Au\}dt + dg_u \\ u(a) = \xi, \end{cases} \tag{12.2.1}$$

where $g_u = G(u)$. As in the preceding section, the meaning of the $\mathfrak{L}^\infty$-solution is that given by Definition 9.1.1.

Definition 12.2.1. The mapping $G : L^\infty(a, b\,; X) \to BV([\,a, b\,]\,; X)$ is *bounded and weakly pointwise continuous* if its range is of equibounded variation and for each sequence $(u_n)_{n\in\mathbb{N}}$ satisfying

$$\begin{cases} (u_n)_{n\in\mathbb{N}} \ \text{ is bounded in } L^\infty(a, b\,; X) \text{ and} \\ \lim\limits_{n\to\infty} u_n = u \ \text{ in } \ L^1(a, b\,; X) \end{cases}$$

we have

$$\lim_{n\to\infty} G(u_n)(t) = G(u)(t)$$

in the $\sigma(X, X^{\odot})$-topology for each $t \in [a,b]$.

The main result in this section is:

Theorem 12.2.1. *Let $A : D(A) \subseteq X \to X$ be the generator of a compact C_0-semigroup of contractions and let $G : L^\infty(a,b;X) \to BV([a,b];X)$ be bounded and weakly pointwise continuous. Then, for each $\xi \in X$, the problem* (12.2.1) *has at least one $\mathcal{L}^\infty$-solution defined on $[a,b]$.*

Proof. Let $M > 0$ be such that

$$\mathrm{Var}\,(g_u, [a,b]) \leq M$$

for each $u \in L^\infty(a,b;X)$. Let $\xi \in X$, let $r = \|\xi\| + M$ and let K be the closed ball with center 0 and radius r in $L^\infty(a,b;X)$. We define the operator $Q : K \to L^\infty(a,b;X)$ by

$$Q(u)(t) = S(t-a)\xi + \int_a^t S(t-s)\,dg_u(s)$$

for each $u \in K$ and each $t \in [a,b]$. Obviously u is an $\mathcal{L}^\infty$-solution of (12.2.1) if and only if u is a fixed point of Q. So, to complete the proof, it suffices to show that Q has at least one fixed point in K. In order to prove this, we will show that Q satisfies the hypotheses of Schauder fixed point theorem A.1.5. First, let us observe that

$$\|Q(u)(t)\| \leq \|S(t-a)\xi\| + \left\| \int_a^t S(t-s)\,dg_u(s) \right\|$$
$$\leq \|\xi\| + \mathrm{Var}\,(g, [a,t]) \leq \|\xi\| + \mathrm{Var}\,(g, [a,b]) \leq r.$$

So Q maps K into itself. Since G is weakly pointwise continuous and bounded, by virtue of Theorem 9.4.2, it follows that $Q(K)$ is relatively compact in $L^1(a,b;X)$ (in fact in any $L^p(a,b;X)$, provided $p \in [1,+\infty)$). Summing up, K, as a subset of $L^1(a,b;X)$, is closed, convex and bounded and $Q(K)$ is relatively compact. To achieve the proof it remains merely to check that Q is continuous from K to K in the norm topology of $L^1(a,b;X)$. But, this follows from the weak pointwise continuity and boundedness of G combined with the relative compactness of $Q(K)$, along with Theorem 9.5.1. Indeed, let $u \in K$ and let $(u_k)_{k\in\mathbb{N}}$ be a sequence in K which converges in $L^1(a,b;X)$ to u. Since G is bounded and weakly pointwise continuous and $Q(K)$ is relatively compact, by Theorem 9.5.1, it follows that

$$\begin{cases} \lim_{k\to\infty} Q(u_k) = v & \text{in } L^1(a,b;X) \text{ and} \\ \lim_{k\to\infty} Q(u_k)(t) = v(t) & \text{in the } \sigma(X,X^{\odot}) - \text{topology on } X, \end{cases}$$

where $v : [a, b] \to X$ is the unique generalized $\mathcal{L}^\infty$-solution of the problem

$$\begin{cases} dv = \{Av\}dt + dg_u \\ v(a) = \xi. \end{cases}$$

Inasmuch as g_u belongs to $BV([a, b]; X)$, the problem above has a unique $\mathcal{L}^\infty$-solution which must coincide with v. So $v = Q(u)$ and Q is continuous from K to K in the norm topology of $L^1(a, b; X)$. Consequently Q satisfies the hypotheses of Schauder fixed point Theorem A.1.5 and this completes the proof. □

A useful application of Theorem 12.2.1 is given below.
Throughout in this chapter, $BV(\mathbb{R}; X)$ denotes the space of all functions $g : \mathbb{R} \to X$ whose restrictions to any interval $[a, b]$ belong to $BV([a, b]; X)$. Let $\mathcal{D}$ be a nonempty and open subset in $\mathbb{R} \times X$, let $f : \mathcal{D} \to X$ be a continuous function and $g \in BV(\mathbb{R}; X)$. We consider the Cauchy problem

$$\begin{cases} du = \{Au + f(t, u)\}dt + dg \\ u(a) = \xi. \end{cases} \tag{12.2.2}$$

The next local existence theorem, extending Pazy's Theorem 11.2.1, is useful in the study of semilinear parabolic problems with distributed measures as we shall see in Section 5. We notice that Pazy's Theorem 11.2.1 refers to the special case $g \equiv 0$.

Theorem 12.2.2. *If $A : D(A) \subseteq X \to X$ is the infinitesimal generator of a compact C_0-semigroup of contractions $\{S(t);\ t \geq 0\}$, $f : \mathcal{D} \to X$ is continuous, and $g \in BV(\mathbb{R}; X)$, then, for each $(a, \xi) \in \mathbb{R} \times X$ satisfying*

$$(a, g(a + 0) - g(a) + \xi) \in \mathcal{D},$$

there exists $b > a$ such that the problem (12.2.2) *has at least one $\mathcal{L}^\infty$-solution defined on $[a, b]$.*

Proof. Let $(a, \xi) \in \mathbb{R} \times X$ with $(a, g(a+0) - g(a) + \xi) \in \mathcal{D}$, and let us denote by $\eta = g(a+0) - g(a) + \xi$. Inasmuch as $\mathcal{D}$ is open and f is continuous, there exist $c > a$, $r > 0$ and $M > 0$, such that $[a, c] \times B(\eta, r) \subset \mathcal{D}$, and

$$\|f(t, u)\| \leq M \tag{12.2.3}$$

for each $(t, u) \in [a, c] \times B(\eta, r)$. Let us define $\rho : X \to X$ by

$$\rho(y) = \begin{cases} y & \text{for } y \in B(\eta, r) \\ \dfrac{r}{\|y - \eta\|}(y - \eta) + \eta & \text{for } y \in X \setminus B(\eta, r). \end{cases}$$

Clearly, ρ maps X to $B(\eta, r)$ and is continuous. Now, let us define the function $f_r : \mathbb{R} \times X \to X$ by

$$f_r(t,u) = \begin{cases} f(a, \rho(u)) & \text{if } t \in (-\infty, a] \quad \text{and } u \in X \\ f(t, \rho(u)) & \text{if } t \in (a, c) \quad \text{and } u \in X \\ f(c, \rho(u)) & \text{if } t \in [c, +\infty) \quad \text{and } u \in X. \end{cases}$$

From (12.2.3), we conclude that f_r is bounded. Since both functions f and ρ are continuous, it follows that f_r is continuous. Finally, let us define the mapping $G : L^\infty(a, c; X) \to BV([a, c]; X)$ by

$$G(u)(t) = g_u(t) = \int_a^t f_r(s, u(s))\, ds + g(t)$$

for each $u \in L^\infty(a, c; X)$ and $t \in [a, c]$. It is easy to see that G is bounded, continuous with respect to the variation seminorm on $BV([a, c]; X)$, and thus weakly pointwise continuous in the $\sigma(X, X^{\odot})$-topology. By virtue of Theorem 12.2.1, it follows that the Cauchy problem

$$\begin{cases} du = \{Au + f_r(t, u)\}dt + dg \\ x(a) = \xi \end{cases}$$

has at least one $\mathcal{L}^\infty$-solution, $u : [a, c] \to X$. Since, $u(a) = \xi$ and, by (9.2.1) in Theorem 9.2.1,

$$u(a + 0) = g(a + 0) - g(a) + u(a) = \eta,$$

there exists $b \in (a, c]$ such that, for each $t \in (a, b]$, $u(t) \in B(\eta, r)$. But in this case $\rho(u(t)) = u(t)$ for each $t \in (a, b]$, and consequently $f_r(s, u(s))$ must coincide with $f(s, u(s))$ for each $s \in (a, b]$. Since $u : [a, b] \to X$ is piecewise continuous and $(t, u(t)) \in (a, c] \times B(\eta, r) \subset \mathcal{D}$, it follows that $(t, u(t + 0)) \in [a, c] \times B(\eta, r) \subset \mathcal{D}$, and thus, by Remark 12.1.3, u is an $\mathcal{L}^\infty$-solution of the problem (12.2.2). This completes the proof. □

12.3. Saturated $\mathcal{L}^\infty$-Solutions

In this section, we shall say a few words about saturated $\mathcal{L}^\infty$-solutions for the semilinear Cauchy problem

$$\begin{cases} du = \{Au + f(t, u)\}dt + dg \\ u(a) = \xi. \end{cases} \tag{12.3.1}$$

Here, as in the preceding sections, $\mathcal{D}$ is a nonempty and open subset in $\mathbb{R} \times X$, $f : \mathcal{D} \to X$ is a continuous function and $g \in BV(\mathbb{R}; X)$.

Definition 12.3.1. An $\mathfrak{L}^\infty$ solution $u : \mathbb{I} \to X$ of (12.3.1), $\mathbb{I} = [\,a, b\,]$, or $\mathbb{I} = [\,a, b)$, is called *continuable* if there exists an $\mathfrak{L}^\infty$ solution $v : [\,a, c\,] \to X$ of (12.3.1), with $C > b$ and such that $u(t) = v(t)$ for all $t \in \mathbb{I}$. A solution which is not continuable is called *saturated*, or noncontinuable.

Remark 12.3.1. In contrast with the case considered Chapter 11, where each saturated C^0-solution were necessarily defined on an interval of the form $[\,a, b)$, here it may exist saturated $\mathfrak{L}^\infty$-solutions of (12.3.1) which are defined on a closed interval $[\,a, b\,]$. Clearly this happens only if the right jump of g at b is such that $(b, g(b+0)-g(b)+u(b)) \notin \mathfrak{D}$. See Theorems 12.1.2 and 12.2.2. Therefore, in all that follows, we confine ourselves only to the study of saturated $\mathfrak{L}^\infty$-solutions defined on intervals of the form $[\,a, b)$.

Lemma 12.3.1. *Let $A : D(A) \subseteq X \to X$ be the generator of a C_0-semigroup of contractions and let $f : \mathfrak{D} \to X$ be continuous. Assume that either the semigroup is compact, or f is locally Lipschitz with respect to its last argument. Let $g \in BV(\mathbb{R}; X)$. Then an $\mathfrak{L}^\infty$-solution $u : [\,a, b) \to X$ of* (12.3.1) *is continuable if and only if there exists*

$$u^* = \lim_{t \uparrow b} u(t) \tag{12.3.2}$$

and

$$(b, g(b+0) - g(b-0) + u^*) \in \mathfrak{D} \tag{12.3.3}$$

Proof. *Necessity.* Assume that u is continuable. Then, in view of (9.2.1) and (9.2.3), we have

$$g(b+0) - g(b-0) + u^* = u(b+0) - u(b-0) + u(b-0) = u(b+0)$$

and, by (i) in Definition 12.1.4, $(b, u(b+0)) \in \mathfrak{D}$. Thus (12.3.3) holds.
Sufficiency. Let us assume that (12.3.2) and (12.3.3) hold. Take

$$\xi = g(b) - g(b-0) + u^*$$

and let us observe that $g(b+0) - g(b) + \xi = g(b+0) - g(b-0) + u^*$. Then, by virtue of (12.3.3), we are in the hypotheses of either Theorem 12.1.2, or Theorem 12.2.2, and accordingly the problem

$$\begin{cases} dv = \{Av + f(t, v)\}dt + dg \\ v(b) = g(b) - g(b-0) + u^* \end{cases}$$

has at least one $\mathfrak{L}^\infty$-solution $v : [\,b, c) \to X$. By Proposition 9.1.1, we conclude that the concatenate function $w : [\,a, c) \to X$, defined by

$$w(t) = \begin{cases} u(t) & \text{for } t \in [\,a, b) \\ v(t) & \text{for } t \in [\,b, c) \end{cases}$$

is an $\mathfrak{L}^\infty$-solution of (12.3.1) on $[\,a, c)$ which coincides with u on $[\,a, b)$. Thus u is continuable and this completes the proof. □

Proposition 12.3.1. *Let $A : D(A) \subseteq X \to X$ be the generator of a C_0-semigroup of contractions which is continuous in the uniform operator topology from $(0,+\infty)$ to $\mathcal{L}(X)$, let $f : \mathfrak{D} \to X$ be continuous and let $g \in BV(\mathbb{R}; X)$. Let $u : [a,b) \to X$ be an $\mathcal{L}^\infty$-solution of* (12.3.1). *If $b < +\infty$ and $t \mapsto f(t,u(t+0))$ belongs to $L^1(a,b;X)$, then there exists*

$$u(b-0) = \lim_{t \uparrow b} u(t).$$

Proof. Since u is given by

$$u(t) = S(t-a)u(a) + \int_a^t S(t-s)f(s,u(s+0))\,ds + \int_a^t S(t-s)\,dg(s)$$

and the first two terms on the right-hand side have limit as t tends to b as the semigroup is continuous and $t \mapsto f(t,u(t+0))$ belongs to $L^1(a,b;X)$, it remains only to show that there exists

$$\lim_{t \uparrow b} \int_a^t S(t-s)\,dg(s).$$

But this is a direct consequence of the second part of Theorem 9.2.1. The proof is complete. □

Remark 12.3.2. The condition $t \mapsto f(t,u(t+0))$ belongs to $L^1(a,b;X)$ in Proposition 12.3.1 above can be replaced with, either $t \mapsto f(t,u(t))$, or $t \mapsto f(t,u(t-0))$ belongs to $L^1(a,b;X)$. This follows from the simple observation that these three mappings coincide a.e. on (a,b) simply because u is piecewise continuous on $[a,b)$.

Theorem 12.3.1. *Let $A : D(A) \subseteq X \to X$ be the generator of a C_0-semigroup of contractions, let $f : \mathfrak{D} \to X$ be continuous and $g \in BV(\mathbb{R}; X)$. Assume that either the semigroup is compact, or f is locally Lipschitz with respect to its last argument. Let $u : [a,b) \to X$ be an $\mathcal{L}^\infty$-solution of* (12.3.1). *Then either u is saturated, or u can be continued up to a saturated one.*

The proof follows the same lines as those of Theorem 11.3.2 and therefore we omit it.

From Theorems 12.1.2, 12.2.2 and 12.3.1 we deduce:

Corollary 12.3.1. *Let $A : D(A) \subseteq X \to X$ be the generator of a C_0-semigroup of contractions, let $f : \mathfrak{D} \to X$ be continuous and $g \in BV(\mathbb{R}; X)$. Assume that either the semigroup is compact, or f is locally Lipschitz with respect to its last argument. Then, for each $(a,\xi) \in \mathbb{R} \times X$ with*

$$(a, g(a+0) - g(a-0) + \xi) \in \mathfrak{D},$$

the problem (12.3.1) *has at least one saturated $\mathcal{L}^\infty$-solution.*

Theorem 12.3.2. *Let $A : D(A) \subseteq X \to X$ be the generator of a C_0-semigroup of contractions, let $f : \mathfrak{D} \to X$ be continuous and $g \in BV(\mathbb{R}; X)$. Assume that f maps bounded subsets in $\mathfrak{D}$ into bounded subsets in X and, either the semigroup is compact, or f is locally Lipschitz with respect to its last argument. Let $u : [\,a, b) \to X$ be a saturated $\mathfrak{L}^\infty$-solution of* (12.3.1). *Then either*

(i) *$t \mapsto u(t+0)$ is unbounded on (a, b) and, if $b < +\infty$, there exists*

$$\lim_{t\uparrow b} \|u(t+0)\| = +\infty, \text{ or}$$

(ii) *$t \mapsto u(t+0)$ is bounded and, either u is global, or*

(iii) *$t \mapsto u(t+0)$ is bounded and u is non-global, case in which there exists*

$$\lim_{t\uparrow b} u(t) = u(b-0) \text{ and}$$
$$(b, g(b+0) - g(b-0) + u(b-0)) \notin \mathfrak{D}.$$

Remark 12.3.3. Unlike the case of semilinear evolution equations already considered in Chapter 11, here, under the circumstances described in (iii), it may happen that, not only $(b, g(b+0)-g(b-0)+u(b-0)) \notin \mathfrak{D}$, but even $(b, g(b+0) - g(b-0) + u(b-0)) \notin \overline{\mathfrak{D}}$. This apparently strange situation is due to the (possible) existence of an "excessive" right jump of g at b. For instance, let us consider the following simple example. Take $X = \mathbb{R}$, $\mathfrak{D} = \mathbb{R} \times (-1, 1)$, $A \equiv 0$, $g : \mathbb{R} \to \mathbb{R}$ defined by

$$g(t) = \begin{cases} 0 & \text{if } t < 0 \\ t^n & \text{if } t \in [\,0, 1\,] \\ 3 & \text{if } t > 1 \end{cases}$$

and $f : \mathfrak{D} \to \mathbb{R}$, $f \equiv 0$. Then one may easily verify that the unique saturated $\mathfrak{L}^\infty$-solution of the Cauchy problem

$$\begin{cases} du = \{Au + f(t,u)\}dt + dg \\ u(0) = 0 \end{cases}$$

is $u : [\,0, 1) \to \mathbb{R}$, given by $u(t) = g(t)$ for each $t \in [\,0, 1)$. But in this case, $(1, g(1+0) - g(1-0) + u(1-0)) = (1, 3) \notin \overline{\mathfrak{D}}$.

We may now pass to the proof of Theorem 12.3.2.

Proof. Up to one point, the proof parallels that of Theorem 11.3.4. So, let us assume first that $t \mapsto u(t-0)$ is bounded and non-global. Since, except for an at most countably subset, $t \mapsto f(t, u(t))$ and $t \mapsto f(t, u(t-0))$ coincide and the latter is bounded, from Proposition 12.3.1, it follows that there exists $\lim_{t\uparrow b} u(t) = u(b-0)$. Thanks to Lemma 12.3.1, we conclude that $(a, g(b+0)-g(b-0)+u(b-0) \notin \mathfrak{D}$ and thus we have (iii). To complete the proof, we have only to check (i). So, let us assume that u is unbounded. To show that $\lim_{t\uparrow b}\|u(t-0)\| = +\infty$, let us assume the contrary, i.e. that, under these circumstances, we have

$$\liminf_{t\uparrow b}\|u(t-0)\| < +\infty.$$

Equivalently, there exist at least one sequence $(t_k)_{k\in\mathbb{N}}$ in (a,b), convergent to b, and $r>0$ such that $\|u(t_k-0)\|\leq r$, for each $k\in\mathbb{N}$.
Let $C=\{v\in X;\ \|v\|\leq r+1\}$. Since f maps bounded subsets in $\mathfrak{D}$ into bounded subsets in X, $b<+\infty$ and C is bounded, there exists $M>0$, such that

$$\|f(\tau,v)\|\leq M, \tag{12.3.4}$$

for each $(\tau,v)\in([\,a,b)\times C)\cap\mathfrak{D}$. Now let us choose $d>0$ satisfying

$$\begin{cases} dM<\dfrac{1}{2}, \\ \operatorname{Var}(g,[\,b-d,b))<\dfrac{1}{2} \end{cases} \tag{12.3.5}$$

and fix $k\in\mathbb{N}$ such that $b-d<t_k<b$. Since $t\mapsto u(t+0)$ is unbounded on $[\,a,b)$, it is necessarily unbounded on $[\,t_k,b)$. Let

$$t^*=\sup\{t\in[\,t_k,b)\,;\ \|u(s+0)\|\leq r+1, \forall s\in[\,t_k,t)\}.$$

Recalling that u is piecewise continuous on $[\,a,b)$ (see Theorem 9.2.1), it readily follows that $t\mapsto u(t+0)$ is continuous from the right on $[\,a,b)$. This clearly implies that $t^*\in(t_k,b)$. By the definition of t^* it follows that there exists a sequence $(t_k^*)_{k\in\mathbb{N}}$ in (t^*,b) with $t_k^*\downarrow t^*$ and such that $r+1<\|u(t_k^*+0)\|$ for each $k\in\mathbb{N}$. Again from the right continuity of the mapping $t\mapsto u(t+0)$, we conclude that

$$r+1\leq\|u(t^*+0)\|.$$

A simple calculation argument shows that

$$r+1\leq\|u(t^*+0)\|\leq\|u(t_k+0)\|+\operatorname{Var}(g,[\,t_k,t^*))+(t^*-t_k)M.$$

But (12.3.4), (12.3.5) and the obvious inequalities $b-d<t_k<t^*<b$, yield

$$r+1\leq\|u(t^*+0)\|\leq\|u(t_k+0)\|+\operatorname{Var}(g,[\,b-d,b))+dM<r+\frac{1}{2}+\frac{1}{2}=r+1.$$

This contradiction can be eliminated only if

$$\lim_{t\uparrow b}\|u(t+0)\|=+\infty$$

and this completes the proof. □

Corollary 12.3.2. *Let us assume that $A : D(A) \subseteq X \to X$ generates a C_0-semigroup of contractions, $f : \mathbb{R}\times X \to X$ is continuous and $g \in BV(\mathbb{R}; X)$. Assume further that f maps bounded subsets in $\mathcal{D}$ into bounded subsets in X and, either the semigroup is compact, or f is locally Lipschitz with respect to its last argument. Then a saturated $\mathcal{L}^\infty$-solution $u : [\,a, b) \to X$ of (12.3.1) is either global, or it is not global, and in this case there exists*

$$\lim_{t\uparrow b} \|u(t+0)\| = +\infty.$$

Proof. If u is global we have nothing to prove. So, let us assume that u is non-global. This means that $b < +\infty$ and thus $g \in BV([\,a, b\,]; X)$. According to Theorem 12.3.3, either $t \mapsto u(t+0)$ is unbounded and the conclusion holds, or $t \mapsto u(t+0)$ is bounded. In the later case, since $g \in BV([\,a, b\,]; X)$ and inasmuch as f maps bounded subsets in $\mathbb{R} \times X$ into bounded subsets in X, we conclude that $t \mapsto f(t, u(t+0))$ is bounded. Thanks to Proposition 12.3.1, there exists

$$\lim_{t\uparrow b} u(t) = u(b-0).$$

By virtue of Theorem 12.3.3, we must have

$$(b, g(b+0) - g(b-0) + u(b-0)) \notin \mathbb{R} \times X$$

which is absurd. This contradiction can be eliminated only if $t \mapsto u(t+0)$ is unbounded and this completes the proof. □

The next corollary is a reformulation of the preceding one.

Corollary 12.3.3. *Let us assume that $A : D(A) \subseteq X \to X$ generates a C_0-semigroup of contractions, $f : \mathbb{R}\times X \to X$ is continuous and $g \in BV(\mathbb{R}; X)$. Assume further that f maps bounded subsets in $\mathcal{D}$ into bounded subsets in X and, either the semigroup is compact, or f is locally Lipschitz with respect to its last argument. Then a necessary and sufficient condition in order that an $\mathcal{L}^\infty$-solution $u : [\,a, b) \to X$ of (12.3.1) be continuable is that $b < +\infty$ and $t \mapsto u(t+0)$ be bounded on $[\,a, b)$.*

We conclude this section with a sufficient conditions on f ensuring the existence of global $\mathcal{L}^\infty$-solutions of (12.3.1).

Theorem 12.3.3. *Let $A : D(A) \subseteq X \to X$ be the generator of a C_0-semigroup of contractions, $f : \mathbb{R} \times X \to X$ a continuous function and $g \in BV(\mathbb{R}; X)$. Let us assume that either the semigroup is compact, or f is locally Lipschitz with respect to its last argument. Further, let us assume that there exist two continuous functions $h, k : \mathbb{R}_+ \to \mathbb{R}_+$ such that*

$$\|f(\tau, v)\| \leq k(\tau)\|v\| + h(\tau), \tag{12.3.6}$$

for each $(\tau, v) \in \mathbb{R}_+ \times X$. *Then, for each* $(a, \xi) \in \mathbb{R}_+ \times X$, (12.3.1) *has at least one global* $\mathcal{L}^\infty$*-solution.*

Proof. Let $u : [a, b) \to X$ be a saturated solution of (12.3.1) whose existence is ensured by Corollary 12.3.1. To complete the proof it suffices to show that $b = +\infty$. Thus, let us assume by contradiction that $b < +\infty$. From the definition of the $\mathcal{L}^\infty$-solution (see (iii) in Definition 12.1.4) and the growth condition on f we deduce

$$\|u(t+0)\| \le \|S(t-a)\xi\|$$

$$+ \int_a^t \|S(t-s)\|_{\mathcal{L}(X)} \|f(s, u(s+0))\| \, ds + \left\| \int_a^{t+0} S(t-s)\, dg(s) \right\|$$

$$\le \|\xi\| + \operatorname{Var}(g, [a, b]) + \int_a^t k(s) \|u(s+0)\| \, ds$$

for each $t \in [a, b)$. By virtue of Lemma 10.2.1, it follows that $t \mapsto u(t+0)$ is bounded on $[a, b)$. From Corollary 12.3.3, we conclude that u is continuable thereby contradicting the hypothesis. This contradiction can be eliminated only if $b = +\infty$ and this achieves the proof. □

Remark 12.3.4. All the results in this section can be reformulated to handle the more general case

$$\begin{cases} du = \{Au + f(t, u)\}dt + dg_u \\ u(a) = \xi, \end{cases}$$

where A and f are as above and, for each interval $[a, b]$, either $u \mapsto g_u$ is locally Lipschitz from $L^\infty(a, b; X)$ to $BV([a, b]; X)$ whenever f is locally Lipschitz, or is bounded and weakly pointwise continuous from $L^\infty(a, b; X)$ to $BV([a, b]; X)$ whenever f is continuous. See Definition 12.2.1. However, we notice that in both cases, we have to assume that $u \mapsto g_u$ is hereditary (see Definition 12.1.1) and there exists a function $h \in BV(\mathbb{R})$ such that, for each $a, b \in \mathbb{R}$ with $a < b$ and each $u \in L^\infty(a, b; X)$, we have

$$\operatorname{Var}(g_u, \mathbb{I}) \le \operatorname{Var}(h, \mathbb{I}),$$

where $\mathbb{I}$ is any of the intervals $[a, b]$, $(a, b]$, $[a, b)$ or (a, b).

12.4. The Case of Spatial Measures

Let $f : \mathbb{R} \times X \to X$ be a continuous function, $g \in BV(\mathbb{R}; X_A)$ and let us consider the Cauchy problem

$$\begin{cases} du = \{Au + f(t, u)\}dt + dg \\ u(a) = \xi_A. \end{cases} \tag{12.4.1}$$

Here X_A is the space of all admissible measures for A, i.e. $X_A = (X^{\odot})^*$, where $X^{\odot}$ is the sun dual of X and $\xi_A \in X_A$. See Sections 9.1 and 9.5.

Theorem 12.4.1. *Let $A : D(A) \subseteq X \to X$ be the infinitesimal generator of a compact C_0-semigroup of contractions $\{S(t)\,;\ t \geq 0\}$ and $f : \mathbb{R} \times X \to X$ a continuous function for which there exists $\alpha, \beta \in \mathcal{L}^1_{\text{loc}}(\mathbb{R})$ satisfying*

$$\|f(t,u)\| \leq \alpha(t)\|u\| + \beta(t) \tag{12.4.2}$$

a.e. for $t \in \mathbb{R}$ and for every $u \in X$. Then, for each $a, b \in \mathbb{R}$ with $a < b$, each $\xi \in X_A$, and $g \in BV([\,a,b\,]; X_A)$ which is the pointwise limit in the $\sigma(X_A, X^{\odot})$ topology of a sequence of functions $(g_n)_{n\in\mathbb{N}}$ in $BV([\,a,b\,]; X)$ with equibounded variation, there exists at least one $\mathcal{L}^\infty$-generalized solution of (12.4.1) *defined on $[\,a,b\,]$.*

Proof. First, let us observe that we may assume without any loss of generality that $g \in BV(\mathbb{R}; X_A)$ and it can be pointwise approximated in the $\sigma(X_A, X^{\odot})$-topology by a sequence of functions $(g_n)_{n\in\mathbb{N}}$ in $BV(\mathbb{R}; X)$ with equibounded variation. Indeed, if $g \in BV([\,a,b\,]; X)$ satisfies the hypotheses of Theorem 12.4.1, then the function $\tilde{g}$, defined by

$$\tilde{g}(t) = \begin{cases} g(a) & \text{if } t < a \\ g(t) & \text{if } t \in [\,a,b\,] \\ g(b) & \text{if } t > b, \end{cases}$$

belongs to $BV(\mathbb{R}; X_A)$, it can be pointwise approximated in the $\sigma(X_A, X^{\odot})$-topology by a sequence of functions $(g_n)_{n\in\mathbb{N}}$ in $BV(\mathbb{R}; X)$ with equibounded variation and coincides with g on $[\,a,b\,]$. Let $(\xi_n)_{n\in\mathbb{N}}$ be a sequence in X which converges in the weak-$\odot$ topology to ξ_A and let $(u_n)_{n\in\mathbb{N}}$ the sequence of saturated solutions of (12.4.1) corresponding to $(\xi_n)_{n\in\mathbb{N}}$ and $(g_n)_{n\in\mathbb{N}}$. We notice that the existence of these solutions is ensured by Corollary 12.3.1. We claim that u_n is global, i.e. it is defined on $[\,a, +\infty)$. Indeed, let us assume by contradiction that, for some $n \in \mathbb{N}$, u_n is defined on $[\,a,b)$ with $b < +\infty$. By (12.4.2) and Lemma 10.2.1, it readily follows that u_n is bounded on $[\,a,b)$. By Corollary 12.3.3, we conclude that u_n is continuable, thereby contradicting the fact that u_n is saturated. This contradiction can be eliminated only if u_n is global. So, for each $b \in \mathbb{R}$, $b > a$ and $n \in \mathbb{N}$, u_n is defined at least on $[\,a,b\,]$. By the uniform boundedness principle, it follows that $(\xi_n)_{n\in\mathbb{N}}$ is bounded in X, while, by hypotheses, $(g_n)_{n\in\mathbb{N}}$ is of equibounded variation on $[\,a,b\,]$. Again by (12.4.2) and Lemma 10.2.1, we conclude that $(u_n)_{n\in\mathbb{N}}$ is uniformly bounded on $[\,a,b\,]$, and therefore the family of functions

$$\left\{t \mapsto \int_a^t f(s, u_n(s))\, ds + g_n(t)\,;\ k \in \mathbb{N}\right\}$$

is of equibounded variation on $[a, b]$. By Theorem 9.4.2, we conclude that, on a subsequence at least, $(u_n)_{n\in\mathbb{N}}$ is convergent in $L^p(a, b; X)$, for each $p \in [1, +\infty)$, to a certain function $u : [a, b] \to X$. Since f is continuous and we may assume without loss of generality that $(u_n)_{n\in\mathbb{N}}$ converges a.e. on $[a, b]$ to u, a simple computational argument shows that u is an $\mathcal{L}^\infty$-generalized solution of (12.4.1) and this completes the proof. $\square$

12.5. Two Examples

We conclude this chapter with two illuminating examples. Let Ω be a bounded domain in $\mathbb{R}^n$ whose boundary Γ is of class C^2. In this section we consider the semi-dynamic elliptic problem

$$\begin{cases} -\Delta u = 0 & \text{in } Q_T \\ du + \{u_\nu + \beta(t,u)\}dt = d\varphi & \text{on } \Sigma_T \\ u(0) = u_0 & \text{on } \Gamma, \end{cases} \tag{12.5.1}$$

as well as the parabolic problem with dynamic boundary conditions

$$\begin{cases} du + \{-\Delta u + \alpha(t,u)\}dt = d\eta & \text{in } Q_T \\ du + \{u_\nu + \beta(t,u)\}dt = d\psi & \text{on } \Sigma_T \\ u(0) = u_0^\Omega & \text{in } \Omega \\ u(0) = u_0^\Gamma & \text{on } \Gamma \end{cases} \tag{12.5.2}$$

where $Q_T = (0,T) \times \Omega$, $\Sigma_T = (0,T) \times \Gamma$, $\alpha, \beta : [0,T] \times \mathbb{R} \to \mathbb{R}$ are continuous, $u_0, u_0^\Gamma \in L^2(\Gamma)$, $u_0^\Omega \in L^2(\Omega)$, u_ν is the conormal derivative of u at points of Γ, $\varphi, \psi \in BV([0,T]; L^2(\Gamma))$ and $\eta \in BV([0,T]; L^2(\Omega))$. We will show next how (12.5.1) and (12.5.2) can be analyzed by means of Theorem 12.2.2. We start with the problem (12.5.1). Namely we have:

Theorem 12.5.1. *Assume that $\beta : [0,T] \times \mathbb{R} \to \mathbb{R}$ is continuous, has sublinear growth and $\varphi \in BV([0,T]; L^2(\Gamma))$. Then, for each $u_0 \in L^2(\Gamma)$, there exists $T_0 \in (0,T)$ such that the problem* (12.5.1) *has at least one solution $u : [0, T_0] \to L^2(\Omega)$ satisfying:*

(i) *for each $t \in [0, T_0)$ and each $s \in (0, T_0]$, we have*

$$\begin{cases} u_{|\Gamma}(t+0) - u_{|\Gamma}(t) = \varphi(t+0) - \varphi(t) \\ u_{|\Gamma}(s) - u_{|\Gamma}(s-0) = \varphi(s) - \varphi(s-0) \end{cases};$$

(ii) *for each interval $\mathbb{I}$ of absolute continuity of φ for which φ' lies in $L^2(\mathbb{I}; L^2(\Gamma))$, we have: $u_{|\Gamma} \in C(\mathbb{I}; H^{1/2}(\Gamma)) \cap W^{1,2}(\mathbb{I}; L^2(\Gamma))$, $u \in C(\mathbb{I}; H^1(\Omega))$ and u satisfies* (12.5.2) *a.e. on $\mathbb{I}$.*

Proof. Take $H = L^2(\Gamma)$ and let us define $A : D(A) \subseteq H \to H$ by

$$D(A) = \{y \in H^{1/2}(\Gamma);\ u_\nu \in L^2(\Gamma)\},$$

where $u \in H^1(\Omega)$ is the unique solution of the elliptic problem

$$\begin{cases} -\Delta u = 0 & \text{in } \Omega \\ u = y & \text{on } \Gamma \end{cases}$$

and $Ay = -u_\nu$ for each $u \in D(A)$. By virtue of Theorem 7.5.1, it follows that A generates a compact analytic C_0-semigroup of contractions on H. Let $f : [\,0,T\,] \times L^2(\Gamma) \to L^2(\Gamma)$ and $g : [\,0,T\,] \to L^2(\Gamma)$ be defined by $f(t,u)(\sigma) = -\beta(t,u(\sigma))$ for $(t,u) \in [\,0,T\,] \times L^2(\Gamma)$ and a.e. for $\sigma \in \Gamma$ and respectively by $g(t) = \varphi(t)$ for each $t \in [\,0,T\,]$. Then (12.5.1) rewrites as an abstract Cauchy problem of the form (12.2.2) in the Hilbert space H, with $\xi = u_0$, and Theorem 12.2.2 applies. An appeal to Theorem 9.2.1 shows that (i) holds. Since (ii) follows from well-known regularity results concerning elliptic equations along with the Green's formula, the proof is complete. □

Theorem 12.5.2. *If $g, \beta : [\,0,T\,] \times \mathbb{R} \to \mathbb{R}$ is continuous, has sublinear growth, $\eta \in BV([\,0,T\,];L^2(\Omega))$ and $\psi \in BV([\,0,T\,];L^2(\Gamma))$, then, for each $u_0^\Omega \in L^2(\Omega)$ and $u_0^\Gamma \in L^2(\Gamma)$, there exists $T_0 \in (0,T)$ such that the problem* (12.5.2) *has at least one solution $u : [\,0,T_0] \to L^2(\Omega)$ satisfying:*

(i) *for each $t \in [\,0,T_0)$ and each $s \in (0,T_0\,]$, we have*

$$\begin{cases} u(t+0) - u(t) = \eta(t+0) - \eta(t) \\ u_{|\Gamma}(t+0) - u_{|\Gamma}(t) = \psi(t+0) - \psi(t) \\ u(s) - u(s-0) = \eta(s) - \eta(s-0) \\ u_{|\Gamma}(s) - u_{|\Gamma}(s-0) = \psi(s) - \psi(s-0)\,; \end{cases}$$

(ii) *for each interval $\mathbb{I}$ of absolute continuity of both η and ψ for which $\eta' \in L^2(\mathbb{I};L^2(\Omega))$ and $\psi' \in L^2(\mathbb{I};L^2(\Gamma))$, we have:*

$$\begin{cases} u \in C(\mathbb{I};H^1(\Omega)) \cap W^{1,2}(\mathbb{I};L^2(\Omega)) \\ u_{|\Gamma} \in C(\mathbb{I};H^{1/2}(\Gamma)) \cap W^{1,2}(\mathbb{I};L^2(\Gamma)) \end{cases}$$

and u, $u_{|\Gamma}$ satisfy (12.5.2) *a.e. on $\mathbb{I}$.*

Proof. Take $H = L^2(\Omega) \times L^2(\Gamma)$ endowed with the inner product

$$\langle (u,v),(\tilde{u},\tilde{v})\rangle_H = \langle u,\tilde{u}\rangle_{L^2(\Omega)} + \langle v,\tilde{v}\rangle_{L^2(\Gamma)},$$

for each $(u,v),(\tilde{u},\tilde{v}) \in H$ and let us define $A : D(A) \subseteq H \to H$ by

$$D(A) = \{(u,v) \in H\,;\ \Delta u \in L^2(\Omega),\ u_\nu \in L^2(\Gamma) \text{ and } u_\nu = v \text{ on } \Gamma\}$$

and $A(u,v) = (\Delta u, -u_\nu)$ for each $(u,v) \in D(A)$. From Theorem 7.4.1, it follows that A is a densely defined, self-adjoint operator which generates a compact, analytic C_0-semigroup of contractions on H. Let us define both $f \colon [\,0,T\,] \times L^2(\Omega) \times L^2(\Gamma) \to L^2(\Omega) \times L^2(\Gamma)$ and $g \colon [\,0,T\,] \to L^2(\Omega) \times L^2(\Gamma)$ by

$f(t,u,v)(x) = -(\alpha(t,u(x)), \beta(t,v(\sigma)))$ for $(t,u,v) \in [0,T] \times L^2(\Omega) \times L^2(\Gamma)$, a.e. for $x \in \Omega$ and a.e. for $\sigma \in \Gamma$, and respectively by $g(t) = (\eta(t), \psi(t))$ for $t \in [0,T]$. Then (12.5.1) can be rewritten as an abstract Cauchy problem of the form (12.3.1) in the Hilbert space H, with $\xi = (u_0^\Omega, u_0^\Gamma)$ and we are in the hypotheses of Theorem 12.2.2. The conclusion follows from Theorem 9.2.1 combined with well-known regularity results concerning parabolic equations. □

12.6. One More Example

In order to show how Theorem 12.4.1 applies, in this section we analyze a simple example of semilinear parabolic problem with distributed measures with respect to both time and space. Let Ω be a nonempty, open and bounded subset in $\mathbb{R}^n$ whose boundary Γ is of class C^1, let $b : \mathbb{R}\times\Omega\times\mathbb{R} \to \mathbb{R}$, let $g \in BV(\mathbb{R}; (C_0(\overline{\Omega}))^*)$, $\xi_A \in (C_0(\overline{\Omega}))^*$ and let us consider the problem

$$\begin{cases} du = \{\Delta u + b(t,x,u)\}dt + dg(t) & \text{in } Q_\infty \\ u = 0 & \text{on } \Sigma_\infty \\ u(0,x) = \xi_A & \text{in } \Omega, \end{cases} \tag{12.6.1}$$

where $Q_\infty = [0,+\infty) \times \Omega$ and $\Sigma_\infty = [0,+\infty) \times \Gamma$. As usual, we rewrite (12.6.1) as an abstract Cauchy problem of the form

$$\begin{cases} du = \{Au + f(t,u)\}dt + dg \\ u(0) = \xi_A, \end{cases} \tag{12.6.2}$$

satisfying the hypotheses of Theorem 12.4.1. To this aim, let us assume that $b : \mathbb{R} \times \Omega \times \mathbb{R} \to \mathbb{R}$ is continuous and there exist two continuous functions $k, h : \mathbb{R} \to \mathbb{R}_+$ such that

$$|b(t,x,u)| \leq k(t)|u| + h(t) \tag{12.6.3}$$

for each $t \in \mathbb{R}$, each $u \in \mathbb{R}$ and a.e. for $x \in \Omega$. Now, take $X = L^1(\Omega)$ and $A : D(A) \subseteq X \to X$, defined by

$$\begin{cases} D(A) = \{u \in W_0^{1,1}(\Omega)\,;\ \Delta u \in L^1(\Omega)\} \\ Au = \Delta u & \text{for } u \in D(A), \end{cases}$$

and let us recall that A generates an analytic and compact C_0-semigroup of contractions. See Theorem 7.2.7. Furthermore, let us recall that in this case $X^\odot = C_0(\overline{\Omega})$ and $X_A = (C_0(\overline{\Omega}))^*$, i.e. the space of all Radon measures concentrated in Ω. See Example 9.5.1. Next, let us define $f : \mathbb{R} \times X \to X$ by

$$f(t,u)(x) = b(t,x,u(x))$$

for each $u \in L^1(\Omega)$, each $t \in \mathbb{R}$ and a.e. for $x \in \Omega$. Thanks to (ii) in Lemma A.6.1, f is well-defined and continuous. At this point it is easy to see that (12.6.1) can be rewritten in the abstract form (12.6.2), where A, f, g and ξ_A are as above. From Theorem 12.4.1 we deduce:

Theorem 12.6.1. *Let us assume that* $b : \mathbb{R} \times \Omega \times \mathbb{R} \to \mathbb{R}$ *satisfies* (12.6.3) *and* $g \in BV(\mathbb{R}; (C_0(\overline{\Omega}))^*)$ *is the pointwise limit in the* $\sigma((C_0(\overline{\Omega}))^*, C_0(\overline{\Omega}))$*-topology of a sequence* $(g_n)_{n\in\mathbb{N}}$ *of equibounded variation in* $BV(\mathbb{R}; L^1(\Omega))$*. Then, for each* $\xi_A \in (C_0(\overline{\Omega}))^*$ *which is the limit in the* $\sigma((C_0(\overline{\Omega}))^*, C_0(\overline{\Omega}))$*-topology of a sequence* $(\xi_n)_{n\in\mathbb{N}}$ *bounded in* $L^1(\Omega)$*, the problem* (12.6.2) *has at least one generalized* $\mathcal{L}^\infty$*-solution* $u : [0, +\infty) \to L^1(\Omega)$.

We describe below a typical instance in which both g and ξ_A satisfy all the hypotheses of Theorem 12.6.1. More precisely, let $x_0 \in \Omega$ and let $\xi_A = \delta(x - x_0)$, i.e. the Dirac delta concentrated at x_0. Then ξ_A is the limit in the $\sigma((C_0(\overline{\Omega}))^*, C_0(\overline{\Omega}))$-topology of a sequence $(\xi_n)_{n\in\mathbb{N}}$ which is bounded in $L^1(\Omega)$. Indeed, let $\varepsilon_0 > 0$ be sufficiently small such that $B(x_0, \varepsilon_0) \subset \Omega$, let $\varepsilon \in (0, \varepsilon_0)$, and let us define the function $\delta_\varepsilon(\cdot - x_0) : \Omega \to \mathbb{R}$ by

$$\delta_\varepsilon(x - x_0) = \begin{cases} \dfrac{1}{\mu(B_\varepsilon)} & \text{if } \|x - x_0\| < \varepsilon \\ 0 & \text{if } x \in \Omega,\ \|x - x_0\| \geq \varepsilon. \end{cases}$$

Here $\mu(B_\varepsilon)$ is the Lebesgue measure of the closed ball in $\mathbb{R}^n$ centered at 0 and having radius ε. For any $\varepsilon \in (0, \varepsilon_0)$, we have $\|\delta_\varepsilon(\cdot - x_0)\|_{L^1(\Omega)} = 1$ and therefore $\{\delta_\varepsilon(\cdot - y_0)\,;\ \varepsilon \in (0, \varepsilon_0)\}$ is bounded in $L^1(\Omega)$. On the other hand, we have

$$\lim_{\varepsilon \downarrow 0} \delta_\varepsilon(x - x_0) = \delta(x - x_0)$$

in the $\sigma((C_0(\overline{\Omega}))^*, C_0(\overline{\Omega}))$-topology. Indeed, this means that

$$\lim_{\varepsilon \downarrow 0} \int_\Omega \delta_\varepsilon(x - x_0)\varphi(x)\, dx = \varphi(x_0)$$

for each $\varphi \in C_0(\overline{\Omega})$, which clearly holds true. So, $\xi_A = \delta(x - x_0)$ satisfies the hypotheses of Theorem 12.6.1. Next, let $t_0 \in \mathbb{R}$ and $y_0 \in \Omega$ and let $g : \mathbb{R} \to (C_0(\overline{\Omega}))^*$ be defined by

$$g(t) = \begin{cases} -\frac{1}{2}\delta(x - y_0) & \text{if } t < t_0 \\ 0 & \text{if } t = t_0 \\ \frac{1}{2}\delta(x - y_0) & \text{if } t > t_0. \end{cases}$$

We shall show that g satisfies the hypohteses of Theorem 12.6.1. It is easy to see that $g \in BV(\mathbb{R}; (C(\overline{\Omega}))^*)$. Let $\varepsilon_0 > 0$ be such that $B(y_0, \varepsilon_0) \subset \Omega$, let

$\varepsilon \in (0, \varepsilon_0)$, and let us define $g_\varepsilon : \mathbb{R} \to C(\overline{\Omega})$ by

$$g_\varepsilon(t) = \begin{cases} -\frac{1}{2}\delta_\varepsilon(x - y_0) & \text{if } t < t_0 \\ 0 & \text{if } t = t_0 \\ \frac{1}{2}\delta_\varepsilon(x - y_0) & \text{if } t > t_0, \end{cases}$$

where $\delta_\varepsilon(x - y_0)$ is defined as above. By the preceding proof, it readily follows that

$$\lim_{\varepsilon \downarrow 0} g_\varepsilon(t) = g(t),$$

for each $t \in \mathbb{R}$, in the $\sigma((C_0(\overline{\Omega}))^*, C_0(\overline{\Omega}))$-topology. Therefore, g fulfils the hypotheses of Theorem 12.6.1, and thus, if $b : \mathbb{R} \times \Omega \times \mathbb{R} \to \mathbb{R}$ satisfies (12.6.3), the semilinear parabolic problem

$$\begin{cases} du = \{\Delta u + b(t, x, u)\}dt + \delta(t - t_0) \times \delta(x - y_0) & \text{in } Q_\infty \\ u = 0 & \text{on } \Sigma_\infty \\ u(0, x) = \delta(x - x_0) & \text{in } \Omega, \end{cases}$$

has at least one global generalized $\mathfrak{L}^\infty$-solution. By a generalized $\mathfrak{L}^\infty$-solution of the problem above, we mean in fact a generalized $\mathfrak{L}^\infty$-solution of the corresponding problem (12.6.2).

Problems

Problem 12.1. Let $A : D(A) \subseteq X \to X$ be the generator of a C_0-semigroup of contractions and let $G : L^\infty(a, b; X) \to BV_A([\,a, b\,]; X)$ be a hereditary locally Lipschitz mapping. Then, for each $\xi \in X$, there exists $c \in (a, b\,]$ such that the problem

$$\begin{cases} du = \{Au\}dt + dg_u \\ u(a) = \xi \end{cases}$$

has a unique $\mathfrak{L}^\infty$-solution defined on $[\,a, c\,]$.

We denote by $BV_A(\mathbb{R}; X)$ the space of all functions $g \in BV(\mathbb{R}; X)$ with the property that, for each $a < b$, $g \in BV_A([\,a, b\,]; X)$.

Problem 12.2. Let $t_0 \in \mathbb{R}$ be fixed and let $h : \mathbb{R} \to \mathbb{R}$ be continuous. Using the result in Problem 12.1, show that, for each $\xi \in L^1(\mathbb{R})$, the problem

$$\begin{cases} u_t = u_x + h(x)\delta(t - t_0) & \text{for } (t, x) \in \mathbb{R} \times \mathbb{R}, \\ u(0, x) = \xi(x) & \text{for } x \in \mathbb{R} \end{cases}$$

has an $\mathfrak{L}^\infty$-solution defined on $\mathbb{R}_+$, although the C_0-group generated by the operator $Au = u'$ is not continuous from $(0, +\infty)$ to $\mathfrak{L}(L^1(\mathbb{R}))$ in the uniform operator topology.

Problem 12.3. Prove that if $A : D(A) \subseteq X \to X$ is the infinitesimal generator of a C_0-semigroup of contractions $\{S(t)\,;\ t \geq 0\}$, $f : \mathbb{R} \times X \to X$ is continuous with $f(\mathbb{R} \times X)$ compact, and $g \in BV_A(\mathbb{R}; X)$ then, for each $(a, \xi) \in \mathbb{R} \times X$ and $c > a$, the problem

$$\begin{cases} du = \{Au + f(t,u)\}dt + dg \\ u(a) = \xi \end{cases}$$

has at least one $\mathfrak{L}^\infty$-solution defined on $[\,a, c\,]$.

Problem 12.4. Let $\mathfrak{D}$ be a nonempty and open subset in $\mathbb{R} \times X$. Prove the following variant of Theorem 12.2.2. If $A : D(A) \subseteq X \to X$ is the infinitesimal generator of a C_0-semigroup of contractions $\{S(t)\,;\ t \geq 0\}$, $f : \mathfrak{D} \to X$ is continuous and $\mathfrak{b}$-compact, and $g \in BV_A(\mathbb{R}; X)$, then, for each $(a, \xi) \in \mathbb{R} \times X$ satisfying

$$(a, g(a+0) - g(a) + \xi) \in \mathfrak{D},$$

there exists $b > a$ such that the problem

$$\begin{cases} du = \{Au + f(t,u)\}dt + dg \\ u(a) = \xi \end{cases}$$

has at least one $\mathfrak{L}^\infty$-solution defined on $[\,a, b\,]$.

Problem 12.5. Let Ω be a nonempty, bounded and open subset in $\mathbb{R}^n$ whose boundary Γ is of class C^2, let $Q_T = (0,T) \times \Omega$ and $\Sigma_T = (0,T) \times \Gamma$. We consider the nonlinear hyperbolic equation

$$\begin{cases} du_t = \{\Delta u + b(t,x,u)\}dt + \delta(t - t_0) & (t,x) \in Q_T \\ u(t,x) = 0 & (t,x) \in \Sigma_T \\ u(0,x) = u_0(x) & x \in \Omega, \\ u_t(0,x) = v_0(x) & x \in \Omega \end{cases}$$

where $b : \mathbb{R}_+ \times \Omega \times \mathbb{R} \to \mathbb{R}$ is a continuous function and $t_0 \in (0, +\infty)$. Prove that, if $b : \mathbb{R}_+ \times \Omega \times \mathbb{R} \to \mathbb{R}$ is a continuous function for which there exist $c > 0$ and $d > 0$ such that

$$|b(t,x,u)| \leq c|u| + d$$

for $(t,x,u) \in \mathbb{R}_+ \times \Omega \times \mathbb{R}$, then, for each $u_0 \in H_0^1(\Omega)$ and each $v_0 \in L^2(\Omega)$, the problem above has at least one global $\mathfrak{L}^\infty$-solution.

Problem 12.6. Let us consider the following parabolic problem

$$\begin{cases} du = \{\Delta u\}dt + \delta(t - t_0) \times \delta(x - x_u) & \text{in } Q_\infty \\ u = 0 & \text{on } \Sigma_\infty \\ u(0,x) = \delta(x - x_0) & \text{in } \Omega, \end{cases}$$

where $t_0 \in (0, +\infty)$, $x_0 \in \Omega$ and $u \mapsto x_u$ is continuous from $L^\infty(a, b; L^1(\Omega))$ respectively to Ω_K, the latter being a compact subset in Ω. Prove that this problem has at least one generalized $\mathcal{L}^\infty$-solution. We notice that the problem above may be interpreted as a closed loop system whose feedback law $u \mapsto \delta(t - t_0) \times \delta(x - x_u)$ takes values in the space of Radon measures over $\mathbb{R} \times \Omega$.

Notes. The existence results concerning semilinear differential equation with distributed measures in Sections 12.1 and 12.2, i.e. Theorems 12.1.1 and 12.2.1 were established by Grosu [**65**]. Theorems 12.1.2 and 12.2.2 are from Vrabie [**131**], but the proofs are new, and due to Grosu [**65**]. The qualitative results referring to continuable and saturated $\mathcal{L}^\infty$-solutions in Section 12.3 are in the same spirit as those in Sections 10.3 and 11.3 and extend the later to the general case of semilinear equations involving distributed measures. Theorem 12.4.1 and the examples in Section 12.5 are from Vrabie [**131**], while the example in Section 12.6 and the problems at the end of this chapter are new.

APPENDIX A

Compactness Results

The aim of this chapter is to present several necessary and sufficient conditions in order that a given subset in a certain function space be relatively compact. One of the most fundamental results in Real Analysis says that each bounded sequence in $\mathbb{R}^n$ has at least one convergent subsequence (Cesàro's lemma). In addition, one knows that a Banach space is finite-dimensional if and only if the class of its relatively compact subsets coincides with the class of its bounded subsets. In other words, Cesàro's lemma is a characteristic of the finite-dimensional frame. On the other hand, the usual function spaces employed in Nonlinear Analysis ($C([a,b];X), L^p(a,b;X)$, with X Banach space) are infinite-dimensional (even if X is finite-dimensional). From this reason at least, it is very important to get appropriate variants of Cesàro's lemma for each one of this function spaces. As we shall see, these variants differ from one type of space to another, and are closely related to the nature of the space in question.

A.1. Compact Operators

In this section we shall present some auxiliary concepts and results we need in that follows.

Definition A.1.1. A subset C of a topological space $(X, \mathcal{T})$ is called:

(i) *relatively compact*, if each generalized sequence in C has at least one generalized convergent subsequence;
(ii) *compact*, if is relatively compact and closed;
(iii) *sequentially relatively compact*, if each sequence in C has at least one convergent subsequence;
(iv) *sequentially compact*, if is sequentially relatively compact, and closed.

If (X, d) is a metric space, $C \subset X$ is called *precompact*, or *totally bounded*, if for each $\varepsilon > 0$ there exists a finite family of closed balls of radius ε whose union includes C.

Remark A.1.1. Inasmuch as each metric space satisfies the first axiom of countability i.e., each point has an at most countable fundamental system of neighborhoods, in such a space, a subset is (relatively) compact if and only if it is sequentially (relatively) compact.

Remark A.1.2. One may easily show that a subset $C \subset X$ is precompact if and only if, for each $\varepsilon > 0$ there exists a finite family of closed balls centered in points of C and having radii ε, whose union includes C.

We recall, without proof, the following result due to Hausdorff.

Theorem A.1.1. *If (X, d) is a complete metric space, then a subset of it is sequentially relatively compact if and only if it is precompact.*

We also recall:

Theorem A.1.2. (Tichonov) *Let $\{(X_\alpha, \mathcal{T}_\alpha);\ \alpha \in \Gamma\}$ be a given family of topological spaces. Then the product space $X = \Pi_{\alpha\in\Gamma}X_\alpha$, endowed with the usual product topology, is compact if and only if, for each $\alpha \in \Gamma$, $(X_\alpha, \mathcal{T}_\alpha)$ is compact.*

See Dunford and Schwartz [**49**], Theorem 5, p. 32.

Theorem A.1.3. (Mazur) *The closed convex hull of a compact subset in a Banach space X is compact.*

See Dunford and Schwartz [**49**], Theorem 6, p. 416.

Theorem A.1.4. (Mazur) *The weak closure of every convex subset in a Banach space coincides with its strong closure.*

See Hille and Phillips [**70**], Theorem 2.9.3, p. 36.
Let $(X_0, \|\cdot\|_0)$ and $(X, \|\cdot\|)$ be two Banach spaces with $X_0 \subset X$. We say that the inclusion $X_0 \subset X$ is compact, or that $(X_0, \|\cdot\|_0)$ is compactly imbedded in $(X, \|\cdot\|)$, if each bounded subset in X_0 is relatively compact in X. We say that the inclusion $X_0 \subset X$ is continuous, or that $(X_0, \|\cdot\|_0)$ is continuously imbedded in $(X, \|\cdot\|)$, if there exists $k > 0$ such that

$$\|x\| \leq k\|x\|_0$$

for each $x \in X$.

Lemma A.1.1. (Lions) *Let $(X_0, \|\cdot\|_0)$, $(X, \|\cdot\|)$, and $(X_1, \|\cdot\|_1)$ be three real Banach spaces with $X_0 \subset X \subset X_1$. If the inclusion $X_0 \subset X$ is compact and $X \subset X_1$ is continuous then, for each $\varepsilon > 0$, there exists $\eta(\varepsilon) > 0$ such that*

$$\|x\| \leq \varepsilon\|x\|_0 + \eta(\varepsilon)\|x\|_1 \tag{A.1.1}$$

for each $x \in X_0$.

Proof. Let us assume by contradiction that this is not the case. Then there exists $\varepsilon > 0$ such that, for each $\eta > 0$ there exists $x_\eta \in X_0$ with the property

$$\|x_\eta\| > \varepsilon\|x_\eta\|_0 + \eta\|x_\eta\|_1.$$

Taking $\eta = n$ with $n \in \mathbb{N}^*$, and denoting x_η by x_n, the last inequality can be rewritten under the form

$$\|x_n\| > \varepsilon\|x_n\|_0 + n\|x_n\|_1$$

for each $n \in \mathbb{N}^*$. Obviously $x_n \neq 0$ and so, we can define

$$y_n = \frac{1}{\|x_n\|_0} x_n$$

for $n \in \mathbb{N}^*$. From the preceding inequality we obtain

$$\|y_n\| > \varepsilon + n\|y_n\|_1$$

for each $n \in \mathbb{N}^*$. As X_0 is compactly imbedded in X and $(y_n)_{n\in\mathbb{N}^*}$ satisfies $\|y_n\|_0 = 1$ for each $n \in \mathbb{N}^*$, we may assume without loss of generality that there exists $y \in X$ such that

$$\lim_{n\to\infty} y_n = y$$

in the norm $\|\cdot\|$. The last inequality yields $\lim_{n\to\infty}\|y_n\|_1 = 0$, which shows that $y = 0$. On the other hand, $\|y_n\| > \varepsilon$ for each $n \in \mathbb{N}^*$ and, thus $\|y\| \geq \varepsilon$, relation which contradicts the preceding one. This contradiction can be eliminated only if $(A.1.1)$ holds, and this completes the proof. □

We shall see in the next sections that Lemma A.1.1 is useful in proving some sufficient conditions of compactness in several function spaces. Therein, the reader will find some examples of Banach spaces X_0, X and X_1 satisfying the hypotheses of Lemma A.1.1.
Let X, Y be two real Banach spaces and let D a nonempty subset in Y.

Definition A.1.2. A (possible nonlinear) operator $Q : D \to X$ is called *compact* if it is continuous and maps bounded subsets in D into relatively compact subsets in X.

Lemma A.1.2. *Let Γ be a nonempty subset in $\overline{\mathbb{R}} = \mathbb{R} \cup \{-\infty\} \cup \{+\infty\}$, let μ be an accumulation point of Γ, and $\{Q_\lambda;\ \lambda \in \Gamma\}$ a family of operators from $D \subset Y$ in X. If, for each $\lambda \in \Gamma$, $Q_\lambda D$ is relatively compact, and*

$$\lim_{\lambda\to\mu} Q_\lambda x = Qx, \tag{A.1.2}$$

uniformly for $x \in D$, then QD is relatively compact.

Proof. Let $\varepsilon > 0$. From $(A.1.2)$ it follows that there exists $\lambda \in \Gamma$ such that

$$\|Q_\lambda x - Qx\| \leq \varepsilon \tag{A.1.3}$$

for each $x \in D$. As $Q_\lambda D$ is precompact, for the very same $\varepsilon > 0$, there exists a finite family $\{x_1, x_2, \ldots, x_{k(\varepsilon)}\}$ in D such that, for each $x \in D$, there exists $i = 1, 2, \ldots, k(\varepsilon)$ with

$$\|Q_\lambda x - Q_\lambda x_i\| \leq \varepsilon.$$

From $(A.1.3)$, and the preceding inequality, we deduce

$$\|Qx - Qx_i\| \leq \|Qx - Q_\lambda x\| + \|Q_\lambda x - Q_\lambda x_i\| + \|Q_\lambda x_i - Qx_i\| \leq 3\varepsilon,$$

which shows that QD is precompact in X. By virtue of Theorem A.1.1, QD is relatively compact in X, and this completes the proof. □

Three useful consequences of Lemma A.1.2 are listed below.

Corollary A.1.1. *Let Γ be a nonempty subset in $\overline{\mathbb{R}} = \mathbb{R} \cup \{-\infty\} \cup \{+\infty\}$, let μ an accumulation point of Γ, and $\{Q_\lambda;\ \lambda \in \Gamma\}$ a family of compact operators from $D \subset Y$ in X. If*

$$\lim_{\lambda \to \mu} Q_\lambda x = Qx,$$

uniformly for x in each subset bounded in D, then $Q : D \to X$ is a compact operator.

Corollary A.1.2. *Let K a bounded subset in X, Γ a nonempty subset in $\overline{\mathbb{R}} = \mathbb{R} \cup \{-\infty\} \cup \{+\infty\}$, μ an accumulation point of Γ, and $\{Q_\lambda;\ \lambda \in \Gamma\}$ a family of compact operators from K to X. If*

$$\lim_{\lambda \to \mu} Q_\lambda x = x,$$

uniformly for x in K, then K is relatively compact.

We recall that, if $T : Y \to X$ is a linear continuous operator, its *operator norm* $\|T\|_{\mathcal{L}(Y,X)}$ is defined by

$$\|T\|_{\mathcal{L}(Y,X)} = \sup_{\|y\| \leq 1} \|Ty\|.$$

Corollary A.1.3. *Let Γ be a nonempty subset in $\overline{\mathbb{R}} = \mathbb{R} \cup \{-\infty\} \cup \{+\infty\}$, let μ an accumulation point of Γ and $\{Q_\lambda;\ \lambda \in \Gamma\}$ a family of linear compact operators from Y to X. If*

$$\lim_{\lambda \to \mu} \|Q_\lambda - Q\|_{\mathcal{L}(Y,X)} = 0$$

then $Q : Y \to X$ is a compact operator.

Lemma A.1.3. *Let K be a compact subset in X and let $\mathcal{F}$ a family of continuous functions from $[a,b]$ to K. Then*

$$\left\{\int_a^b f(t)\,dt;\ f\in\mathcal{F}\right\}$$

is relatively compact in X.

Proof. Using Riemann sums, if follows that

$$\left\{\int_a^b f(t)\,dt;\ f\in\mathcal{F}\right\}\subset(b-a)\overline{\text{conv K}},$$

which, by virtue of Mazur's Theorem A.1.3, completes the proof. □

Lemma A.1.4. *Let D be nonempty in X, $Q:D\to X$ and $G:D\to X$. If G is compact, and there exists an increasing function $\alpha:\mathbb{R}_+\to\mathbb{R}_+$ with*

$$\lim_{r\downarrow 0}\alpha(r)=0,$$

and such that

$$\|Qv-Qw\|\leq\alpha(\|Gv-Gw\|)$$

for each $v,w\in D$, then Q is compact.

Proof. Each finite ε-net $\{Gv_1,Gv_2,\ldots,Gv_k\}$ of $G(E)$ defines a finite $\alpha(\varepsilon)$-net $\{Qv_1,Qv_2,\ldots,Qv_k\}$ of $Q(E)$. Since $\lim_{r\downarrow 0}\alpha(r)=0$, this achieves the proof. □

Theorem A.1.5. (Schauder) *Let K be nonempty, convex and bounded in a Banach space X and let $Q:K\to K$ be compact. Then Q has at least one fixed point in K, i.e. there exists at least one $u\in K$ such that $Q(u)=u$.*

A.2. Compactness in $C([a,b];X)$

Let X be a real Banach space, and let $[a,b]$ be an interval. In that follows, we denote by $C([a,b];X)$ the space of all continuous functions from $[a,b]$ to X. Endowed with the sup-norm $\|\cdot\|_\infty$, defined by

$$\|f\|_\infty=\sup\{\|f(t)\|;\ t\in\ [a,b]\},$$

for each $f\in C([a,b];X)$, this is a real Banach space.

Definition A.2.1. A family $\mathcal{F}$ in $C([a,b];X)$ is *equicontinuous* at t in $[a,b]$, if for each $\varepsilon>0$ there exists $\delta(\varepsilon,t)>0$ such that, for each $s\in[a,b]$ with $|t-s|<\delta(\varepsilon,t)$, we have

$$\|f(t)-f(s)\|<\varepsilon,$$

uniformly with respect to $f\in\mathcal{F}$.

The family $\mathcal{F}$ is *equicontinuous* on $[a,b]$, if it is equicontinuous at each point $t \in [a,b]$, in the sense mentioned above.
The family $\mathcal{F}$ is *uniformly equicontinuous* on $[a,b]$, if it is equicontinuous on $[a,b]$, and $\delta(\varepsilon,t)$ can be chosen independent of $t \in [a,b]$.

Remark A.2.1. We leave to the reader the proof of the fact that a family $\mathcal{F}$ in $C([a,b];X)$ is equicontinuous on $[a,b]$ if and only if it is uniformly equicontinuous on $[a,b]$.

Theorem A.2.1. (Arzelà-Ascoli) *A subset $\mathcal{F}$ in $C([a,b];X)$ is relatively compact if and only if:*

(i) *$\mathcal{F}$ is equicontinuous on $[a,b]$;*
(ii) *there exists a dense subset D in $[a,b]$ such that, for each $t \in D$, $\mathcal{F}(t) = \{f(t);\ f \in \mathcal{F}\}$ is relatively compact in X.*

Proof. We begin by showing the necessity of (i) and (ii). To this end, let $\mathcal{F}$ be a relatively compact subset in $C([a,b];X)$ and let $\varepsilon > 0$. As $\mathcal{F}$ is relatively compact, it is precompact, and therefore there exists a finite family $\{f_1, f_2, \dots, f_{n(\varepsilon)}\}$ included in $\mathcal{F}$, such that, for each $f \in \mathcal{F}$, there exists $i \in \{1, 2 \dots, n(\varepsilon)\}$ with

$$\|f - f_i\|_\infty < \varepsilon. \tag{A.2.1}$$

Let $t \in [a,b]$. Obviously, as $\{f_1, f_2, \dots, f_{n(\varepsilon)}\}$ is finite, it is equicontinuous. Hence there exists $\delta(\varepsilon) > 0$ such that, for each $s \in [a,b]$ with $|t-s| < \delta(\varepsilon)$, we have

$$\|f_i(t) - f_i(s)\| < \varepsilon,$$

uniformly with respect to $i \in \{1, 2, \dots, n(\varepsilon)\}$. Then

$$\begin{aligned}\|f(t) - f(s)\| &\leq \|f(t) - f_i(t)\| + \|f_i(t) - f_i(s)\| + \|f_i(s) - f(s)\| \\ &\leq 2\|f - f_i\|_\infty + \|f_i(t) - f_i(s)\| < 3\varepsilon,\end{aligned}$$

for each $s \in [a,b]$ with $|t - s| < \delta(\varepsilon)$, where f_i is associated with f such that $(A.2.1)$ holds. As the last inequality shows that $\mathcal{F}$ is equicontinuous at t, arbitrarily chosen in $[a,b]$, it follows that $\mathcal{F}$ satisfies (i).
We shall prove next that $\mathcal{F}$ satisfies (ii) as well, with $D = [a,b]$. To this aim, let $t \in [a,b]$ and let $(f_n(t))_{n\in\mathbb{N}}$ be a sequence in $\mathcal{F}(t)$. As $\mathcal{F}$ is relatively compact, the sequence $(f_n)_{n\in\mathbb{N}}$ has at least one subsequence (denoted for simplicity again by $(f_n)_{n\in\mathbb{N}}$), uniformly convergent on $[a,b]$ to some function $f \in C([a,b];X)$. Obviously, this subsequence is pointwise convergent on $[a,b]$, also to f. Hence $(f_n(t))_{n\in\mathbb{N}}$ is convergent to $f(t)$ and therefore $\mathcal{F}(t)$ is relatively compact in X, which completes the proof of the necessity of both (i) and (ii).

We may now proceed to the proof of the sufficiency. Let $\mathcal{F}$ be a family in $C([a,b];X)$ satisfying (i) and (ii). By virtue of Hausdorff's Theorem A.1.1, it suffices to check that $\mathcal{F}$ is precompact. Let $\varepsilon > 0$. As $\mathcal{F}$ is equicontinuous on $[a,b]$, in view of Remark A.2.1, it is uniformly equicontinuous. Hence, there exists $\delta(\varepsilon) > 0$ such that, for each $t, s \in [a,b]$ with $|t-s| < \delta(\varepsilon)$, we have

$$\|f(t) - f(s)\| < \varepsilon,$$

for all $f \in \mathcal{F}$.
Let now $t_0 < t_1 < \cdots < t_k$ in $[a,b]$ with $t_i \in D$ for $i = 1,2,\ldots,k$, and satisfying

$$t_0 - a < \delta(\varepsilon), \quad b - t_k < \delta(\varepsilon), \quad t_{i+1} - t_i < \delta(\varepsilon) \tag{A.2.2}$$

for $i = 1,2,\ldots,k$. Let us define the set

$$\mathcal{K} = \{(f(t_1), f(t_2), \ldots, f(t_k));\ f \in \mathcal{F}\}.$$

Obviously, $\mathcal{K}$ is relatively compact being included in the product space

$$\mathcal{F}(t_1) \times \mathcal{F}(t_2) \times \cdots \times \mathcal{F}(t_k),$$

which, by virtue of (ii) and Tichonov's Theorem A.1.2, is compact inasmuch as all the factors are compact. Let us observe that the product topology on X^k can be equivalently defined by means of the norm $\|\cdot\|_p$, given by

$$\|(x_1, x_2, \ldots, x_k)\|_p = \max\{\|x_1\|, \|x_2\|, \ldots, \|x_k\|\}.$$

From Hausdorff's Theorem A.1.1, $\mathcal{K}$ is precompact in X^k. So, for $\varepsilon > 0$ as above, there exists a finite family $\{f^1, f^2, \ldots, f^{m(\varepsilon)}\}$ in $\mathcal{F}$, so that, for each $f \in \mathcal{F}$ there exists $j \in \{1,2,\ldots,m(\varepsilon)\}$ such that

$$\|f(t_i) - f^j(t_i)\| < \varepsilon \tag{A.2.3}$$

for each $i = 1,2,\ldots,k$. We have

$$\|f(t) - f^j(t)\| \leq \|f(t) - f(t_i)\| + \|f(t_i) - f^j(t_i)\| + \|f^j(t_i) - f^j(t)\|$$

for each $t \in [a,b]$, where $i \in \{1,2,\ldots,k\}$ is chosen such that $t \in [t_i, t_{i+1}]$, and $j \in \{1,2,\ldots,m(\varepsilon)\}$ is associated to f by means of $(A.2.3)$. From this inequality, $(A.2.2)$ and $(A.2.3)$, it follows that, for each $\varepsilon > 0$, there exists a finite family $\{f^1, f^2, \ldots, f^{m(\varepsilon)}\}$ in $\mathcal{F}$ such that, for each $f \in \mathcal{F}$ there exists $j \in \{1,2,\ldots,m(\varepsilon)\}$ with

$$\|f - f^j\|_\infty < 3\varepsilon.$$

Hence $\mathcal{F}$ is precompact in $C([a,b];X)$ and this achieves the proof. □

Remark A.2.2. As we have seen from the proof of the necessity, if $\mathcal{F}$ is relatively compact in $C([a,b];X)$, (ii) holds with $D = [a,b]$. In fact, in this case, even a stronger condition holds as the next consequence shows.

Corollary A.2.1. *If $\mathcal{F} \subset C([a,b];X)$ is relatively compact, then the set*

$$\mathcal{F}([a,b]) = \{f(t);\ f \in \mathcal{F},\ t \in [a,b]\} \tag{A.2.4}$$

is relatively compact in X.

Proof. Let $(f_n(t_n))_{n\in\mathbb{N}}$ be a sequence in $\mathcal{F}([a,b])$. As both $[a,b]$ and $\mathcal{F}$ are compact in $\mathbb{R}$, and respectively in $C([a,b];X)$, we may assume without loss of generality that, on a subsequence at least, we have

$$\lim_{n\to\infty} t_n = t,$$

and

$$\lim_{n\to\infty} f_n = f,$$

uniformly on $[a,b]$. Let us observe that

$$\begin{aligned}\|f_n(t_n) - f(t)\| &\le \|f_n(t_n) - f(t_n)\| + \|f(t_n) - f(t)\| \\ &\le \|f_n - f\|_\infty + \|f(t_n) - f(t)\|\end{aligned}$$

for each $n \in \mathbb{N}$. Clearly, the first term on the right-hand side of the last inequality goes to 0 for n tending to ∞. Moreover, as f is continuous, the second term tends to 0 too for n tending to ∞, and thus we have

$$\lim_{n\to\infty} f(t_n) = f(t).$$

So $\mathcal{F}([a,b])$ is relatively compact in X and this completes the proof. □

Corollary A.2.2. *Let U be nonempty and closed in X, $g : [a,b] \times U \to X$ a continuous function,*

$$\mathcal{U} = \{u \in C([a,b]),\ u(t) \in U \text{ for } t \in [a,b]\}$$

and let $G : \mathcal{U} \to C([a,b];X)$ the superposition operator associated to the function g, i.e.

$$G(u)(t) = g(t, u(t))$$

for each $u \in \mathcal{U}$ and $t \in [a,b]$. Then G is continuous from $\mathcal{U}$ in $C([a,b];X)$, both the domain and range being endowed with the norm topology $\|\cdot\|_\infty$.

Proof. Let $(u_n)_{n\in\mathbb{N}}$ be a sequence in $\mathcal{U}$ convergent to $u \in \mathcal{U}$ in the norm $\|\cdot\|_\infty$. Obviously, $\{u_n\,;\ n \in \mathbb{N}\}$ is relatively compact in $C([a,b];X)$. Then, according to Corollary A.2.1, the set

$$K =: \overline{\{u_n(t)\,;\ n \in \mathbb{N}, t \in [a,b]\}}$$

is compact in X and consequently $[a,b] \times K$ is compact in $\mathbb{R} \times X$. Then the restriction of the function g to $[a,b] \times K$ is uniformly continuous. This means that, for each $\varepsilon > 0$, there exists $\delta(\varepsilon) > 0$ such that, for each $(t,x),(s,y) \in [a,b] \times K$ with $|t-s| + \|x-y\| \leq \delta(\varepsilon)$ we have

$$\|g(t,x) - g(t,y)\| \leq \varepsilon.$$

Let $\varepsilon > 0$ and let $\delta(\varepsilon) > 0$ as above. As $(u_n)_{n\in\mathbb{N}}$ converges uniformly to u on $[a,b]$, there exists $n(\varepsilon) \in \mathbb{N}$ such that

$$\|u_n(t) - u(t)\| \leq \delta(\varepsilon)$$

for each $n \in \mathbb{N}$ satisfying $n \geq n(\varepsilon)$ and each $t \in [a,b]$. Therefore,

$$\|g(t,u_n(t)) - g(t,u(t))\| \leq \varepsilon$$

for each $n \geq n(\varepsilon)$ and each $t \in [a,b]$. Taking the supremum both sides of the above inequality for $t \in [a,b]$, we obtain

$$\|G(u_n) - G(u)\|_\infty \leq \varepsilon$$

for each $n \geq n(\varepsilon)$. Thus $\lim_{n\to\infty} G(u_n) = G(u)$ in the norm $\|\cdot\|_\infty$, and this completes the proof. □

In order to state the next consequence of Theorem A.2.1, some notations and definitions are needed.
If X is a real Banach space, we denote by X_w the set X endowed with the weak topology, and by $C([a,b];X_w)$ the set of all continuous functions from $[a,b]$ to X_w.

Definition A.2.2. A family $\mathfrak{F}$ in $C([a,b];X_w)$ is *weakly equicontinuous* at a point $t \in [a,b]$, if for each $x^* \in X^*$ and each $\varepsilon > 0$, there exists $\delta(\varepsilon,x^*,t) > 0$ such that, for each $s \in [a,b,]$ with $|t-s| < \delta(\varepsilon,x^*,t)$ we have

$$|x^*(f(t) - f(s))| < \varepsilon,$$

uniformly with respect to $f \in \mathfrak{F}$, i.e $\{x^* \circ f;\ f \in \mathfrak{F}\}$ is equicontinuous at t.
The family $\mathfrak{F}$ is called *weakly equicontinuous* on $[a,b]$, if it is weakly equicontinuous at each $t \in [a,b]$, in the sense mentioned in above.
The family $\mathfrak{F}$ is called *uniformly weakly equicontinuous* on $[a,b]$, if it is weakly equicontinuous on $[a,b]$, and $\delta(\varepsilon,x^*,t)$ can be chosen independent of $t \in [a,b]$.

Corollary A.2.3. *The family $\mathfrak{F}$ in $C([a,b];X_w)$ is included and relatively compact in $C([a,b];X)$ if and only if it is weakly equicontinuous on $[a,b]$, and*

$$\mathfrak{F}([a,b]) = \{f(t);\ f \in \mathfrak{F}, t \in [a,b]\}$$

is relatively compact in X.

Proof. The necessity is a direct consequence of Corollary A.2.1. For the sufficiency, let us observe that, by virtue of Theorem A.2.1, it would be enough to show that $\mathcal{F}$ is equicontinuous on $[a, b]$. To this aim, let us assume by contradiction that this is not the case. Then there exist $t \in [a, b]$, $\varepsilon > 0$, $(f_n)_{n\in\mathbb{N}}$ in $\mathcal{F}$ and $(t_n)_{n\in\mathbb{N}}$ in $[a, b]$, convergent to t, such that

$$\|f_n(t_n) - f_n(t)\| \geq \varepsilon \tag{A.2.5}$$

for each $n \in \mathbb{N}$. As $\mathcal{F}([a, b])$ is relatively compact, we may assume without loss of generality (by extracting a subsequence if necessary), that there exists $x \in X$ such that

$$\lim_{n\to\infty} (f_n(t_n) - f_n(t)) = x$$

strongly in X. But $\mathcal{F}$ is weakly equicontinuous, and consequently

$$\lim_{n\to\infty} (f_n(t_n) - f_n(t)) = 0$$

in the weak topology of X. This implies that $x = 0$, which contradicts $(A.2.5)$. This contradiction can be eliminated only if $\mathcal{F} \subset C([a, b]; X)$ and is equicontinuous on $[a, b]$ in the sense of Definition A.2.1, thereby completing the proof. □

Theorem A.2.2. *Let X_0, X, and X_1 be three real Banach spaces with the inclusion $X_0 \subset X$ compact and $X \subset X_1$ continuous. If $\mathcal{F}$ is a family of functions from $[a, b]$ to X_0, which is uniformly bounded in the norm $\|\cdot\|_0$, and equicontinuous in the norm $\|\cdot\|_1$, then $\mathcal{F}$ is relatively compact in $C([a, b]; X)$.*

Proof. As $\mathcal{F}$ is uniformly bounded in the norm $\|\cdot\|_0$, there exists $M > 0$ such that

$$\|f(t)\|_0 \leq M$$

for each $f \in \mathcal{F}$ and each $t \in [a, b]$. From the preceding inequality and the compactness of the inclusion $X_0 \subset X$, it follows that, for each $t \in [a, b]$, $\mathcal{F}(t) = \{f(t);\ f \in \mathcal{F}\}$ is relatively compact in X. Hence $\mathcal{F}$ satisfies (ii) in Theorem A.2.1.
Let $\varepsilon > 0$. According to Lemma A.1.1, there exists $\eta = \eta(\varepsilon)$ such that

$$\|x\|_0 \leq \varepsilon\|x\| + \eta(\varepsilon)\|x\|_1$$

for each $x \in X_0$. Inasmuch as $\mathcal{F}$ is equicontinuous from $[a, b]$ to X_1, for each $t \in [a, b]$ there exists $\delta(t, \varepsilon) > 0$ such that

$$\|f(t) - f(s)\|_1 \leq \frac{\varepsilon}{\eta(\varepsilon)}$$

for each $s \in [a,b]$ satisfying $|t-s| \le \delta(t,\varepsilon)$. We then have

$$\begin{aligned}\|f(t)-f(s)\| &\le \varepsilon\|f(t)-f(s)\|_0 + \eta(\varepsilon)\|f(t)-f(s)\|_1 \\ &\le 2M\varepsilon + \eta(\varepsilon)\frac{\varepsilon}{\eta(\varepsilon)} \le (2M+1)\varepsilon\end{aligned}$$

for each $s \in [a,b]$, $|t-s| \le \delta(t,\varepsilon)$. Hence $\mathfrak{F}$ is equicontinuous from $[a,b]$ to X, and therefore it satisfies (i) in Theorem A.2.1. Consequently $\mathfrak{F}$ is relatively compact in $C([a,b];X)$, and the proof is complete. □

We conclude this section with two examples of triples X_0, X, X_1 satisfying the general condition in Theorem A.2.2.

Example A.2.1. Let $X_0 = C^1([0,1])$, $X = C([0,1])$, $X_1 = L^1(0,1)$. In view of Theorem A.2.1, it follows that each bounded subset with respect to the norm $\|\cdot\|_{C^1([0,1])}$, defined by

$$\|f\|_{C^1([0,1])} = \sup_{x\in[0,1]} |f(t)| + \sup_{x\in[0,1]} |f'(t)|$$

for each $f \in C^1([0,1])$, is relatively compact in $C([0,1])$ endowed with the norm $\|\cdot\|_\infty$. In addition, each convergent sequence in $C([0,1])$ is convergent in $L^1(0,1)$. Therefore X_0 is compactly included in X which, in its turn, is continuously included in X_1.

Example A.2.2. Let $X_0 = H_0^1(\Omega)$, $X = L^2(\Omega)$ and let $X_1 = H^{-1}(\Omega)$, where Ω is a nonempty, bounded and open subset in $\mathbb{R}^n$ whose boundary Γ is of class C^1. Then, by virtue of Theorem 1.5.4, it follows that X_0 is compactly included in X, and this is continuously (in fact even compactly) included in X_1.

A.3. Compactness in $C([a,b];X_w)$

In this section we shall present a variant of Theorem A.2.1 referring to the space $C([a,b];X_w)$ of all continuous functions from $[a,b]$ to X. On this space we define a topology as follows. Let $\mathrm{Fin}\,(X^*)$ be the class of all nonempty and finite subsets in X^*, let $F^* \in \mathrm{Fin}\,(X^*)$ and let us define $\|\cdot\|_{F^*} : C([a,b];X_w) \to \mathbb{R}_+$ by

$$\|f\|_{F^*} = \sup_{t\in[a,b]} \sup_{x^*\in F^*} |x^*(f(t))|$$

for each $f \in C([a,b];X_w)$. One may easily see that $\{\|\cdot\|_{F^*};\ F^* \in \mathrm{Fin}\,(X^*)\}$ is a family of seminorms on $C([a,b];X_w)$ which defines a topology of a locally convex, separated space, called *the uniform weak convergence topology.* We emphasize that this topology (except for the case in which X is finite-dimensional) is not metrisable. Therefore, the main result in this section referring to the sequential relative compactness of a subset in $C([a,b];X_w)$ is rather surprising because it cannot be obtained directly by means of the equivalence between relative, and sequentially relative compactness.

Theorem A.3.1. (Arzelà-Ascoli) *Let X be a sequentially weakly complete*[1] *Banach space. A family $\mathfrak{F}$ in the space $C([a,b];X_w)$, endowed with the uniform weak convergence topology, is sequentially relatively compact if and only if:*

(i) *$\mathfrak{F}$ is weakly equicontinuous on $[a,b]$;*
(ii) *there exists a dense subset D in $[a,b]$ such that, for each $t \in D$, the section $\mathfrak{F}(t) = \{f(t);\ f \in \mathfrak{F}\}$ is sequentially weakly relatively compact in X.*

Proof. *Necessity.* Let $\mathfrak{F} \subset C([a,b,];X_w)$ be sequentially relatively compact. In order to show that it satisfies (i), we proceed by contradiction. So, let us assume that there exist $t \in [a,b]$, $\varepsilon > 0$, $(t_n)_{n\in\mathbb{N}}$ in $[a,b]$ convergent to some t, $(f_n)_{n\in\mathbb{N}}$ in $\mathfrak{F}$, and $x^* \in X^*$ such that

$$|x^*(f_n(t_n) - f_n(t))| \geq \varepsilon \tag{A.3.1}$$

for each $n \in \mathbb{N}$. As $\mathfrak{F}$ is sequentially relatively compact, we may assume without loss of generality that there exists f in $C([a,b];X_w)$, such that

$$\lim_{n\to\infty} f_n = f$$

in $C([a,b];X_w)$. In particular, we have

$$\lim_{n\to\infty} |x^*(f_n(s) - f(s))| = 0, \tag{A.3.2}$$

uniformly for $s \in [a,b]$. Let us remark that

$$\varepsilon \leq |x^*(f_n(t_n) - f_n(t))|$$
$$\leq |x^*(f_n(t_n) - f(t_n))| + |x^*(f(t_n) - f(t))| + |x^*(f(t) - f_n(t))|$$

for each $n \in \mathbb{N}$. In view of $(A.3.2)$ and the weak continuity of the function f, it follows that, for $n \in \mathbb{N}$ large enough, we have

$$|x^*(f_n(t_n) - f(t_n))| < \frac{\varepsilon}{3}, \quad |x^*(f(t) - f_n(t))| < \frac{\varepsilon}{3},$$

and

$$|x^*(f(t_n) - f(t))| < \frac{\varepsilon}{3}.$$

But these inequalities, and $(A.3.1)$, lead to $\varepsilon < \varepsilon$. This contradiction can be eliminated only if the family $\mathfrak{F}$ satisfies (i). To complete the proof of

[1]Each reflexive Banach space is sequentially weakly complete, but there are non-reflexive spaces enjoying this property. For instance, $L^1(\Omega)$, with Ω measurable in $\mathbb{R}^n$, although non-reflexive, is sequentially weakly complete.

the necessity it suffices to show that (ii) holds with $D = [a,b]$. To this aim, take $t \in [a,b]$ and $(f_n(t))_{n\in\mathbb{N}}$ in $\mathcal{F}(t)$. By hypothesis, there exist f_∞ in $C([a,b];X_w)$ and a subsequence $(f_{n_k})_{k\in\mathbb{N}}$ of $(f_n)_{n\in\mathbb{N}}$ with the property that $\lim_{k\to\infty} f_{n_k} = f_\infty$ in $C([a,b];X_w)$. Therefore, for each $x^* \in X^*$, we have $\lim_{k\to\infty} x^*(f_{n_k}(t)) = x^*(f_\infty(t))$. Thus (ii) holds and this completes the proof of the necessity.

Sufficiency. Inasmuch as $[a,b]$ is compact, one may easily see that a family $\mathcal{F}$ is weakly equicontinuous on $[a,b]$ if and only if it is uniformly weakly equicontinuous on $[a,b]$. See Definition A.2.2. Let $\mathcal{F}$ be a family which satisfies (i) and (ii). First, let us observe that, since $[a,b]$ is separable, we may always assume that D is countable (by taking a subset if necessary). So, let $D = \{t_k;\ k \in \mathbb{N}\}$ be the dense set in $[a,b]$ given by (ii). Let $(f_n)_{n\in\mathbb{N}}$ be a sequence in $\mathcal{F}$. Using (ii) and a standard diagonal procedure, we deduce that there exists $f_\infty : D \to X$ such that

$$\lim_{n\to\infty} f_n(t_k) = f_\infty(t_k) \tag{A.3.3}$$

weakly in X for each $t_k \in D$.

We shall show next that $(f_n)_{n\in\mathbb{N}}$ is a Cauchy sequence in $C([a,b];X_w)$. To this aim, let $\varepsilon > 0$ and $F^* \in \mathrm{Fin}(X^*)$. At this point, let us observe that we may assume without loss of generality that F^* is a singleton, i.e. $F = \{x^*\}$. By virtue of (i), there exists $\delta(\varepsilon) > 0$ such that, for each $f \in \mathcal{F}$ and each $t, s \in [a,b]$ with $|t-s| < \delta(\varepsilon)$, we have

$$|x^*(f(t) - f(s))| < \varepsilon.$$

Let $a \le t_1 < t_2 < \cdots < t_{k(\varepsilon)} \le b$ be such that $t_i \in D$, $t_1 - a < \delta$, $b - t_{k(\varepsilon)} < \delta$ and $|t_{i+1} - t_i < \delta(\varepsilon)$ for $i = 1, 2, \ldots, k(\varepsilon) - 1$. Let $t \in [a,b]$ and let us choose $i \in \{1, 2, \ldots, k(\varepsilon)\}$ so that $|t - t_i| < \delta(\varepsilon)$. Since $(x^*(f_n(t_i)))_{n\in\mathbb{N}^*}$ are Cauchy sequences, $i = 1, 2, \ldots, k(\varepsilon)$, there exists $n(\varepsilon) \in \mathbb{N}^*$ such that, for each $n, m \in \mathbb{N}$ with $n, m \ge n(\varepsilon)$ and each $i \in \{1, 2, \ldots, k(\varepsilon)\}$, we have $|x * (f_n(t_i) - f_m(t_i))| < \varepsilon$. So, for each $t \in [a,b]$, with i and n, m as above, we have

$$|x^*(f_n(t) - f_m(t))| \le |x^*(f_n(t) - f_n(t_i))| + |x^*(f_n(t_i) - f_m(t_i))|$$
$$+|x^*(f_m(t_i) - f_m(t))| < 3\varepsilon.$$

Thus $(f_n)_{n\in\mathbb{N}^*}$ is a Cauchy sequence in $C([a,b];X_w)$. Inasmuch as X is sequentially weakly complete, we deduce that there exists $f \in C([a,b];X_w)$ such that

$$\lim_{n\to\infty} \|f_n - f\|_{x^*} = 0,$$

and this completes the proof. □

We conclude this section with a weak variant of Theorem A.2.2.

Theorem A.3.2. *Let X_0, X and X_1 be three real Banach spaces with the inclusion $X_0 \subset X$ compact and $X \subset X_1$ continuous. If $\mathfrak{F}$ is a family of functions from $[a,b]$ to X_0 which is uniformly bounded with respect to $\|\cdot\|_0$, and weakly equicontinuous from $[a,b]$ to X_1, then $\mathfrak{F}$ is relatively compact in $C([a,b];X)$.*

Proof. As $\mathfrak{F}$ is uniformly bounded from $[a,b]$ to X_0, X_0 is compactly imbedded in X, and the latter, in its turn, is continuously imbedded in X_1, it follows that the set $\mathfrak{F}([a,b]) = \{f(t);\ f \in \mathfrak{F},\ t \in [a,b]\}$ is relatively compact in X_1. By virtue of Theorem A.3.1, it follows that $\mathfrak{F}$ is relatively compact in $C([a,b];X_1)$, and from Theorem A.1.1, we deduce that $\mathfrak{F}$ is equicontinuous with respect to the norm $\|\cdot\|_1$. By Lions' Lemma A.1.1 it follows that, for each $\varepsilon > 0$ there exists $\eta(\varepsilon) > 0$ such that

$$\|f(t) - f(s)\| \leq \varepsilon\|f(t) - f(s)\|_{X_0} + \eta(\varepsilon)\|f(t) - f(s)\|_{X_1}$$

for each $f \in \mathfrak{F}$, and all $t, s \in [a,b]$. Since $\mathfrak{F}$ is relatively compact in X_0 and thus bounded, and $\mathfrak{F}$ is equicontinuous from $[a,b]$ to X_1, this inequality shows that $\mathfrak{F}$ is equicontinuos from $[a,b]$ to X. An appeal to Theorem A.2.1 achieves the proof. □

A.4. Compactness in $L^p(a,b;X)$

In this section we shall prove several necessary and/or sufficient conditions in order for a given subset in $L^p(a,b;X)$ to be relatively compact. We begin by introducing some notions we need in that follows. Let X be a real Banach space and $p \in [1,+\infty)$.

Definition A.4.1. A family $\mathfrak{F} \subset L^p(a,b;X)$ is *p-equiintegrable* if

$$\lim_{h\downarrow 0} \int_a^{b-h} \|f(t+h) - f(t)\|^p dt = 0,$$

uniformly for $f \in \mathfrak{F}$.

Remark A.4.1. Each finite set in $L^p(a,b;X)$ is p-equiintegrable. This follows from the fact that $C([a,b];X)$ is dense in $L^p(a,b;X)$, and each finite subset in $C([a,b];X)$ has this property.

Remark A.4.2. A subset $\mathfrak{F}$ in $L^p(a,b;X)$ is p-equiintegrable if and only if

$$\lim_{h\downarrow 0} \int_{a+h}^{b} \|f(t-h) - f(t)\|^p dt = 0$$

uniformly for $f \in \mathfrak{F}$. Indeed, the condition above follows from the p-equiintegrability condition by a simple change of variable, and conversely.

We can now proceed to the first important result in this section.

Theorem A.4.1. (Kolmogorov-Riesz-Weil) *A subset $\mathcal{F}$ in $L^p(a,b;X)$ is relatively compact if and only if:*

(i) *$\mathcal{F}$ is p-equiintegrable;*

(ii) *for each interval $[\alpha,\beta]$ included in $[a,b]$, the set*

$$\left\{\int_\alpha^\beta f(t)\,dt;\ f\in\mathcal{F}\right\}$$

is relatively compact in X.

Proof. We begin by showing the necessity of the conditions (i) and (ii). Let $\mathcal{F}$ be a relatively compact subset in $L^p(a,b;X)$, and let $\varepsilon>0$. As $\mathcal{F}$ is precompact, it has a finite ε-net $\mathcal{F}_\varepsilon=\{f_1,f_2,\ldots,f_{m(\varepsilon)}\}$. According to Remark A.4.1, $\mathcal{F}_\varepsilon$ is p-equiintegrable. Therefore, there exists $\delta(\varepsilon)>0$ such that, for each $h\in(0,\delta(\varepsilon)]$ and $i\in\{1,2,\ldots,n(\varepsilon)\}$, we have

$$\left(\int_a^{b-h}\|f_i(t+h)-f_i(t)\|^p dt\right)^{1/p}<\varepsilon. \tag{A.4.1}$$

As $\mathcal{F}_\varepsilon$ is an ε-net for $\mathcal{F}$, for each $f\in\mathcal{F}$, there exists $i\in\{1,2,\ldots,n(\varepsilon)\}$, such that

$$\left(\int_a^{b}\|f(t)-f_i(t)\|^p dt\right)^{1/p}<\varepsilon. \tag{A.4.2}$$

From $(A.4.1)$ and $(A.4.2)$, it follows that

$$\left(\int_a^{b-h}\|f(t+h)-f(t)\|^p dt\right)^{1/p}$$

$$\leq\left(\int_a^{b-h}\|f(t+h)-f_i(t+h)\|^p dt\right)^{1/p}$$

$$+\left(\int_a^{b-h}\|f_i(t+h)-f_i(t)\|^p dt\right)^{1/p}+\left(\int_a^{b-h}\|f_i(t)-f(t)\|^p dt\right)^{1/p}$$

$$\leq 2\varepsilon+\varepsilon$$

for each $h\in(0,\delta(\varepsilon)]$. Obviously, this inequality shows that $\mathcal{F}$ satisfies (i). In order to prove (ii), let $[\alpha,\beta]$ be any interval included in $[a,b]$, and let $(f_n)_{n\in\mathbb{N}}$ be a sequence in $\mathcal{F}$. As $\mathcal{F}$ is relatively compact, we may assume without loss of generality that this sequence is convergent in $L^p(a,b;X)$ to some element f. Then

$$\lim_{n\to\infty}\int_\alpha^\beta f_n(t)\,dt=\int_\alpha^\beta f(t)\,dt,$$

which shows that $\mathcal{F}$ satisfies (ii).

We can now proceed to the proof of sufficiency. Let $\mathcal{F}$ be a family satisfying (i) and (ii). Let $f \in \mathcal{F}$, $\lambda > 0$ with $a < b - \lambda$, and let $f_\lambda : [a, b-\lambda] \to X$ be defined by

$$f_\lambda(t) = \frac{1}{\lambda}\int_t^{t+\lambda} f(s)\, ds.$$

We shall show next that the family $\mathcal{F}_\lambda = \{f_\lambda;\ f \in \mathcal{F}\}$ is relatively compact in $C([a, b-\lambda]; X)$, and that

$$\lim_{\lambda \downarrow 0} \|f_\lambda - f\|_{L^p(a,b-\lambda;X)} = 0 \tag{A.4.3}$$

uniformly for $f \in \mathcal{F}$.

From (ii) it follows that, for each $t \in [a, b-\lambda]$ the t-section of the family $\mathcal{F}_\lambda$, $\mathcal{F}_\lambda(t) = \{f_\lambda(t);\ f \in \mathcal{F}\}$, is relatively compact in X. In addition, for each $t, s \in [a, b-\lambda]$, with $t < s$ and $s - t \leq \lambda$, we have

$$f_\lambda(t) - f_\lambda(s) = \frac{1}{\lambda}\int_t^{t+\lambda} f(\tau)\, d\tau - \frac{1}{\lambda}\int_s^{s+\lambda} f(\tau)\, d\tau.$$

The change of variable $\tau = \theta + s - t$ in the last integral, leads to

$$f_\lambda(t) - f_\lambda(s) = \frac{1}{\lambda}\int_t^{t+\lambda} f(\theta)\, d\theta - \frac{1}{\lambda}\int_t^{t+\lambda} f(\theta + s - t)\, d\theta.$$

Therefore

$$\|f_\lambda(t) - f_\lambda(s)\| \leq \frac{1}{\lambda}\int_t^{t+\lambda} \|f(\theta) - f(\theta + s - t)\| d\theta$$

$$\leq \frac{1}{\lambda}\int_a^{b+t-s} \|f(\theta) - f(\theta + s - t)\| d\theta$$

$$\leq \frac{1}{\lambda}(b - a + t - s)^{1/q}\left(\int_a^{b+t-s} \|f(\theta) - f(\theta + t - s)\|^p d\theta\right)^{1/p},$$

where $1/p + 1/q = 1$. From this inequality, and (i), it follows that $\mathcal{F}_\lambda$ satisfies the hypotheses of Arzèla-Ascoli's Theorem A.2.1, and therefore it is relatively compact in $C([a, b-\lambda]; X)$. In order to check $(A.4.3)$, let us observe that

$$\|f - f_\lambda\|^p_{L^p(a,b-\lambda;X)} = \int_a^{b-\lambda} \|f(t) - f_\lambda(t)\|^p dt$$

$$= \int_a^{b-\lambda} \left\| f(t) - \frac{1}{\lambda}\int_t^{t+\lambda} f(s)\, ds \right\|^p dt$$

$$= \frac{1}{\lambda^p}\int_a^{b-\lambda} \left\| \int_t^{t+\lambda} f(t)\, ds - \int_t^{t+\lambda} f(s)\, ds \right\|^p dt.$$

From this relation, by changing the variable $s = t + h$ in the last integral, using Hölder's Inequality and changing the order of integration, we obtain

$$\|f - f_\lambda\|^p_{L^p(a,b-\lambda;X)} = \frac{1}{\lambda^p}\int_a^{b-\lambda}\left\|\int_0^\lambda f(t)\,ds - \int_0^\lambda f(t+h)\,dh\right\|^p dt$$

$$\leq \frac{1}{\lambda^p}\int_a^{b-\lambda}\lambda^{p/q}\left(\int_0^\lambda \|f(t) - f(t+h)\|^p dh\right) dt$$

$$= \frac{1}{\lambda}\int_0^\lambda\int_a^{b-\lambda}\|f(t) - f(t+h)\|^p dt\,dh.$$

But, this inequality, along with (i), proves $(A.4.3)$. Finally, we shall show that, for each $c \in (a,b)$, the set $\mathcal{F}|_{[a,c]}$, of restrictions to $[a,c]$ of elements in $\mathcal{F}$, is relatively compact in $L^p(a,c;X)$. From the preceding arguments, and $(A.4.3)$, we deduce that the family of operators

$$\{Q_\lambda;\ \lambda \in (0, b-c)\},$$

$Q_\lambda : \mathcal{F} \to C([a,c];X)$, defined by

$$Q_\lambda f = f_\lambda$$

for each $\lambda \in (0, b-c)$ and $f \in \mathcal{F}$, satisfies the hypotheses of Lemma A.1.2. Indeed, for each $\lambda \in (0, b-c)$, $(Q_\lambda\mathcal{F})|_{[a,c]}$ is relatively compact in $L^p(a,c;X)$ being relatively compact in $C([a,c];X)$, and

$$\lim_{\lambda\downarrow 0}\|Q_\lambda f - f\|_{L^p(a,c;X)} = 0$$

uniformly for $f \in \mathcal{F}$. In view of Lemma A.1.2, we conclude that $\mathcal{F}|_{[a,c]}$ is precompact in $L^p(a,c;X)$. Finally, using Remark A.2.1, and reasoning as above, we deduce that, for each $c \in (a,b)$, $\mathcal{F}|_{[c,b]}$ is relatively compact in $L^p(c,b;X)$. Therefore $\mathcal{F}$ is relatively compact in $L^p(a,b;X)$, and this achieves the proof. □

Theorem A.4.2. (Aubin) *Let $1 \leq p < +\infty$, and X_0, X and X_1 be three real Banach spaces with X_0 compactly included in X and X continuously included in X_1. If $\mathcal{F}$ is bounded in $L^p(a,b;X_0)$ and p-equiintegrable in $L^p(a,b;X_1)$, then it is relatively compact in $L^p(a,b;X)$.*

Proof. Let $[\alpha,\beta] \subset [a,b]$. Since $\mathcal{F}$ is bounded in $L^p(a,b;X_0)$, it follows that the set

$$\left\{\int_\alpha^\beta f(t)\,dt;\ f \in \mathcal{F}\right\}$$

is bounded in X_0, and therefore relatively compact in X. Hence $\mathcal{F}$ satisfies (ii) in Theorem A.4.1. Let $\varepsilon > 0$. According to Lemma A.1.1, there exists $\eta(\varepsilon) > 0$ such that

$$\|x\| \leq \varepsilon\|x\|_0 + \eta(\varepsilon)\|x\|_1$$

for each $x \in X_0$. Then, for each $f \in \mathcal{F}$ and each $h > 0$, we have

$$\left(\int_a^{b-h} \|f(t+h) - f(t)\|^p dt\right)^{1/p} \leq \varepsilon\left(\int_a^{b-h} \|f(t+h) - f(t)\|_0^p dt\right)^{1/p}$$

$$+\eta(\varepsilon)\left(\int_a^{b-h} \|f(t+h) - f(t)\|_1^p dt\right)^{1/p}.$$

As $\mathcal{F}$ is bounded in $L^p(a,b;X_0)$, there exists $M > 0$, such that

$$\left(\int_a^{b-h} \|f(t+h) - f(t)\|^p dt\right)^{1/p}$$

$$\leq 2M\varepsilon + \eta(\varepsilon)\left(\int_a^{b-h} \|f(t+h) - f(t)\|_1^p dt\right)^{1/p}.$$

But $\mathcal{F}$ is p-equiintegrable in $L^p(a,b;X_1)$, and consequently there exists $\delta(\varepsilon) > 0$, such that, for each $f \in \mathcal{F}$ and each $h \in (0, \delta(\varepsilon))$, we have

$$\left(\int_a^{b-h} \|f(t+h) - f(t)\|_1^p dt\right)^{1/p} \leq \frac{\varepsilon}{\eta(\varepsilon)}.$$

From this inequality and the preceding one, we deduce

$$\left(\int_a^{b-h} \|f(t+h) - f(t)\|^p dt\right)^{1/p} \leq (2M+1)\varepsilon$$

for each $f \in \mathcal{F}$ and each $h \in (0, \delta(\varepsilon))$. Therefore $\mathcal{F}$ is p-equiintegrable in $L^p(a,b;X)$, and, according to Theorem A.4.1, it is relatively compact in this space. The proof is complete. □

A.5. Compactness in $L^p(a,b;X)$. Continued

In this section we shall present a variant of Theorem A.4.1 which proves much easier to check in many practical situations.

Definition A.5.1. A family $\mathcal{F}$ in $L^p(a,b;X)$ is called *uniformly integrable* if for each $\varepsilon > 0$ there exists $\delta(\varepsilon) > 0$ such that, for each measurable subset E in $[a,b]$ whose Lebesgue measure $\mu(E) < \delta(\varepsilon)$, we have

$$\int_E \|f(t)\| dt < \varepsilon$$

for all $f \in \mathcal{F}$.

Remark A.5.1. As $[a,b]$ can be decomposed into a finite union of sets of arbitrary small Lebesgue measure, it follows that each uniformly integrable subset is bounded. The converse statement is not true, as we can see from the example below.

Example A.5.1. Let $\mathfrak{F} = \{f_n;\ n \in \mathbb{N}^*\}$, where $f_n : [-1,1] \to \mathbb{R}$ is defined by

$$f_n(t) = \begin{cases} 0, & \text{for } t \in [-1,-1/n] \cup [1/n,1] \\ \dfrac{n}{2}, & \text{for } t \in [-1/n,1/n] \end{cases} \qquad n \in \mathbb{N}^*.$$

Obviously $\mathfrak{F}$ is bounded in $L^1(-1,1;\mathbb{R})$ (all elements having the norm 1), but it is not uniformly integrable "near 0". Namely, we have

$$\int_{-1/n}^{1/n} |f_n(t)|\, dt = 1$$

for each $n \in \mathbb{N}^*$, relation which shows that the family $\mathfrak{F}$ cannot satisfy the condition in Definition A.5.1.

Remark A.5.2. If $p > 1$ and $\mathfrak{F}$ is bounded in $L^p(a,b;X)$, then it is uniformly integrable. Indeed, by virtue of Hölder's Inequality, it follows

$$\int_E \|f(t)\|\, dt \leq (\mu(E))^{1/q} \left(\int_E \|f(t)\|^p\, dt \right)^{1/p}$$

for each $f \in \mathfrak{F}$ and each measurable subset E in $[a,b]$, where q is the conjugate of p, i.e. $1/p + 1/q = 1$. From this inequality we easily deduce that $\mathfrak{F}$ satisfies the condition in Definition A.5.1.

Remark A.5.3. Let $\mathfrak{F} \subset L^1(a,b;X)$ and let us assume that there exists $g \in L^1(a,b)$, such that $\|f(t)\| \leq g(t)$ for each $f \in \mathfrak{F}$ and a.e. $t \in (a,b)$. Then $\mathfrak{F}$ is uniformly integrable. Indeed, in this case, $\int_E \|f(t)\|\, dt \leq \int_E g(t)\, dt$ for each $f \in \mathfrak{F}$ and each measurable subset E in $[a,b]$. As g is integrable, from this inequality, it follows that $\mathfrak{F}$ is uniformly integrable.

The main result in this section is:

Theorem A.5.1. (Gutman) *A uniformly integrable family $\mathfrak{F}$ in $L^p(a,b;X)$ is relatively compact if and only if:*

(j) *$\mathfrak{F}$ is p-equiintegrable;*
(jj) *for each $\varepsilon > 0$, there exists a compact subset C_ε in X, such that, for each $f \in \mathfrak{F}$ there exists a measurable subset $E_{\varepsilon,f}$ in $[a,b]$ with $\mu(E_{\varepsilon,f}) \leq \varepsilon$, and such that $f(t) \in C_\varepsilon$ for each $t \in [a,b] \setminus E_{\varepsilon,f}$.*

Proof. *Necessity.* Let $\mathcal{F} \subset L^p(a,b;X)$ be relatively compact. From Theorem A.4.1 it follows that $\mathcal{F}$ satisfies (j). To prove (jj), let $\varepsilon \in (0,1)$, and let $E(\varepsilon) \subset C([a,b];X)$ be a finite ε-net of $\mathcal{F}$ with respect to the norm $\|\cdot\|_{L^p(a,b;X)}$. Let $n \in \mathbb{N}^*$, and let

$$E_n(\varepsilon) = E(\varepsilon^{(n+1)/p}/2^{n/p}).$$

Then, for each $n \in \mathbb{N}^*$ and each $f \in \mathcal{F}$, there exists $f_n \in E_n(\varepsilon)$ such that

$$\|f - f_n\|_{L^p(a,b;X)} \leq \varepsilon.$$

Let us define

$$E^n_{\varepsilon,f} = \{t \in [a,b]\,;\ \|f(t) - f_n(t)\| > \varepsilon^n\}.$$

We deduce that $\mu(E^n_{\varepsilon,f}) \leq \varepsilon/2^n$. Let now

$$E_{\varepsilon,f} = \bigcup_n E^n_{\varepsilon,f},$$

and let us observe that, for each $t \in [a,b] \setminus E_{\varepsilon,f}$, and each $n \in \mathbb{N}^*$, we have

$$\|f(t) - f_n(t)\| \leq \varepsilon^n.$$

Finally, let

$$C_{\varepsilon,f} = \{f(t)\,;\ t \in [a,b] \setminus E_{\varepsilon,f}\} \text{ and } C_\varepsilon = \bigcup_{f \in \mathcal{F}} C_{\varepsilon,f}.$$

We shall prove that C_ε is totally bounded. To this aim, let $\delta > 0$, and let us fix $n \in \mathbb{N}^*$ satisfying $\varepsilon^n < \delta/2$. We can always do this, because $\varepsilon \in (0,1)$. As $E_n(\varepsilon)$ is a finite family of continuous functions, the set

$$E_n(\varepsilon)([a,b]) = \{h(t)\,;\ h \in E_n(\varepsilon),\ t \in [a,b]\}$$

is relatively compact. Hence it has a finite $\delta/2$-net $\{x_1, x_2, \ldots, x_{k(\delta/2)}\}$. Then, for each $f_n \in E_n(\varepsilon)$ and $t \in [a,b]$, there exists $i \in \{1, 2, \ldots, k(\delta/2)\}$, such that we have

$$\|f_n(t) - x_i\| < \frac{\delta}{2}.$$

From this inequality, recalling that $\mu(E^n_{\varepsilon,f}) \leq \varepsilon/2^n$, we deduce

$$\|f(t) - x_i\| \leq \|f(t) - f_n(t)\| + \|f_n(t) - x_i\| < \frac{\delta}{2} + \frac{\delta}{2} = \delta,$$

which shows that C_ε is precompact and this completes the proof of the necessity.

Sufficiency. Let $\mathcal{F}$ be a uniformly integrable family satisfying (j) and (jj) and let $[\alpha, \beta]$ be any interval included in $[a,b]$. By Theorem A.4.1, to complete the proof, it suffices to show that $\mathcal{F}^\beta_\alpha = \left\{\int_\alpha^\beta f(t)\,dt\,;\ f \in \mathcal{F}\right\}$ is relatively compact in X. To this aim, let $\varepsilon > 0$ and let us define the operator $Q_\varepsilon : \mathfrak{F}_\alpha^\beta \to X$ by

$$Q_\varepsilon \left(\int_\alpha^\beta f(t)\,dt \right) = \int_{[\,\alpha,\beta\,]\setminus E_{\varepsilon,f}} f(t)\,dt,$$

where $E_{\varepsilon,f}$ is the subset whose existence is ensured by (jj). As $f(t) \in C_\varepsilon$ for each $t \in [\,\alpha,\beta\,] \setminus E_{\varepsilon,f}$ and C_ε is compact, from Lemma A.1.3, it follows that $Q_\varepsilon(\mathfrak{F}_\alpha^\beta)$ is relatively compact. Let us observe that

$$\left\| Q_\varepsilon \int_\alpha^\beta f(t)\,dt - \int_\alpha^\beta f(t)\,dt \right\| \leq \int_{E_{\varepsilon,f}} \|f(t)\|\,dt,$$

for each $f \in \mathfrak{F}$. Since $\mathfrak{F}$ is uniformly integrable, this inequality shows that

$$\lim_{\varepsilon \downarrow 0} Q_\varepsilon x = x$$

uniformly for $x \in \mathfrak{F}_\alpha^\beta$. According to Corollary A.2.1, it follows that $\mathfrak{F}_\alpha^\beta$ is relatively compact. Hence $\mathfrak{F}$ satisfies the conditions in Theorem A.4.1, and this completes the proof of Theorem A.5.1. □

We conclude this section with a variant of Theorem A.4.2.

Theorem A.5.2. *Let $1 \leq p < +\infty$, and let X_0, X and X_1 be real Banach spaces with X_0 compactly imbedded in X, and X continuously imbedded in X_1. Let $\mathfrak{F}$ be subset in $L^p(a,b;X_0)$ such that $\{\|f\|^p\,;\ f \in \mathfrak{F}\}$ is uniformly integrable and which satisfies:*

(l) *$\mathfrak{F}$ is p-equiintegrable in $L^p(a,b;X_1)$;*

(ll) *for each $\varepsilon > 0$ there exists $M_\varepsilon > 0$ such that, for each $f \in \mathfrak{F}$ there exists a measurable subset $E_{\varepsilon,f}$ in $[\,a,b\,]$ with $\mu(E_{\varepsilon,f}) \leq \varepsilon$, and such that*

$$\|f(t)\|_0 \leq M_\varepsilon$$

for each $t \in [\,a,b\,] \setminus E_{\varepsilon,f}$.

Then $\mathfrak{F}$ is relatively compact in $L^p(a,b;X)$.

Proof. As X_0 is compactly imbedded in X, from (ll), it follows that $\mathfrak{F}$ satisfies (jj) in Theorem A.5.1. Let $\varepsilon > 0$. Since $\{\|f\|^p\,;\ f \in \mathfrak{F}\}$ is uniformly integrable, there exists $\delta = \delta(\varepsilon) > 0$ such that

$$\int_E \|f(t)\|^p\,dt \leq \varepsilon^p$$

for each $f \in \mathfrak{F}$, and each measurable subset E in $[\,a,b\,]$ with $\mu(E) \leq 2\delta$. From (ll), we deduce that, for $\delta > 0$, there exists $M_\delta > 0$ such that, for each $f \in \mathfrak{F}$, there exists a measurable subset $E_{\delta,f}$ in $[\,a,b\,]$, with $\mu(E_{\delta,f}) \leq \delta$, and $\|f(t)\|_0 \leq M_\delta$ for each $t \in [a,b] \setminus E_{\delta,f}$. Let us observe that, for each $h > 0$, we can choose

$$E_{\delta,f(\cdot+h)} = E_{\delta,f} + h.$$

Let us denote by $E = E_{\delta,f} \cup E_{\delta,f(\cdot+h)}$ and $G = [a, b-h] \setminus E$. We then have

$$\left(\int_a^{b-h} \|f(t+h) - f(t)\|^p dt\right)^{1/p} \leq \left(\int_E \|f(t+h) - f(t)\|^p dt\right)^{1/p}$$
$$+\left(\int_G \|f(t+h) - f(t)\|^p dt\right)^{1/p} \leq 2\varepsilon + \left(\int_G \|f(t+h) - f(t)\|^p dt\right)^{1/p}. \tag{A.5.1}$$

Next, let $\lambda > 0$ with $2\lambda M_\delta \leq \varepsilon$, and let $\eta(\lambda) > 0$ be given by Lemma A.1.1. We recall that $\eta(\lambda)$ satisfies

$$\|x\| \leq \lambda \|x\|_0 + \eta(\lambda)\|x\|_1 \tag{A.5.2}$$

for each $x \in X_0$. As $\mathfrak{F}$ is p-equiintegrable in $L^p(a,b;X_1)$, there exists $\gamma(\varepsilon) > 0$, such that

$$\eta(\lambda)\left(\int_a^{b-h} \|f(t+h) - f(t)\|^p dt\right)^{1/p} \leq \varepsilon \tag{A.5.3}$$

for each $f \in \mathfrak{F}$ and each $h > 0$ with $h \leq \gamma(\varepsilon)$. From $(A.5.1)$, $(A.5.2)$ and $(A.5.3)$, it follows

$$\left(\int_a^{b-h} \|f(t+h) - f(t)\|^p dt\right)^{1/p} \leq 2\varepsilon + \left(\int_G \|f(t+h) - f(t)\|^p dt\right)^{1/p}$$
$$\leq 2\varepsilon + 2\lambda M_\delta + \eta(\lambda)\left(\int_G \|f(t+h) - f(t)\|_1^p dt\right)^{1/p}$$
$$\leq 2\varepsilon + \varepsilon + \eta(\lambda)\left(\int_a^{b-h} \|f(t+h) - f(t)\|_1^p dt\right)^{1/p} \leq 4\varepsilon,$$

for each $f \in \mathfrak{F}$, and each $h > 0$, with $h \leq \gamma(\varepsilon)$. Hence $\mathfrak{F}$ is p-equiintegrable in $L^p(a,b;X)$, and we are in the hypotheses of Theorem A.5.1. The proof is complete. □

A.6. The Superposition Operator

In this section we shall prove several sufficient conditions in order that the operator obtained by composing a fixed continuous real function by elements in a given space of functions be well-defined and continuous. In the sequel, Ω is a bounded domain in $\mathbb{R}^n$, $n \geq 1$, $g : \mathbb{R}_+ \times \overline{\Omega} \times \mathbb{R} \to \mathbb{R}$, and $p \in [1, +\infty]$.

Definition A.6.1. The function $f_p : \mathbb{R}_+ \times D_p \subset \mathbb{R}_+ \times L^p(\Omega) \to L^1(\Omega)$ defined by

$$f_p(t,u)(x) = g(t,x,u(x))$$

for each $t \in \mathbb{R}_+$, $u \in D_p$, and a.e. $x \in \Omega$, where

$$D_p = \{u \in L^p(\Omega);\ g(t,\cdot,u(\cdot)) \in L^1(\Omega) \text{ for each } t \in \mathbb{R}_+\}$$

is called *the superposition operator on $L^p(\Omega)$ associated to the function g.*

We say that the superposition operator on $L^p(\Omega)$ associated to g is *well-defined on $L^p(\Omega)$*, if $D_p = L^p(\Omega)$, and f_p maps $\mathbb{R}_+ \times D_p$ in $L^p(\Omega)$. We say that f_p is *well-defined on $C(\overline{\Omega})$*, if $C(\overline{\Omega}) \subset D_p$ and f_p maps $\mathbb{R}_+ \times C(\overline{\Omega})$ in $C(\overline{\Omega})$.

We recall without proof the following result due to Vitali.

Theorem A.6.1. *Let X be a Banach space, let (Ω, Σ, μ) be a σ-finite and complete measure space, and let $p \in [1,+\infty)$. Let $(u_n)_{n\in\mathbb{N}}$ be a sequence in $L^p(\Omega,\mu;X)$ and $u : \Omega \to X$. Then $u \in L^p(\Omega,\mu;X)$ and $\lim_{n\to\infty} u_n = u$ in $L^p(\Omega,\mu;X)$ if and only if*

(i) *$\lim_{n\to\infty} u_n = u$ in measure;*
(ii) *the set $\{\|u_n\|^p;\ n \in \mathbb{N}\}$ is uniformly integrable in $L^1(\Omega,\mu;\mathbb{R})$;*
(iii) *for each $\varepsilon > 0$, there exists $\Omega_\varepsilon \in \Sigma$, with $\mu(\Omega_\varepsilon) < +\infty$, and*

$$\int_{\Omega\setminus\Omega_\varepsilon} \|u_n(\omega)\|^p \leq \varepsilon$$

for each $n \in \mathbb{N}$.

For the proof see Dunford and Schwartz [**49**], Theorem 6, p. 122.

The next lemma gives some sufficient conditions in order that $D_p = L^p(\Omega)$, or even that f_p be well-defined, either on $L^p(\Omega)$, or on $C(\overline{\Omega})$.

Lemma A.6.1. *Assume that $g : \mathbb{R}_+ \times \overline{\Omega} \times \mathbb{R} \to \mathbb{R}$ satisfies:*

(g_1) *for each $(t,x) \in \mathbb{R}_+ \times \overline{\Omega}$, $u \mapsto g(t,x,u)$ is continuous;*
(g_2) *for each $u \in \mathbb{R}$, $(t,x) \mapsto g(t,x,u)$ is measurable;*
(g_3) *for each $T > 0$, and $r > 0$, the restriction of g to $[0,T]\times\Omega\times[-r,r]$ is bounded.*

Let $p \in [1,+\infty]$, and let $f_p : \mathbb{R}_+ \times D_p \subset \mathbb{R}_+ \times L^p(\Omega) \to L^1(\Omega)$ be the superposition operator on $L^p(\Omega)$ associated to g. Then

(i) *$L^\infty(\Omega) \subset D_p$. If $p = +\infty$, $D_\infty = L^\infty(\Omega)$, and f_∞ is well-defined on $L^\infty(\Omega)$. In addition, for each $t \in \mathbb{R}_+$, the function $u \mapsto f_\infty(t,u)$ is continuous from $L^\infty(\Omega)$ into itself and, for each $u \in L^\infty(\Omega)$, the function $t \mapsto f_\infty(t,u)$ is strongly measurable. Also in this case,* *i.e.* $p=+\infty$, *if* g *is jointly continuous, then* f_∞ *is well-defined on* $C(\overline{\Omega})$ *and continuous from* $\mathbb{R}_+\times C(\overline{\Omega})$ *to* $C(\overline{\Omega})$.

(ii) *If* $p\in[1,+\infty)$, *and for each* $T>0$ *there exists* $a_T>0$ *and* $b_T\in\mathbb{R}$, *such that*

$$|g(t,x,u)|\leq a_T|u|^p+b_T \tag{A.6.1}$$

for each $(t,x,u)\in[0,T]\times\overline{\Omega}\times\mathbb{R}$, *then* $D_p=L^p(\Omega)$, *and, for each* $t\in\mathbb{R}_+$, *the function* $u\mapsto f_p(t,u)$ *is continuous from* $L^p(\Omega)$ *to* $L^1(\Omega)$. *In addition, for each* $u\in L^p(\Omega)$, *the function* $t\mapsto f_p(t,u)$ *is strongly measurable.*

(iii) *If* $p=1$, *and* g *satisfies* (*A.6.1*) *then, for each* $r\in[1,+\infty]$, *the restriction of* f_1 *to* $\mathbb{R}_+\times L^r(\Omega)$ *coincides with* f_r *which is well-defined on* $L^r(\Omega)$. *In addition, for each* $t\in\mathbb{R}_+$, *the function* $u\mapsto f_1(t,u)$ *is continuous from* $L^r(\Omega)$ *into itself and, for each* $u\in L^r(\Omega)$, *the function* $t\mapsto f_1(t,u)$ *is strongly measurable from* $\mathbb{R}_+$ *to* $L^r(\Omega)$. *Finally, for each* $T>0$, *we have*

$$\|f_1(t,u)\|_{L^r(\Omega)}\leq a_T\|u\|_{L^r(\Omega)}+|b_T|\mu(\Omega)^{1/r} \tag{A.6.2}$$

for each $t\in[0,T]$, *and* $u\in L^r(\Omega)$, *where* a_T *and* b_T *are given by* (ii), *and* $\mu(\Omega)$ *is the Lebesgue measure of* Ω.

Proof. (i) By virtue of (g_3), it follows that g maps bounded subsets in $\mathbb{R}_+\times\overline{\Omega}\times\mathbb{R}$ into bounded subsets in $\mathbb{R}$. Therefore, for each $u\in L^\infty(\Omega)$ and $t\in\mathbb{R}_+$, $x\mapsto g(t,x(u(x))$ is bounded on $\overline{\Omega}$. In addition, from (g_1) and (g_2), the function above is measurable. Thus $L^\infty(\Omega)\subset D_p$ for each $p\in[1,+\infty]$, and f_∞ is well-defined on $L^\infty(\Omega)$. If $u\in L^\infty(\Omega)$ and $(u_k)_{k\in\mathbb{N}}$ is a sequence with $\lim_{k\to\infty}u_k=u$ in $L^\infty(\Omega)$, then $(u_k)_{k\in\mathbb{N}}$ is a.e. uniform convergent to u on Ω. By (g_1) we have $\lim_{k\to\infty}f(t,u_k)=f_\infty(t,u)$ in $L^\infty(\Omega)$ for each $t\in\mathbb{R}_+$. As g can be approximated, uniformly on bounded subsets in $\mathbb{R}_+\times\overline{\Omega}\times\mathbb{R}$, by countably-valued functions, the strong measurability of $t\mapsto f_\infty(t,u)$, for each u in $L^\infty(\Omega)$, follows by using standard arguments. If g is jointly continuous then, obviously f_∞ is well-defined on $C(\overline{\Omega})$, and is continuous from $\mathbb{R}_+\times C(\overline{\Omega})$ to $C(\overline{\Omega})$.

(ii) Clearly, for each $(t,u)\in\mathbb{R}_+\times L^p(\Omega)$, $x\mapsto g(t,x,u(x))$ is measurable. On the other hand, in view of $(A.6.1)$, the function above is in $L^1(\Omega)$. So $D_p=L^p(\Omega)$. In order to prove that, for each $t\in\mathbb{R}_+$, the function $u\mapsto f_p(t,u)$ is continuous from $L^p(\Omega)$ to $L^1(\Omega)$, we shall make use of Vitali's Theorem A.6.1. To this aim, let $(u_k)_{k\in\mathbb{N}}$ be a sequence which converges in $L^p(\Omega)$ to u. For the beginning, we shall show that $(g(t,\cdot,u_k(\cdot)))_{k\in\mathbb{N}}$ converges in measure to $g(t,\cdot,u(\cdot))$. Indeed, if we assume the contrary, then there exists $\varepsilon>0$ and a subsequence of $(g(t,\cdot,u_k(\cdot)))_{k\in\mathbb{N}}$, denoted for simplicity again by $(g(t,\cdot,u_k(\cdot)))_{k\in\mathbb{N}}$, such that

$$\mu\left(\{x\in\Omega;\ |g(t,x,u_k(x))-g(t,x,u(x))|\geq\varepsilon\}\right)\geq\varepsilon,$$

for each $k\in\mathbb{N}$, where μ is the Lebesgue measure on $\mathbb{R}^n$. On the other hand, $(u_k)_{k\in\mathbb{N}}$ has at least one subsequence which converges a.e. on Ω to u. From (g_1) it follows that $(g(t,\cdot,u_k(\cdot)))_{k\in\mathbb{N}}$ has at least a subsequence convergent a.e. on Ω to $g(t,\cdot,u(\cdot))$. Consequently, in view of Lebesgue theorem, $(g(t,\cdot,u_k(\cdot)))_{k\in\mathbb{N}}$ converges in measure on that subsequence to $g(t,\cdot,u(\cdot))$, thereby contradicting the last inequality. This contradiction can be eliminated only if $(g(t,\cdot,u_k(\cdot)))_{k\in\mathbb{N}}$ is convergent in measure on Ω to $g(t,\cdot,u(\cdot))$. Next, we shall show that $\{g(t,\cdot,u_k(\cdot));\ k\in\mathbb{N}\}$ is uniformly integrable in $L^1(\Omega)$, i.e. for each $\varepsilon>0$ there exists $\delta>0$ such that

$$\int_\omega |g(t,x,u_k(x))|dx\leq\varepsilon$$

for each measurable subset ω in Ω with $\mu(\omega)\leq\delta$, and uniformly for $k\in\mathbb{N}$. To this aim, let $T>0$ and and let us observe that, by virtue of $(A.6.1)$, we have

$$\int_\omega |g(t,x,u_k(x))|dx\leq a_T\int_\omega |u_k(x)|^p dx+|b_T|\mu(\omega)$$

for each $k\in\mathbb{N}$, and each measurable subset ω in Ω. As $(u_k)_{k\in\mathbb{N}}$ is convergent in $L^p(\Omega)$ to u, from Theorem A.6.1, we know that the set $\{|u_k|^p;\ k\in\mathbb{N}\}$ is uniformly integrable in $L^1(\Omega)$. Then, from the last inequality, it follows that $\{g(t,\cdot,u_k(\cdot));\ k\in\mathbb{N}\}$ is uniformly integrable too. Consequently, by virtue of Theorem A.6.1, we conclude that $(g(t,\cdot,u_k(\cdot)))_{k\in\mathbb{N}}$ is convergent in $L^1(\Omega)$ to $g(t,\cdot,u(\cdot))$.
(iii) The proof of the first part of (iii) follows the same lines as those in the proof of (ii), and therefore we shall omit it. In order to check $(A.6.2)$, it suffices to observe that $(A.6.1)$ with $p=1$, along with Minkowski's inequality, leads to

$$\left(\int_\Omega |g(t,x,u(x)|^r dx\right)^{1/r}\leq a_T\left(\int_\Omega |u(x)|^r dx\right)^{1/r}+|b_T|\left(\int_\Omega dx\right)^{1/r}$$

for each $t\in[\,0,T\,]$ and $u\in L^r(\Omega)$, relation which completes the proof. □

Problems.

A family $\mathfrak{X}$ in c_0 is called *uniformly convergent* if, for each $\varepsilon>0$, there exists $n(\varepsilon)\in\mathbb{N}^*$ such that, for each $x\in\mathfrak{X}$ and each $n\geq n(\varepsilon)$, we have $|x_n|\leq\varepsilon$.

Problem A.1. Prove that a subset $\mathcal{X}$ in c_0 is relatively compact if and only if it is bounded and uniformly convergent.

Let X be a real Banach space and $c_0(X)$ the space of all sequences with elements in X vanishing at ∞. This, endowed with the norm $\|\cdot\|_\infty$, defined by $\|(x_n)_{n\in\mathbb{N}^*}\|_\infty = \sup_{n\in\mathbb{N}^*}\|x_n\|$ for each $(x_n)_{n\in\mathbb{N}^*} \in c_0$, is a real Banach space.

Problem A.2. Using the model offered by Problem A.1, state and prove a necessary and sufficient condition in order that a subset in $c_0(X)$ be relatively compact.

A family $\mathcal{X}$ in l_p is called *uniformly p-summable* if, for each $\varepsilon > 0$, there exists $n(\varepsilon) \in \mathbb{N}^*$, such that, for each $x \in \mathcal{X}$ and each $n \geq n(\varepsilon)$, we have $\left(\sum_{k=n}^{\infty} |x_k|^p\right)^{1/p} \leq \varepsilon$.

Problem A.3. Prove that a subset $\mathcal{X}$ in l_p is relatively compact if and only if it is bounded and uniformly p-summable.

Let X be a real Banach space, let $p \in [1, +\infty)$, and let $l_p(X)$ the space of all sequences $(x_n)_{n\in\mathbb{N}^*}$ with elements in X satisfying $\sum_{n=1}^{\infty}\|x_n\|^p < +\infty$. This, endowed with the norm $\|\cdot\|_p$, defined by $\|(x_n)_{n\in\mathbb{N}^*}\|_p = \left(\sum_{n=1}^{\infty}|x_n|^p\right)^{1/p}$ for each $(x_n)_{n\in\mathbb{N}^*} \in l_p(X)$, is a real Banach space.

Problem A.4. Using the model offered by Problem A.3, state and prove a necessary and sufficient condition in order that a subset in $l_p(X)$ be relatively compact.

Problem A.5. Prove that a family $\mathcal{F}$ in the linear space $C_\ell(\mathbb{R})$, of all continuous functions from $\mathbb{R}$ to $\mathbb{R}$, having finite limits at both $+\infty$ and $-\infty$, endowed with the norm $\|\cdot\|_\infty$, is relatively compact if and only if it is uniformly bounded and equicontinuous on $\mathbb{R}$.

Problem A.6. Prove that a family $\mathcal{F}$ in the linear space $C_\ell(\mathbb{R}; X)$ of all continuous functions from $\mathbb{R}$ to X, having limits at both $+\infty$ and $-\infty$, endowed with the norm $\|\cdot\|_\infty$, is relatively compact if and only if:

(i) $\mathcal{F}$ is equicontinuous on $\mathbb{R}$;

(ii) there exists a dense subset D in $\mathbb{R}$ such that, for each $t \in D$, the set $\mathcal{F}(t) = \{f(t);\ f \in \mathcal{F}\}$ is relatively compact in X.

Problem A.7. Prove that a family $\mathcal{F}$ in the linear space $C_\ell(\mathbb{R}; X)$, endowed with the norm $\|\cdot\|_\infty$, is relatively compact if and only if:

(i) $\mathcal{F}$ is weakly equicontinuous on $\mathbb{R}$;

(ii) for each $[a, b] \subset \mathbb{R}$, $\mathcal{F}([a, b]) = \{f(t);\ f \in \mathcal{F},\ t \in [a, b]\}$ is relatively compact in X.

Problem A.8. Prove that a family $\mathfrak{F}$ in $L^p(\mathbb{R}; X)$ is relatively compact if and only if:

(i) $\mathfrak{F}$ is p-equiintegrable;

(ii) for each interval $[\alpha, \beta]$ the set $\left\{\int_\alpha^\beta f(t)\,dt;\ f \in \mathfrak{F}\right\}$ is relatively compact in X;

(iii) for each $\varepsilon > 0$ there exists $\delta(\varepsilon) > 0$ such that

$$\left(\int_{-\infty}^{-\delta(\varepsilon)} \|f(t)\|^p dt + \int_{\delta(\varepsilon)}^{+\infty} \|f(t)\|^p dt\right)^{1/p} \leq \varepsilon,$$

uniformly for $f \in \mathfrak{F}$.

Problem A.9. Prove that a family $\mathfrak{F}$ in $L^p(\mathbb{R}; X)$ is relatively compact if and only if:

(j) $\mathfrak{F}$ is p-equiintegrable;

(jj) for each $\varepsilon > 0$ there exists a compact subset C_ε in X such that, for each $f \in \mathfrak{F}$, there exists a measurable subset $E_{\varepsilon,f}$ in $\mathbb{R}$ with $\mu(E_{\varepsilon,f}) \leq \varepsilon$, and such that $f(t) \in C_\varepsilon$ for each $t \in \mathbb{R} \setminus E_{\varepsilon,f}$;

(jjj) for each $\varepsilon > 0$, there exists $\delta(\varepsilon) > 0$ such that

$$\left(\int_{-\infty}^{-\delta(\varepsilon)} \|f(t)\|^p dt + \int_{\delta(\varepsilon)}^{+\infty} \|f(t)\|^p dt\right)^{1/p} \leq \varepsilon,$$

uniformly for $f \in \mathfrak{F}$.

Problem A.10. Let X_0, X, and X_1 be three real Banach spaces with X_0 compactly imbedded in X, and X continuously imbedded in X_1. Let $\mathfrak{F}$ be a subset in $L^p(\mathbb{R}; X_0)$ with $\{\|f\|^p;\ f \in \mathfrak{F}\}$ uniformly integrable and which, in addition, satisfies:

(l) $\mathfrak{F}$ is p-equiintegrable in $L^p(\mathbb{R}; X_1)$;

(ll) for each $\varepsilon > 0$, there exists $M_\varepsilon > 0$ such that, for each $f \in \mathfrak{F}$ there exists a measurable subset $E_{\varepsilon,f}$ in $\mathbb{R}$ with $\mu(E_{\varepsilon,f}) \leq \varepsilon$, and such that

$$\|f(t)\|_0 \leq M_\varepsilon$$

for each $t \in \mathbb{R} \setminus E_{\varepsilon,f}$;

(lll) for each $\varepsilon > 0$ there exists $\delta(\varepsilon) > 0$ such that

$$\left(\int_{-\infty}^{-\delta(\varepsilon)} \|f(t)\|^p dt + \int_{\delta(\varepsilon)}^{+\infty} \|f(t)\|^p dt\right)^{1/p} \leq \varepsilon,$$

uniformly for $f \in \mathfrak{F}$.

Prove that $\mathfrak{F}$ is relatively compact in $L^p(\mathbb{R}; X)$.

Notes The results in section 1 are classical. Theorem A.2.1 is an infinite-dimensional version of the famous Arzelà-Ascoli's Compactness Theorem, and Theorem A.2.2 is essentially due to Aubin [**9**]. Theorems A.3.1 and A.3.2 are in fact weak variants of Theorems A.2.1 and respectively A.3.2. Theorem A.4.1 is an infinite-dimensional generalization of a well-known compactness result in L^p spaces and, in the form here presented, is due to Simon [**115**]. For a similar variant see Vrabie [**127**]. Theorem A.4.2 is an extension and, at the same time, a generalization of some compactness results established by Aubin [**9**], as well as by Brezis and Browder [**30**]. Theorem A.5.1 is from Gutman [**66**], and Theorem A.5.2 is new. Finally, Lemma A.6.1, well-known in the mathematical folklore, is undertaken in this form from Vrabie [**127**]. The problems included at the end of the chapter are classical.

Solutions

Chapter 1

Problem 1.1. Let $n \in \mathbb{N}^*$, let $\mathcal{P}_n : a = t_0^n < t_1^n \cdots < t_n^n = b$ be such that $t_{i+1}^n - t_i^n = (b-a)/n$, $i = 1, 2, \dots, n-1$, and let $x_n : [a, b] \to X$, defined by $x_n(t) = x_n(t_{i+1}^n)$ for $t \in (t_i^n, t_{i+1}^n]$, $i = 1, \dots, n-1$ and $x_n(t) = x_n(t_1^n)$ for each $t \in [t_0^n, t_1^n]$. Obviously, for every $x^* \in X^*$ and every $t \in [a, b]$, we have $\lim_{n\to\infty}(x_n(t), x^*) = (x(t), x^*)$. Therefore x is weakly measurable. On the other hand, $x([a, b])$ is included in the weak closure of the convex hull of the set $\{x(t_{i+1}^n)\,;\ n \in \mathbb{N}^*,\ i = 0, 1, \dots, n-1\}$ which, by virtue of Theorem A.1.4, coincides with the strong closure of the same convex hull. As the later is obviously separable, it follows that x is almost separably valued and, thanks to Theorem 1.1.3, we deduce that x is strongly measurable. The conclusion still holds true in the general case specified, under the extra-condition that Ω is separable.

Problem 1.2. First, we shall prove that, for each $p \in [1, +\infty]$, $W^{1,p}(0, \pi)$ is continuously imbedded in $C([0, \pi])$. As the case $p = \infty$ is evident, we shall assume that $p < +\infty$. Let $\mathcal{F}$ be a subset in $W^{1,p}(0, \pi)$ for which there exists $M > 0$ such that

$$\|f\|_{L^p(0,\pi)} + \|f'\|_{L^p(0,\pi)} \le M \tag{$*$}$$

for each $f \in \mathcal{F}$. Obviously, for each $f \in \mathcal{F}$, we have

$$\operatorname{essinf}\{|f(t)|\,;\ t \in [0, \pi]\} = \sup\{\alpha;\ |f(t)| \ge \alpha \text{ a.e. } t \in [0, \pi]\} \le M. \tag{$**$}$$

(Otherwise, we should have $\|f\|_{L^p(0,\pi)} > M$, inequality in contradiction with $(*)$.) As f is continuous, there exists $t_f \in [0, \pi]$ such that

$$\inf\{|f(t)|\,;\ t \in [0, \pi]\} = |f(t_f)|.$$

In view of Leibniz-Newton formula, $(*)$, $(**)$, and Hölder's inequality, we have

$$|f(t)| \le |f(t_f)| + \left|\int_{t_f}^{t} |f(s)|ds\right|$$

$$\le |f(t_f)| + \int_0^{\pi} |f(s)|ds \le M + \pi^{1/q}\|f\|_{L^p(0,\pi)},$$

for each $f \in \mathcal{F}$, where $1/p + 1/q = 1$. Hence $\mathcal{F}$ is uniformly bounded on $[0, \pi]$, and consequently $W^{1,p}(0, \pi)$ is continuously imbedded in $C([0, \pi])$.

Let now $t, s \in [0,\pi]$, and $p \in (1,+\infty)$. Again by Leibniz-Newton formula, Hölder's inequality, and $(*)$, we have

$$|f(t) - f(s)| \leq \left| \int_s^t |f(\tau)|\, d\tau \right| \leq M|t-s|^{1/q},$$

for each $f \in \mathcal{F}$. Hence $\mathcal{F}$, which is uniformly bounded, is equicontinuous on $[0,\pi]$. By virtue of Theorem A.2.1 it follows that it is relatively compact in $C([0,\pi])$. If $p = 1$, the set $\mathcal{F} = \{f_n\,;\ n \in \mathbb{N}^*\}$ with $f'_n = g_n$, and $f_n(0) = 0$, where

$$g_n(t) = \begin{cases} n & \text{for } t \in [0, 1/n] \\ 0 & \text{for } t \in (1/n, \pi], \end{cases}$$

although bounded in $W^{1,1}(0,\pi)$, is not relatively compact in $C([0,\pi])$ as it is not equicontinuous at $t = 0$.

Problem 1.3. We have

$$\begin{aligned} \langle Au, v\rangle_{L^2(0,\pi)} &= \int_0^\pi u''(x)v(x)\,dx = u'(x)v(x)\,\big|_0^\pi - \int_0^\pi u'(x)v'(x)\,dx \\ &= -u(x)v'(x)\,\big|_0^\pi + \int_0^\pi u(x)v''(x)\,dx = \langle u, Av\rangle_{L^2(0,\pi)} \end{aligned}$$

for each $u, v \in D(A)$, and therefore A is symmetric. Let us denote by $H = L^2(0,\pi)$. From the definition of the adjoint of an operator, we have

$$D(A^*) = \{v \in H\,;\ \exists C > 0,\ |\langle Au, v\rangle_H \leq C\|u\|_H,\ \forall u \in D(A)\}.$$

Let us observe that $H^2(0,\pi) \subseteq D(A^*)$ because, for each $v \in H^2(0,\pi)$, we have

$$|\langle Au, v\rangle| \leq \|v''\|_{L^2(0,\pi)}\|u\|_{L^2(0,\pi)}$$

for each $u \in D(A)$. In addition, from the symmetry, we have $A^*v = v''$ for each $v \in H^2(0,\pi)$. To conclude, it suffices to show that $D(A^*) \subseteq H^2(0,\pi)$. As $A^*_{|_{H^2(0,\pi)}}$ is a closed operator and $D(A^*)$ is dense in $H^2(0,\pi)$, we deduce that $D(A^*) = H^2(0,\pi)$. Thus $A \neq A^*$.

Problem 1.4. We have $D(A^*) \subseteq H^2(0,\pi)$. Then, for each $u \in D(A)$, and each $v \in H^2(0,\pi)$, we have

$$\begin{aligned} \langle Au, v\rangle_{L^2(0,\pi)} &= \int_0^\pi u''(x)v(x)\,dx = u'(x)v(x)\,\big|_0^\pi - \int_0^\pi u(x)'v'(x)\,dx \\ &= -u'(0)v(0) + u(\pi)v'(\pi) + \int_0^\pi u(x)v''(x)\,dx. \end{aligned}$$

If $v(0) \neq 0$, or $v'(\pi) \neq 0$, then v does not satisfy the condition in the definition of $D(A^*)$, because for each $n \in \mathbb{N}^*$ there exists $u_n \in D(A)$ with $|\langle Au_n, v\rangle_{L^2(0,\pi)}| > n\|u_n\|_{L^2(0,\pi)}$. Indeed, if $v(0) \neq 0$ then each function

$u_n \in D(A)$, with $\|u_n\|_{L^2(0,\pi)} = 1$ and $-u'_n(0)v(0) > \|v''\|_{L^2(0,\pi)} + n$, enjoys the previously specified property. As the case $v'(\pi) \neq 0$ follows analogously, the proof is complete.

Problem 1.5. We have $D(A^*) \subseteq W^{1,2}(0,\pi)$. For each $u \in D(A)$, and $v \in W^{1,2}(0,\pi)$, it follows that

$$\langle Au, v\rangle = \int_0^\pi u'(x)v(x)\,dx = -u(\pi)v(\pi) - \int_0^\pi u(x)v'(x)\,dx.$$

Let us observe that, $v \in D(A^*)$ if and only if $v \in W^{1,2}(0,\pi)$, and $v(\pi) = 0$. Hence $D(A^*) = \{v \in W^{1,2}(0,\pi)\,;\ v(\pi) = 0\}$, and $A^*v = v'$ for $v \in D(A^*)$.

Chapter 2

Problem 2.1. A simple calculation shows that $\{S(t)\,;\ t \geq 0\}$ is a semigroup of linear bounded operators. In addition

$$\|S(t)(x_n)_{n\in\mathbb{N}^*}\|_p \leq \|(x_n)_{n\in\mathbb{N}^*}\|_p,$$

for each $t \geq 0$, and each $(x_n)_{n\in\mathbb{N}^*} \in l_p$, which shows that the semigroup contains only nonexpansive operators. Let us observe that, for each $t \geq 0$, each $(x_n)_{n\in\mathbb{N}^*} \in l_p$, and each $m \in \mathbb{N}^*$, we have

$$\|S(t)(x_n)_{n\in\mathbb{N}^*} - (x_n)_{n\in\mathbb{N}^*}\|_p^p = \sum_{k=1}^{m} \left(1 - e^{-a_k t}\right)^p |x_k|^p$$

$$+ \sum_{k=m+1}^{\infty} \left(1 - e^{-a_k t}\right)^p |x_k|^p$$

$$\leq \max_{k\in\{1,2,\ldots,m\}} \left(1 - e^{-a_k t}\right)^p \|(x_n)_{n\in\mathbb{N}^*}\|_p^p + \sum_{k=m+1}^{\infty} |x_k|^p.$$

Let $\varepsilon > 0$. Let us fix $m = m_\varepsilon > 0$ such that

$$\sum_{k=m+1}^{\infty} |x_k|^p \leq \frac{\varepsilon^p}{2},$$

and let us choose $\delta = \delta(\varepsilon) > 0$ such that, for each $t \in (0, \delta(\varepsilon))$, we have

$$\max_{k\in\{1,2,\ldots,m\}} \left(e^{-a_k t} - 1\right)^p \leq \frac{\varepsilon^p}{2\,\|(x_n)_{n\in\mathbb{N}^*}\|_p^p}.$$

From the last three inequalities, we deduce that the semigroup is of class C_0. Next we shall prove that

$$D(A) = \{(x_n)_{n\in\mathbb{N}^*} \in l_p\,;\ (a_n x_n)_{n\in\mathbb{N}^*} \in l_p\},$$

and

$$A(x_n)_{n\in\mathbb{N}^*} = -(a_k x_k)_{k\in\mathbb{N}^*}.$$

Indeed, let $(y_n)_{n\in\mathbb{N}^*} \in l_p$ be such that

$$\lim_{t\downarrow 0}\left\|\frac{1}{t}\left(S(t)-I\right)(x_n)_{n\in\mathbb{N}^*} - (y_n)_{n\in\mathbb{N}^*}\right\|_p = 0,$$

i.e.

$$\lim_{t\downarrow 0}\sum_{k=0}^{\infty}\left|\frac{1}{t}\left(e^{-a_k t}-1\right)x_k - y_k\right|^p = 0.$$

This implies that

$$y_n = \lim_{t\downarrow 0}\frac{1}{t}\left(e^{-ant}-1\right)x_n = -a_n x_n$$

for each $n\in\mathbb{N}^*$, and $(a_n x_n)_{n\in\mathbb{N}^*}\in l_p$. Hence

$$D(A)\subseteq\{(x_n)_{n\in\mathbb{N}^*}\in l_p\,;\ (a_n x_n)_{n\in\mathbb{N}^*}\in l_p\},$$

and

$$A(x_n)_{n\in\mathbb{N}^*} = (-a_k x_k)_{k\in\mathbb{N}^*}$$

for each $(x_n)_{n\in\mathbb{N}^*}\in D(A)$. Let $(x_n)_{n\in\mathbb{N}^*}\in l_p$ be such that $(a_n x_n)_{n\in\mathbb{N}^*}\in l_p$. Obviously, we have

$$\left\|\frac{1}{t}\left(S(t)-I\right)(x_n)_{n\in\mathbb{N}^*} + (a_n x_n)_{n\in\mathbb{N}^*}\right\|_p^p$$

$$= \sum_{k=1}^{m}\left|\frac{1}{t}\left(e^{-a_k t}-1\right)+a_k\right|^p |x_k|^p + \sum_{k=m+1}^{\infty}\left|\frac{1}{t}\left(e^{-a_k t}-1\right)+a_k\right|^p |x_k|^p.$$

Since $\left(1-e^{-at}\right)/t\leq a$ for each $t\geq 0$, and each $a\geq 0$, we have

$$\sum_{k=m+1}^{\infty}\left|\frac{1}{t}\left(e^{-a_k t}-1\right)+a_k\right|^p x_k^p \leq 2^p\sum_{k=m+1}^{\infty} a_k^p x_k^p.$$

Consequently,

$$\left\|\frac{1}{t}\left(S(t)-I\right)(x_n)_{n\in\mathbb{N}^*} + (a_n x_n)_{n\in\mathbb{N}^*}\right\|_p^p$$

$$\leq \frac{1}{t}\left\|(x_n)_{n\in\mathbb{N}^*}\right\|_p^p \max_{i\in\{1,2,\ldots,m\}}\left\{\left|\left(e^{-a_i t}-1\right)+a_i\right|^p\right\} + 2^p\sum_{k=m+1}^{\infty} a_k^p x_k^p.$$

Let $\varepsilon>0$. As $(a_n x_n)_{n\in\mathbb{N}^*}\in l_p$, there exists $m=m_\varepsilon>0$ with

$$\sum_{k=m+1}^{\infty} a_k^p x_k^p \leq \frac{\varepsilon^p}{2^{p+1}}.$$

For m so fixed, there exists $\delta = \delta(\varepsilon) > 0$ such that

$$\frac{1}{t} \max_{i \in \{1,2,\ldots,m\}} \left\{ \left| \left(e^{-a_i t} - 1\right) + a_i \right|^p \right\} \leq \frac{\varepsilon^p}{2 \left\| (x_n)_{n \in \mathbb{N}^*} \right\|_p^p}$$

for each $t \in (0, \delta)$. From the last three inequalities, we deduce that $(x_n)_{n\in\mathbb{N}^*} \in D(A)$. Finally, if $(a_n)_{n\in\mathbb{N}^*}$ is bounded, it follows that $D(A) = l_p$, and A is continuous. Hence A generates a uniformly continuous semigroup. If $(a_n)_{n\in\mathbb{N}^*}$ is unbounded, then there exists $(n_k)_{k\in\mathbb{N}^*}$ with $n_k > k$, and $a_{n_k} > k^\alpha$ for each $k \in \mathbb{N}^*$, where $\alpha = 1$ for $p > 1$, and $\alpha > 1$ is fixed for $p = 1$. Then the sequence $(x_n)_{n\in\mathbb{N}^*}$, defined by

$$x_n = \begin{cases} \dfrac{1}{k^\alpha} & \text{if } n = n_k \\ 0 & \text{if } n \neq n_k \ (\forall) k \in \mathbb{N}^*, \end{cases}$$

is in l_p, but $(a_n x_n)_{n\in\mathbb{N}^*} \notin l_p$. Hence in this case $D(A) \neq l_p$, and therefore A cannot generate a uniformly continuous semigroup.

Problem 2.2. Obviously, $\{S(t)\,;\ t \geq 0\}$ is a semigroup of nonexpansive linear operators. Let $(x_n)_{n\in\mathbb{N}^*}$, $t > 0$, and $m \in \mathbb{N}^*$. We have

$$\left\| S(t)(x_n)_{n\in\mathbb{N}^*} - (x_n)_{n\in\mathbb{N}^*} \right\|_\infty$$

$$\leq \max \left\{ \max_{k\in\{1,2,\ldots,m\}} \left| 1 - e^{-a_k t} \right| \left\| (x_n)_{n\in\mathbb{N}^*} \right\|_\infty, \sup_{k \geq m} |x_k| \right\}.$$

Let $\varepsilon > 0$. Let us fix $m > 0$ satisfying

$$\sup_{k\geq m} |x_k| \leq \varepsilon.$$

This is always possible because $(x_n)_{n\in\mathbb{N}^*} \in c_0$. For m fixed as before, there exists $\delta = \delta(\varepsilon) > 0$, such that

$$\max_{k\in\{1,2,\ldots,m\}} \left| e^{-a_k t} - 1 \right| \leq \frac{\varepsilon}{\left\| (x_n)_{n\in\mathbb{N}^*} \right\|_\infty},$$

and therefore the semigroup is of class C_0. We shall prove next that

$$D(A) = \{(x_n)_{n\in\mathbb{N}^*} \in c_0\,;\ (a_n x_n)_{n\in\mathbb{N}^*} \in c_0\}$$

and $A(x_n)_{n\in\mathbb{N}^*} = -(a_k x_k)_{k\in\mathbb{N}^*}$. Indeed, let $(y_n)_{n\in\mathbb{N}^*} \in c_0$ be such that

$$\lim_{t\downarrow 0} \left\| \frac{1}{t} \left(S(t) - I\right)(x_n)_{n\in\mathbb{N}^*} - (y_n)_{n\in\mathbb{N}^*} \right\|_\infty = 0,$$

i.e.

$$\lim_{t\downarrow 0} \frac{1}{t} \sup_{k\geq 1} \left| \left(e^{-a_k t} - 1\right) x_k - y_k \right| = 0.$$

This implies

$$y_n = \lim_{t \downarrow 0} \frac{1}{t}\left(e^{-ant} - 1\right) x_n = -a_n x_n$$

for each $n \in \mathbb{N}^*$, and $(a_n x_n)_{n\in\mathbb{N}^*} \in c_0$. Hence

$$D(A) \subseteq \{(x_n)_{n\in\mathbb{N}^*} \in c_0\,;\ (a_n x_n)_{n\in\mathbb{N}^*} \in c_0\}$$

and

$$A(x_n)_{n\in\mathbb{N}^*} = (-a_k x_k)_{k\in\mathbb{N}^*}$$

for each $(x_n)_{n\in\mathbb{N}^*} \in D(A)$. Let $(x_n)_{n\in\mathbb{N}^*} \in c_0$ with $(a_n x_n)_{n\in\mathbb{N}^*} \in c_0$. For each $m \in \mathbb{N}^*$, we have

$$\left\| \frac{1}{t}\left(S(t) - I\right)(x_n)_{n\in\mathbb{N}^*} + (a_n x_n)_{n\in\mathbb{N}^*} \right\|_\infty = \max\{M_1, M_2\},$$

where

$$\begin{cases} M_1 = \max_{k\in\{1,2,\ldots,m\}} \left|\frac{1}{t}\left[\left(e^{-a_k t} - 1\right) + a_k\right] x_k\right| \\ M_2 = \sup_{k\geq m} \left|\frac{1}{t}\left[\left(e^{-a_k t} - 1\right) + a_k\right] x_k\right|. \end{cases}$$

Since $\left(1 - e^{-at}\right)/t \leq a$ for each $t \geq 0$, and each $a \geq 0$, we have

$$\sup_{k\geq m} \left|\frac{1}{t}\left[\left(e^{-a_k t} - 1\right) + a_k\right] x_k\right| \leq 2\sup_{k\geq m} |a_k x_k|.$$

Consequently

$$\left\| \frac{1}{t}\left(S(t) - I\right)(x_n)_{n\in\mathbb{N}^*} + (a_n x_n)_{n\in\mathbb{N}^*} \right\|_\infty$$

$$\leq \max\left\{ \max_{k\in\{1,2,\ldots,m\}} \left|\frac{1}{t}\left[\left(e^{-a_k t} - 1\right) + a_k\right] x_k\right|, 2\sup_{k\geq m} |a_k x_k| \right\}.$$

Let $\varepsilon > 0$. As $(a_n x_n)_{n\in\mathbb{N}^*} \in c_0$, there exists $m = m_\varepsilon > 0$ with

$$\sup_{k\geq m} |a_k x_k| \leq \frac{\varepsilon}{2}.$$

For m fixed as above, there exists $\delta = \delta(\varepsilon) > 0$, such that

$$\max_{k\in\{1,2,\ldots,m\}} \left|\frac{1}{t}\left[\left(e^{-a_k t} - 1\right) + a_k\right] x_k\right| \leq \frac{\varepsilon}{\|(x_n)_{n\in\mathbb{N}^*}\|_\infty},$$

for each $t \in (0, \delta)$. From the last three inequalities, we deduce that $(x_n)_{n\in\mathbb{N}^*} \in D(A)$. Finally, if $(a_n)_{n\in\mathbb{N}^*}$ is bounded, we get that $D(A) = c_0$ and A is continuous. So A generates a uniformly continuous semigroup.

If $(a_n)_{n\in\mathbb{N}^*}$ is unbounded, then there exists $(n_k)_{k\in\mathbb{N}^*}$ with $n_k > k$, and $a_{n_k} > k$ for each $k \in \mathbb{N}^*$. Then the sequence $(x_n)_{n\in\mathbb{N}^*}$, defined by

$$x_n = \begin{cases} \dfrac{1}{k} & \text{if } n = n_k \\ 0 & \text{if } n \neq n_k \ (\forall) k \in \mathbb{N}^*, \end{cases}$$

is in c_0, but $(a_n x_n)_{n\in\mathbb{N}^*} \notin c_0$. Hence, in this case, $D(A) \neq c_0$, and therefore A cannot generate a uniformly continuous semigroup.

Problem 2.3. Clearly $\{S(t)\,;\ t \geq 0\}$ is a semigroup of nonexpansive, linear operators. Let us observe that $\lim_{t\downarrow 0} S(t)f = f$ if and only if f is uniformly continuous on $\mathbb{R}$. Since $C_b(\mathbb{R})$ contains functions which are not uniformly continuous (as for instance $f(x) = \sin x^2$), the semigroup is not of class C_0. A simple calculation shows that $D(A)$ coincides with the space of all functions which are uniformly differentiable at the right on $\mathbb{R}$, and whose right derivatives are continuous, and bounded on $\mathbb{R}$, and $Af = f'$ for each $f \in D(A)$. As the closure of $D(A)$ coincides with the space of all uniformly continuous and bounded functions from $\mathbb{R}$ to $\mathbb{R}$, and the later is a proper closed subspace in $C_b(\mathbb{R})$, it follows that $D(A)$ is not dense in X.

Problem 2.4. In order to solve the problem, it suffices to check that, for each $f \in X$, the function $u : \mathbb{R} \to X$, $u(t)(x) = [G(t)f](x)$ for each $t \in \mathbb{R}$, and each $x \in \mathbb{R}$, is differentiable on $\mathbb{R}$, and satisfies

$$\begin{cases} u' = Au \\ u(0) = f \end{cases}$$

To this aim, let us observe that, for each $f \in X$, each $t \geq 0$, each $h > 0$, and each $x \in \mathbb{R}$, we have

$$\frac{1}{h}[u(t+h)(x) - u(t)(x)] - [Au(t)](x)$$

$$= e^{-\lambda t}\frac{1}{h}\left\{\left[e^{-\lambda h}\sum_{k=0}^{\infty}\frac{\lambda^k (t+h)^k}{k!}f(x-k\delta) - \sum_{k=0}^{\infty}\frac{(\lambda t)^k}{k!}f(x-k\delta)\right]\right.$$

$$\left. -\lambda\sum_{k=0}^{\infty}\frac{\lambda^k t^k}{k!}f(x-(k+1)\delta) + \lambda\sum_{k=0}^{\infty}\frac{\lambda^k t^k}{k!}f(x-k\delta)\right\}$$

$$= e^{-\lambda t}\left[\frac{e^{-\lambda h}-1}{h} + \lambda\right]\sum_{k=0}^{\infty}\frac{\lambda^k t^k}{k!}f(x-k\delta)$$

$$+e^{-\lambda t}\sum_{k=0}^{\infty}\left[\frac{(t+h)^k - t^k}{h} - kt^{k-1}\right]\frac{\lambda^k}{k!}f(x-k\delta).$$

We then have

$$\lim_{h\downarrow 0}\left\|\frac{1}{h}[u(t+h)-u(t)]-Au(t)\right\|_{\infty}=0.$$

As $\|G(t)f\|_\infty = \|f\|_\infty$ for each $f \in X$, it follows that $\{G(t)\,;\ t\in\mathbb{R}\}$ is a group of isometries, and this achieves the proof.

Problem 2.5. Since $\{e^{-t\mathcal{A}}\,;\ t\in\mathbb{R}\}$ is a uniformly continuous group on $\mathbb{R}^n$, it follows that $\{G(t)\,;\ t\in\mathbb{R}\}$ is a group of linear bounded operators on $L^p(\mathbb{R}^n)$. As for each $f\in L^p(\mathbb{R}^n)$, $\{f\}$ is p-equiintegrable, it follows that the group is of class C_0. Finally, let us observe that its infinitesimal generator is defined by

$$D(A)=\{f\in L^p(\mathbb{R}^n)\,;\ x\mapsto\langle\mathcal{A}x,\nabla f(x)\rangle \text{ belongs to } L^p(\mathbb{R}^n\}$$

and

$$[Af](x)=-\langle\mathcal{A}x,\nabla f(x)\rangle$$

for each $f\in D(A)$, and a.e. for $x\in\mathbb{R}^n$. If, in addition, $\sum_{i=1}^n a_{ii}=0$, from Liouville theorem (see Corduneanu [**39**], Theorem 4.4, p. 63) it follows that $\|e^{-t\mathcal{A}}\|_{\mathcal{L}(L^p(\mathbb{R}_n))}=1$ for each $t\in\mathbb{R}$, while from the formula of change of variable in multiple integrals, we have $\|G(t)f\|_{L^p(\mathbb{R}^n)}=\|f\|_{L^p(\mathbb{R}^n)}$, which shows that $G(t)$ is an isometry.

Problem 2.6. A simple computation shows that $\{G(t)\,;\ t\in\mathbb{R}\}$ is a group on X. This is not a C_0-group simply because $\lim_{t\downarrow 0}\|e^{-t\mathcal{A}}x-x\|=0$ is not uniform for $x\in\mathbb{R}^n$.

Problem 2.7. One simply has to repeat step by step all the arguments in the proof of Theorem 2.3.1, with the remark that, in this case, the pointwise boundedness of the set $\{S(t_n)x\,;\ n\in\mathbb{N}^*\}$ follows from the fact that $\lim_{n\to\infty}S(t_n)x=x$ in the weak topology on X.

Problem 2.8. Let $Y=\{x\in X\,;\ \lim_{t\downarrow 0}\|S(t)x-x\|=0\}$. We observe that Y is nonempty ($0\in Y$) and closed in X. To complete the proof, it suffices to show that Y is dense in X. To this aim, let us observe that, for each $x\in X$, the function $t\mapsto S(t)x$ is weakly continuous from the right on $\mathbb{R}_+$. So, according to Problem 1.1, this function is strongly measurable on $\mathbb{R}_+$. From Problem 2.7, it follows that there exists $M\geq 1$, and $\omega\in\mathbb{R}$, such that $\|S(t)\|_{\mathcal{L}(X)}\leq Me^{t\omega}$ for each $t\geq 0$. Consequently $S(\,\cdot\,)$ is locally integrable on $\mathbb{R}_+$. Let now $x\in X$, and let $\varphi:\mathbb{R}\to\mathbb{R}_+$ be a function of class C^∞ for which there exists $[\,a,b\,]\subset(0,+\infty)$ such that $\varphi(t)=0$ for each $t\notin[\,a,b\,]$. Let us define

$$x(\varphi)=\int_0^{+\infty}\varphi(t)S(t)x\,dt$$

and let us observe that, by Lebesgue dominated convergence theorem, we deduce

$$\begin{aligned}
&\lim_{h\downarrow 0}[S(h)x(\varphi) - x(\varphi)] \\
&= \lim_{h\downarrow 0}\int_0^\infty \left(\int_0^\infty \varphi(t)S(t+h)x\,dt - \int_0^\infty \varphi(t)S(t)x\,dt\right) \\
&= \lim_{h\downarrow 0}\int_0^\infty [\varphi(\tau - h) - \varphi(\tau)]S(\tau)x\,d\tau = 0
\end{aligned}$$

strongly in X, because φ is continuous, and $S(\cdot)$ is measurable and bounded on the support of φ. Therefore $x(\varphi) \in Y$. To complete the proof, it suffices to show that $\{x(\varphi)\,;\ x \in X,\ \varphi \in \mathcal{D}((0,+\infty))\}$ is dense in X. To this aim, let $\varphi \in \mathcal{D}((0,+\infty))$ be positive, with $\int_0^\infty \varphi(t)\,dt = 1$, and $\operatorname{supp}\varphi \subset [\,a, b\,]$, let $\varepsilon > 0$, and let us define $\varphi_\varepsilon : \mathbb{R} \to (0,+\infty)$ by

$$\varphi_\varepsilon(t) = \frac{1}{\varepsilon}\varphi\left(\frac{t}{\varepsilon}\right).$$

Clearly φ_ε is of class C^∞, $\varphi_\varepsilon(t) = 0$ for $t \notin [\,\varepsilon a, \varepsilon b\,]$, and $\int_0^{+\infty} \varphi_\varepsilon(t)\,dt = 1$. Let us observe that, for each $x^* \in X^*$, we have

$$\begin{aligned}
\lim_{\varepsilon\downarrow 0}|(x(\varphi_\varepsilon) - x, x^*)| &= \left|\int_0^\infty \varphi_\varepsilon(t)(S(t)x - x, x^*)\,dt\right| \\
&\leq \lim_{\varepsilon\downarrow 0}\ \sup_{t\in[\,\varepsilon a, \varepsilon b\,]} |(S(t)x - x, x^*)| = 0.
\end{aligned}$$

Hence the weak closure of Y coincides with X. As Y is a convex set, in view of Mazur's Theorem A.1.4, it follows that the strong closure of Y coincides with X, and this achieves the proof.

Chapter 3

Problem 3.1 The idea is to show that A satisfies (i) and (ii) in Hille-Yosida Theorem 3.1.1. Clearly $D(A)$ is dense in l_2 because the set of all sequences whose terms are 0 except for as finite number of indices is contained in $D(A)$ and dense in l_2. Next, let $\lambda > 0$, let $y = (y_n)_{n\in\mathbb{N}^*}$ and let us consider the equation $(\lambda I - A)x = y$. One may easily verify that this equation has a unique solution x which is given by

$$x_n = \frac{1}{\lambda + a_n} y_n$$

for each $n \in \mathbb{N}^*$. So, $\|x\|_{l_2} \leq \frac{1}{\lambda}\|y\|_{l_2}$ and therefore $\lambda I - A$ is invertible, with continuous inverse and $\|(\lambda I - A)^{-1}\| \leq \frac{1}{\lambda}$. Consequently A satisfies (ii) in Theorem 3.1.1 and therefore A is closed. Summing up, we conclude that A

satisfies (i) and (ii) in Theorem 3.1.1, and thus it generates a C_0-semigroup of contractions.

Problem 3.2 The proof follows the same lines as those in the preceding proof, with the sole exception that instead of the norm of l_2 we have to use the norm of c_0.

Problem 3.3 Since $C_0^\infty(0,\pi)$ is included in $D(A)$ and is dense in $L^2(0,\pi)$, we conclude that A is densely defined. Let $\lambda > 0$, $f \in L^2(0,\pi)$ and let us consider the equation $(\lambda I - A)u = f$ which rewrites equivalently as

$$\begin{cases} \lambda u - u'' = f & \text{in } (0,\pi) \\ u(0) = u'(\pi) = 0. \end{cases} \tag{$*$}$$

From the variation of constants formula we conclude that this problem has a unique solution u in $L^2(0,\pi)$, and so $\lambda I - A$ is invertible. Multiplying both sides of the first equation in $(*)$ by u, integrating by parts and taking into account the boundary conditions, we get

$$\lambda\|u\|^2 + \|u'\|^2 = \langle f, u\rangle,$$

where $\langle\cdot,\cdot\rangle$ and $\|\cdot\|$ are the inner product and respectively the norm of $L^2(0,\pi)$. Thanks to the Cauchy-Schwarz inequality, this implies that $\lambda\|u\| \leq \|f\|$, or equivalently $\|(\lambda I - A)^{-1}\| \leq \frac{1}{\lambda}$. Therefore Theorem 3.1.1 applies and accordingly A generates a C_0-semigroup of contractions.

Problem 3.4 Since $C_0^\infty(0,\pi)$ is included in $D(A)$ and is dense in $C_\pi(0,\pi)$, we conclude that A is densely defined. Let $\lambda > 0$, $f \in C_\pi(0,\pi)$ and let us consider the equation $(\lambda I - A)u = f$. This rewrites as

$$\begin{cases} \lambda u + u' = f & \text{in } (0,\pi) \\ u(0) = u(\pi). \end{cases}$$

By the variation of constants formula we have

$$u(x) = e^{-\lambda x}u(0) + \int_0^x e^{-\lambda(x-y)} f(y)\,dy.$$

Imposing the periodicity condition $u(0) = u(\pi)$, we get

$$u(0) = \frac{1}{1 - e^{-\lambda\pi}} \int_0^\pi e^{-\lambda(\pi-y)} f(y)\,dy$$

and so $(\lambda I - A)u = f$ has a unique solution. Consequently $\lambda I - A$ is invertible. A simple calculation leads to

$$|u(x)| \leq \frac{1}{\lambda}\left|e^{-\lambda x}e^{-\lambda\pi} + e^{-\lambda x}(1 - e^{-\lambda x})\right| \|f\| \leq \frac{1}{\lambda}\|f\|,$$

where $\|\cdot\|$ stands for the norm of $C_\pi(0,\pi)$. Hence $\|(\lambda I - A)^{-1}\| \leq \frac{1}{\lambda}$. From Theorem 3.1.1, we get that A generates a C_0-semigroup of contractions.

Problem 3.5 In view of (ii) in Lumer-Phillips Theorem 3.4.2, it suffices to show that $I - A$ is surjective. To this aim let us observe that $R(I - A)$ is closed because A is closed and dissipative. Indeed, let $(y_n)_{n\in\mathbb{N}^*}$ be in $R(A)$ with $\lim_{n\to\infty} y_n = y$. By Theorem 3.4.1 we have $\|y_n - y_m\| \leq \|x_n - x_m\|$ for each $n, m \in \mathbb{N}^*$, where $x_n \in D(A)$ is such that $y_n = x_n - Ax_n$, for $n \in \mathbb{N}^*$. So there exists $\lim_{n\to\infty} x_n = x$, $x \in D(A)$ and $x - Ax = y$. Next, if we assume that $R(I - A) \neq X$, there exists $x^* \in X^*$, $x^* \neq 0$ and such that $(x^*, x - Ax) = 0$ for each $x \in D(A)$. This implies $x^* - A^*x^* = 0$ and since A^* is also dissipative, we must have $x^* = 0$. This contradiction can be eliminated only if $R(I - A) = X$. So A satisfies the hypotheses of Lumer-Phillips Theorem 3.4.2, and accordingly A generates a C_0-semigroup of contractions on X.

Problem 3.6 The first part, (i) has been already proved in Theorem 3.4.2. To check (ii), let $x \in D(\overline{A})$ and $y = \overline{A}x$. Let $((x_n, y_n))_{n\in\mathbb{N}^*}$ be a sequence in graph$\,(A)$ such that $\lim_{n\to\infty} x_n = x$ and $\lim_{n\to\infty} y_n = y$. By virtue of Theorem 3.4.1 we have $\|\lambda x_n - Ax_n\| \geq \lambda\|x_n\|$ for each $n \in \mathbb{N}^*$ and $\lambda > 0$, and thus $\|\lambda x - Ax\| \leq \lambda\|x\|$ for each $\lambda > 0$ and $x \in D(\overline{A})$. By Theorem 3.4.1, it follows that $\overline{A}$ is dissipative and this proves (ii). Finally, let us assume that although $D(A)$ is dense in X, A is not closable. Thus, there exists $(x_n)_{n\in\mathbb{N}^*}$ with $\lim_{n\to\infty} x_n = 0$ and $\lim_{n\to\infty} Ax_n = y$, with $\|y\| = 1$. Using once again Theorem 3.4.1, we conclude that, for each $t > 0$, each $n \in \mathbb{N}^*$ and $x \in D(A)$ we have

$$\left\|\left(x + \frac{1}{t}x_n\right) - tA\left(x + \frac{1}{t}x_n\right)\right\| \geq \left\|x + \frac{1}{t}x_n\right\|.$$

Letting successively $n \to \infty$ and $t \to 0$, we get $\|x - y\| \geq \|x\|$ for each $x \in D(A)$. Since $\|y\| = 1$ this shows that y cannot be approximated with elements in $D(A)$ thereby contradicting the density of $D(A)$. This contradiction can be eliminated only if A is closable, and this proves (iii).

Problem 3.7 Since $D(A)$ is a subspace in X, it remains to show that if $x^* \in X^*$ is such that $(x^*, x) = 0$ for every $x \in D(A)$, then $x^* = 0$. Recalling that $I - A$ is surjective, to prove that $x^* = 0$, it suffices to show that $(x^*, x - Ax) = 0$ for every $x \in D(A)$ which, in view of the supposition that $(x^*, x) = 0$, is equivalent to $(x^*, Ax) = 0$ for every $x \in D(A)$. Let $x \in D(A)$. By virtue of (i) in Problem 3.6, it follows that, for each n in $\mathbb{N}^*$ there exists $x_n \in D(A)$ such that $x = x_n - (1/n)Ax_n$. On the other hand $Ax_n = n(x_n - x) \in D(A)$ and therefore $x_n \in D(A^2)$ and

$Ax = Ax_n - (1/n)A^2x_n$, or $Ax_n = (I-(1/n)A)^{-1}Ax_n$. From Theorem 3.4.1 we deduce that $\|(I-(1/n)A)^{-1}\| \leq 1$ and consequently $\|Ax_n\| \leq \|Ax\|$ for every $n \in \mathbb{N}^*$. Hence $\|x_n - x\| \leq (1/n)\|Ax_n\| \leq \|Ax\|$ and accordingly $\lim_{n\to\infty} x_n = x$. Since X is reflexive and $(Ax_n)_{n\in\mathbb{N}^*}$ is bounded, we may assume without loss of generality that there exists $\lim_{n\to\infty} Ax_n = y$ weakly in X. On the other hand, by Theorem 3.4.2 A generates a C_0-semigroup of contractions and thus, by Theorem 3.1.1, it is closed. So $x \in D(A)$ and $Ax = y$, because the strong$\times$strong closure of graph(A) coincides with its weak$\times$weak closure and thus with its strong$\times$weak closure. See Mazur Theorem A.1.4 in Appendix. Inasmuch as $(x^*, z) = 0$ for every $z \in D(A)$, we have $(x^*, Ax_n) = n(x^*, x_n - x) = 0$. Letting $n \to \infty$ we get $(x^*, Ax) = 0$ for every $x \in D(A)$. Thus $x^* = 0$ and therefore $D(A)$ is dense in X.

Problem 3.8 We begin by observing that, for each $\lambda > 0$ and $f \in C([0,1])$, the equation $\lambda u - Au = f$ has a unique solution given by

$$u(x) = \int_0^x e^{-\lambda(x-y)} f(y)\,dy \tag{$*$}$$

and thus $\lambda I - A$ is surjective. Moreover, from $(*)$ we deduce that

$$\lambda|u(x)| \leq \left(1 - e^{-\lambda x}\right)\|f\| \leq \|\lambda u - Au\|$$

and therefore

$$\lambda\|u\| \leq \|\lambda u - Au\|.$$

Hence A is dissipative. Nevertheless, $\overline{D(A)} = \{u \in C([0,1])\,;\ u(0) = 0\}$ which does not coincide with $C([0,1])$.

Chapter 4

Problem 4.1 We shall use Lumer-Phillips Theorem 3.4.2. Since $C^2([0,1])$ is included in $D(A)$ and is dense in X, A is densely defined. Moreover, let us observe that, for each $\lambda > 0$, we have $|u(\theta) - \theta(1-\theta)u''(\theta)| \geq |u(\theta)|$ where $\theta \in [0,1]$ is such that $|u(\theta)| = \|u\|_{C([0,1])}$. Therefore,

$$\|u\|_{C([0,1])} \leq \|u - \lambda Au\|_{C([0,1])}$$

for all $u \in D(A)$ and $\lambda > 0$, and thus A is dissipative. We shall show next that $I - A$ is surjective. To this aim, let us remark that $u_1(s) = s$ and $u_2(s) = 1 - s$ belong to $D(A)$ and, for every $\lambda > 0$, we have

$$(\lambda I - A)u_i = \lambda u_i, \tag{$*$}$$

for $i = 1, 2$. Thus, it suffices to prove only that $\lambda I - A_0$ is surjective, where A_0 is the part of A in the closed subspace $X_0 = \{f \in X\,;\ f(0) = f(1) = 0\}$

with $D(A_0) = \{f \in X_0 \cap C^2(0,1);\ \lim_{s\to 0,1} s(1-s)f''(s) = 0\}$. At this point let us observe that A_0 is in fact injective and its inverse is given by

$$[A_0^{-1}f](s) = \int_0^1 G(s,t)\frac{f(t)}{t(1-t)}dt,$$

where the kernel G is defined by

$$G(s,t) = \begin{cases} s(t-1) & \text{if } 0 \le s \le t \le 1 \\ t(s-1) & \text{if } 0 \le t \le s \le 1. \end{cases}$$

So $0 \in \rho(A_0)$ and since A_0 is still dissipative, one has $[0,+\infty) \subseteq \rho(A_0)$. By virtue of $(*)$, it then follows that $(0,+\infty) \subseteq \rho(A)$ and thus A satisfies the hypotheses of Lumer-Phillips Theorem 3.4.2.

Problem 4.2. The Problem is an "$L^2(0,\pi)$" variant of Problem 2.1. From this reason, we shall use very similar arguments. More precisely, let us observe that $\{S(t)\,;\ t \ge 0\}$ is a semigroup of linear bounded operators. In addition

$$\|S(t)\xi\|_{L^2(0,\pi)} \le \|\xi\|_{L^2(0,\pi)},$$

for each $t \ge 0$, and each $\xi \in L^2(0,\pi)$, which shows that the semigroup contains only nonexpansive operators. We recall that, for each $\xi \in L^2(0,\pi)$,

$$\|\xi\|^2_{L^2(0,\pi)} = \sum_{k=1}^{\infty} a_k^2(\xi)$$

and let us observe that, for each $t \ge 0$, each $\xi \in L^2(0,\pi)$, and each $m \in \mathbb{N}^*$, we have

$$\begin{aligned}\|S(t)\xi - \xi\|^2_{L^2(0,\pi)} &= \sum_{k=1}^{m}\left(e^{-k^2t}-1\right)^2 a_k^2(\xi) \\ &\quad + \sum_{k=m+1}^{\infty}\left(e^{-k^2t}-1\right)^2 a_k^2(\xi) \\ &\le \max_{k\in\{1,2,\dots,m\}}\left(e^{-k^2t}-1\right)^2 \|\xi\|^2_{L^2(0,\pi)} + 2^2\sum_{k=m+1}^{\infty} a_k^2(\xi).\end{aligned}$$

Let $\varepsilon > 0$, let us fix $m = m_\varepsilon > 0$, such that

$$\sum_{k=m+1}^{\infty} a_k^2(\xi) \le \frac{\varepsilon^2}{2^3},$$

and let us choose $\delta = \delta(\varepsilon) > 0$, such that

$$\max_{k\in\{1,2,\dots,m\}}\left(e^{-k^2t}-1\right)^2 \le \frac{\varepsilon^2}{2\,\|\xi\|^2_{L^2(0,\pi)}}.$$

From the last three inequalities, we deduce that the semigroup is of class C_0. Next, we shall prove that

$$D(A) = \left\{ \xi \in L^2(0,\pi)\,;\ \sum_{k=1}^{\infty} k^4 a_k^2(\xi) < +\infty \right\}, \tag{$*$}$$

and

$$[A\xi](x) = -\sum_{k=1} k^2 a_k(\xi) \sin kx. \tag{$**$}$$

Indeed, let $\eta \in L^2(0,\pi)$ be such that

$$\lim_{t\downarrow 0} \left\| \frac{1}{t}\left(S(t) - I\right)\xi - \eta \right\|_{L^2(0,\pi)} = 0,$$

i.e.

$$\lim_{t\downarrow 0} \sum_{k=1}^{\infty} \left[\frac{1}{t}\left(e^{-k^2 t} - 1\right) a_k(\xi) - a_k(\eta) \right]^2 = 0.$$

This implies that

$$a_n(\eta) = \lim_{t\downarrow 0} \frac{1}{t}\left(e^{-n^2 t} - 1\right) a_n(\xi) = -n^2 a_n(\xi)$$

for $n \in \mathbb{N}^*$, and $\sum_{k=1}^{\infty} k^4 a_k^2(\xi) < +\infty$. Hence $(*)$ and $(**)$ hold. Let $\xi \in L^2(0,\pi)$ with $\sum_{k=1}^{\infty} k^4 a_k^2(\xi) < +\infty$, and let $\eta = -\sum_{k=1}^{\infty} k^2 a_k(\xi) \sin kx$. Obviously, we have

$$\left\| \frac{1}{t}\left(S(t) - I\right)\xi - \eta \right\|^2_{L^2(0,\pi)}$$

$$= \sum_{k=1}^{m} \left[\frac{1}{t}\left(e^{-k^2 t} - 1\right) + k^2 \right]^2 a_k^2(\xi) + \sum_{k=m+1}^{\infty} \left[\frac{1}{t}\left(e^{-k^2 t} - 1\right) + k^2 \right]^2 a_k^2(\xi).$$

Since $\left(1 - e^{-at}\right)/t \le a$ for each $t \ge 0$ and each $a \ge 0$, we have

$$\sum_{k=m+1}^{\infty} \left[\frac{1}{t}\left(e^{-k^2 t} - 1\right) + k^2 \right]^2 a_k^2(\xi) \le 2^2 \sum_{k=m+1}^{\infty} k^4 a_k^2(\xi).$$

Consequently

$$\left\| \frac{1}{t}\left(S(t) - I\right)\xi - \eta \right\|^2_{L^2(0,\pi)}$$

$$\le \frac{1}{t} \|\xi\|^2_{L^2(0,\pi)} \max_{i\in\{1,2,\dots,m\}} \left\{ \left| \left(e^{-i^2 t} - 1\right) + i^2 \right|^2 \right\} + 2^2 \sum_{k=m+1}^{\infty} a_k^4 a_k^2(\xi).$$

Let $\varepsilon > 0$. As $\sum_{k=1}^{\infty} k^4 a_k^2(\xi) < +\infty$, there exists $m = m_\varepsilon > 0$ with

$$\sum_{k=m+1}^{\infty} k^4 a_k^2(\xi) \leq \frac{\varepsilon^2}{2^3}.$$

For m fixed as before, there exists $\delta = \delta(\varepsilon) > 0$, such that

$$\frac{1}{t} \max_{i \in \{1,2,\ldots,m\}} \left\{ \left| \left(e^{-i^2 t} - 1 \right) + i^2 \right|^2 \right\} \leq \frac{\varepsilon^2}{2 \|\xi\|_{L^2(0,\pi)}^2}$$

for each $t \in (0, \delta)$. From the last three inequalities, we get $\xi \in D(A)$. Finally, we have $\eta \in D(A)$ if and only if $\eta \in H^2(0,\pi) \cap H_0^1(0,\pi)$, and $A\eta = \eta''$. In addition, if $\xi \in L^2(0,\pi)$, we have $S(t)\xi \in D(A)$ for each $t > 0$, and consequently (iii) holds.

Problem 4.3. Let $\mathcal{S}$ be the vector space of all functions $f : \mathbb{R} \to \mathbb{C}$ of class C^∞ with the property that, for each $m, k \in \mathbb{N}$, the function $x \mapsto x^m f^{(k)}$ is bounded on $\mathbb{R}$. Since $C_0^\infty(\mathbb{R})$ is included in $\mathcal{S}$, and is dense in $L^p(\mathbb{R})$, it follows that $\mathcal{S}$ is dense in $L^p(\mathbb{R})$. Let us denote by

$$G_t(x) = \frac{1}{\sqrt{4\pi t}} e^{-\frac{x^2}{4t}}$$

and let us observe that $[S(t)f](x) = G_t(x) * f(x)$. From Brezis [**29**], Théorème IV.15, p. 66, we know that, for each $p \in [1, +\infty)$, and each $f \in L^p(\mathbb{R})$, we have

$$\|S(t)f\|_{L^p(\mathbb{R})} \leq \|G_t\|_{L^1(\mathbb{R})} \|f\|_{L^p(\mathbb{R})},$$

which shows that $S(t)$ maps $L^p(\mathbb{R})$ into itself and

$$\|S(t)\|_{\mathcal{L}(L^p(\mathbb{R}))} \leq 1.$$

Since the convolution is associative, we have

$$S(s)S(t)f = G_s * (G_t * f) = (G_s * G_t) * f$$

for each $f \in L^p(\mathbb{R})$. Let us we denote by $\exp\{a\} = e^a$, and let us observe that

$$(G_s * G_t)(x) = \frac{1}{4\pi\sqrt{st}} \int_{-\infty}^{+\infty} \exp\left\{ -\frac{(x-y)^2}{4s} \right\} \exp\left\{ -\frac{y^2}{4t} \right\} dy$$

$$= \frac{1}{4\pi\sqrt{st}} \int_{-\infty}^{+\infty} \exp\left\{ -\frac{(tx-y)^2 + sy^2}{4st} \right\} dy$$

$$= \frac{1}{4\pi\sqrt{st}} \int_{-\infty}^{+\infty} \exp\left\{ -\frac{(s+t)y^2 - 2txy + tx^2}{4st} \right\} dy$$

$$= \frac{1}{4\pi\sqrt{st}} \int_{-\infty}^{+\infty} \exp\left\{ -\frac{(s+t)y^2}{4s}\left(y - \frac{tx}{s+t}\right)^2 \right.$$

$$\left. + \left(\frac{t}{s+t} - \frac{t^2}{(s+t)^2}\right) x^2 \right\} dy$$

$$= \frac{1}{4\pi\sqrt{st}} \int_{-\infty}^{+\infty} \exp\left\{ -\frac{(s+t)^2}{4s} y^2 \right\} \exp\left\{ -\frac{(s+t)}{4st} \frac{st}{(s+t)^2} x^2 \right\} dy$$

$$= \frac{1}{2\pi\sqrt{s+t}} \int_{-\infty}^{+\infty} \exp\left\{ -\frac{x^2}{4(s+t)} \right\} \int_{-\infty}^{+\infty} \exp\left\{ -z^2 \right\} dz$$

$$= \frac{1}{2\pi\sqrt{s+t}} \exp\left\{ -\frac{x^2}{4(s+t)} \right\} \sqrt{\pi} = G_{s+t}(x),$$

where

$$z = \sqrt{\frac{s+t}{4st}}\, y.$$

Hence $S(s)S(t)f = S(s+t)f$ for each $f \in \mathcal{S}$, and each $t, s \geq 0$. So $\{S(t)\,;\ t \geq 0\}$ is a semigroup of linear bounded operators on $L^p(\mathbb{R})$. In addition,

$$\left(\int_{-\infty}^{+\infty} |[S(t)f](x) - f(x)|^p dx \right)^{1/p}$$

$$= \frac{1}{\sqrt{4\pi t}} \left(\int_{-\infty}^{+\infty} \left| \int_{-\infty}^{+\infty} \exp\left\{ -\frac{(x-y)^2}{4t} \right\} [f(y) - f(x)]\, dy \right|^p dx \right)^{1/p}$$

$$= \frac{1}{\sqrt{\pi}} \left(\int_{-\infty}^{\infty} \left| \int_{-\infty}^{+\infty} \exp\left\{ -\tau^2 \right\} \left[f(y) - f(y + 2\sqrt{t}\tau) \right] dy \right|^p d\tau \right)^{1/p}.$$

As

$$\lim_{t \downarrow 0} \int_{-\infty}^{\infty} |f(y) - f(y + 2\sqrt{t}\tau)|\, dy = 0$$

a.e. for $\tau \in \mathbb{R}$, in view of Vitali's Theorem A.6.1, it follows that

$$\lim_{t \downarrow 0} \left(\int_{-\infty}^{+\infty} |[S(t)f](x) - f(x)|^p dx \right)^{1/p} = 0.$$

Hence the semigroup is of class C_0, and the proof is complete.

Problem 4.4. The fact that u satisfies the partial differential equation in question follows from the differentiation rule of a convolution product. The initial condition $u(0,x) = f(x)$ reflects in fact the following continuity property of the solution u

$$\lim_{t \downarrow 0} \|u(t,\cdot) - f\|_{L^p(\mathbb{R})} = 0.$$

Problem 4.5. Let us denote by

$$P_t(x) = \frac{t}{\pi(t^2 + x^2)}$$

and let us observe that $[S(t)f](x) = P_t(x) * f(x)$, where the convolution is considered with respect to the variable x. Moreover, from Brezis [**29**], Théorème IV.15, p. 66, we know that for each $p \in [1, +\infty)$, and each $f \in L^p(\mathbb{R})$, we have

$$\|S(t)f\|_{L^p(\mathbb{R})} \leq \|P_t\|_{L^1(\mathbb{R})}\|f\|_{L^p(\mathbb{R})},$$

which shows that $S(t)$ maps $L^p(\mathbb{R})$ into itself and

$$\|S(t)\|_{\mathcal{L}(L^p(\mathbb{R}^n))} \leq 1.$$

Using the Residues theorem to compute the Fourier transform $\widetilde{P_t}$ of P_t with respect to the variable x, we obtain

$$\widetilde{P_t}(x) = e^{-t|x|}.$$

Consequently

$$\widetilde{S(t)f}(x) = e^{-t|x|}\widetilde{f}(x),$$

which shows that $\{S(t);\ t \geq 0\}$ satisfies $S(t+s) = S(t)S(s)$ for each $t, s \geq 0$. Finally, let us observe that

$$\|S(t)f - f\|_{L^p(\mathbb{R})}$$

$$= \frac{1}{\pi}\left(\int_{-\infty}^{+\infty}\left|\int_{-\infty}^{+\infty}\frac{t}{t^2 + (x-y)^2}[f(y) - f(x)]\,dy\right|^p dx\right)^{1/p}$$

$$= \frac{1}{\pi}\left(\int_{-\infty}^{+\infty}\left|\int_{-\infty}^{+\infty}\frac{t}{t^2 + \theta^2}[f(y) - f(y+\theta)]\,dy\right|^p d\theta\right)^{1/p}$$

$$= \frac{1}{\pi}\left(\int_{-\infty}^{+\infty}\left|\int_{-\infty}^{+\infty}\frac{1}{1 + \tau^2}[f(y) - f(y+t\tau)]\,dy\right|^p d\tau\right)^{1/p}.$$

As

$$\lim_{t \downarrow 0}\int_{-\infty}^{+\infty}\frac{1}{1+\tau^2}|f(y) - f(y+t\tau)|\,dy = 0$$

a.e. for $\tau \in \mathbb{R}$, in view of Vitali's Theorem A.6.1, it follows that

$$\lim_{t \downarrow 0}\|S(t)f - f\|_{L^p(\mathbb{R})} = 0$$

for each $f \in L^p(\mathbb{R})$, which shows that the semigroup is of class C_0.

Problem 4.6. By a simple differentiation under the integral sign and by virtue of Fubini theorem, it follows that u is harmonic on the half-plane $\{(t,x)\,;t>0,\ x\in\mathbb{R}\}$. The condition $u(0,x)=f(x)$ signifies the fact that

$$\lim_{t\downarrow 0}\|u(t,\cdot)-f\|_{L^p(\mathbb{R})}=0.$$

Problem 4.7. Let

$$R_t(x)=\begin{cases}\dfrac{x^{t-1}}{\Gamma(t)} & \text{for } t>0 \text{ and } x\in[\,0,1\,]\\[2ex] 0 & \text{for } t>0 \text{ and } x\in\mathbb{R}\setminus[\,0,1\,].\end{cases}$$

We observe that, for each $t>0$ and each $f\in L^p(0,1)$, we have

$$[S(t)f](x)=R_t(x)*f(x),$$

where the convolution is considered with respect to the variable x. On the other hand, from Brezis [**29**], Théorème IV.15, p. 66, it follows

$$\|S(t)f\|_{L^p(0,1)}$$

$$=\frac{1}{\Gamma(t)}\left(\int_0^1\left|\int_0^x(x-y)^{t-1}f(y)\,dy\right|^p dx\right)^{1/p}\leq\frac{1}{\Gamma(t+1)}\|f\|_{L^p(0,1)},$$

which shows that $\{S(t)\,;\ t\geq 0\}$ contains only nonexpansive operators. In addition

$$[S(s)S(t)f](x)=\frac{1}{\Gamma(s)}\int_0^x(x-u)^{s-1}\left(\frac{1}{\Gamma(t)}\int_0^u(u-v)^{t-1}f(v)\,dv\right)du$$

$$=\frac{1}{\Gamma(s)\Gamma(t)}\int_0^x f(v)\left(\int_v^x(x-u)^{s-1}(u-v)^{t-1}du\right)dv$$

$$=\frac{1}{\Gamma(s)\Gamma(t)}\int_0^x(x-v)^{s+t-1}f(v)\left(\int_0^1\theta^{s-1}(1-\theta)^{t-1}d\theta\right)dv$$

$$\frac{1}{\Gamma(s)\Gamma(t)}\frac{\Gamma(s)\Gamma(t)}{\Gamma(s+t)}\int_0^x(x-v)^{s+t-1}f(v)\,dv,$$

where

$$\theta=\frac{x-u}{x-v}.$$

Hence $\{S(t)\,;\ t\geq 0\}$ is a semigroup of linear bounded operators. Finally, again from in Brezis [**29**], Théorème IV.15, p. 66, we have

$$\|S(t)f-f\|_{L^p(0,1)}\leq\left|\frac{1}{\Gamma(t+1)}-1\right|\|f\|_{L^p(0,1)},$$

which shows that $\{S(t)\,;\ t \geq 0\}$ is of class C_0.

Problem 4.8. We remark that in both cases, the infinitesimal generator is self-adjoint. In the case of Problem 4.3 this follows by observing that the infinitesimal generator is symmetric on C_0^∞ which is dense in $L^2(\mathbb{R})$, and using Remark 6.4.1. Therefore, the duals coincide with the initial semigroups.

Problem 4.9. A simple calculation shows that $\{G(t)\,;\ t \geq 0\}$ is a C_0-group of unitary operators whose infinitesimal generator is defined by

$$D(A) = H^1(\mathbb{R}) \times H^1(\mathbb{R}) \quad \text{and} \quad A(u,v) = \left(\frac{\partial v}{\partial x}, \frac{\partial u}{\partial x}\right)$$

for each$(u,v) \in D(A)$.

Problem 4.10. We observe that $x : (-\infty, T\,] \to X$ is an almost everywhere solution of the integro-differential equation with delay if and only if the function $u : [\,0, T\,] \to L^1(-\infty, 0; X)$, defined by $u(t)(\theta) = x(t+\theta)$ for each $t \in [\,0, T\,]$, and a.e. for $\theta \in (-\infty, 0\,]$, is a strong solution of the problem $u' = \mathcal{A}u$, $u(0) = \varphi$.

Chapter 5

Problem 5.1. Let $(f_n)_{n\in\mathbb{N}^*}$ be a sequence of Cantor functions from $[\,0,\pi\,]$ to $[\,0,\pi\,]$ with $\lim_{n\to\infty} f_n(x) = x$ for each $x \in [\,0,\pi\,]$. As $f_n{}'(x) = 0$ a.e. for $x \in [\,0,\pi\,]$, it follows that, for each $n \in \mathbb{N}^*$, $f_n \in D(A)$, and $Af_n = 0$. Let us denote by $f(x) = x$, $g(x) = 1$, and $h(x) = 0$ for each $x \in [\,0,\pi\,]$, and let us observe that both (f,g) and (f,h) are in the closure of the graph of A. Therefore this closure cannot be the graph of a single-valued function, and consequently A is not closable.

Problem 5.2. (i) The operator is closable. The domain of the closure is

$$D(\bar{A}) = \{(x_n)_{n\in\mathbb{N}^*} \in l_2\,;\ (2^n x_n)_{n\in\mathbb{N}^*} \in l_2\}.$$

(ii) The operator is closable. The domain of the closure is $D(\bar{A}) = l_1 \cap l_2$.
(iii) The operator is closable. The domain of the closure is $D(\bar{A}) = l_2$.
(iv) The operator is not closable, because there exists at least one sequence $(u_n)_{n\in\mathbb{N}^*}$ convergent la 0 in $L^2(0,1)$, and such that $u_n(1/2) = 1$ for each $n \in \mathbb{N}^*$.
(v) The operator is closable. The domain of the closure is $D(\bar{A}) = L^2(0,1)$.

Problem 5.3. From Lemma 3.7.3 we know that A^{**} is densely defined and closed. The conclusion follows from the fact that $D(A) \subseteq D(A^{**})$, and $Au = A^{**}u$ for each $u \in D(A)$.

Problem 5.4 Thanks to (3.2.3), $(A_n)_{n\in\mathbb{N}^*}$ is pointwise convergent to A on $D(A)$. Taking $Y_0 = D(A)$, we are in the hypotheses of Theorem 5.1.2 and the conclusion follows.

Problem 5.5 From the semigroup property it follows that A_n commutes with $S(t)$ and thus (i) holds. Furthermore, since $A_n \in \mathcal{L}(X)$ and

$$\left\|e^{tA_n}\right\|_{\mathcal{L}(X)} = \left\|e^{nt(S(1/n)-I)}\right\|_{\mathcal{L}(X)} \leq e^{-nt}e^{nt\|S(1/n)\|_{\mathcal{L}(X)}} = 1,$$

it generates a uniformly continuous semigroup of contractions. Hence (ii) holds too. From the definition of the infinitesimal generator we get (iii). So, we are in the hypotheses of Theorem 5.1.2, and accordingly

$$f(t+s) = \lim_{n\to\infty}\left(e^{tA_n}f\right)(s) = \lim_{n\to\infty}\sum_{k=0}^{\infty}\frac{t^k}{k!}(A_n^k)(f)(s)$$

for each $f \in D(A)$ and uniformly for $s \in \mathbb{R}$ and $t \in [0,1]$. Take $s = 0$ and let us observe that, for each $f \in X$, there exists a sequence $(m_n)_{n\in\mathbb{N}^*}$ of natural numbers with $\lim_{n\to\infty} m_n = \infty$ and such that

$$f(t) = \lim_{n\to\infty}\sum_{k=0}^{m_n}\frac{t^k}{k!}(A_n^k)(f)(0).$$

Since each term on the right-hand side is a polynomial, this is nothing else than the Weierstrass approximation theorem.

Chapter 6

Problem 6.1. Let us observe that, for each $t > 0$, each $h \in \mathbb{R}_+$, and each $(x_n)_{n\in\mathbb{N}^*} \in l_p$ with $\|(x_n)_{n\in\mathbb{N}^*}\|_{l_p} \leq 1$, we have

$$\|S(t+h)(x_n)_{n\in\mathbb{N}^*} - S(t)(x_n)_{n\in\mathbb{N}^*}\|_p^p = \sum_{k=1}^{\infty}\left|e^{-a_k^2(t+h)} - e^{-a_k^2 t}\right|^p |x_k|^p$$

$$\leq \sum_{k=1}^{\infty}\left|1 - e^{-a_k^2 h}\right|^p e^{-pa_k^2 t}|x_k|^p.$$

From this point we distinguish between two cases: $(a_n)_{n\in\mathbb{N}^*}$ bounded, or $(a_n)_{n\in\mathbb{N}^*}$ unbounded. In the first case, by virtue of Problem 2.1, we know that the semigroup is uniformly continuous and thus equicontinuous. So we will consider only the second case. More than this, we may assume with no loss of generality that $\lim_{n\to\infty} a_n = +\infty$. Otherwise, we can split the sequence $(a_n)_{n\in\mathbb{N}^*}$ into two complementary subsequences, one bounded and the other one tending to $+\infty$, and we reduce the problem to these two cases. Thus, let us assume that $\lim_{n\to\infty} a_n = +\infty$. We then have

$$\|S(t+h)(x_n)_{n\in\mathbb{N}^*} - S(t)(x_n)_{n\in\mathbb{N}^*}\|_p^p \leq \sum_{k=1}^{\infty} \left|1 - e^{-a_k^2 h}\right|^p |x_k|^p$$

$$\leq \max_{k\in\{1,2,\ldots,m\}} \left\{\left|1 - e^{-a_k^2 h}\right|\right\} + \sum_{k=m+1}^{\infty} e^{-pa_k^2 t}|x_k|^p.$$

Let $\varepsilon > 0$. Let us fix $m = m(\varepsilon) \in \mathbb{N}^*$ such that

$$\sum_{k=m+1}^{\infty} e^{-pa_k^2 t} \leq \frac{\varepsilon^p}{2}$$

and let us choose $\delta = \delta(\varepsilon) > 0$ such that

$$\max_{k\in\{1,2,\ldots,m\}} \left|1 - e^{-a_k^2 h}\right|^p \leq \frac{\varepsilon^p}{2}$$

for each $h \in \mathbb{R}_+$ with $h \leq \delta$. From the last three inequalities, it follows that

$$\|S(t+h) - S(t)\|_{\mathcal{L}(c_0)} \leq \varepsilon$$

for each $h \in \mathbb{R}$ with $h \leq \delta$, which shows that the semigroup is uniformly continuous "from the right". The uniform continuity from the left follows from the simple observation that, for each $t > 0$ and each $h \in \mathbb{R}$ with $|h| < t/2$, we have $\|S(t+h) - S(t)\|_{\mathcal{L}(X)} \leq \|S(t/2 - h) - S(t/2)\|_{\mathcal{L}(X)}$. If $a_n = n^2$, we know form Problem 2.1 that the semigroup is not uniformly continuous because its infinitesimal generator is not defined on the whole space l_p.

Problem 6.2. Let us observe that, for each $t > 0$, each $h \in \mathbb{R}$ with $|h| < t/2$, and each $(x_n)_{n\in\mathbb{N}^*} \in c_0$ with $\|(x_n)_{n\in\mathbb{N}^*}\|_{c_0} \leq 1$, we have

$$\|S(t+h)(x_n)_{n\in\mathbb{N}^*} - S(t)(x_n)_{n\in\mathbb{N}^*}\|_\infty = \sup_{k\in\mathbb{N}^*} \left|e^{-a_k^2(t+h)} - e^{-a_k^2 t}\right| |x_k$$

$$\leq \sup_{k\in\mathbb{N}^*} \left|e^{-a_k^2(t+h)} - e^{-a_k^2 t}\right|$$

$$= \max\left\{\sup_{k\in\{1,2,\ldots,m\}} \left|e^{-a_k^2(t+h)} - e^{-a_k^2 t}\right|, \sup_{k\geq m+1} \left|e^{-a_k^2(t+h)} - e^{-a_k^2 t}\right|\right\}$$

$$\leq \max\left\{\max_{k\in\{1,2,\ldots,m\}} \left|e^{-a_k^2(t+h)} - e^{-a_k^2 t}\right|, 2\sup_{k\geq m+1} e^{-\frac{pa_k^2 t}{2}}\right\}.$$

Let $\varepsilon > 0$, let us fix $m = m(\varepsilon) \in \mathbb{N}^*$, such that

$$\sup_{k \geq m+1} e^{-\frac{pa_k^2 t}{2}} \leq \frac{\varepsilon}{2}$$

and let us choose $\delta = \delta(\varepsilon) \in (0, t/2)$, such that

$$\max_{k \in \{1,2,\ldots,m\}} \left| e^{-a_k^2(t+h)} - e^{-a_k^2 t} \right| \leq \varepsilon$$

for each $h \in \mathbb{R}$ with $|h| \leq \delta$. From the last three inequalities it follows that

$$\|S(t+h) - S(t)\|_{\mathcal{L}(c_0)} \leq \varepsilon$$

for each $h \in \mathbb{R}$ with $|h| \leq \delta$, which shows that the semigroup is of class C_0. If $a_n = n^2$, we know from Problem 2.2 that the semigroup is not uniformly continuous because its infinitesimal generator is not defined on the whole space c_0.

Problem 6.3. Let us observe that if the semigroup is compact, then it cannot be uniformly continuous, because l_p is infinite-dimensional. (Otherwise, one should have that $S(0) = I$ is a compact operator). From Problem 2.1, it follows that $\lim_{n\to\infty} a_n = \infty$. Conversely, if $\lim_{n\to\infty} a_n = \infty$, then, according to Problem 6.1 the semigroup is equicontinuous. Moreover, for each $\lambda > 0$, $(\lambda I - A)^{-1}$ is a compact operator. Indeed,

$$(x_n)_{n\in\mathbb{N}^*} = (\lambda I - A)^{-1}(y_k)_{k\in\mathbb{N}^*}$$

if and only if

$$x_n = \frac{1}{\lambda + a_n^2} y_n$$

for each $n \in \mathbb{N}^*$. On the other hand, if $\mathcal{Y}$ is a bounded subset in l_p, from the relation above and from the condition $\lim_{n\to\infty} a_n = \infty$, we deduce that $(\lambda I - A)^{-1}\mathcal{Y}$ is bounded and uniformly p-summable in l_p. In view of Problem A.2, it follows that $(\lambda I - A)^{-1}$ is a compact operator, and from Theorem 6.2.1 we deduce that the semigroup generated by A is compact.

Problem 6.4. Let us observe that if the semigroup is compact then it cannot be uniformly continuous because c_0 is infinite-dimensional. (Otherwise, one should have that $S(0) = I$ is a compact operator). From Problem 2.2 it follows that $\lim_{n\to\infty} a_n = \infty$. Conversely, if $\lim_{n\to\infty} a_n = \infty$, then, tanks to Problem 6.2, the semigroup is equicontinuous. In addition, for each $\lambda > 0$, $(\lambda I - A)^{-1}$ is a compact operator. Indeed,

$$(x_n)_{n\in\mathbb{N}^*} = (\lambda I - A)^{-1}(y_k)_{k\in\mathbb{N}^*}$$

if and only if

$$x_n = \frac{1}{\lambda + a_n^2} y_n$$

for each $n \in \mathbb{N}^*$. On the other hand, if $\mathcal{Y}$ is a bounded subset in c_0, from the relation above and from the condition $\lim_{n\to\infty} a_n = \infty$, we deduce that $(\lambda I - A)^{-1}\mathcal{Y}$ is bounded and uniformly convergent in c_0. From Problem A.1 it follows that $(\lambda I - A)^{-1}$ is a compact operator, and from Theorem 6.2.1, we deduce that the semigroup generated by A is compact.

Problem 6.5. For each $t \geq 0$ we have $S(t) = T(t)G(t) = G(t)T(t)$, where

$$T(t)(x_n)_{n\in\mathbb{N}^*} = \left(e^{-k^2 t} x_k\right)_{k\in\mathbb{N}^*} \quad \text{and} \quad G(t)(x_n)_{n\in\mathbb{N}^*} = \left(e^{ie^{k^4} t} x_k\right)_{k\in\mathbb{N}^*}.$$

From Problem 6.3, we know that $\{T(t)\,;\ t \geq 0\}$ is a compact semigroup. As $\{G(t)\,;\ t \geq 0\}$ comes from a C_0-group, we deduce that $\{S(t)\,;\ t \geq 0\}$ is a C_0-compact semigroup. Its infinitesimal generator is defined by

$$D(A) = \left\{(x_k)_{k\in\mathbb{N}^*}\,;\ \left((-k^2 + ie^{k^4})x_k\right)_{k\in\mathbb{N}^*} \in l_2\right\},$$

and

$$(A(x_n)_{n\in\mathbb{N}^*})_{k\in\mathbb{N}^*} = \left((-k^2 + ie^{k^4})x_k\right)_{k\in\mathbb{N}^*}$$

for each $(x_n)_{n\in\mathbb{N}^*} \in D(A)$. On the other hand

$$(AS(t)(x_n)_{n\in\mathbb{N}^*})_{k\in\mathbb{N}^*} = \left((-k^2 + ie^{k^4})e^{-k^2 t + ie^{k^4} t} x_k\right)_{k\in\mathbb{N}^*}$$

for each $(x_n)_{n\in\mathbb{N}^*} \in D(A)$. In order to conclude the proof it suffices to observe that, for $t > 0$,

$$\left((-k^2 + ie^{k^4})e^{-k^2 t + ie^{k^4} t} \frac{1}{k}\right)_{k\in\mathbb{N}^*} \notin l_2,$$

although $\left(\frac{1}{n}\right)_{n\in\mathbb{N}^*} \in l_2$. Hence the semigroup is not differentiable.

Problem 6.6. Let us observe that each Hilbert-Schmidt operator is the uniform limit on the unit ball of operators with finite-dimensional range, and thus compact. Indeed, if T is a Hilbert-Schmidt operator, then

$$\lim_{n\to\infty} T_n x = Tx$$

uniformly for $x \in B(0,1)$, where, for each $n \in \mathbb{N}^*$, T_n is defined by

$$T_n x = T\left(\sum_{k=1}^{n} \langle x, e_k\rangle e_k\right)$$

for each $x \in H$, $x = \sum_{k=1}^{\infty} \langle x, e_k \rangle e_k$.

Problem 6.7. Let $\{e_n\,;\ n \in \mathbb{N}^*\}$ be the canonic orthonormal system in l_2, i.e. the system defined by $e_n = (e_k^n)_{k \in \mathbb{N}^*}$, where

$$e_k^n = \begin{cases} 0 & \text{for } n \neq k \\ 1 & \text{for } n = k. \end{cases}$$

Then, for each $t > 0$, we have

$$\sum_{n=1}^{\infty} \|S(t)e_n\|_{l_2}^2 = \sum_{n=1}^{\infty} e^{-2n^2 t} < +\infty,$$

which proves that $S(t)$ is a Hilbert-Schmidt operator. We complete the proof with the help of Problem 6.6.

Problem 6.8. Let $\mathcal{S} = \{e_k\,;\ k \in \mathbb{N}^*\}$, where for each $k \in \mathbb{N}^*$,

$$e_k(x) = \sqrt{\frac{2}{\pi}} \sin kx$$

a.e. for $x \in (0, \pi)$. One knows that $\mathcal{S}$ is a complete orthonormal system in $L^2(0, \pi)$. Let us observe that, for each $t > 0$, we have

$$\sum_{n=1}^{\infty} \|S(t) \sin nx\|_{L^2(0,\pi)}^2 = \sum_{n=1}^{\infty} e^{-n^2 t} < +\infty,$$

which shows that $S(t)$ is a Hilbert-Schmidt operator. We complete the proof with the help of Problem 6.6.

Problem 6.9. If a C_0-group is uniformly continuous then, its infinitesimal generator is in $\mathcal{L}(X)$, and therefore it is uniformly differentiable, and hence differentiable. See Theorem 2.2.1. Conversely, if a C_0-group $\{G(t)\,;\ t \in \mathbb{R}\}$ is differentiable, then it is differentiable at $t = 0$ because, for each $x \in X$, we have

$$\lim_{h \downarrow 0} \frac{1}{h} (G(h)x - x) = G^{-1}(t) \lim_{h \downarrow 0} \frac{1}{h} (G(t+h)x - G(t)x),$$

where $t > 0$ is arbitrary. Therefore $D(A) = X$, and, according to the closed graph theorem, A is continuous. From Theorem 2.2.1 it follows that the group is uniformly continuous, and hence uniformly differentiable.

Chapter 7

Problem 7.1. If a group has an analytic extension to a sector in the complex plane then it is uniformly differentiable and, according to Problem 6.9, it is uniformly continuous. In view of Theorem 2.2.1, its infinitesimal generator is bounded.

Problem 7.2. The eigenvalues of A are $\lambda_k = k^2$ with $k \in \mathbb{N}^*$ and the corresponding set of eigenfunctions, which is a complete orthonormal basis in $L^2(0,\pi)$, is

$$\left\{\sqrt{\frac{2}{\pi}}\sin kx\,;\ k \in \mathbb{N}^*\right\}.$$

So, according to Example 7.6.1, we have

$$D(A^\alpha) = \left\{u \in L^2(0,\pi)\,;\ u(x) = \sum_{k=1}^{\infty} k^{2\alpha} u_k \sin kx\right\},$$

where

$$u_k = \sqrt{\frac{2}{\pi}}\int_0^\pi u(y)\sin ky\,dy$$

for $k \in \mathbb{N}^*$. Therefore, if $\alpha \in [\,\frac{1}{2}, 1\,]$, $D(A) = H^{2\alpha}(0,\pi) \cap H_0^1(0,\pi)$.

Problem 7.3. Let $f \in C_0^\infty(0,+\infty)$. We have

$$(A^{-\alpha}f)(s) = \frac{1}{\Gamma(\alpha)}\int_0^\infty t^{\alpha-1}[S(t)f](s)\,dt$$

$$= \frac{1}{\Gamma(\alpha)}\int_0^\infty t^{\alpha-1}f(t+s)\,dt = \frac{1}{\Gamma(\alpha)}\int_s^\infty (u-s)^{\alpha-1}f(u)\,du = (K^\alpha f)(s).$$

We notice that, for $f \in C_0^\infty(0,+\infty)$, the integral from 0 to $+\infty$, which defines $A^{-\alpha}$, is convergent for each $\alpha > 0$. Finally, let us observe that, for each $f, g \in C_0^\infty(0,+\infty)$, and each $\alpha > 0$, we have

$$\int_0^\infty f(s)(K^\alpha g)(s)\,ds = \int_0^{+\infty} f(s)\left(\frac{1}{\Gamma(\alpha)}\int_s^\infty (u-s)^{\alpha-1}g(u)\,du\right)ds$$

$$= \int_0^\infty g(u)\left(\frac{1}{\Gamma(\alpha)}\int_0^u (u-s)^{\alpha-1}f(s)\,ds\right)du = \int_0^\infty (J^\alpha f)(u)g(u)\,du.$$

Problem 7.4 Let us recall that $D(A) \subseteq D(A_{1/2})$. See (ii) in Theorem 7.6.1. Let $f \in D(A)$, and let $\{S_{1/2}(t)\,;\ t \geq 0\}$ be the semigroup generated by $A_{1/2}$. Let us observe that the function $u(t) = S_{1/2}(t)f$ satisfies

$$\begin{cases} u'' = \left(A_{1/2}\right)^2 f = -Af \\ u(0) = f. \end{cases}$$

Recalling that A is the Laplace operator on $L^p(\mathbb{R})$ (see Problem 4.4), we deduce that u is the solution of the elliptic equation

$$\begin{cases} u_{tt} + u_{xx} = 0 \\ u(0,x) = f(x). \end{cases}$$

An appeal to Problem 4.6 shows that $\{S_{1/2}(t)\,;\ t \geq 0\}$ coincides with the Poisson semigroup. Finally, the analytic expression of $A_{1/2}f$ follows directly from Definition 2.1.2 of the infinitesimal generator.

Chapter 8

Problem 8.1. We begin by observing that $D(\mathcal{A})$ is dense in $\mathcal{X}$ because $D(A)$ is dense in X, and $W^{1,1}(0,+\infty;X)$ is dense in $L^1(0,+\infty;X)$. Moreover, the operator $\mathcal{A}$ is closed because both A, and the differentiation operator are closed. Let $\lambda > 0$. It is easy to see that

$$(\lambda\mathcal{I} - \mathcal{A})(\xi, f) = (\eta, g)$$

if and only if

$$\begin{cases} \lambda\xi - A\xi = f(0) + \eta \\ \lambda f - f' = g. \end{cases} \tag{$*$}$$

The second equation has the general solution given by

$$f(t) = f(0)e^{\lambda t} - \int_0^t e^{\lambda(t-s)} g(s)\,ds,$$

where $f(0) \in X$. In order that f belong to $L^1(0,+\infty;X)$, we must have

$$\liminf_{t\to+\infty} \|f(t)\| = 0.$$

Let us observe that this condition is satisfied if and only if

$$f(0) = \int_0^\infty e^{-\lambda s} g(s)\,ds.$$

As A is the infinitesimal generator of a C_0-semigroup of contractions, from Theorem 3.1.1, it follows that, for $f(0)$ as above, the first equation in $(*)$ has a unique solution ξ. In view of Theorem 3.1.1, to conclude the proof, it suffices to show that, for each $(\eta, g) \in \mathcal{X}$, we have

$$\|(\xi, f)\|_{\mathcal{X}} \leq \frac{1}{\lambda}\|(\eta, g)\|_{\mathcal{X}}. \tag{$**$}$$

To this aim, let us observe that

$$\|f\|_{L^1(0,+\infty;X)} \leq \int_0^\infty e^{\lambda t}\left(\int_t^\infty e^{-\lambda s}\|g(s)\|\,ds\right)dt$$

$$= \frac{1}{\lambda} e^{\lambda t} \int_t^{\infty} e^{-\lambda s} \|g(s)\| \, ds \big|_0^{\infty} + \frac{1}{\lambda} \int_0^{\infty} \|g(t)\| \, dt$$

$$= -\frac{1}{\lambda} \|f(0)\| + \frac{1}{\lambda} \|g\|_{L^1(0,+\infty;X)}.$$

On the other hand, again from Theorem 3.1.1, we deduce that

$$\|\xi\| \leq \frac{1}{\lambda} \|f(0)\| + \frac{1}{\lambda} \|\eta\|.$$

From the last two inequalities, we deduce

$$\|\xi\| + \|f\|_{L^1(0,+\infty;X)} \leq \frac{1}{\lambda} \left(\|\eta\| + \|g\|_{L^1(0,+\infty;X)} \right),$$

inequality which is equivalent to $(**)$.

Problem 8.2. Theorem 4.4.1 implies that $A : D(A) \subseteq L^p(\mathbb{R}^3) \to L^p(\mathbb{R}^3)$, defined by

$$\begin{cases} D(A) = \{u \in X; \, a \cdot \nabla u \in X\} \\ Au = -a \cdot \nabla u = -\displaystyle\sum_{i=1}^{n} a_i \frac{\partial u}{\partial x_i} \quad \text{for each} \quad u \in D(A), \end{cases}$$

generates a C_0-group of isometries $\{G(t)\,;\ t \in \mathbb{R}\}$, $[G(t)f](x) = f(x - ta)$ for each $f \in L^p(\mathbb{R}^3)$, each $t \in \mathbb{R}$, and a.e. $x \in \mathbb{R}^3$. Therefore, the unique C^0-solution of the problem considered is given by

$$u(t, x) = u_0(x - ta) + \int_0^t f\left(s, x - (t - s)a\right) ds.$$

Problem 8.3. In view of Problem 2.5, the unique C^0-solution of the Cauchy problem considered is given by

$$u(t, x) = u_0\left(e^{-tA} x\right) + \int_0^t f\left(s, e^{-(t-s)A} x\right) ds.$$

Problem 8.4. We have $(\lambda I - A)^{-1} f = u$ if and only if $u, u' \in L^2_{2\pi}(\mathbb{R})$, and $\lambda u - u' = f$. This means that

$$u(t) = u(0) e^{\lambda t} - \int_0^t e^{\lambda(t-s)} f(s) \, ds.$$

From the periodicity condition, $u(0) = u(2\pi)$, it follows

$$u(0) = \frac{e^{2\pi\lambda}}{e^{2\pi\lambda} - 1} \int_0^{2\pi} e^{-\lambda s} f(s) \, ds.$$

Hence, if $\mathcal{F}$ is bounded in $L^2_{2\pi}(\mathbb{R})$, the set

$$\left\{ \left[(\lambda I - A)^{-1} f \right](0)\,;\ f \in \mathcal{F} \right\}$$

is bounded in $\mathbb{R}$. Consequently $(\lambda I - A)^{-1}(\mathfrak{F})$ is bounded in $W^{1,2}(0, 2\pi)$. From Theorem 1.5.4 it follows that$(\lambda I - A)^{-1}(\mathfrak{F})$ is relatively compact in $L^2_{2\pi}(\mathbb{R})$. Let us observe that, for each $n \in \mathbb{N}^*$, $Qf_n = t \cdot \sin(n(t+x))$, where $f_n = \sin(n(t+x))$. On the other hand, one can easily see that the family $\{t \cdot \sin(n(t+x))\,;\ n \in \mathbb{N}^*\}$ is not relatively compact in $C([\,0,1\,]; L^2_{2\pi}(\mathbb{R}))$, because it is not equicontinuous on $[\,0,1\,]$ except for $t = 0$.

Problem 8.5 First, let us observe that, in view of the resolvent equation (1.7.3), $R(\lambda; A) = (\lambda I - A)^{-1}$ is compact for all $\lambda > 0$. On the other hand, we have

$$\|\lambda R(\lambda; A)u(t) - u(t)\| = \left\| \lambda \int_0^\infty e^{-\lambda\tau}(S(\tau)u(t) - u(t))\, d\tau \right\|$$

$$\leq \lambda \int_0^\infty e^{-\lambda\tau} \|S(\tau)u(t) - u(t+\tau)\|\, d\tau + \lambda \int_0^\infty e^{-\lambda\tau} \|u(t+\tau) - u(t)\|\, d\tau.$$

But

$$\|S(\tau)u(t) - u(t+\tau)\| \leq \int_t^{t+\tau} \|f(\theta)\|\, d\theta$$

and

$$\lim_{\tau\downarrow 0} \|u(t+\tau) - u(t)\| = 0$$

uniformly for $t \in \mathbb{R}_+$. So, we have

$$\lim_{\lambda\to\infty} \lambda R(\lambda; A)u(t) = u(t)$$

uniformly for $t \in \mathbb{R}_+$. The conclusion follows from Corollary A.1.2.

Chapter 9

Problem 9.1 Let $\eta \in (C_{ub}(\mathbb{R}))^*$ and let us observe that

$$(f(s), [S(t)^*\eta](s)) = ([S(t)f](s), \eta(s)) = \int_{\mathbb{R}} f(s-t)\, d\eta(s)$$

$$= \int_{\mathbb{R}} f(s)\, d\eta(s+t) = (f(s), \eta(s+t)).$$

Thus, $[S(t)^*\eta](s) = \eta(s+t)$, i.e. the translation of the Radon measure η. Consequently, $X^\odot$ coincides with the space of all Radon measures on $\mathbb{R}$ whose right translations are continuous with respect to the variation, i.e. the space of all $\eta^\odot$ for which $\lim_{t\downarrow 0} \mathrm{Var}\,((\eta^\odot(\cdot + t) - \eta^\odot(\,\cdot\,)), \mathbb{R}) = 0$. We note that the Lebesgue measure is in $X^\odot$, but the Dirac delta is not.

Problem 9.2 Let $\Delta : a = t_0 < t_1 \cdots < t_n = t$ and let $\tau_i \in [\,t_{i-1}, t_i\,]$ for $i = 1, 2, \ldots n$. If either $a \leq t < 0$, or $0 < a < t$, the Riemann-Stieltjes sum over $[a,t]$ of $s \mapsto S(t-s)$ with respect to g is 0. So, it suffices to analyze the remaining cases. If $a < 0 = t$, we have

$$\sigma_{[a,t]}(\Delta, S, g, \tau_i) = S(t-\tau_n)(g(t_n) - g(t_{n-1})),$$

simply because, on any other interval of Δ, g is constant. Therefore

$$\sigma_{[a,t]}(\Delta, S, g, \tau_i) = S(t-\tau_n)x$$

and hence, in this case $\int_a^t S(t-s)\,dg(s) = x$. The cases $a < 0 < t$ and $0 = a < t$ follow similarly.

Problem 9.3 Let $\Delta : a = t_0 < t_1 \cdots < t_n = t$ and let $\tau_i \in [t_{i-1}, t_i]$ for $i = 1, 2, \dots n$. We have $\sigma_{[a,t]}(\Delta, S, g, \tau_i) = \sum_{i=1}^n (h(t_i) - h(t_{i-1}))S(t-\tau_i)x$. But the later sum is the Riemann-Stieltjes sum over $[a,t]$ of the vector valued continuous function $s \mapsto S(t-s)x$ with respect to g. Since, by classical arguments, it follows that $s \mapsto S(t-s)x$ is Riemann-Stieltjes integrable on $[a,t]$ with respect to g, we easily conclude that there exists $\int_a^t S(t-s)\,dg(s) = \int_a^t S(t-s)x\,dh(s)$.

Problem 9.4 The conclusion follows from Problem 9.3 by a simple induction argument.

Problem 9.5 Let $(g_n)_{n\in\mathbb{N}^*}$ be a sequence in $BV_A([a,b];X)$ such that

$$\lim_{n\to\infty} \operatorname{Var}(g_n - g, [a,b]) = 0.$$

Then

$$\lim_{n\to\infty} \|\sigma_{[a,t]}(\Delta, S, g_n, \tau_i) - \sigma_{[a,t]}(\Delta, S, g, \tau_i)\| = 0$$

uniformly with respect to $\Delta \in \mathcal{D}[a,t]$ and $\tau_i \in [t_{i-1}, t_i]$. See Section 9.1 for the notations above. But this relation shows that whenever $s \mapsto S(t-s)$ is Riemann-Stieltjes integrable on $[a,t]$ with respect to g_n in the norm of X for $n \in \mathbb{N}^*$, then $s \mapsto S(t-s)$ is Riemann-Stieltjes integrable on $[a,t]$ with respect to g too, of course, in the norm of X.

Problem 9.6 Since $g \in BV([a,b];X)$ satisfies $\operatorname{Var}(g,[c,d]) \leq \ell(d-c)$ for each $[c,d] \subset [a,b]$, it follows that g is continuous on $[a,b]$. Hence, in view of Theorem 9.2.1, for each $\xi \in X$, the unique $\mathcal{L}^\infty$-solution $u = Q(\xi, g)$ is continuous on $[a,b]$. From now on, we adapt the proof of Theorem 8.4.1 to the case here considered. Thanks to Theorem A.2.1, the necessity is obvious. In order to prove the sufficiency, we shall make use again of Theorem A.2.1. To this aim, we shall show that $Q(\mathcal{D}, \mathcal{G})$ is equicontinuous on $(a,b]$. Let $\varepsilon > 0$. Since the function ℓ in the definition of the equi-absolute continuity is continuous at $t = 0$ and $\ell(0) = 0$, for each $\varepsilon > 0$

there exists $\gamma(\varepsilon) > 0$ such that

$$\ell(r) \leq \varepsilon$$

for each $r \in [0, \gamma(\varepsilon)$ So, for each $[c, d] \subseteq [a, b]$ with $d - c \leq \gamma(\varepsilon)$, we have

$$\operatorname{Var}(g, [c, d]) \leq \varepsilon$$

uniformly for $g \in \mathcal{G}$. Let $t \in (a, b]$ and let us fix $\lambda = \lambda(\varepsilon) > 0$ such that $t - \lambda > a$, $2\lambda \leq \gamma(\varepsilon)$ and $t - \lambda \in D$. This is always possible because D is dense in $[a, b]$. Inasmuch as $Q(\mathcal{D}, \mathcal{G}))(t - \lambda)$ is relatively compact in X, for each $\varepsilon > 0$, there exists a finite family $\{(\xi_1, g_1), (\xi_2, g_2), \ldots (\xi_k, g_{k(\varepsilon)})\}$ in $\mathcal{D} \times \mathcal{G}$ such that, for each $(\xi, g) \in \mathcal{D} \times \mathcal{G}$, there exists $i \in \{1, 2, \ldots k(\varepsilon)\}$ with the property

$$\|Q(\xi, g)(t - \lambda) - Q(\xi_i, g_i)(t - \lambda)\| \leq \varepsilon.$$

On the other hand, the family $\{Q(\xi_1, g_1), Q(\xi_2, g_2), \ldots Q(\xi_{k(\varepsilon)}, g_{k(\varepsilon)})\}$ is equicontinuous at t, being a finite family of continuous functions on $[a, b]$. Therefore, for the very same $\varepsilon > 0$, there exists $\delta(\varepsilon) \in (0, \lambda]$, such that

$$\|Q(\xi_i, g_i)(t + h) - Q(\xi_i, g_i)(t)\| \leq \varepsilon$$

for each $i = 1, 2, \ldots k(\varepsilon)$, and each $h \in \mathbb{R}$ with $|h| \leq \delta(\varepsilon)$. Let us observe that

$$Q(\eta, g)(t + h) = S(\lambda + h)Q(\eta, g)(t - \lambda) + \int_{t-\lambda}^{t+h} S(t + h - s)\, dg(s),$$

For the sake of simplicity, let us denote $u = Q(\xi, g)$ and $u_i = Q(\xi, g_i)$. Since

$$\begin{aligned} u(t + h) - u(t) &= [S(\lambda + h)u(t - \lambda) - S(\lambda + h)u_i(t - \lambda)] \\ &+ \left[\int_{t-\lambda}^{t+h} S(t + h - s)\, dg(s)\right] + [S(\lambda + h)u_i(t - \lambda) - u_i(t + h)] \\ &+ [u_i(t + h) - u_i(t)] + [u_i(t) - u(t)], \end{aligned}$$

we have

$$\begin{aligned} \|u(t + h) - u(t)\| &\leq \|S(\lambda + h)u(t - \lambda) - S(\lambda + h)u_i(t - \lambda)\| \\ &+ \left\|\int_{t-\lambda}^{t+h} S(t + h - s)\, dg(s)\right\| + \left\|\int_{t-\lambda}^{t+h} S(t + h - s)\, d(g_i)(s)\right\| \\ &+ \|u_i(t + h) - u_i(t)\| + \|S(\lambda)u_i(t - \lambda) - S(\lambda)u(t - \lambda)\| \\ &+ \left\|\int_{t-\lambda}^{t} S(t - s)\, dg(s)\right\| + \left\|\int_{t-\lambda}^{t} S(t - s)\, dg_i(s)\right\| \end{aligned}$$

$$\leq \|u(t - \lambda) - u_i(t - \lambda)\| + \operatorname{Var}(g, [t - \lambda, t + h]) + \operatorname{Var}(g_i, [t - \lambda, t + h])$$
$$+ \|u_i(t+h) - u_i(t)\| + \|u_i(t-\lambda) - u(t-\lambda)\| + \operatorname{Var}(g_i, [t-\lambda, t]) + \operatorname{Var}(g, [t-\lambda, t])$$

for each $(\xi, g) \in \mathcal{D} \times \mathcal{G}$ and each $h \in (-\delta(\varepsilon), \delta(\varepsilon))$. From this inequality and from the manner of choice of both $\lambda > 0$ and $\delta(\varepsilon) > 0$, it follows

$$\|Q(\xi, g)(t+h) - Q(\xi, g)(t)\| \leq 7\varepsilon$$

for each $(\xi, g) \in \mathcal{D} \times \mathcal{G}$ and each $h \in \mathbb{R}$ with $|h| \leq \delta(\varepsilon)$. Therefore $Q(\mathcal{D}, \mathcal{G})$ is equicontinuous on $(a, b]$, and, by virtue of Theorem A.2.1, it is relatively compact in $C([c, b]; X)$ for each $c \in (a, b)$. If $a \in D$, then $Q(\mathcal{D}, \mathcal{G})$ is equicontinuous at a as well, and this achieves the proof.

Problem 9.7 Thanks to Problem 9.6, it suffices to check that, for each $t \in (a, b)$, $Q(\mathcal{D}, \mathcal{F}))(t)$ is relatively compact in X. Let $t \in (a, b)$ and let $\lambda > 0$ with $t - \lambda > a$. We have

$$Q(\xi, f)(t) = S(t-a)\xi + S(\lambda) \int_a^{t-\lambda} S(t-s) dg(s) + \int_{t-\lambda}^t S(t-s)\, dg(s).$$

Since $S(\lambda)$ is compact, it follows that the operator $P_\lambda : Q(\mathcal{D}, \mathcal{G})(t) \to X$, defined by

$$P_\lambda Q(\xi, g)(t) = S(t-a)\xi + S(\lambda) \int_a^{t-\lambda} S(t-\lambda-s) f(s)\, dg(s),$$

maps the set $Q(\mathcal{D}, \mathcal{G})(t)$ into a relatively compact subset in X. In addition, from the preceding relation and from the equi-absolute continuity of $\mathcal{G}$, it follows that

$$\lim_{\lambda \downarrow 0} \|P_\lambda Q(\xi, g)(t) - Q(\xi, g)(t)\| \leq \limsup_{\lambda \downarrow 0} \operatorname{Var}(g, [t-\lambda, t]) = 0$$

uniformly for $g \in \mathcal{G}$. From Lemma A.1.2 it follows that $Q(\mathcal{D}, \mathcal{G})(t)$ is relatively compact in X for each $t \in [a, b]$. An appeal to Problem 9.5 completes the proof.

Chapter 10

Problem 10.1. We have only to repeat the proof of Theorem 10.1.1, noticing that, first of all we have to state and prove an appropriate extension of Lemma A.1.3 from the case f continuous, to the case f integrable. In order to obtain the latter result, one approximates the functions f in $L^1(a, b; X)$ by continuous functions (in fact of class C^∞) taking values in the closed convex hull of the range of f (see Theorem 1.3.3), and then one uses Mazur's

Theorem A.1.3. For passing to the limit in the approximating equations one uses Lebesgue dominated convergence theorem.

Problem 10.2. We have only to repeat the proof Theorem 10.2.1, using the extension of Lemma A.1.3 suggested in the proof of Problem 10.1. We notice that, here, we need the following "local" variant of Lemma 10.2.2:
Lemma *Let* $g : \mathbb{I} \times X \to X$ *be a function satisfying* (i), (ii) *in* Problem 10.1, *and* (iv) *in* Problem 10.2. *Then, for each* $a \in \mathbb{I}$, *and each* $\xi \in X$, *there exists* $\delta > 0$ *such that* $[\,a, a+\delta\,] \subset \mathbb{I}$, *and, for each* h *in* $L^1(a, a+\delta; X)$, *the problem*

$$\begin{cases} u' = g(t,u) + h(t) \\ u(a) = \xi, \end{cases} \tag{$*$}$$

has a unique solution defined on $[\,a, a+\delta\,]$. *In addition, for each* a, ξ, *and* δ *fixed as above, the mapping* $h \mapsto \mathcal{S}(h)$*-the unique solution of the problem* ($*$) *corresponding to* h*-satisfies*

$$\|\mathcal{S}(h_1) - \mathcal{S}(h_2)\|_{C([\,a,a+\delta\,];X)} \leq e^{\int_a^{a+\delta} \ell(t)\,dt} \|H_1 - H_2\|_{C([\,a,a+\delta\,];X)},$$

where ℓ *is given by* (iv), *and*

$$H_i(t) = \int_a^t h_i(s)\,ds$$

for $i = 1, 2$, *and* $t \in [\,a, a+\delta\,]$. In order to prove this lemma, let us choose $\delta > 0$, sufficiently small, such that

$$q = \int_a^{a+\delta} \ell(t)\,dt < 1. \tag{$**$}$$

Using mathematical induction, we show that, for each $n \in \mathbb{N}^*$, and each $u, v \in C([\,a, a+\delta\,]; X)$, we have (with the same notations as in the proof of Lemma 10.2.2)

$$\|Q^n u - Q^n v\|_{C([\,a,a+\delta\,];X)} \leq q^n \|u - v\|_{C([\,a,a+\delta\,];X)}. \tag{$***$}$$

Chapter 11

Problem 11.1. We have to repeat the proof Theorem 11.2.1, with the special mention that, in the process of passing to the limit one has to use Lebesgue dominated convergence theorem.

Problem 11.2. Consider first $\Omega = X$. We choose $\delta > 0$ satisfying ($**$) in the proof of Problem 10.2, and we define $Q : C([\,a, a+\delta\,]; X) \to C([\,a, a+\delta\,]; X)$ by

$$(Qu)(t) = S(t-a)\xi + \int_0^t S(t-s) f(s, u(s))\,ds$$

for each $u \in C([a, a+\delta]; X)$. We then show that this operator satisfies $(***)$ in the proof of Problem 10.2 and therefore, according to a consequence of Banach fixed point theorem, Q has a unique fixed point. For the general case, in which Ω is strictly included in X, we use the method of proof of Theorem 11.2.1.

Problem 11.3. We apply Problem 11.2 and then, using Theorem 11.3.2, we show that each local solution can be extended up to one defined on $\mathbb{R}_+$.

Problem 11.4. We apply Problem 8.3 combined with Problem 11.2, and with the suggestion made for the preceding proof.

Problem 11.5. Let us observe that the problem in question can be rewritten as an abstract ordinary differential equation of the form

$$\begin{cases} u' = Au + f(t, u) \\ u(0) = \xi \end{cases}$$

in the space $X = C_{ub}(\mathbb{R} \times [-1, 1])$, where $A : D(A) \subseteq X \to X$ is defined by

$$D(A) = \{u \in X\,;\ yu_x \in X\},$$

$$Au = -yu_x$$

for each $u \in X$, and $f : \mathbb{R}_+ \times X \to X$ is defined by

$$f(t, u)(x, y) = \frac{\varphi(x)}{2} \int_{-1}^{1} u(x, z)\, dz$$

for each $(t, u) \in \mathbb{R}_+ \times X$, and each $(x, y) \in \mathbb{R} \times [-1, 1]$. We can easily see that we are in the hypotheses of Problem 11.2, with $\ell = c$.

Chapter 12

Problem 12.1 We have to repeat the very same routine as that in the proof Theorem 12.1.1 with the sole special mention that here, the existence of the operator Q is ensured by the definition of $BV_A([a, b]; X)$, and not by Theorem 9.1.1 which might not be applicable in this case.

Problem 12.2 We observe that the problem above rewrites equivalently as

$$\begin{cases} du = \{Au\}dt + dg \\ u(0) = \xi, \end{cases}$$

where $X = L^1(\mathbb{R})$, $A : D(A) \subseteq X \to X$ is defined by

$$\begin{cases} D(A) = \{u \in X\,;\ u' \in X\} \\ Au = u' \end{cases} \qquad \text{for } u \in D(A),$$

and $g : \mathbb{R} \to X$ is given by

$$g(t) = \begin{cases} -\frac{1}{2}h & \text{if } t < t_0 \\ 0 & \text{if } t = t_0 \\ \frac{1}{2}h & \text{if } t > 0. \end{cases}$$

Repeating the arguments in Problem 9.1, we easily see that $g \in BV_A(\mathbb{R}; X)$ and thus the conclusion follows from Problem 1.2.1.

Problem 12.3 Let $(a, \xi) \in \mathbb{R} \times X$ and let $c > a$. We define the operator $Q : L^\infty(a, c; X) \to L^\infty(a, c; X)$ by

$$Q(v)(t) = S(t-a)\xi + \int_a^t S(t-s)f(s, v(s))\, ds + \int_a^t S(t-s)\, dg(s)$$

for each $v \in L^\infty(a, c; X)$ and $t \in [a, b]$. Since $f(\mathbb{R} \times X)$ is compact, there exists $M > 0$ such that $\|f(t, \eta)\| \leq M$ for all $(t, \eta) \in \mathbb{R} \times X$. Let $r = M + \operatorname{Var}(g, [a, c])$ and let

$$K = \{v \in L^\infty(a, c; X);\ \|v(t)\| \leq r \text{ a.e. for } t \in [a, c]\}.$$

In fact K is the closed ball of radius r and centered at 0 in $L^\infty(a, c; X)$. We consider K as a subset of $L^1(a, c; X)$ and we notice that it is nonempty, closed, convex and bounded. Moreover it is easy to see that Q maps K into itself. So, in order to apply Schauder fixed point theorem A.1.5, we have merely to show that Q is continuous from K to K in the norm topology of $L^1(a, c; X)$ and $Q(K)$ is relatively compact in the same topology. Since for every $v, w \in K$ we have

$$\|Q(v)(t) - Q(w)(t)\| \leq \int_a^t \|f(s, v(s)) - f(s, w(s))\|\, ds$$

and f is continuous and bounded, by Lebesgue dominated convergence theorem, it follows that Q is continuous from K to K when both domain and range are endowed with the norm topology of $L^1(a, c; X)$. To check the compactness of $Q(K)$, let $t \in [a, b]$ and let us observe that, inasmuch as $(s, \eta) \mapsto S(t-s)\eta$ is continuous from $[a, t] \times \overline{f(\mathbb{R} \times X)}$ to X and the former is relatively compact, $\{S(t-s)f(s, v(s));\ v \in K,\ s \in [a, t]\}$ is relatively compact in X. From this remark and Lemma A.1.3, we readily conclude that, for each $t \in [a, c]$,

$$\left\{\int_a^t S(t-s)f(s, v(s))\, ds;\ v \in K\right\}$$

is relatively compact in X. Let us denote by $G(t) = \int_a^t S(t-s)\, dg(s)$ and let us observe that, by virtue of Theorem 8.4.1, we conclude that the set $\{Q(v) - G;\ v \in K\}$ is relatively compact in $C([a, c]; X)$. On the other hand $Q(K) = \{Q(v) - G\,; v \in K\} + G$, and thus it is relatively compact in $L^1(a, c\,; X)$ (in fact, it is relatively compact even in the uniform convergence topology on $[\,a, c\,]$). By Schauder fixed point theorem A.1.5, Q has at least one fixed point in K which obviously is an $\mathfrak{L}^\infty$-solution of the problem considered on $[\,a, c\,]$, as claimed.

Problem 12.4 Let us observe first that, if $g \in BV_A(\mathbb{R}\,; X)$ and $h \in C(\mathbb{R}\,; X)$, then $t \mapsto \tilde{g}(t) = \int_a^t h(s)\,ds + g(t)$ belongs to $BV_A(\mathbb{R}\,; X)$ too. From now on, we shall mimic the proof of Theorem 12.2.2. Namely, let $(a, \xi) \in \mathbb{R} \times X$ with $(g(a+0) - g(a) + \xi) \in \mathfrak{D}$, and let us denote by $\eta = g(a+0) - g(a) + \xi$. Inasmuch as $\mathfrak{D}$ is open and f is b-compact, there exist $c > a$ and $r > 0$, such that $[\,a, c\,] \times B(\eta, r) \subset \mathfrak{D}$, and $K = \overline{f([\,a, c\,] \times B(\eta, r))}$ is relatively compact in X. In particular, there exists $M > 0$ such that

$$\|f(t, u)\| \leq M$$

for each $(t, u) \in [\,a, c\,] \times B(\eta, r)$. Let us define $\rho : X \to X$ by

$$\rho(y) = \begin{cases} y & \text{for } y \in B(\eta, r) \\ \dfrac{r}{\|y - \eta\|}(y - \eta) + \eta & \text{for } y \in X \setminus B(\eta, r). \end{cases}$$

We may easily see that ρ maps X to $B(\eta, r)$ and is continuous Now, let us define the function $f_r : \mathbb{R} \times X \to X$ by

$$f_r(t, u) = \begin{cases} f(a, \rho(u)) & \text{if } t \in (-\infty, a\,] & \text{and } u \in X \\ f(t, \rho(u)) & \text{if } t \in (a, c) & \text{and } u \in X \\ f(c, \rho(u)) & \text{if } t \in [\,c, +\infty) & \text{and } u \in X. \end{cases}$$

Clearly f_r is continuous and $f_r(\mathbb{R} \times X)$ is relatively compact. By virtue of Problem 12.3, it follows that the Cauchy problem

$$\begin{cases} du = \{Au + f_r(t, u)\}dt + dg \\ x(a) = \xi \end{cases}$$

has at least one $\mathfrak{L}^\infty$-solution, $u : [\,a, c\,] \to X$. Since, $u(a) = \xi$ and, by (9.2.1) in Theorem 9.2.1, $u(a + 0) = g(a + 0) - g(a) + u(a) = \eta$, there exists $b \in (a, c\,]$ such that, for each $t \in (a, b\,]$, $u(t) \in B(\eta, r)$. But in this case $\rho(u(t)) = u(t)$ for each $t \in (a, b\,]$, and consequently $f_r(s, u(s))$ must coincide with $f(s, u(s))$ for each $s \in (a, b\,]$. Since $u : [\,a, b\,] \to X$ is piecewise continuous and $(t, u(t)) \in (a, c\,] \times B(\eta, r) \subset \mathfrak{D}$, it follows that $(t, u(t + 0)) \in [\,a, c\,] \times B(\eta, r) \subset \mathfrak{D}$, and thus, by Remark 12.1.3, u is an $\mathfrak{L}^\infty$-solution of the problem (12.2.2).

Problem 12.5 Set

$$H = H_0^1(\Omega) \times L^2(\Omega)$$

which, endowed with the inner product $\langle \cdot, \cdot \rangle$, defined by

$$\langle (u,v), (\tilde{u}, \tilde{v}) \rangle = \int_\Omega u'(x)\tilde{u}'(x)\,dx + \int_\Omega v(x)\tilde{v}(x)\,dx$$

for each $(u,v), (\tilde{u}, \tilde{v}) \in H$, is a real Hilbert space. We define the operator $A : D(A) \subset H \to H$ by

$$\begin{cases} D(A) = (H^2(\Omega) \cap H_0^1(\Omega)) \times H_0^1(\Omega) \\ A(u,v) = (v, \Delta u) \quad \text{for each } (u,v) \in D(A). \end{cases}$$

Furthermore, let us define $f : \mathbb{R}_+ \times D(f) \to H$ by

$$\begin{cases} D(f) = \{(u,v) \in H\,;\ b(t,\,\cdot\,,u(\,\cdot\,)) \in L^2(\Omega) \text{ for each } t \in [0,+\infty)\} \\ f(t,(u,v))(x) = (0, b(t,x,u(x))) \end{cases}$$

for each $t \in [0,+\infty)$, each $(u,v) \in D(f)$ and a.e. for $x \in \Omega$. At this point, let us observe that the problem considered can be rewritten under the equivalent form

$$\begin{cases} dz = \{Az + f(t,z)\}dt + dg(t) \\ z(a) = \xi, \end{cases}$$

where A and f are as above, $z(t)(x) = (u(t,x), v(t,x))$ a.e. for (t,x) in $(0,T) \times \Omega$, $\xi = (u_0, v_0)$ and $g : \mathbb{R} \times H \to H$ is defined by

$$g(t)(x) = \begin{cases} -(0,\frac{1}{2}) & \text{if } t < t_0 \text{ and a.e. for } x \in \Omega \\ (0,0) & \text{if } t = t_0 \text{ and a.e. for } x \in \Omega \\ (0,\frac{1}{2}) & \text{if } t < t_0 \text{ and a.e. for } x \in \Omega. \end{cases}$$

In order to prove that A, f and g satisfy the conditions in Problem 12.4, let us remark first that, by virtue of Theorem 4.6.2, A generates a C_0-group of isometries. As concerns f, from Theorem 1.5.4 and Lemma A.6.1, it follows that $D(f) = H$ and f is continuous on $\mathbb{R} \times H$. Finally, again from Theorem 1.5.4, we know that $H_0^1(\Omega)$ is compactly imbedded in $L^2(\Omega)$ and thus f is compact. Using similar arguments as those in the proof of Problem 9.2, we show that $g \in BV_A(\mathbb{R}; H)$ and therefore we are in the hypotheses of Problem 12.4.

Problem 12.6 We rewrite the problem in the abstract form

$$\begin{cases} du = \{Au\}dt + dg_u \\ u(0) = \xi_A, \end{cases}$$

To this aim, take $X = L^1(\Omega)$ and $A : D(A) \subseteq X \to X$, defined by

$$\begin{cases} D(A) = \{u \in W_0^{1,1}(\Omega)\,;\ \Delta u \in L^1(\Omega)\} \\ Au = \Delta u \qquad\qquad \text{for } u \in D(A), \end{cases}$$

and let us recall that A generates an analytic and compact C_0-semigroup of contractions. See Theorem 7.2.7. We also recall that in this case $X^{\odot} = C_0(\overline{\Omega})$ and $X_A = (C_0(\overline{\Omega}))^*$, i.e. the space of all Radon measures concentrated in Ω. See Example 9.5.1. Next, take $T > 0$ with $t_0 \in (0,T)$ and let us define $G : L^\infty(0,T;X) \to BV([0,T];(C_0(\overline{\Omega}))^*)$ by

$$G(u)(t) = \begin{cases} -\frac{1}{2}\delta(x - x_u) & \text{for } t < t_0 \\ 0 & \text{for } t = t_0 \\ \frac{1}{2}\delta(x - x_u) & \text{for } t > t_0. \end{cases}$$

Since $u \mapsto x_u$ takes values in a compact subset Ω_K in Ω, there exists $r > 0$ such that, for each $x \in \Omega_K$, $B(x,r) \subset \Omega$. Next, take $\varepsilon \in (0,r)$ and let us define $G_\varepsilon : L^\infty(0,T;L^1(\Omega)) \to BV([0,T];L^1(\Omega))$ by

$$G_\varepsilon(u)(t) = \begin{cases} -\frac{1}{2}\delta_\varepsilon(x - x_u) & \text{for } t < t_0 \\ 0 & \text{for } t = t_0 \\ \frac{1}{2}\delta_\varepsilon(x - x_u) & \text{for } t > t_0, \end{cases}$$

where

$$\delta_\varepsilon(x - x_u) = \begin{cases} \frac{1}{\mu B(x_u,\varepsilon)} & \text{if } x \in B(x_u,\varepsilon) \\ 0 & \text{if } x \in \Omega \setminus B(x_u,\varepsilon), \end{cases}$$

$\mu(B(x_n,\varepsilon))$ being the Lebesgue measure of $B(x_n,\varepsilon)$. We may easily verify that $\{G_\varepsilon;\varepsilon \in (0,r)\}$ is of equibounded variation in $BV([0,T];L^1(\Omega))$. Moreover, for each $\varepsilon \in (0,r)$, G_ε satisfies the conditions in Theorem 12.2.1 because it is bounded and pointwise continuous in the $\sigma(L^1(\Omega), C_0(\overline{\Omega}))$ topology. The boundedness was already been proved. From Theorem 12.2.1 it then follows that the ε-approximating problem obtained from the initial one by substituting $\delta(t - t_0) \times \delta(x - x_u)$ by $\delta_\varepsilon(t - t_u) \times \delta_\varepsilon(x - x_u)$, has at least one $\mathfrak{L}^\infty$-solution u_ε. By virtue of Theorem 9.4.2, it follows that $\{u_\varepsilon;\ \varepsilon \in (0,r)\}$ is relatively compact in $L^1(0,T;L^1(\Omega))$. Moreover, by the Helly-Bray Theorem 1.4.6, we conclude that

$$\lim_{\varepsilon \downarrow 0} G_\varepsilon(u_\varepsilon)(t) = g(t)$$

pointwise in the $\sigma((C_0(\overline{\Omega}))^*, L^1(\Omega))$-topology. Again by the continuity of the mapping $x \mapsto x_u$, it follows that $g = G(u)$, and this completes the proof.

Appendix

Problem A.1. *Necessity.* Let $\mathfrak{X}$ be a relatively compact family in c_0, and let $\varepsilon > 0$. In view of Hausdorff's Theorem A.1.1, $\mathfrak{X}$ is precompact. So, it has a finite $\varepsilon/2$-net $\mathfrak{X}_\varepsilon = \{(x_n^i)_{n\in\mathbb{N}^*}\,;\ i = 1,2,\ldots,k(\varepsilon)\}$. Since $\mathfrak{X}_\varepsilon$ is finite, it is uniformly convergent, and thus, for the same $\varepsilon > 0$, there exists $n(\varepsilon) \in \mathbb{N}^*$ such that

$$|x_n^i| \leq \frac{\varepsilon}{2}$$

for each $i = 1, 2, \ldots, k(\varepsilon)$, and each $n \in \mathbb{N}^*$, $n \geq n(\varepsilon)$. In order to conclude the proof of the necessity, let us observe that, for each $(x_k)_{k\in\mathbb{N}^*} \in \mathcal{X}$, and each $n \in \mathbb{N}^*$, $n \geq n(\varepsilon)$, we have

$$|x_n| \leq |x_n - x_n^i| + |x_n^i| \leq \frac{\varepsilon}{2} + \frac{\varepsilon}{2} = \varepsilon.$$

Sufficiency. Let $\mathcal{X}$ be a bounded and uniformly convergent family, and let $\varepsilon > 0$. Let $n(\varepsilon)$ in $\mathbb{N}^*$ such that, for each $(x_n)_{n\in\mathbb{N}^*}$, and each $n \in \mathbb{N}^*$, $n \geq n(\varepsilon)$, we have

$$|x_n| \leq \frac{\varepsilon}{2}.$$

As $\left\{(x_1, x_2, \ldots, x_{n(\varepsilon)})\,;\ (x_n)_{n\in\mathbb{N}^*} \in \mathcal{X}\right\}$ is bounded in $\mathbb{R}^{n(\varepsilon)}$, it is precompact and thus it has a finite $\varepsilon/2$-net $\left\{(x_1^i, x_2^i, \ldots, x_{n(\varepsilon)}^i)\,;\ i = 1, 2, \ldots, k(\varepsilon)\right\}$. Let us observe that the set $\left\{(x_n^i)_{n\in\mathbb{N}^*}\,;\ i = 1, 2, \ldots, k(\varepsilon)\right\}$ is a finite ε-net for $\mathcal{X}$, because

$$\max_{n\in\mathbb{N}^*} |x_n - x_n^i| \leq \max\left\{\max_{n\leq n(\varepsilon)} |x_n - x_n^i|,\ \sup_{n\geq n(\varepsilon)} |x_n - x_n^i|\right\}$$

$$\leq \sup\left\{\max_{n\leq n(\varepsilon)} |x_n - x_n^i|,\ \sup_{n\geq n(\varepsilon)} |x_n| + \sup_{n\geq n(\varepsilon)} |x_n^i|\right\} \leq \frac{\varepsilon}{2} + \frac{\varepsilon}{2} = \varepsilon.$$

Problem A.2. By analogy with the case $X = \mathbb{R}$, we introduce the notion of uniformly convergent family in $c_0(X)$. Using similar arguments as those in the proof of Problem A.1 (one replaces $|\cdot|$ by $\|\cdot\|$), we obtain the following compactness criterion:

A family $\mathcal{X}$ in $c_0(X)$ is relatively compact if and only if it is uniformly convergent and, for each $i \in \mathbb{N}^*$, the section of $\mathcal{X}$ at i, $\{x_i\,;\ (x_n)_{n\in\mathbb{N}^*} \in \mathcal{X}\}$, is relatively compact in X.

Problem A.3. *Necessity.* Let $\mathcal{X} \subset l_p$ be relatively compact, and let $\varepsilon > 0$. In view of Hausdorff's Theorem A.1.1, $\mathcal{X}$ is precompact, and thus it has a finite $\varepsilon/2$-net $\mathcal{X}_\varepsilon = \left\{(x_n^i)_{n\in\mathbb{N}^*}\,;\ i = 1, 2, \ldots, k(\varepsilon)\right\}$. Since $\mathcal{X}_\varepsilon$ is finite, it is uniformly p-summable, and therefore, for the very same $\varepsilon > 0$, there exists $n(\varepsilon) \in \mathbb{N}^*$ such that

$$\left(\sum_{k=n}^{\infty} |x_k^i|^p\right)^{1/p} \leq \frac{\varepsilon}{2}$$

for each $i = 1, 2, \ldots, k(\varepsilon)$, and each $n \in \mathbb{N}^*$, $n \geq n(\varepsilon)$. In order to conclude the proof of the necessity, let us observe that, for each $(x_k)_{k\in\mathbb{N}^*} \in \mathcal{X}$, and each $n \in \mathbb{N}^*$, $n \geq n(\varepsilon)$, we have

$$\left(\sum_{k=n}^{\infty} |x_k|^p\right)^{1/p} \leq \left(\sum_{k=n}^{\infty} |x_k - x_k^i|^p\right)^{1/p} + \left(\sum_{k=n}^{\infty} |x_k^i|^p\right)^{1/p} \leq \frac{\varepsilon}{2} + \frac{\varepsilon}{2} = \varepsilon.$$

Sufficiency. Let $\mathcal{X}$ be a bounded and uniformly p-summable family, and let $\varepsilon > 0$. Let $n(\varepsilon) \in \mathbb{N}^*$ be such that, for each $(x_k)_{k\in\mathbb{N}^*}$ and each $n \in \mathbb{N}^*$ with $n \geq n(\varepsilon)$, we have

$$\left(\sum_{k=n}^{\infty} |x_k|^p\right)^{1/p} \leq \frac{\varepsilon}{4}.$$

Inasmuch as $\left\{(x_1, x_2, \ldots, x_{n(\varepsilon)})\,;\ (x_n)_{n\in\mathbb{N}^*} \in \mathcal{X}\right\}$ is bounded in $\mathbb{R}^{n(\varepsilon)}$, it is precompact and therefore it has a finite $\varepsilon/2$-net (with respect to any one of the equivalent norms on $\mathbb{R}^{n(\varepsilon)}$), $\left\{(x_1^i, x_2^i, \ldots, x_{n(\varepsilon)}^i)\,;\ i = 1, 2, \ldots, k(\varepsilon)\right\}$. Let us observe that the set $\left\{(x_n^i)_{n\in\mathbb{N}^*}\,;\ i = 1, 2, \ldots, k(\varepsilon)\right\}$ is a finite ε-net for $\mathcal{X}$, because

$$\left(\sum_{k=1}^{\infty} |x_k - x_k^i|^p\right)^{1/p} \leq \left(\sum_{k=1}^{n(\varepsilon)} |x_k - x_k^i|^p\right)^{1/p} + \left(\sum_{k=n(\varepsilon)+1}^{\infty} |x_k - x_k^i|^p\right)^{1/p}$$

$$\leq \left(\sum_{k=n(\varepsilon)}^{\infty} |x_k - x_k^i|^p\right)^{1/p} + \left(\sum_{k=n(\varepsilon)+1}^{\infty} |x_k|^p\right)^{1/p} + \left(\sum_{k=n(\varepsilon)+1}^{\infty} |x_k^i|^p\right)^{1/p}$$

$$\leq \frac{\varepsilon}{2} + \frac{\varepsilon}{4} + \frac{\varepsilon}{4} = \varepsilon.$$

Problem A.4. We mimic the case $X = \mathbb{R}$, and we define the concept of uniformly p-summable family in $l_p(X)$. Similar arguments as those used in the proof of Problem A.3 (one replaces $|\cdot|$ by $\|\cdot\|$), lead to the following compactness criterion:

A family $\mathcal{X} \subset l_p(X)$ is relatively compact if and only if it is uniformly p-summable and, for each $i \in \mathbb{N}^*$, the section of $\mathcal{X}$ at i, $\{x_i\,;\ (x_n)_{n\mathbb{N}^*} \in \mathcal{X}\}$, is relatively compact in X.

Problem A.5. The function $g : \mathbb{R} \cup \{-\infty, +\infty\} \to [-\frac{\pi}{2}, \frac{\pi}{2}]$

$$g(t) = \begin{cases} -\dfrac{\pi}{2} & \text{for } t = -\infty \\ \arctan t & \text{for } t \in \mathbb{R} \\ \dfrac{\pi}{2} & \text{for } t = +\infty, \end{cases}$$

defines a homeomorphism $\mathcal{B} : C_\ell(\mathbb{R};\,\mathbb{R}) \to C([-\frac{\pi}{2}, \frac{\pi}{2}];\,\mathbb{R})$, by

$$[\mathcal{B}(f)](t) = \begin{cases} \lim_{\tau\to-\infty} f(\tau) & \text{for } t = -\infty \\ f(\tan t) & \text{for } t \in \mathbb{R} \\ \lim_{\tau\to+\infty} f(\tau) & \text{for } t = +\infty. \end{cases}$$

Since this homeomorphism preserves the property of being equicontinuous of any subset in $C_\ell(\mathbb{R})$, the conclusion follows from Theorem A.2.1.

Problem A.6. The proof follows the same lines as those used in solving Problem A.5.

Problem A.7. The proof follows the same lines as those used in solving Problem A.5, with the mention that, instead of Theorem A.2.1, we have to use Corollary A.2.3.

Problem A.8. *Necessity.* Let $\mathcal{F}$ be a relatively compact family in $L^p(\mathbb{R}; X)$. Obviously, $\mathcal{F}$ is bounded. Let $\varepsilon > 0$, and let $\mathcal{F}_\varepsilon = \left\{f_1, f_2, \ldots, f_{n(\varepsilon)}\right\}$ be a finite $\varepsilon/2$-net. As $\mathcal{F}_\varepsilon$ is finite, it is p-integrable. Accordingly, there exists $\eta(\varepsilon) > 0$ such that

$$\left(\int_{-\infty}^{+\infty} \|f_i(t+h) - f_i(t)\|^p\,dt\right)^{1/p} \leq \frac{\varepsilon}{4}$$

for each $i = 1, 2, \ldots, n(\varepsilon)$, and each $h \in (0, \eta(\varepsilon))$. We have then

$$\left(\int_{-\infty}^{+\infty} \|f(t+h) - f(t)\|^p\,dt\right)^{1/p}$$

$$\leq \left(\int_{-\infty}^{+\infty} \|f(t+h) - f_i(t+h)\|^p\,dt\right)^{1/p}$$

$$+ \left(\int_{-\infty}^{+\infty} \|f_i(t+h) - f_i(t)\|^p\,dt\right)^{1/p}$$

$$+ \left(\int_{-\infty}^{+\infty} \|f_i(t) - f(t)\|^p\,dt\right)^{1/p} \leq \frac{\varepsilon}{2} + \frac{\varepsilon}{4} + \frac{\varepsilon}{4} = \varepsilon$$

for each $f \in \mathcal{F}$ and $h \in (0, \eta(\varepsilon))$. So the family is p-equiintegrable. Let $[\alpha, \beta]$ be an arbitrary compact interval and let $\left(\int_\alpha^\beta f_n(t)\,dt\right)_{n\in\mathbb{N}^*}$ be a sequence in $\left\{\int_\alpha^\beta f(t)\,dt;\ f \in \mathcal{F}\right\}$. As $\mathcal{F}$ is relatively compact, the sequence $(f_n)_{n\in\mathbb{N}^*}$ has at least one subsequence, convergent to some element f in

$L^p(\mathbb{R};X)$. We denote for simplicity this subsequence again by $(f_n)_{n\in\mathbb{N}^*}$, and we observe that

$$\left\|\int_\alpha^\beta f_n(t)\,dt-\int_\alpha^\beta f(t)\,dt\right\|\leq\int_\alpha^\beta\|f_n(t)-f(t)\|\,dt$$

$$\leq(\beta-\alpha)^{1/q}\left(\int_\alpha^\beta\|f_n(t)-f(t)\|^p dt\right)^{1/p}$$

$$\leq(\beta-\alpha)^{1/q}\left(\int_{-\infty}^{+\infty}\|f_n(t)-f(t)\|^p dt\right)^{1/p},$$

for each $n\in\mathbb{N}^*$. From this inequality, we deduce that

$$\lim_{n\to\infty}\int_\alpha^\beta f_n(t)\,dt=\int_\alpha^\beta f(t)\,dt,$$

and consequently (ii) holds. Finally, for the very same $\varepsilon>0$, there exists $\delta(\varepsilon)>0$ such that

$$\left(\int_{-\infty}^{\delta(\varepsilon)}\|f_i(t)\|^p\,dt+\int_{\delta(\varepsilon)}^{+\infty}\|f_i(t)\|^p\,dt\right)^{1/p}\leq\frac{\varepsilon}{2}$$

for $i=1,2,\ldots,n(\varepsilon)$. Consequently

$$\left(\int_{-\infty}^{\delta(\varepsilon)}\|f(t)\|^p\,dt+\int_{\delta(\varepsilon)}^{+\infty}\|f(t)\|^p\,dt\right)^{1/p}$$

$$\leq\left(\int_{-\infty}^{\delta(\varepsilon)}\|f(t)-f_i(t)\|^p\,dt+\int_{\delta(\varepsilon)}^{+\infty}\|f(t)-f_i(t)\|^p\,dt\right)^{1/p}$$

$$+\left(\int_{-\infty}^{\delta(\varepsilon)}\|f_i(t)\|^p\,dt+\int_{\delta(\varepsilon)}^{+\infty}\|f_i(t)\|^p\,dt\right)^{1/p}\leq\frac{\varepsilon}{2}+\frac{\varepsilon}{2}=\varepsilon.$$

Therefore (iii) holds, and this completes the proof of the necessity.
Sufficiency. Let $\mathcal{F}$ be a family satisfying (i), (ii) and (iii), let $\varepsilon>0$, and let $\delta=\delta(\varepsilon/4)$ be given by (iii). From Theorem A.4.1, it follows that the restriction of the family $\mathcal{F}$ to $L^p(-\delta,\delta;X)$ is relatively compact in this space. Therefore, for the very same $\varepsilon>0$, it has a finite $\varepsilon/2$-net, $\{f_1,f_2,\ldots,f_{n(\varepsilon)}\}$. We have

$$\left(\int_{-\infty}^{+\infty}\|f(t)-f_i(t)\|^p\,dt\right)^{1/p}\leq\left(\int_{-\infty}^{-\delta}\|f(t)\|^p\,dt++\int_\delta^\infty\|f(t)\|^p\,dt\right)^{1/p}$$

$$+\left(\int_{-\infty}^{-\delta}\|f_i(t)\|^p\,dt+\int_\delta^\infty\|f_i(t)\|^p\,dt\right)^{1/p}+\left(\int_{-\delta}^{\delta}\|f(t)-f_i(t)\|^p\,dt\right)^{1/p}$$

$$\leq \frac{\varepsilon}{4} + \frac{\varepsilon}{4} + \frac{\varepsilon}{2},$$

which proves that $\{f_1, f_2, \ldots, f_{n(\varepsilon)}\}$ is a finite ε-net for $\mathfrak{F}$. Therefore $\mathfrak{F}$ is precompact and the proof is complete.

Problem A.9. The proof follows the same lines as those used in solving Problem A.8, with the mention that, instead of Theorem A.4.1, one has to use Theorem A.5.1.

Problem A.10. The proof follows the same lines as those used in solving Problem A.8, with the mention that, instead of Theorem A.4.1, one has to use Theorem A.5.2.

Bibliography

[1] Adams, R. A. *Sobolev spaces*, Academic Press, Boston San Diego New York London Sidney Tokyo Toronto, 1978.

[2] Ahmed, N. U. *Semigroup theory with applications to systems and control*, Pitman Research Notes in Mathematics Series **246**, Longman Scientific and Technical, 1991.

[3] Ahmed, N. U. Some remarks on the dynamics of impulsive systems in Banach spaces, *Journal of Dynamics of Continuous, Discrete and Impulsive Systems, Ser. A Math. Anal.*, **8**(2001), 261-274.

[4] Aizicovici, S. On a nonlinear integrodifferential equation, *J. Math. Anal. Appl.* **63**(1978), 385-395.

[5] Amann, H. *Linear and quasilinear parabolic problems, Volume I Abstract theory*, Birchäuser Verlag, Basel Boston Berlin, 1995.

[6] Amann, H. Linear parabolic problems involving measures, *RACSAM, Rev. R. Acad. Cien. Serie A. Mat.*, **95**(2001), 85-119.

[7] Aniţa, S. *Analysis and control of age-dependent population dynamics*, Mathematical Modelling. Theory and Applications, **11**, Kluwer Academic Publishers, Dordrecht / Boston / London, 2000.

[8] Arendt, W., Vector-valued Laplace transforms and Cauchy problems, *Israel J. Math.*, **59**(1987), 327-351.

[9] Aubin, J. P. Un théorème de compacité, *C. R. Acad. Sci. Paris*, **256**(1963), 5042-5044.

[10] Balakrishnan, A. V. An operator calculus for infinitesimal generators of semi-groups, *Trans. Amer. Math. Soc.*, **91**(1959), 330-353.

[11] Balakrishnan, A. V. Fractional powers of closed operators and the semigroups generated by them, *Pacific J. Math.*, **10**(1960), 419-437.

[12] Balakrishnan, A. V. *Applied functional analysis*, Applications of Mathematics Vol. **3**, Springer Verlag, New York, 1976.

[13] Baras, P. Compacité de l'opérateur definissant la solution d'une équation d'évolution non linéaire $(du/dt) + Au \ni f$, *C. R. Acad. Sci. Sér. I Math.*, **286**(1978), 1113-1116.

[14] Baras, P., Hassan, J. C. and Veron, L. Compacité de l'opérateur definissant la solution d'une équation d'évolution non homogène, *C. R. Acad. Sci. Paris Sér. I Math.*, **284**(1977), 779-802.

[15] Barbu, V. Sur un problème aux limites pour une classe d'équations différentielles nonlineaires abstraites du deuxiême ordre en t, *C. R. Acad. Sci. Paris*, **274**(1972), 459-462.

[16] Barbu, V. A class of boundary problems for second-order abstract differential equations, *J. Fac. Sci. Univ. Tokyo Sect. I.*, **19**(1972), 295-319.

[17] Barbu, V. *Nonlinear semigroups and differential equations in Banach spaces*, Noordhoff, Leyden, 1976.

[18] Barbu, V. Constrained control problems with convex cost in Hilbert spaces, *J. Math. Anal. Appl.*, **56**(1976), 502-528.

[19] Barbu, V. A semigroup approach to an infinite delay equation in Hilbert spaces, *Abstract Cauchy problems and functional differential equations*, Kappel, F. and Schappacker, W. Editors, Research Notes in Mathematics **48**, Pitman Boston, London, Melbourne, 1981.

[20] Barbu, V. *Analysis and control of nonlinear infinite dimensional systems*, Academic Press, 1993.

[21] Barbu, V. *Boundary value problems for partial differential equations*, Editura Academiei Române, Bucureşti, 1993 (Romanian).

[22] Barbu, V. and Precupanu, Th. *Convexity and optimization in Banach spaces*, Second Edition, Editura Academiei Bucureşti, D. Reidel Publishing Company, 1986.

[23] Bejenaru, I., Diaz, J. I. and Vrabie, I. I. An abstract approximate controllability result and applications to elliptic and parabolic systems with dynamic boundary conditions, *Electronic Journal of Differential Equations*, *Electronic Journal of Differential Equations*, Vol. 2001(2001), No. 50, 1-19.

[24] Belleni-Morante, A. *Applied semigroups and evolution equations*, Oxford Mathematical Monographs, Oxford University Press, Oxford, 1979.

[25] Bochner, S. Diffusion equations and sctochastic processes, *Proc. Nat. Acad. Sci. U.S.A.*, **35**(1949), 368-370.

[26] Bochner, S. and Taylor, A. E. Linear functionals on certain spaces of abstractly-valued functions, *Ann. of Math.* **39**(1938), 913-944.

[27] Bogachev, V. I. Deterministic and stochastic differential equations in infinite dimensional spaces, *Acta Appl. Math.*, **20**(1995), 25-93.

[28] Brezis, H. *Operateurs maximaux monotones et semi-groupes de contractions dans les espaces de Hilbert*, North-Holland Mathematics Studies **5**, North-Holland/American Elsevier, 1973.

[29] Brezis, H. *Analyse fonctionelle Théorie et applications*, Masson Paris Milan Barcelone Bonn 1992.

[30] Brezis, H. and Browder, F. E. Strongly nonlinear parabolic initial-boundary value problems, *Proc. Nat. Acad. Sci. USA*, **76**(1979), 38-40.

[31] Brezis, H. and Cazenave, T. *Linear semigroups of contractions, the Hille Yosida theory and some applications*, Publications du Laboratoire d'Analyse Numerique, Université Pierre et Marie Curie, A92004, 1993.

[32] Butzer, P. L. and Berens, H. *Semi-groups of operators and approximation*, Springer Verlag, New York, 1967.

[33] Cârjă, O. *Some methods of nonlinear analysis*, MATRIXROM, Bucureşti, 2003 (Romanian).

[34] Cârjă, O. and Vrabie, I. I. Some new viability results for semilinear differential inclusions, *NoDEA, Nonlinear Differential Equations Appl.*, **4**, (1997), 401-424.

[35] Casas, E. Pontryagin's principle for state-constraint boundary control; problems of semilinear parabolic equations, *SIAM J. Control Optim.*, **35**(1997), 1297-1327.

[36] Cazenave, T. and Haraux, A. *An introduction to semilinear evolution equations*, Clarendon Press · Oxford, 1998.

[37] Chernoff, P. R. Note on product formuals for operator semigrous, *J. Funct. Anal.*, **2**(1968), 238-242.

[38] Constantin, P. and Foiaş, C. *Navier-Stokes equations*, The University of Chicago Press, Chicago and London, 1988.

[39] Corduneanu, C., *Principles of differential and integral equations*, Chelsea Publishing Company The Bronx, New York, Second Edition, 1977.
[40] Corduneanu, C. *Integral equations and applications*, Cambridge University Press Cambridge New York Port Chester Melbourne Sydney, 1991.
[41] Courant, R. and Hilbert, D. *Methods of mathematical physics*, vol. 2, Interscience, John Wiley, 1962.
[42] Crandall, M. G., Londen, S.-O. and Nohel, J. A. An abstract nonlinear Volterra integrodifferential equation, *J. Math. Anal. Appl.* **64**(1978), 701-735.
[43] Crandall, M. G., Pazy, A. and Tartar, L. Remarks on generators of analytic semigroups, *Israel J. Math.*, **32**(1979), 363-374.
[44] Dafermos, C. Contraction semigroups and trend to equilibrium in continuum mechanics, *Applications of methods of functional analysis to problems in mechanics*, Germain, P. and Nayroles, B. editors, Lecture Notes in Mathematics **503**, Springer Verlag, Berlin • Heidelberg • New York, 1976, 295-306.
[45] Davies, E. B. *One-parameter semigroups*, Academic Press, London, 1980.
[46] Diestel, J. and Uhl, J. J. Jr, *Vector measures*, American Mathematical Society, Mathematical Surveys Number 15, 1977.
[47] Dinculeanu, N. *Measure theory and real functions*, Editura Didactică şi Pedagogică, Bucureşti, 1964 (Romanian).
[48] Dinculeanu, N. *Vector measures*, Pergamon Press, New York, 1967.
[49] Dunford, N. and Schwartz, J. T. *Linear operators part I: general theory*, Interscience Publishers, Inc. New York, Interscience Publishers Ltd. London, 1958.
[50] Edwards, R. E. *Functional analysis theory and applications*, Holt, Rinehart and Winston, New York • Chicago • San Francisco • Toronto • London, 1965.
[51] Engel, K.-J and Nagel, R. et al *One-parameter semigroups for linear evolution equations*, Graduate Text in Mathematics, **194**, Springer 2000.
[52] Fattorini, H. O. Boundary control systems, *SIAM J. Control Optimization*, **6**(1968), 349-385.
[53] Feller, W. The parabolic differential equations and the associated semi-groups of transformations, *Ann. of Math.*, **55**(1952), no. 2, 468-519.
[54] Feller, W. Semi-groups of transformations in general weak topologies, *Ann. of Math.*, **57**(1953), 287-308.
[55] Feller, W. On the generation of unbounded semi-gorups of bounded linear operators, *Ann. of Math.*, **58**(1953), 166-174.
[56] Feynman, R. P. Space-time approach to non-relativistic quantum mechanics, *Rev. Modern Phys.*, **20**(1948), 367-387.
[57] Frigon, M. and O'Reagan, D. Existence results for initial value problems in Banach spaces, *Differential Equations and Dynamical Systems*, **2**(1994), 41-48.
[58] Fujita, H. and Kato, T. On the Navier-Stokes initial-value problem, *Arch. Rational Mech. Anal.* **16**(1964), 269-315.
[59] Gilbarg, D. and Trudinger, N. S. *Elliptic partial differential equations of second order*, Springer Verlag, Berlin, 1983.
[60] Goldstein, J. A. Semigroup theoretic proofs of the central limit theorem and other theorems of analysis, Semigroup Forum, **12**(1976), 189-206, 388.
[61] Goldstein, J. A. *Semi-groups of operators and abstract Cauchy problems*, Cambridge University Press, 1985.
[62] Goldstein, J. A., Radin, C. and Showalter, R. E. Convergence rates of ergodic limits for semigroups and cosine functions, *Semigroup Forum*, **16**(1978), 89-95.

[63] Graves, L. M. *The theory of functions of real variable*, McGraw-Hill Company, New York, 1946.

[64] Grisvard, P. *Elliptic problems in nonsmooth domains*, Monographs and Studies in Mathematics **24**, Pitman Advanced Publishing Program, Boston London Melbourne, Pitman Publishing Inc. XIV 1985.

[65] Grosu, G Existence results for semilinear evolution equations involving measures, submitted.

[66] Gutman, S. Compact perturbations of m-accretive operators in general Banach spaces, *SIAM J. Math. Anal.* **13**(1982), 789-800.

[67] Hale, J. *Functional differential equations*, Applied Mathematical Sciences **3**, Springer Verlag New • York • Heidelberg • Berlin, 1971.

[68] Haraux, A. *Linear semi-groups in Banach spaces*, Pitman Research Notes in Mathematics, **152**, Longman, 1986, 93-135.

[69] Hille, E. *Functional analysis and semi-groups*, Amer. Math. Soc. Colloq. Publ., **31**, New York, 1948.

[70] Hille, E. and Phillips, R. S. *Functional analysis and semi-groups*, Amer. Math. Soc. Colloquium Publications, **31**, Fourth Printing of Revised Edition, 1981.

[71] Hrusa, W. J. A nonlinear functional-differential equation in Banach space with application to materials with fading memory, *Arch. Rational Mech. Anal.* **84**(1984), 99-137.

[72] Kato, T. Remarks on pseudo-resolvents and infinitesimal generators of semigroups, *Proc. Japan Acad.*, **35**(1959), 467-468.

[73] Kellermann, H. and Hieber, M., Integrated semigroups, *J. Funct. Anal.*, **84**(1989), 160-180.

[74] Kisýsinski, J. A proof of the Trotter-Kato theorem on the approximation of semigroups, *Colloq. Math.*, **18**(1967), 181-184.

[75] Komatsu, H. Fractional powers of operators, *Pacific J. Math.*, **19**(1966), 285-346.

[76] Komatsu, H. Fractional powers of operators II Interpolation spaces, *Pacific J. Math.*, **21**(1967), 89-111.

[77] Komatsu, H. Fractional powers of operators III Negative powers, *J. Math. Soc. Japan*, **21**(1966), 205-220.

[78] Komatsu, H. Fractional powers of operators IV Potential operators, *J. Math. Soc. Japan*, **21**(1969), 221-228.

[79] Komatsu, H. Fractional powers of operators V Dual operators, *J. Fac. Sci. Univ. of Tokyo*, **17**(1970), 373-396.

[80] Krasnoselskii, M., A. Two remarks on the method of successive approximations, *Uspehi Mat. Nauk*, **10**(1955), 123-127 (Russian).

[81] Laha, R. G. and Rohatgi, V. K. *Probability theory*, John Wiley & Sons, New York Chichester Brisbane Toronto, 1979.

[82] Lasiecka, I. Boundary control of parabolic systems: regularity of optimal solutions, *Appl. Math. Optim.*, **4**(1978), 301-327.

[83] Lasiecka, I. State constrained control problems for parabolic systems: regularity of optimal solutions, *Appl. Math. Optim.*, **6**(1980), 1-29.

[84] Lasiecka, I. and Triggiani, R. *Control theory for partial differential equations: continuous and approximation theories I abstract parabolic systems*, Encyclopedia of Mathematics and its Applications **74**, Cambridge University Press, 2000.

[85] Lions, J. L. and Magenes, E. *Problèmes aux limites non homogènes et applications*, Volume 1, Dunod Paris, 1968.

[86] Liu, K. A characterization of strongly continuous groups of linear operators on a Hilbert space, *Bull. London. Math. Soc*, **32**(2000), 54-62.

[87] Lumer, G. and Phillips, R. S. Dissipative operators in a Banach space, *Pacific J. Math.*, **11**(1961), 679-698.

[88] Maz'ya, V. G. *Some estimates for solutions of elliptic second-order equations*, Doklady Acad. Nauk SSSR, **137**(1961), 1057-1059 (Russian).

[89] McBride, A. C. *Semigroups of linear operators: an introduction*, Pitman Research Notes in Mathematics Series, **156**, Longman Scientific&Technical, 1987.

[90] Mitidieri, E. and Vrabie, I. I. Existence for nonlinear functional differential equations, *Hiroshima Math. J.*, **17**(1987), 627-649.

[91] Miyadera, I. Generation of strongly continuous semi-groups of operators, *Tôhoku Math. J.*, **4**(1952), 109-114.

[92] Nagumo, M. Einige analytische Untersuchungen in liniaren metrische Ringen, *Japan J. Math.*, **13**(1936), 61-80.

[93] Nathan, D. S. One parameter groups of transformations in abstract vector spaces, *Duke Math. J.*, **1**(1935), 518-526.

[94] Nečas, J. *Les méthodes directes dans la théorie des équations elliptiques*, Academia, Éditions de l'Academie Tchécoslovaque des Sciences, Prague, 1967.

[95] Neveu, J. Théorie des semi-groupes de Markov, *Univ. Calif. Publ. Statistics*, **2**(1958), 319-394.

[96] Nikolsky, S. M. *A Course of mathematical analysis*, Volume 1, Mir Publishers Moscow, 1981.

[97] Pavel, N. H. *Evolution equations and nonlinear semigroups*, Lecture Notes in Mathematics, **1260**, Springer Verlag, 1987.

[98] Pazy, A. On the differentiability and compactness of semi-groups of linear operators, *J. Math. and Mech.*, **17**(1968), 1131-1141.

[99] Pazy, A. A class of semi-linear equations of evolution, *Israel J. Math.*, **20**(1975), 23-36.

[100] Pazy, A. A perturbation theorem for linear m-dissipative operators, *Memorias de Matematica da U.F.R.J.*, **131**(1981).

[101] Pazy, A. *Semigroups of linear operators and applications to partial differential equations*, Applied Mathematical Sciences, **44**, Springer Verlag, New York Berlin Heidelberg Tokyo, 1983.

[102] Peano, G. Integrazione per le serie delle equazioni differenziali lineari, *Atti Reale Acc. Sci. Torino*, **22**(1887), 293-302.

[103] Peano, G. Intégration par séries des équations différentielles linéaires, *Math. Ann.*, **32** (1888), 450-456.

[104] Phillips, R. S. On the generation of semi-groups of linear operators, *Pacific J. Math.*, **2**(1952), 343-369.

[105] Phillips, R. S. The adjoint semi-group, *Pacific J. Math.*, **5**(1955), 269-283.

[106] Phillips, R. S. Dissipative operators and hyperbolic systems of partial differential equations, *Trans. Amer. Math. Soc.* **90**(1959), 193-254.

[107] Raymond, J. P. Nonlinear boundary control of semilinear parabolic equations with pointwise state constraints, *Disc. Cont. Dyn. Syst.*, **3**(1997), 341-370.

[108] Raymond, J. P. and Tröltzsch, F. Second order sufficient optimality conditions for nonlinear parabolic problems with state constraints, *Discrete Contin. Dynam. Systems*, **6**(2000), 431-450.

[109] Raymond, J. P. and Zidani, H. Time-optimal problems with boundary controls, *Differential Integral Equations*, **13**(2000), 1039-1072.

[110] Richtmyer, R. D. *Principles of advanced mathematical physics*, Volume I, Texts and Monographs in Physics, Springer Verlag, New York Heidelberg Berlin, 1978.

[111] Segal, I. Nonlinear semi-groups, *Ann. of Math.*, **78**(1963), 339-364.

[112] Showalter, R. E. *Monotone operators in Banach space and nonlinear partial differential equations*, Mathematical Surveys and Monographs Volume **49**, American Mathematical Society, 1997.

[113] Showalter, R. E. and Ting, T. W. Pseudoparabolic partial differential equations, *SIAM J. Math. Anal.*, **1**(1970), 1-26.

[114] Simon, B. *Functional integration and quantum physics*, Academic Press, New York, 1980.

[115] Simon, J. Compact sets in the space $L^p(0,T;X)$, *Ann. Mat. Pura. Appl.* **146**(1987), 65-96.

[116] Stein, E. M. and Weiss, G. *Introduction to Fourier analysis on euclidean spaces*, Princeton University Press, Princeton, New Jersey, 1971.

[117] Stone, M. H. On one-parameter groups of unitary operators in Hilbert spaces, *Ann. of Math.*, **33**(1932), 643-648.

[118] Sulem, C. and Sulem, P.-L. *The nonlinear Schrödinger equation Self-focusing and wave collapse*, Applied Mathematical Sciences **139**, Springer, 1999.

[119] Taylor, M. S. *Partial differential equations basic theory*, Text in Applied Mathematics, Springer, 1996.

[120] Thieme, H. R., "Integrated semigroups" and integrated solutions to abstract Cauchy problems, *J. Math. Anal. Appl.*, **152**(1990), 416-447.

[121] Thieme, H. R., Positive perturbation of operator semigroups: growth bounds, essential compactness, and asynchronous exponential growth, *Discrete and Continuous Dynamical Systems*, **4**(1998), 735-764.

[122] Trotter, H. F. Approximation of semi-groups of operators, *Pacific J. Math.*, **8**(1958), 887-919.

[123] Trotter, H. F. On the product of semi-groups of operators, *Proc. Amer. Math. Soc.*, **10**(1959), 545-551.

[124] Vrabie, I. I. The nonlinear version of Pazy's local existence theorem, *Israel J. Math.*, **32**, (1979), 225-235.

[125] Vrabie, I. I. A compactness criterion in $C(0,T;X)$ for subsets of solutions of nonlinear evolution equations governed by accretive operators, *Rend. Sem. Mat. Univ. e Politec. Torino*, **45**(1985), 149-157.

[126] Vrabie, I. I. A new characterization of generators of linear compact semigroups, *An Ştiinţ. Univ. Al. I. Cuza Iaşi, Secţ. I a Mat.* **35**(1989), 145-151.

[127] Vrabie, I. I. *Compactness methods for nonlinear evolutions*, Pitman Monographs and Surveys in Pure and Applied Mathematics, Second Edition, **75**, Addison Wesley and Longman, 1995.

[128] Vrabie, I. I. Compactness in L^p of the set of solutions to a nonlinear evolution equation, *Qualitative problems for differential equations and control theory*, Corduneanu, C. Editor, World Scientific, 1995, 91-101.

[129] Vrabie, I. I. *Differential equations*, MATRIXROM, Bucharest 1999 (Romanian).

[130] Vrabie, I. I. A new characterization of generators of differentiable semigroups, *RACSAM, Rev. R. Acad. Cien. Serie A. Mat.*, **95**(2001), 297-302.

[131] Vrabie, I. I. Compactness of the solution operator for a linear evolution equation with distributed measures, *Trans. Amer. Math. Soc.*, **354**(2002), 3181-3205.

[132] Webb, G. *Theory of nonlinear age-dependent population dynamics*, Marcel Dekker, New York 1985.

[133] Widder, D. V. *The Laplace transform*, Princeton University Press, London: Humphrey Milford Oxford University Press, 1946.

[134] Yosida, K. On the group embedded in the metrical complete ring, *Japan J. Math.*, **13**(1936), 7-26, 459-472.

[135] Yosida, K. On the differentiability and representation of one parameter semi-groups of linear operators, *J. Math. Soc. Japan*, **1**(1948), 15-21.

[136] Yosida, K. *Functional analysis*, Third Edition, Springer Verlag Berlin Heidelberg New York, 1971.

List of Symbols

$A^{k,p}(a,b;X)$ — the set of all $u:[a,b]\to X$, whose m^{th}-order derivatives $u^{(m)}$ are absolutely continuous on $[a,b]$ for $m=0,1,\ldots,k-1$, and belong to $L^p(a,b;X)$, for $m=0,1,\ldots,k$

A^* — the adjoint of the operator $A:D(A)\subseteq X\to X$

$a\cdot\nabla u$ — the directional derivative of u along a

$\approx$ — isomorphic

$BV([a,b])$ — the space of functions of bounded variation from $[a,b]$ to $\mathbb{R}$

$BV([a,b];X)$ — the space of functions of bounded variation from $[a,b]$ to X

$BV(\mathbb{R})$ — the space of functions $f:\mathbb{R}\to\mathbb{R}$ whose restrictions on any interval $[a,b]$ are of bounded variation

$BV(\mathbb{R};X)$ — the space of functions $f:\mathbb{R}\to X$ whose restrictions on any interval $[a,b]$ are of bounded variation

$BV_A([a,b];X)$ — the space of all $g\in BV([a,b];X)$ such that, for each $t\in[a,b]$, there exists the integral $\int_0^t S(t-s)\,dg(s)$ in the norm topology of X

$BV_A(\mathbb{R};X)$ — the space of all functions $g\in BV(\mathbb{R};X)$ with the property that, for each $a<b$, $g\in BV_A([a,b];X)$

$C(a,b)$ — the space of continuous functions from (a,b) to $\mathbb{R}$

$C([a,b])$ — the space of continuous functions from $[a,b]$ to $\mathbb{R}$

$C([a,b];X)$ — the space of continuous functions from $[a,b]$ to X

$C([a,b];X_w)$ — the space of continuous functions from $[a,b]$ to X_w

$C^k(a,b)$ — the space of functions of class C^k from (a,b) to $\mathbb{R}$

$C^k([a,b])$ — the space of functions of class C^k from $[a,b]$ to $\mathbb{R}$

c_0 — the space of sequences of real numbers tending to 0 for $n\to\infty$

$c_0(X)$ — the space of sequences in X tending to 0 for $n\to\infty$

$C_0^\infty(\Omega)$ — the space of indefinite differentiable functions from Ω to $\mathbb{R}$, with compact support in Ω, i.e. $\mathcal{D}(\Omega)$

$C_{ub}(D)$ — the space of uniformly continuous and bounded functions from $D\subseteq\mathbb{R}^n$ to $\mathbb{R}$

$D(A)$ — the domain of A

Δ — the Laplace operator

$\mathcal{D}'(a,b;X)$ — the set of linear continuous operators from $\mathcal{D}(a,b)$ to X

$f^{(k)}$ — the kth-order derivative of f in the sense of X-valued distributions on (a,b)

$\mathcal{D}(\Omega)$	—	the space of indefinite differentiable functions from Ω to $\mathbb{R}$, with compact support in Ω, i.e $C_0^\infty(\Omega)$
$\mathcal{D}^\alpha u$	—	the derivative of order α of u in the sense of distributions on $\mathcal{D}(\Omega)$
$\mathcal{D}'(\Omega)$	—	the space of distributions on $\mathcal{D}(\Omega)$, i.e. of linear continuous functionals from $\mathcal{D}(\Omega)$ to $\mathbb{R}$
$\hat{f}$	—	the Fourier transform of f, i.e. $\hat{f} = \frac{1}{(2\pi)^{n/2}} \int_{\mathbb{R}^n} e^{-i\langle\xi,\omega\rangle} f(\omega)\, d\omega$
graph (A)	—	the graph of A
$H^m(\Omega)$	—	the space $W^{m,2}(\Omega) = H^{m,2}(\Omega)$
$H^{m,p}(\Omega)$	—	the completion of the space $\{u \in C^m(\Omega);\ \|\|u\|\|_{m,p} < +\infty\}$ with respect to $\|\| \cdot \|\|_{m,p}$
$H^{-1}(\Omega)$	—	the dual of $H_0^1(\Omega)$
$H^s(\Gamma)$	—	the space of all functions $u \in L^2(\Gamma)$ with the property that $u_i \in H^s(\mathbb{R}^{n-1})$ for each $i \in \mathcal{I}$, where $\{\chi_i;\ i \in \mathcal{I}\}$ is a family of local charts on Γ, $\{\theta_i;\ i \in \mathcal{I}\}$ a subordinated finite partition of the unity and $u_i = \chi_i \circ \theta_i u$
$H^s(\Omega)$	—	the set of restrictions to Ω of all elements in $H^s(\mathbb{R}^n)$
$H^s(\mathbb{R}^n)$	—	is the space of all $u \in L^2(\mathbb{R}^n)$ with $(1 + \|\|\xi\|\|^2)^{s/2}\hat{u}(\xi) \in L^2(\mathbb{R}^n)$
$\mathrm{Im}\,\lambda$	—	the imaginary part of the complex number λ
$\ker(A)$	—	the kernel of A, i.e. $\ker(A) = \{x \in D(A);\ Ax = 0\}$
$\mathcal{L}(X)$	—	the space of all linear bounded operators from X to X
l_p	—	the space of sequences $(x_n)_{n\in\mathbb{N}^*}$ in $\mathbb{R}$ with $\sum_{n=1}^\infty \|x_n\|^p < +\infty$.
$L^p(a,b;X)$	—	the space $L^p((a,b),\mu;X)$ with μ the Lebesgue measure on (a,b)
$\mathcal{L}^p(\Omega)$	—	the set of all functions $f : \Omega \to \mathbb{R}$ with the property that f is strongly measurable on $\Omega \subseteq \mathbb{R}^n$ and $\|\|f\|\|^p$ is integrable on Ω with respect to the Lebesgue measure
$\mathcal{L}^p(\Omega,\mu;X)$	—	the set of all functions $f : \Omega \to X$ with the property that f is strongly measurable on Ω and $\|\|f\|\|^p$ is integrable on Ω with respect to μ
$L^p(\Omega,\mu;X)$	—	the quotient space $\mathcal{L}^p(\Omega,\mu;X)/\sim$, $\sim$ being the almost everywhere μ-equality on Ω
$l_p(X)$	—	the space of sequences $(x_n)_{n\in\mathbb{N}^*}$ in X with $\sum_{n=1}^\infty \|\|x_n\|\|^p < +\infty$.
$\mathcal{L}^1_{\mathrm{loc}}(\mathbb{R}_+)$	—	the space of measurable functions $f : \mathbb{R}_+ \to \mathbb{R}$ satisfying $f \in \mathcal{L}^1(0,b)$ for each $b > 0$
$\mathbb{N}$	—	the set of nonnegative integers
$\mathbb{N}^*$	—	the set of positive integers
∇	—	the nabla operator, i.e. $\nabla\varphi = \mathrm{grad}\,\varphi$, $\nabla \cdot \mathbf{F} = \mathrm{div}\,\mathbf{F}$ and $\nabla \times \mathbf{F} = \mathrm{curl}\,\mathbf{F}$
ν		the unit exterior normal at Γ at the current point
$\|\| \cdot \|\|_{D(A)}$	—	the norm on $D(A)$, $\|\|x\|\|_{D(A)} = \|\|x\|\| + \|\|Ax\|\|$ for $x \in D(A)$
$\| \cdot \|_{D(A)}$	—	the norm on $D(A)$, $\|x\|_{D(A)} = \|\|x - Ax\|\|$ for $x \in D(A)$
$\|\| \cdot \|\|_{\mathcal{L}(X)}$	—	the operator norm on $\mathcal{L}(X)$, i.e. $\|\|U\|\|_{\mathcal{L}(X)} = \sup_{\|\|x\|\|\le 1} \|\|Ux\|\|$

$\|\cdot\|_{m,p}$ — the norm on $W^{m,p}(\Omega)$, defined by

$$\|u\|_{m,p} = \begin{cases} \left(\sum_{0\le|\alpha|\le m} \|\mathcal{D}^\alpha u\|^p_{L^p(\Omega)} \right)^{1/p} & \text{if } 1 \le p < +\infty \\ \max_{0\le|\alpha|\le m} \|\mathcal{D}^\alpha u\|_{L^\infty(\Omega)} & \text{if } p = \infty \end{cases}$$

$\mathcal{P}$ — a partition of the interval $[a,t]$
$\mathcal{P}[a,t]$ — the set of all partitions of the interval $[a,t]$
$\mathbb{R}$ — the set of real numbers
$\mathbb{R}^*$ — the set of real numbers excepting 0
$\mathbb{R}_+$ — the set of nonnegative real numbers
$\mathbb{R}^*_+$ — the set of positive real numbers
$R(A)$ — the range of A, i.e. $R(A) = A(D)$
$\operatorname{Re}\lambda$ — the real part of the complex number λ
$R(\lambda; A)$ — the operator $(\lambda I - A)^{-1}$
$\rho(A)$ — the resolvent set of the operator $A : D(A) \subseteq X \to X$
$r_\sigma(A)$ — the spectral radius of the operator A
$\sigma(A)$ — the spectrum of the operator $A : D(A) \subseteq X \to X$
$\sim$ — equivalence relation
$\operatorname{supp}\varphi$ — the set $\overline{\{x \in \Omega\,;\ \varphi(x) \neq 0\}}$
$\{S(t)^\odot;\ t \ge 0\}$ — the sun dual of the semigroup $\{S(t)\,;\ t \ge 0\}$
u_ν — the normal derivative of u on Γ, i.e. $u_\nu = \nu \cdot \nabla u$
u_t — the partial derivative of u with respect to t
u_{xx} (u_{xxx}) — the second (third) order partial derivative of u with respect to x
$\mathbf{u} \times \mathbf{v}$ — the vector product of $\mathbf{u}$ by $\mathbf{v}$
$W^{m,p}(a,b;X)$ — the set of all X-valued distributions f on (a,b) with f^m in $L^p(a,b;X)$ for each $m = 0,1,\ldots,k$
$W^{m,p}(\Omega)$ — the set of all $u \in L^p(\Omega)$ with $\mathcal{D}^\alpha u \in L^p(\Omega)$ for $0 \le |\alpha| \le m$
$W^{m,p}_0(\Omega)$ — the closure of $\mathcal{D}(\Omega)$ in $W^{m,p}(\Omega)$
$W^{-m,q}(\Omega)$ — the set of all distributions $u \in \mathcal{D}'(\Omega)$, $u = \sum_{|\alpha|\le m} \mathcal{D}^\alpha f_\alpha$ with $f_\alpha \in L^q(\Omega)$, the dual of $W^{m,p}(\Omega)$
X_A — the space $(X^\odot)^*$
$X^\odot$ — the sun dual of the space X with respect to the operator $A : D(A) \subseteq X \to X$, i.e., $X^\odot = \overline{D(A^*)}$
$X_1 \oplus X_2$ — the direct sum of the spaces X_1, X_2
X^* — the topological dual of X, i.e. the space of linear continuous functionals from X to $\mathbb{R}$, or $\mathbb{C}$
x_t — the function $x_t : [-r,0] \to \mathbb{R}^n$ defined by $x_t(\theta) = x(t+\theta)$,
X^* — where $x : [-r,+\infty) \to \mathbb{R}^n$
X_w — the space X endowed with the weak topology

Subject Index

www.ingramcontent.com/pod-product-compliance
Ingram Content Group UK Ltd.
Pitfield, Milton Keynes, MK11 3LW, UK
UKHW020222290926
15470UKWH00001B/321

9 780444 512888